D1556683

LIVERPOOL JMU LIBRARY
3 1111 01441 6919

The Law and Management of Building Subcontracts

The Law and Management of Building Subcontracts

Second edition

John McGuinness BSc, MSc, FCIOB

Blackwell Publishing editorial offices:
Blackwell Publishing Ltd, 9600 Garsington Road, Oxford OX4 2DQ, UK
Tel: +44 (0)1865 776868
Blackwell Publishing Inc., 350 Main Street, Malden, MA 02148-5020, USA
Tel: +1 781 388 8250
Blackwell Publishing Asia Pty Ltd, 550 Swanston Street, Carlton, Victoria 3053, Australia
Tel: +61 (0)3 8359 1011

First edition published 2004 by Athena Press
Second edition published 2007 by Blackwell Publishing Ltd

ISBN: 978-1-4051-6102-2

Library of Congress Cataloging-in-Publication Data

McGuinness, John.
The law and management of building subcontracts / John McGuinness.
– 2nd ed.
p. cm.
Includes bibliographical references and index.
ISBN-13: 978-1-4051-6102-2 (hardback : alk. paper)
ISBN-10: 1-4051-6102-7 (hardback : alk. paper)
1. Construction contracts–England. 2. Subcontracting–England.
3. Construction contracts–Wales. 4. Subcontracting–Wales. 5. Construction industry–Management. I. Title.
KD1641.M39 2007
346.4202′2–dc22

2006029934

A catalogue record for this title is available from the British Library
Set in 9.5/11.5 Palatino
by SNP Best-set Typesetter Ltd., Hong Kong
Printed and bound in Great Britain
by TJ International, Padstow, Cornwall

The publisher's policy is to use permanent paper from mills that operate a sustainable forestry policy, and which has been manufactured from pulp processed using acid-free and elementary chlorine-free practices. Furthermore, the publisher ensures that the text paper and cover board used have met acceptable environmental accreditation standards.

For further information on Blackwell Publishing, visit our website:
www.blackwellpublishing.com/construction

Permission to reproduce relevant conditions from the following are acknowledged: General Conditions of Contract (MF/1), Form of Sub-contract, by The Institution of Engineering and Technology; CIC Novation Agreement, by The Construction Industry Council; and CLL Standard Form of Novation, by the City of London Law Society.

Contents

Definitions

Contractor	employing contractor or subcontractor
Employer	ultimate employer under the head contract
Employer's agent	any person named in the main contract to represent the employer
Head contract	contract between the ultimate contractor and the employer
Main contract	contract between the contractor and another
Subcontract	contract between the contractor and the subcontractor
Subcontractor	employed party

CDM	Construction (Design and Management) Regulations 1994
CECA/6th	Civil Engineering Contractors Association Form of Subcontract July 1998
CIC/Nov Agr	Construction Industry Council Novation Agreement, for use where the appointment of a consultant is to be novated from an employer to a design and build contractor
CLL	City of London Law Society standard novation agreement
DBSub/A	JCT Design and Build Sub-Contract Agreement
DBSub/C	JCT Design and Build Sub-Contract Conditions
DOM/1	The subcontract conditions for use with the Domestic subcontract DOM/1 Articles of Agreement
GC/Works/1	General Conditions of Government Contracts for Building and Civil Engineering Works
HGCRA	Housing Grants, Construction and Regeneration Act 1996
IC	Intermediate Building Contract for works of simple content, issued by the Joint Contracts Tribunal Ltd
ICSub/A	Intermediate Sub-Contract Agreement
ICSub/C	Intermediate Sub-Contract Conditions
ICSub/D/A	Intermediate Sub-Contract with subcontractor's design Agreement
ICSub/D/C	Intermediate Subcontract with subcontractor's design Conditions
ICSub/NAM/A	Intermediate Named Sub-Contract Agreement
ICSub/NAM/C	Intermediate Named Sub-Contract Conditions
ICSub/NAM/IT	Intermediate Named Sub-Contract Invitation to Tender
ICSub/NAM/T	Intermediate Named Sub-Contract Tender
JCT	The Joint Contracts Tribunal Limited
LADs	Liquidated and ascertained damages
LPCDIA	Late Payment of Commercial Debts (Interest) Act 1998
MF/1	Model Form of General Conditions of Contract, issued by the Joint IMechE/IEE Committee

NEC/Sub	The Engineering and Construction Sub-Contract
RTPA	Contracts (Rights of Third Parties) Act 1995
SBC	Standard Building Contract, issued by the Joint Contracts Tribunal Ltd
SBCSub/A	JCT Standard Building Sub-Contract Agreement
SBCSub/C	JCT Standard Building Sub-Contract Conditions
SBCSub/D/A	JCT Standard Building Sub-Contract with Sub-Contractor's Design Agreement
SBCSub/D/C	JCT Standard Building Sub-Contract with Sub-Contractor's Design Conditions
SCDP	Sub-Contractor's Design Portion
Scheme	Scheme for Construction Contracts (England and Wales) Regulations
ShortSub	JCT Short Form of Sub-Contract
Sub/MPF04	Major Project Form of Sub-Contract
Sub/Sub	JCT Standard Form of Sub-subcontract
Works Contract/1	The Standard Form of Works Contract for Works Contractors
Section 1	Invitation to Tender
Section 2	Tender by Works Contractor
Section 3	Articles of Agreement
Works Contract/2	The Standard Form of Works Contract for Works Contractors with whom the Management Contractor contracts in accordance with clause 8.2.1 of the JCT Management Contract 1998 Edition

Preface

This book has been written to be of interest to all those professionally concerned with the subcontracting of building work. It has not therefore been written about specific forms of subcontract, but about the wider issues of subcontracting in the industry. It does, however, relate the more general issues to the specific terms of the subcontract for use with the JCT Standard Building Contract, SBCSub, and similar forms for use on projects of differing size and complexity.

While the primary aim of this book is to set out good practice for those procuring and carrying out subcontract work, whether working as general contractors or for a specialist trade contractor, the secondary objective has been to identify practical difficulties, which the author has personally encountered over several decades of professional involvement, and present solutions as regards both the practical and legal issues. Inevitably this book challenges many established perceptions, and in particular those incorporated from main contract practice, which on examination have been found to be inappropriate to the subcontract situation.

On a conservative estimate there are probably 20 to 25 subcontracts for each major building contract. However, relatively few books explore the problems particular to the relationship between the contractor and its subcontractors. What is more, the number of decided cases relating to subcontract issues is equally disproportionate to those concerning main construction contracts. As a result, in the author's view, this lack of authority and commentary on subcontracts has led to a failure by those preparing both standard and non-standard forms of subcontract, as well as those compiling Sub-Contract Documents and formal agreements, to create workable subcontract arrangements.

One of the consequences of the statutory provision for adjudication is that independent third parties are deciding a much larger number of contractor/subcontractor disputes, as a means of resolving differences on construction projects. At the same time the courts are deciding fewer disputes and as a consequence the law is not developing alongside the cases being decided.

Historically it has been predominantly main contractor and employer disputes that have been decided by the courts, and such decisions may or may not be relevant to the contractor/subcontractor relationship. While certain issues, such as the provision for liquidated damages, may not be generally applicable to subcontracting, other issues such as the obligation as to time, and issues concerning the completion of the work, require a different approach from that under the Standard Building Contract.

A major purpose of reporting the decided cases of the courts has been to ensure consistency of approach between judges. One consequence of the lack of published records of disputes decided by adjudicators, and indeed arbitrators, is a failure to have a consistent approach. One aim of this book is to develop a common understanding of subcontracting issues by discussing the most common matters that lead to disputes between contractors and subcontractors and by identifying the relevant law, thereby leading to a common approach.

Traditionally, subcontracts have been considered and written to reflect the obligations of the contractor under the main contract so as to be a 'minor' or 'secondary' version of it, with the aim of creating 'back to back' obligations. However, the contractual relationship between a contractor and a subcontractor, with the associated integration of the work of a number of individual subcontractors, is significantly more complex than exists in a main contract. This is because, in a normal contractual situation, there is only one party who supplies goods and services to another, with no associated contractor interfaces. However, for many building subcontractors, in addition to having an on-site interface with other trades, there may be a significant amount of their work performed off-site, both of which restrict the subcontractor's freedom of operation.

The results of these differences are discussed following the experiences of the author over many years of working in the construction process and in a large number of disputes referred to adjudication and other dispute resolution processes.

In preparing this edition I have considered and referred to material in my possession in the summer of 2006.

Chapter 1
Tenders

Introduction

A tender for subcontracted work will normally be initiated by a request from the contractor for such an offer. Rarely in subcontracting will the concept of an open offer, such as commonly arises for the supply of materials, apply. An open offer is one where, for example, a supplier advises a number of prospective clients that it can supply materials of a certain type for a specific price or under certain terms and conditions. A specific exception may be the offer to subcontract for the supply of labour and/or plant and equipment on an hourly rate basis.

Tenders for the supply of labour and materials will normally be job specific and will be offered following a request to tender from the contractor. Such requests may arise either when the contractor itself is tendering for work or after the award of the main contract. The method of enquiry may be very formal, with a letter of tender to be completed by the tenderer and accompanied by forms of contract, amendments, health and safety plans, drawings, specification, outline programme and other documents. Or the request may be very informal, requiring a price for a simple schedule of work or for work of specific nature shown on a drawing or drawings.

However the enquiry may be set out, the offer is only what it is said to be in the tenderer's response or offer. The terms of enquiry will only be incorporated into the offer if the offer expressly says so. If A is asked to price for building in bricks but prices to build in blocks and says so, then he has made an offer to build in blocks and not bricks.

In theory, a subcontractor's offer is open for acceptance at any time unless it has a time restriction on it or is withdrawn before acceptance. Many requests to tender will require the offer to be open for acceptance for a specific period and generally tenderers will be happy to comply. However, in the event that a tenderer wishes to withdraw its offer at any time before acceptance, it will normally be able to do so without sanction. For example, if it discovers an error in its tender, then a tenderer can withdraw its offer either entirely or withdraw it and replace it with a revised bid.

Many offers may be made in such a way that they do not in fact comprise a true offer open to acceptance without further discussion and/or agreement. A tender made in the form 'If this tender is of interest we would wish to discuss further . . .' is not an offer that is capable of acceptance, or if accepted it will leave such matters undecided and liable to conflict. Alternatively, the offer may state that it has not complied with all the conditions of the enquiry. Common exclusions may relate to a requirement for the subcontractor to have visited the site, considered the terms of the main contract, etc. The offer may be made subject to a visit to site, review of the main contract requirements, etc. Such offers enable a subcontractor to price a project with the minimum of effort and leave matters not easily valued for consideration only if its bid is otherwise attractive.

Contract of tender

Generally, subcontractors tendering for work under English law will not have any entitlement to be paid for their work in preparing a tender. The common law does however provide the subcontractor with some rights to have its offer properly considered. In the Court of Appeal case of *Blackpool & Fylde Aero Club Ltd* v. *Blackpool Borough Council* [1990] 3 All ER 25, CA, Lord Justice Bingham said:

> 'A tendering procedure of this kind is, in many respects, heavily weighted in favour of the invitor. He can invite tenders from as many or as few parties as he chooses. He need not tell any of them who else, or how many others, he has invited. The invitee may often, although not here, be put to considerable labour and expense in preparing a tender, originally without recompense if he is unsuccessful. The invitation to tender may itself, in a complex case, although again not here, involve time and expense to prepare, but the invitor does not commit himself to proceed with the project, whatever it is; he need not accept the highest tender; he need not accept any tender; he need not give reasons to justify his acceptance or rejection of any tender received. The risk to which the tenderer is exposed does not end with the risk that his tender may not be the highest (or as the case may be, lowest). But where, as here, tenders are solicited from selected parties all of them known to the invitor, and where a local authority's invitation prescribes a clear, orderly and familiar procedure (draft contract conditions available for inspection and plainly not open to negotiation, a prescribed common form of tender, the supply of envelopes designed to preserve the absolute anonymity of tenderers and clearly to identify the tender in question and an absolute deadline) the invitee is in my judgment protected at least to this extent: if he submits a conforming tender before the deadline he is entitled, not as a matter of mere expectation but of contractual right, to be sure that his tender will after the deadline be opened and considered in conjunction with all other conforming tenders or at least that his tender will be considered if others are. Had the club, before tendering, enquired of the Council whether it could rely on any timely and conforming tender being considered along with others, I feel quite sure that the answer would have been "of course". The law would, I think, be defective if it did not give effect to that.'

European competition tendering law applies to public bodies and therefore does not directly refer to subcontracts. However, as it is likely that subcontractors will be required to tender on the basis that they have knowledge of, and accept, the terms of the main contract, it could be argued that there is implied into the invitation to tender to the subcontractor, the principles of the European competitive tendering rules. Further, on many large and/or prestigious projects, significant subcontract packages may be procured by some procedure involving the employer, when it is likely that the public procurement directive may either apply or be implied *into* the procedures.

Specialist contractors' tenders, frequently by their nature, make competitive tendering on a price-only basis unsuitable. Such tenders may involve patented or licensed products, which unless they are preselected by competition, will require selection on merit as well as cost. In the case of *Harmon CFEM Facades (UK) Ltd* v. *The Corporate Officer of the House of Commons* (2000) 67 Con LR 1, the correctness of the selection of the cladding contractor for Portcullis House, the new building for Members of Parliament, was considered in a very lengthy judgment.

The body inviting tenders needed to balance the requirement for confidentiality and for fairness by advising all tenderers as to the priorities for selection, and those requirements were derived from an implied contract within the tendering process. Judge Humphrey Lloyd said:

> 'In my judgment, even though all tenderers accepted that they would not be entitled to see alternatives of detail which were considered to be commercially confidential to a given tenderer, the House of Commons, in soliciting new or revised tenders under the European public works regime, impliedly undertook towards any tenderer which submitted a tender that its submission would be treated as an acceptance of that offer or undertaking and:
>
> (a) that the alternative submitted by any tenderer would be considered alongside a compliant revised tender from that tenderer; and:
> (b) that any alternative would be one of detail and not design;
> (c) that tenderers who responded to that invitation would be treated equally and fairly.
>
> These contractual obligations derive from a contract to be implied from the procurement regime required by European directives, as interpreted by the European Court, whereby the principles of fairness and equality form part of a preliminary contract of the kind that I have indicated. Emery [author of *Administration Law*] shows that such a contract may exist at common law against a statutory background which might otherwise provide the exclusive remedy. I consider that it is now clear in English law that in the public sector where competitive tenders are sought and responded to, a contract comes into existence whereby the prospective employer impliedly agrees to consider all tenderers fairly.'

Later the judge considered the necessary requirements for fairness in making the selection on a basis of 'overall value for money' or that which is 'most economically advantageous' and put his decision in these words:

> 'The principles of transparency and fairness require a tenderer to know, without doubt, what objective criteria are going to be applied and, as Regulation 20 makes clear, their order of importance. All the criteria have to be stated – this is obvious but it is now made clear by Article 30(2) of Directive 93/97 – and to be identifiable as such, either because they are grouped in the same place or because they are clearly marked out. Indeed, the requirement that they should be stated in descending order of importance is perhaps the most significant pointer to the need for the criteria to be clearly identifiable as such.'

Obligation post-tender

In certain circumstances there may be an agreement, either express or implied, as to the binding nature of a quotation. There is no reason why a major subcontractor, or indeed other subcontractors, may not agree to be bound by their offer to a contractor, if at tender stage, as suitable consideration, it is agreed that the subcontractor's offer will be accepted if the contractor is awarded the main contract.

While this is not a situation that often exists in the British contracting industry it appears to be common in Canada.

The case of *Northern Construction Co Ltd* v. *Gloge Heating & Plumbing Ltd* (1984) 6 DLR (4th) 450, although phrased by the judge as being about 'how may acceptance become communicated so as to be binding upon the tenderer', is about the potentially binding nature of a quotation when offered. In that case the subcontractor wished to avoid the subcontract because it had underpriced the work, but, it is submitted, the principle would have been the same had the contractor sought to use a cheaper price from another bidder. Among the facts found by the judge, at page 452, was that:

> 'It was agreed that the custom of the industry would then dictate that any subcontractor interested in bidding to the general contractor would arrange with the owner to procure a set of plans, specifications and tender documents from which it would prepare its prices and make up its bid. It would also know from the documents how long the owner had to accept a bid after tenders had been opened.
>
> It was acceptable practice in the industry for some subcontractors to hold back bids until the last possible moment and that the general contractor would not make up its final tender price and tender until a matter of an hour or less before the closing time.
>
> The supposed rationale behind it was the fear on the part of some subcontractors that if they submitted their bids earlier, there was a risk that somehow their competitors would find out the numbers and adjust their later bids accordingly. The practice was referred to as "bid shopping".'

Under this procedure Gloge submitted a low bid shortly before the closing time. Northern then challenged Gloge to confirm their bid, which the judge recorded, at page 453, as follows:

> 'Bernard [Estimator for Northern, contractor] said that he then called Simons [estimator for Gloge, subcontractor] back on the telephone and informed him of these facts without mentioning specific figures. He also asked Simons if he had used the same figures in Gloge's bids to other general contractors and, if so, was he prepared to commit Gloge to the figures submitted. It is Bernard's clear recollection that Simons said yes to both questions.'

The tender acceptance procedure was found by the judge to be:

> 'According to the tender documents, the Department of Transport had 30 days after the opening of tenders to scrutinise the tenders for the purpose of ensuring that the low bid was indeed lowest and that everything else was in order. During this period and one 30-day extension allowed, the Department of Transport was entitled to hold all tenders open for acceptance and would then confirm in writing that a particular tender was accepted. Upon acceptance, the general contractor would then confirm the contract with its subcontractors.'

The judge then reviewed a number of Canadian cases covering the law of both mistake and offer and acceptance, citing the decision in *The Queen in right of Ontario et al.* v. *Ron Engineering & Construction Eastern Ltd* [1981] 1 SCR 111, which considered that under the tendering arrangements used between those parties there was a two-stage contractual agreement which the judge, at page 461, described as follows:

> 'First, contract A comes into being upon the submission of a bid in response to an invitation to tender. Contract A is a contract collateral to contract B, the formal construction contract. Under contract A, the tenderer agrees to the terms and conditions set out in the call for tenders and undertakes to keep his tender open for the period of time set in that call. Contract B is a formal document consistent with the terms worked out in contract A.'

What *Northern Construction* shows is that contrary to normal English practice it is possible, by agreement, to conclude a binding subcontract at some stage in the main contract tendering process, which binds the contractor and subcontractor without further negotiation. Such an agreement might make the subcontractor's offer binding either on the submission of the contractor's own offer or at the time the contractor could no longer withdraw its offer. Any such arrangements should be reciprocal but need not be so.

Another form of tender agreement may relate to the availability of production capacity. In such circumstances a specialist subcontractor is given the opportunity to tender on an undertaking that the necessary factory capacity will be held open until the result of the tender for the main contract is known.

A major factor in post-tender negotiations may involve the availability of production capacity. A specialist subcontractor with spare capacity may be prepared to offer a very competitive price on the basis that production will be carried out in a specified period, which either becomes a condition of the subcontract or may be considered as a reason for damages if the work is not made available until later than anticipated.

Conversely a specialist subcontractor negotiating with several contractors, in connection with different projects may, before negotiations are concluded, fill its capacity and have either to advise the contractor that its offer is no longer open or that, owing to changed circumstances, it must increase its price to cover for overtime working and/or extend the delivery time.

Many specialist subcontractors are dependent on the availability of materials and/or subcomponents from others. In some cases late procurement may mean that the materials are no longer available within the required time-scale, while in other situations they only become available through stockists at premium rates. It is for the subcontractor to qualify its tender as necessary to safeguard its position.

Specialist trades know the specific difficulties of their branch of the industry and should make their bid safeguarding, as far as possible, the risks set out above. It will be for the contractor to recognise these and plan its procurement with these problems in mind.

Where the employer, or the employer's design consultant, wishes to make a selection as to a subcontractor to carry out specific parts of the contract Works, he may, either before the appointment of the main contractor or shortly after, invite tenders from the proposed subcontractor or prospective subcontractors. The intention here is that the selected subcontractor will become a subcontractor of the contractor, either by assignment, by novation or under specific arrangements such as nomination or naming under the ICSub/NAM form.

Under the ICSub/NAM form the subcontractor, by way of its offer and by completing form ICSub/NAM/T, undertakes, depending on the circumstances at the time of tender, to conclude a subcontract with the contractor, either on notification that the employer and the contractor have entered into the main contract or on receipt of a copy of the architect/contract administrator's instruction in relation to the relevant provisional sum.

Under these arrangements both the contractor and the subcontractor make undertakings to the employer that they will conclude a subcontract on terms defined by the employer with the company selected by the employer to be contractor and subcontractor. Only by reason of any specific terms of their offers will either party not be bound, in the event of their offer being accepted by the employer, to contract with the other.

Nature of the tender

Where a contractor sends out a detailed enquiry stating the obligations it requires the prospective subcontractor to price and allow for in its offer, including the undertaking to enter into a formal agreement in stated terms, then in the absence of any statement to the contrary it is likely that any tender submitted in response to such an enquiry will be deemed to incorporate the requirements of the enquiry.

Exceptions to this general presumption may include agreement to any particularly onerous or unusual terms not expressly drawn to the tenderer's attention, the incorporation of documents which are not in general use and not provided to the subcontractor, and matters where the law requires the express agreement of the parties.

Conversely, where the prospective subcontractor's tender refers to other terms and conditions, then those terms and conditions will be the basis of that offer. Complications will however arise where the tenderer incorporates or excludes from its offer certain matters either listed or referred to in the enquiry or provided by the subcontractor within its offer, but makes no reference to the other documents or conditions within the enquiry. It will be for the courts or other tribunal to decide whether and to what extent other terms can be incorporated by reference. The most likely situation will be that, where the offer is stated to be based on express provisions this will be the limit of the express terms, while where the offer makes exclusions the remainder of those matters in the tender enquiry will stand.

Defining the extent of work included within an enquiry and/or offer is much more problematic than it appears. Phrases such as 'everything necessary for the installation' or 'necessary for the proper functioning and operation' of the subcontract works, will in practice frequently not create the certainty sought. Where the design is complete prior to the enquiry and/or offer, then reference to specific components shown on detail drawings will normally fully define the scope of the subcontract work. However, such details are often not available at the time of the subcontractor's offer; as many specialist subcontractors will have a design or detailing function, there will only be outline or schematic drawings available at the time of tender.

Subcontractors and/or contractors frequently seek to clarify the scope of work by listing either the items expressly included or excluded in the enquiry or offer. While such lists may help to clarify matters, a list may be deemed to be complete and in the case of exclusions to suggest that everything else is included. Inevitably the construction of the resultant subcontract will be defined by the particular circumstances and express words of the offer and/or enquiry.

Reference is frequently made to the case of *Williams* v. *Fitzmaurice* (1858) 3 H & N 844, where the judge had to consider the meaning of the requirement of the specification that the house should be 'completed dry and fit for occupation'. The court found that, despite there being no reference to fitting floor boards in the specification, floor boards were necessary for the house to be occupied and were therefore not to be considered an extra to the contract. The judge said, at page 849:

> 'The main question in this case is whether the plaintiff was bound by the contract to put in the floors. He contracted that for [£1,100] the house should be "completed dry and fit for occupation by the 1st of August, 1858". It is clear that the house would not be complete or fit for occupation without floors.'

While this decision may be of assistance in deciding whether the subcontract has or has not included subcomponents, it is less useful in deciding interface activities, which can be a major cause of dispute. Such items as the supply and installation of cover flashings or movement joints between components are frequently ill defined in subcontracts. The supply and installation of building services can lead to a wide range of disputes including the extent of building work, such as the forming and making good of holes and chases and/or the provision of fixings and supports for both plant and distribution pipes and ducts.

Amended tenders and other pre-contract requests

Specialist subcontractors and suppliers are often approached either to provide assistance and guidance on the basis that they will ultimately be invited to tender for the work involved, or post-tender are requested to provide alternative samples, proposals and tenders on the expectation that they will be given the subcontract for the work. Generally such work will not be extensive and the prospective subcontractor will be content to provide it on a gratuitous basis.

While the law recognises that where an offer is made in response to an enquiry to carry out construction work, it will be presumed that such offer is made without charge, in other circumstances where there is a request to perform a service, it will be presumed that on completion of that service the party will be entitled to a reasonable sum for carrying it out. Where such service is provided on the understanding that the party will be given the opportunity to tender, or post-tender there is an indication that it will be awarded the subcontract for the work, then it may be considered that payment for such further work will be recovered from the anticipated profit that the subcontractor will make from the work.

This situation was considered in the case of *William Lacey (Hounslow) Ltd* v. *Davis* [1957] 1 WLR 932. While this case to an extent reflected the unusual circumstances of post-war reconstruction and the requirements of the War Damage Commission, the judge held that, in the expectation of being given the contract for building work, the builder had at considerable expense provided a number of revised tenders and other details for various schemes for the same development. Mr Justice Barry said, at page 934:

> 'Mr Daniel [Counsel for William Lacey] rightly conceded that if a builder is invited to tender for certain work, either in competition or otherwise, there is no implication that he

will be paid for the work – sometimes the very considerable amount of work – involved in arriving at his price: he undertakes this work as a gamble, and its cost is part of the overhead expenses of his business which he hopes will be met out of the profits of such contracts as are made as a result of tenders which prove to be successful. This generally accepted usage may also – and I think does also – apply to amendments of the original tender necessitated by bona fide alterations in the specification and plans . . . It may also happen – as it certainly did happen in the present case – that when a builder is told that his tender is the lowest and led to believe that the building contract is to be given to him, he, the builder, is prepared to perform other incidental services at the request of the building owner without any intention of charging for them as such. He is not – [Counsel for William Lacey] suggests – rendering these services gratuitously, but is content to be recompensed for them out of the profit which he will make under the contract.'

And at page 935:

'Now, on this evidence, I am quite satisfied that the whole of the work covered by the schedule fell right outside the normal work which a builder, by custom and usage, normally performs gratuitously, when invited to tender for the erection of a building.'

The judge made it clear that the contractor's entitlement to payment was largely because it had been led to believe that the contract would be awarded to it, and that it could expect to recover its costs via the contract sum, at page 939:

'I am unable to see any valid distinction between work done which was to be paid for under the terms of a contract erroneously believed to be in existence, and work done which was to be paid for out of the proceeds of a contract which both parties erroneously believed was about to be made. In neither case was the work to be done gratuitously, and in both cases the party from whom payment was sought requested the work and obtained the benefit of it. In neither case did the parties actually intend to pay for the work otherwise than under the supposed contract, or as part of the total price which would become payable when the expected contract was made. In both cases, when the beliefs of the parties were falsified, the law implied an obligation – and, in this case, I think the law should imply an obligation – to pay a reasonable price for the services which had been obtained . . . In my judgment, the proper inference from the facts proved in this case is not that this work was done in the hope that this building might possibly be reconstructed and that the plaintiff company might obtain the contract, but that it was done under a mutual belief and understanding that this building was being reconstructed and that the plaintiff company was obtaining the contract.'

The judge concluded with the words, at page 940:

'The old-fashioned purpose of an estimate may become an almost subsidiary one, and builders may be called upon to perform all kinds of services and provide all kinds of information quite unconnected with the submission of a tender. I find it difficult to think that any injustice will result if building owners, who obtain the benefit of all these services upon the understanding that a contract is to be given, should be required to make some payment for them, if they subsequently decide that the contract is withheld.'

Where subcontractors are being requested to carry out alternative pricing and/or provide details for consideration by the contractor, they should make it clear to the contractor the basis on which such work is being carried out and the likely timing for provision of such information, together with any requirement for payment in the event that they are not given a subcontract for the work. Likewise, a contractor seeking alternative offers should make it clear when it requires such further information and should obtain agreement that the timetable will be met. It must be acknowledged that specialist subcontractors become increasingly reluctant to provide unpaid information, and a contractor may find it beneficial to have a formalised agreement with consideration for the provision of such additional details.

Chapter 2
Contract

Formation

A contract can be said to exist where there has been an offer from one party, which has been accepted by the other, provided there is provision for consideration from the one party to the other for the benefit obtained. The contract must also be between parties legally capable of forming a contract, who having reached an agreement intended to create a legally binding arrangement.

A binding contract need not be in writing to be enforceable, but in the event of a dispute a decision may be difficult without any suitable written evidence. Where work is carried out at the request of one party and/or with its knowledge, the other party may be granted the value of the service in restitution.

The Housing Grants, Construction and Regeneration Act (HGCRA) applies only to construction contracts in writing. The HGCRA sets out at section 104 what is meant by a construction contract, and at section 105 the definition of 'construction operations' and a list of operations not 'construction operations'. Section 106 makes the HGCRA not applicable to construction contracts with residential occupiers, but it will apply to subcontractors to such contracts. Section 107 defines a contract in writing for the purposes of the HGCRA. For full details of these sections see Chapter 19, on statutes.

The definitions of contracts in writing, as defined in section 107 (2) to (5), show that these may include situations where there is not strictly a contract at law. The provisions of this section have been tested in the courts in the cases referred to in the section 'Failure to conclude an agreement' later in this chapter.

Work contracted for

The subcontractor, by giving a 'quotation' or 'estimate', makes an offer in which it sets out what it is prepared to provide and the price it requires for doing so. Such an offer will generally result from a request from the contractor, which may require the offer to comply with a substantial and detailed set of enquiry documents. Unless specifically stated to comply with these documents, the subcontractor's offer will be for whatever it says it is for and under the terms and conditions stated in that offer.

A period of negotiation will often follow, during which the detailed arrangements of the subcontract will be negotiated between the parties until full agreement is reached, or in the absence of full agreement until some other action creates a contractual situation.

Agreement between the parties may be recorded either by an acceptance letter or by articles of agreement signed by both parties. Either way the agreement will list the relevant documents comprising that agreement.

SBCSub/C at clause 2.1 sets out the subcontractor's obligations and requires it to carry out and complete the subcontract works in compliance with the Sub-Contract Documents, and specifically refers to the health and safety plan and statutory requirements. In addition, the subcontractor is to conform to all directions and reasonable requirements of the contractor that regulate the carrying out of the main contract. The works are to be carried out in a proper and workmanlike manner. Finally, the subcontractor is to give all statutory notices relevant to the subcontract work.

ICSub/C incorporates an identical clause, and SBCSub/D/C, ICSub/D/C and DBSub/C all contain additional sections outlining the subcontractor's obligations for design, to the effect that the subcontractor is to comply with instructions from the contractor for the integration of the subcontractor's design with the design of the main contract Works.

The documents are annexed to the Sub-Contract Agreement (SBCSub/A) and scheduled at section 15 of the Sub-Contract Particulars. It must be stressed that the SBCSub/C Conditions do not in themselves define the obligations of the parties; without the clear incorporation of other documents, which define matters such as scope of works, specification, price and time, these essential elements of the subcontract will be uncertain.

Under SBCSub/C clause 2.4, quality and workmanship depend on the type and standards described in the Sub-Contract Documents, so far as procurable and to the reasonable satisfaction of the architect/contract administrator, or of a standard appropriate to the Sub-Contract Works. Clause 2.4 has five sub-clauses defining the requirements for the quality of the works:

1. The materials and goods for the subcontract works are to be as described within the Sub-Contract Documents, but this requirement is qualified as being 'so far as procurable'. However, there is provision for the subcontractor to seek consent from the contractor to substitute other goods and materials.
2. Workmanship is also to be of the standard described in the Sub-Contract Documents.
3. Goods, materials and workmanship are required to be to the reasonable satisfaction of the architect where described as being to his approval. Where there is no standard specified, then the quality is to be 'appropriate' to the subcontract works.
4. The contractor may require, and the subcontractor is to provide, reasonable proof that the materials and goods used are compliant with the subcontract.
5. The subcontractor is also required to take positive action to encourage its employees to be qualified under the Construction Skills Certification Scheme.

ICSub/C and ICSub/D/C incorporate at clauses 2.3.1 and 2.3.2 the same provisions as SBCSub/C clauses 2.4.3 and 2.4.5. There is no obvious reason for omitting the other three clauses, especially the requirements of clauses 2.4.1 and 2.4.2, unless it is considered that they contain an implied obligation in which case there is no need for an express provision within SBCSub/C.

Sub/MPF04 sets out the subcontractor's general obligations at clause 1, which has two sub-clauses:

1. A general obligation to carry out the completion of the design work and the construction work in accordance with the subcontract.
2. The subcontractor undertakes that it has the competence to do the work and will provide the necessary resources, including design staff to satisfy the requirements of the CDM Regulations.

Sub/MPF04, by the use of the phrase 'in accordance with the Sub-Contract', places more reliance on the Sub-Contract Documents than either the SBCSub or ICSub forms.

In contrast, ShortSub has a list of six short sub-clauses under the general heading of 'Subcontractor's obligations'. Sub-clauses 1 to 3 can be compared with clauses 2.4.1 to 2.4.3 and 2.4.5 of SBCSub/C. These sub-clauses provide that:

1. The subcontractor is to use due diligence and carry out the work in a good and workmanlike manner to completion.
2. As with SBCSub, the goods, materials and workmanship are to be to the standard stated, and where none is stated they are to be of satisfactory quality.
3. Again as with SBCSub, the subcontractor is to encourage its employees to obtain qualifications under the Construction Skills Certification Scheme.

Contractors frequently rely on the details of their enquiry documents and seek to relate the subcontractor's tender to those. As mentioned above, the subcontractor's tender or offer will be for what it says it is, regardless of what the contractor has requested or what the subcontractor has priced for and/or incorporated in its sum. Thus, where the subcontractor's offer is for 'all work as necessary' or 'as detailed on the drawings and in the specification', the subcontractor will be unlikely to be able to claim additional sums if, at a later date, it finds it has failed to allow for certain items covered by such a description. On the other hand, where the subcontractor's tender includes a detailed schedule of the work and it is subsequently found that certain work, required by the contractor, has not been included in that schedule, then the subcontractor will be entitled to payment for such work as an addition.

In the Court of Appeal case of *Boynton and Another* v. *Willers* [2003] EWCA Civ 904, it was found that the contractor's quotation listed a number of items of work with prices against each. It was claimed that the contractor had not included for the bay windows, which were not separately listed. Further, although the contractor constructed the bays it failed to include them in its invoices, which the employer suggested implied that they were not in fact additional work. Lord Justice Potter, rejecting the appeal, said at paragraph 28:

> '28. Assuming that the finding was justified on the evidence (and, despite encouragement to the defendants at the stage leave to appeal was granted, no transcript of the relevant parts of the evidence has been obtained) it sounds the death knell to Mr Smith's [Counsel for Boynton's] submission upon construction. Even if the evidence did not justify a finding of actual agreement, i.e. a common subjective contractual intention that the bays be excluded from the quotation, the judge was in my view right to conclude that the absence

of any reference to such construction of the bays in the quotation, viewed against the contractual background, meant that they were indeed excluded. So far as that background was concerned, the judge found that, in contrast to his view of the defendants, the claimant was in all respects an honest, reasonable and accurate historian. It was the claimant's evidence that, at the time the quotation was given, the parties had not decided upon the precise type of detailing of the bays to be constructed, in which circumstance it is clear that accurate assessment of the costs of the bays would have been impossible. Given that they were a major item in respect of which the necessity for a quoted price would have been apparent in the interests of both parties, and given that not even a p.c. [prime cost] sum was provided for in the quotation, the reasonable and businesslike construction of the quotation was that it did not cover the construction of the bays.'

In the subcontract situation it will normally be for the contractor to define the work to be carried out by specific subcontractors. Even where there is no uncertainty in the contractor's invitation to tender as to the work it requires the subcontractor to price, the subcontractor's offer will only include for the work that its offer states is included.

In post-tender negotiations the contractor may seek to obtain agreement from the subcontractor that its offer has included for the work identified within the enquiry document. It will require very clear evidence that the subcontractor has agreed to provide for work excluded from its accepted offer if there has been no consequential increase in the subcontract sum. Such a situation existed in the Court of Appeal case of *Carillion Construction Limited (trading as Crown House Engineering)* v. *Ballast plc (formerly Ballast Wiltshire plc)* [2001] EWCA Civ 1098, where Lord Justice Aldous said, at paragraph 39:

'39. I will come to the evidence as to what happened at the meeting on 8 November, but it seems improbable that Crown House would warrant that their works, set out in the revised specification, were the same in extent as that required by the employer. They knew that they were not and that [representative of employer] had suggested that the two did not equate. Further, it seems highly unlikely that Crown House would agree to complete all the employer's requirements for the price tendered for carrying out the works in the revised specification. The whole purpose of the revised specification was to indicate the works they would do for the price quoted. Their estimators had worked upon the basis of the revised specification and to agree to different works for the same price without detailed consideration would appear unlikely.'

In this he was supporting the judge at first instance, who he quoted at paragraph 51, as follows:

'I preferred the evidence of Mr Fox [an employee of Carillion] of the claimant because having very carefully priced their tender on a specific basis, it seemed to me wholly implausible that experienced and competent businessmen such as those representing the claimant would already [readily] alter the basis of their written tender without any corresponding alterations, or recalculation or [of] price. I was not impressed with the evidence of Mr Peat [employee of Ballast] for the defendants on this point, which, as I have said, was repetitive on the subject of employer's requirements and whose evidence smacked of the benefit of hindsight.'

In situations such as this, contractors often rely on phrases such as 'the work will comply with the employer's requirements' or 'as specified'. Such phrases are often held by the contractor to mean that the subcontractor will undertake all the work required by the employer. Without clear wording it is more likely that such a phrase will only mean that the work priced will be to the employer's requirements or to the specification, and not that the scope of the works is for more than stated in the offer.

In the *Carillion* case, the subcontractor had defined in its offer the work it was prepared to undertake, and had stated:

> 'Only those systems and services described in the foregoing paragraphs have been included in our tender.'

In his judgment in *Carillion* Lord Justice Aldous makes this distinction, at paragraph 37:

> '37. The documents make it quite clear that as of 22 October Crown House's tender was based on carrying out work set out in the revised specification subject to the exclusions that they had pointed out. No doubt their work had to be carried out to the standard required under the main contract. However, they were not agreeing to carry out all the electrical and mechanical work required by the employers.'

Further difficulties can arise where the subcontract is let under a design and build contract and the contractor has the job of completing the design produced by the employer's design consultants. During the design development it may be found that components have been underdesigned or that a saving made in one element leads to increased costs in another. Similarly, coordination of different elements such as services may require a change in a component size or in its routing through the building. It will require express wording in the subcontract to relieve the contractor from the obligation to pay for such changes. In general terms, the liability or benefit for any change in cost will rest with the party making the change.

In the Scottish case of *Miller Construction Ltd* v. *Trent Concrete Cladding Ltd* (4 August 1995, unreported), Trent had contracted to design a reinforced concrete structure that incorporated precast elements which Trent were to supply and install, and *in situ* concrete work the design and construction of which they subcontracted to Miller. This concrete work included the structural topping to the precast double-T floor slabs.

A number of interface disputes arose for which Miller sought additional payment by way of variations to their subcontract. These areas included the connection between the *in situ* stair and service cores and the treatment given to the double-T slabs prior to the concrete topping. Lord Penrose had to consider Trent's basic argument that where there was an obligation to design and build, the obligation was to do all that was necessary to provide a complete structure. As Lord Penrose put it, at page 86:

> 'The defender's argument relied, in the first place, on the proposition, vouched by Hudson [author of *Hudson's Building and Engineering Contracts*], that a contractor may be obliged to carry out, within a lump sum contract price, work which is not mentioned specifically in any of the technical documents incorporated in the contract but which the general description of the scope of the works nevertheless indicates must have been contemplated by the contract as part of those works.'

Miller, on the other hand, claimed that they only had a liability for that work expressly included in the concluded agreement. In considering the effect of subletting parts of a project, Lord Penrose agreed that it was for the management or main contractor to select what parts of its obligation it passed on to its subcontractors, and that consequently the subcontractor will only have the responsibility to provide a complete element, within the element sublet to it. Lord Penrose put it in these words, at page 87:

'But the issue in the present case is different: the obligation to the management contractor would not necessarily, of itself, give any indication of which of the primary and secondary subcontractors owed the other a contractual duty to procure that result under the secondary subcontract. The sub-subcontracting of elements of the work of its very nature involves the disintegration of the works package project, for domestic purposes of the primary and secondary subcontractors. The description of the secondary subcontract works in the present case is brief and uninformative: to carry out "in situ concrete works and blockwork in connection with the precast concrete superstructure works". As a matter of language it provides little specification of the division of labour at the interface between the in situ cast and precast concrete components of the superstructure. This is an area which might depend to some extent on proof of the indispensable necessity of some component of work as a factor in the operations of one or other of the contractors. But more generally one might expect it to depend on negotiation and agreement of the split between them of the total operations required.'

Lord Penrose goes on to describe what happened in that case:

'In the present case it is accepted that it was the defenders who took the "design" decision that the interconnection between the cores and the double-T slabs would make use of welded steel components, and that the interconnection between the cores and the beams would employ varieties of the Trent Connector, supplied exclusively by them. Some means of connection was admittedly indispensably necessary to the performance of the defenders' contract with THCM [management contractor], and it was the defenders' decision to employ these particular means. That involved the pursuers in the work of fitting the components necessarily embodied in the cores. The defenders were entitled to adopt that solution: the pursuers were bound to accommodate the solution in their works. But there were other solutions which might have been adopted, and which would not have required the accommodation of such components. Typically the cores might have provided tie bars as part of the reinforcing, for connection to the beams and double-T units as reinforcement for in situ cast infill bridges.'

He goes on to say:

'The pursuers' claim for payment is based on the proposition that the selection by the defenders of the particular design solution involved them in work which was not anticipated in their instructions as expressed in the drawings incorporated into the secondary subcontract. To the extent that this issue turns on the content of the drawings, enquiry is necessary. The whole contract drawings may be relevant. The distinction between the connections at first floor level and upper levels must be explored to determine whether it has any substance. Any distinction which turns on the treatment of the welded connectors and the Trent Connectors requires investigation.'

Thus the outcome as to how much, if any, additional payment was allowed to Miller would depend on the facts to be determined at a full trial. It was also argued by Trent that Miller had an obligation to make any necessary enquiries at tender stage to enable them to understand the full extent of the work they were to do if successful, and that in any event there had been an opportunity post-tender, pre-contract to do so. Again, the same duty did not necessarily exist between the primary contractor and the management contractor as between the primary contractor and the subcontractor. Lord Penrose said, on page 91:

> 'It is a matter of concession that this applies as between the parties. But it does not follow that a duty which the parties may in one sense share towards the management contractor to enquire as to the requirements for due performance of the primary subcontract can be taken to include a duty as between the primary and secondary subcontractors binding on the secondary subcontractor to enquire as to design decisions of the primary subcontractor which may have a bearing on the definition of the interface between their respective sub-packages. *Prima facie* it is for the primary subcontractor to specify the work which he requires the secondary subcontractor to perform. In my opinion Mr Taylor's [Counsel for Trent Concrete Cladding's] argument that in any event the responsibility to discover and accommodate the design decisions of the defenders lay on the pursuers is not correct in this case, and can form no part of the decision as to the pursuers' claim for payment.'

In other words, at least on the facts of that case, it was the duty of Trent to make it clear exactly what it required Miller to price and, it is suggested, to satisfy itself that they did so. In this respect merely obtaining an assurance that Miller's tender complied with Trent's enquiry documents would be unlikely to override any ambiguity, inconsistency or incompleteness in those documents. This view is supported by the words of Lord Penrose, on page 92:

> 'Similarly it appears to me that Mr Taylor's [Counsel for Trent Concrete Cladding's] argument that the pursuers were under an implied obligation to make the defenders' design solution effective by providing for the incorporation of the components in the cores is incorrect. The authorities . . . would support the view that "of course" a specialist subcontractor would have regard to the need for some form of connection with the precast elements of the structure in designing the cores. But it would not follow that he would have regard to the particular as distinct from the general need to be aware of the requirement and adapt to it when it was specified.'

However, matters relating to the need of the primary contractor to define its requirements at interfaces and/or clashes between the installation of the subcontractor and the work of others must be distinguished from a failure to describe fully the components necessary to complete the subcontractor's installation. This will be of particular relevance where the subcontractor is installing work to a proprietary system, such as a system for roofing or cladding or internally one for partitioning or ceiling work. In such cases the general law requiring the subcontractor to provide all that is necessary, even if imprecisely defined, is likely to apply.

In the historic case of *Williams* v. *Fitzmaurice* (1858) 3 H & N 844, which concerned a contract to build a house, the floor joists were fully specified but there was no mention of the associated flooring. When he came to fit the flooring the contractor stopped work and sought payment for the flooring as an extra. However, the judge found that the contractor had contracted to provide a house 'complete, dry and fit for occupation' and not just to execute the work as specified. Lord Justice Crowder said, at page 849:

> 'He contracted that for £1100 the house should be "completed dry and fit for occupation by the 1st August, 1858". It is clear that the house would not be complete or fit for occupation without floors. Though the specification sets forth many particulars as to the wood work, it must not be taken that the intention was that those things alone which were specified are to be provided, but that, while the specification directs that certain things shall be of the quality, dimensions and materials named, the flooring was to be of the ordinary character and such as would be reasonably sufficient for the house.'

This decision was supported by the Court of Appeal and Baron Channell said, at page 852:

> 'The contract was that the *house* should be completed and fit for occupation by the 1 August 1858, not that the *works* therein before mentioned should be completed by that day. I think that, looking at the terms of the contract, it would not be reasonable to read it as if it excluded all work not specifically mentioned. The plaintiff contracted to do the entire work in the various characters of bricklayer, carpenter, plumber, &c, for the sum of [£1,100]; and it is not the less a contract to do the whole, because it is specified that certain parts of the building shall be constructed in a particular way.'

The conclusion to be drawn from these decisions is that it was possible to contract to provide either a whole item, in this case a house complete, dry and fit for occupation, or work comprising a schedule of components. It follows that depending on the words of the subcontract, so the liability for anything omitted or overlooked may rest with the contractor or the subcontractor.

Discrepancies between documents

Where the subcontract incorporates a number of documents, some specific to the project and others of general use, then it is very likely that there will be discrepancies between them. This situation is dealt with in SBCSub/C in two ways: first by giving priority to the main documents at clause 2.7, which deals with the contract conditions, and secondly at clause 2.10, which requires the subcontractor to seek a direction from the contractor who should instruct as necessary, should the subcontractor discover any departure, error, omission or inadequacy or any discrepancy or divergence between:

1. the Sub-Contract Documents;
2. the main contract;
3. any direction issued by the contractor under these conditions (save in so far as it is a direction requiring a variation).

Both SBCSub/D/C and DBSub/C add an additional sub-clause 2.10.4 relating to the details to be provided by the subcontractor under its design obligations.

It must be noted that it is not for the subcontractor to seek out such discrepancies; it may be said that it is for the contractor to have made certain what it required.

Although this clause requires the subcontractor to give the contractor notice, in practice it is often the contractor or others, under the main contract, who discovers a discrepancy.

ICSub/C at clause 2.8 has a similar provision but interestingly does not refer to 'If the Subcontractor finds any such . . .', but merely requires the contractor to 'issue directions in regard to any such . . .'. It cannot be the intention under the SBCSub forms that the contractor is only to issue instructions if the subcontractor finds the departures, errors, omissions or inadequacies.

Sub/MPF04 sensibly, under the heading of conflict and discrepancy at clause 5.1, states that 'if either party identifies . . .', then action is to be taken as set out in the succeeding clauses 5.2 to 5.4, which require action as follows:

1. There is an obligation on both parties on finding a discrepancy in the documents, or as a result of a change in statutory requirements, to notify the other party. Such notification is to be given immediately.
2. If the discrepancy is in the documents provided to the subcontractor, then the subcontractor is to advise the contractor of the alternative it elects to adopt, and should the contractor require a different option the contractor is to instruct the subcontractor by way of a change.
3. However, where the discrepancy is in the Sub-Contractor's Proposals, then it is the contractor who elects the alternative and instructs the subcontractor, and such instruction will not be a change.
4. Where the discrepancy concerns a statutory requirement, subject to any instruction complying with the statutory requirements, the contractor is to issue a change instruction to the subcontractor indicating which provision it requires.

The effect of clauses 5.2 and 5.3 is to give the party that has not caused the conflict or discrepancy the right to select the provision it intends or requires to be adopted. In clause 5.2, should the contractor require the subcontractor to adopt another provision, then it will be regarded as giving rise to a change.

ShortSub makes no specific provision for the resolution of discrepancies. However, it is likely that in the event of a discrepancy being identified, a similar provision to ICSub/C would be implied by way of trade practice or business efficacy.

Whoever discovers the discrepancy, so far as the subcontract is concerned, it is ultimately for the contractor to decide the action to be taken and issue directions accordingly. Such directions may require a hold to be put on the work until the matter is resolved and directions can be given as to how to proceed. Under certain circumstances, such as with subcontractor designed work, there may be a prerequisite for the subcontractor to make new proposals prior to an instruction or direction being given by the contractor.

Conclusion of a contract

Agreement of a contract between the parties may be reached after prolonged negotiations, which will have started with the original enquiry followed by the tendered offer. There may be an exchange of letters and/or notes of meetings which may be held to be counter-offers or just a record of agreement as to certain terms. The outcome may be recorded either by an acceptance letter or by a formal agreement signed by both parties.

A subcontract may be concluded by an oral agreement without either the offer or acceptance being in writing. In such a situation it will be difficult for either party, in the event of a dispute arising, to demonstrate the precise nature of the agreement. It is probably for this reason that the HGCRA requires a construction contract, implying the provisions of the Act, to be in writing. The HGCRA defines a 'contract in writing' and there have been various cases defining the law in this respect. In the case of *Connex South Eastern Ltd* v. *MJ Building Services Group plc* [2004] BLR 333, Judge Harvey QC had to consider whether a written offer accepted orally constituted a contract in writing for the purpose of implying in the contract the adjudication procedures of the HGCRA. He found it did, at paragraph 24:

> '[Counsel for Connex South Eastern] drew my attention to the case of *RJT Consulting Engineers* v. *DM Engineering (Northern Ireland) Limited* [2002] 1 WLR 2344, CA. He submitted that the complete agreement (per Ward and Robert Walker LJ at paragraphs 19 and 20 respectively of the judgments), alternatively at least all material terms (per Lord Justice Auld at paragraph 24), must be in writing in order to fulfil the requirements of section 107. [Counsel for MJ Building Services] submitted, and [Counsel for Connex South Eastern] agreed, that whichever approach is adopted, it was manifestly not the intention of Parliament to exclude from the jurisdiction of an adjudicator an agreement solely because it contains implied terms. I accept that very reasonable proposition. Mr Ashton has not identified any express terms of the agreement that are not in writing. His point is that there was no written acceptance of MJ's tender. In my judgment, that is irrelevant. It is not suggested that there was an oral acceptance containing terms. But in any case, there is a brief reference in the minutes of the meeting held on 15 September 2000 to the effect that Connex had given an instruction that the project was to be carried out immediately. In the context, the conclusion is irresistible that that instruction constitutes an acceptance of MJ's tender. Since the minutes were written with the authority of the parties, they constitute evidence, falling within section 107(4) of the Act, of the acceptance. I conclude that the contract was in writing, within the meaning of section 107 of the Act.'

There will however be many situations where work starts before full agreement between the parties is reached. In some cases it will be clear from the documents the state of negotiations and an agreement may well exist, which the parties, at that time, regard as being temporary. For the concept of a temporary contract see the judgment of Judge Thornton QC in *Hall & Tawse South Limited* v. *Ivory Gate Limited* (1996) 62 Con LR 117, in which he states early on in the judgment:

> 'I propose, therefore, to refer to this contract as "the provisional contract". It envisages that the parties will continue to negotiate about the form of contract which would govern the work and which would supersede the provisional contract once all those terms have been agreed.'

This is frequently the situation with letters of intent, where an offer is made for a subcontractor to commence work and to be paid in a certain way until either a deadlock in negotiations occurs and the temporary agreement is terminated, or a full agreement is reached and a finalised subcontract concluded.

The parties, during their pre-subcontract negotiations, may agree that a binding subcontract will come into existence on the happening of some event with no direct connection to the subcontract, such as the conclusion of the main contract agreement. In the Court of Appeal case of *Stent Foundations Limited* v. *Carillion Construction (Contracts) Limited* (2000) 78 Con LR 199, where the piling subcontract was to be a works contract in conjunction with a management contract, the initial enquiry and negotiations had been carried out by the employer's project manager before the appointment of the management contractor. There was a significant period after the selection of both contractors when further negotiations were taking place, before the conclusion of those contracts. The subcontractor commenced work under a letter of intent from the employer's project manager.

Lady Justice Hale, at paragraph 28 of her judgment, found that after the subcontract negotiations were complete, it was recorded that finalisation of the subcontract was dependent on the placing of the management contract as follows:

> '28. Meanwhile, meetings continued between WCM [management contractor] and Stent. At a pre-construction meeting on 5 October 1988 it was recorded under "Order":
>
> "Yet to be placed. Remains dependent on the placing of the Management order with WCM. An AI [Architect's Instruction] and order will then be raised. Work proceeding against the letter of intent from ECHPS [employer's project manager] dated the 12 August 1988." '

However, by the time the management contract was finalised, the piling work was nearly complete and the parties never concluded their agreement. Subsequently, the employer became insolvent and to avoid further payment to the subcontractor, the management contractor claimed that the subcontract arrangements had never progressed beyond the letter of intent with the employer. Lady Justice Hale disagreed and found a subcontract had existed between the management contractor and the piling subcontractor once the management contract was concluded, as follows:

> '47. For my part, therefore, it seems to me quite clear that the learned judge was entirely correct in this case to hold that this work had been done under a contract between Stent and WCM which came into being as soon as WCM concluded a contract with Wiggins [the employer] in January 1989, and I would dismiss this appeal.'

Acceptance by signature

Generally, where one party signs a form agreeing to the terms of the other party, this will conclude a subcontract on those terms. This will be true whether the form is prepared and offered by the contractor or the subcontractor. Unless such a form is endorsed as subject to the contents of a covering letter, it is unlikely that such a covering letter putting forward alternative terms will be regarded as a counter-offer. Even where it is considered to be a counter-offer, that party will need to demonstrate that those terms were accepted by the other party.

In the case of *Butler Machine Tool Co Ltd* v. *Ex-Cell-O Corporation (England) Ltd* [1979] 1 WLR 401, the Court of Appeal was required to consider such a situation, where the purchaser of a machine tool had sent the supplier an order with a return slip to indicate acceptance. The supplier returned the slip signed but with a covering letter stating that delivery was 'in accordance with our revised quotation of

May 23'. The court did not accept that this statement did any more than confirm the price, nor that the letter acted as a further counter-offer re-stating the supplier's terms in preference to those of the purchaser. Mr Justice Lawton put the situation in the following words:

> 'By letter dated 5 June 1969, the sales office manager at the plaintiff's Halifax factory completed that tear-off slip and sent it back to the buyers.
>
> It is true, as Mr Scott [Counsel for Butler Machine Tool] has reminded us, that the return of that printed slip was accompanied by a letter which had this sentence in it: "This is being entered in accordance with our revised quotation of May 23 for delivery in 10/11 months." I agree with Lord Denning MR that, in business sense, that refers to the quotation as to the price and the identity of the machine, and it does not bring into the contract the small print conditions on the back of the quotation. Those small print conditions had disappeared from the story. That was when the contract was made. At that date it was a fixed price contract without a price escalation clause.
>
> As I pointed out in the course of argument to Mr Scott, if the letter of 5 June which accompanied the form acknowledging the terms which the buyers had specified had amounted to a counter-offer, then in my judgment the parties never were *ad idem*. It cannot be said that the buyers accepted the counter-offer by reason of the fact that ultimately they took physical delivery of the machine. By the time they took physical delivery of the machine, they had made it clear by correspondence that they were not accepting that there was any price escalation clause in any contract which they had made with the plaintiffs.'

It is suggested that where the other party signs an acceptance form, then unless there is the clearest evidence that it did not, by signing, accept the contractual arrangements described, then that document will form the subcontract between them. It may be that if the accompanying letter expressly identifies some matters that are unacceptable to it and it can show some evidence of acceptance of this by the other party, such qualifications may be effective. But in any event, such a situation will be decided on the facts in each case.

Acceptance by conduct

Many subcontracts are concluded under terms which are deemed to have been accepted, because the subcontractor started work or was allowed to start work before an agreement was concluded. The so-called 'battle of the forms' is where each party seeks to make an acceptance of the proposals of the other, subject either to its standard terms and conditions or to further terms and conditions.

It cannot be stressed too strongly that once a subcontractor accepts an offer by commencing work, there will generally come into existence at that time, where there was not a pre-existing agreement, a binding subcontract. After that the terms of that subcontract, however uncertain and/or unacceptable to either party, can only be changed by a supplementary agreement accepted by both parties. It can happen that a subcontractor well into the subcontract works is offered 'an agreement' for signature. Such a document will often contain provisions which, in retrospect, the contractor wishes to incorporate into the subcontract. There is no requirement at law for the subcontractor to conclude such an agreement.

Such post-commencement agreement may be issued with a covering letter advising the subcontractor that payment will not be made until the subcontractor has completed and returned the Sub-Contract Agreement. Such a condition might be effective in avoiding contractual arrangements being created where no subcontract exists. However, it is suggested that once a subcontract is in existence, whether formed by conduct or any other informal procedure, it can only be amended by agreement. Once a subcontractor commences work following instructions, there will be a general presumption that it will have a right to be paid for the goods and services provided.

Acceptance by conduct of the terms and conditions in a counter-offer can apply in situations other than by the subcontractor starting work, such as by the contractor admitting the subcontractor to site. Such a situation is likely when the negotiations between the parties have been lengthy and, while significant agreement has been reached, the contractor has yet to conclude an agreement. Such conditions are likely to arise where the contractor fails to integrate the planning of procurement with the on-site activities and progress. Thus if, at a late date, it becomes apparent that work is required on site, there may be a last minute request for the subcontractor to commence its work. In these circumstances the subcontractor may confirm its willingness to commence and in addition state its terms and conditions for doing so. Again, by admitting the subcontractor to site the contractor will probably, at law, be accepting those terms as the terms of the subcontract.

In the case of *Chichester Joinery Ltd* v. *John Mowlem & Co plc* (1987) 42 BLR 100, Judge James Fox-Andrews QC found that a counter-offer made by the subcontractor was accepted when Mowlem took delivery of the joinery. In his judgment, at page 104, he warns against parties seeking to impose their own conditions by the words:

> 'This case highlights the risks facing parties who seek to impose their own respective conditions rather than using some well established form of contract commonly used for the supply of goods to a main contractor for incorporation in construction work. Unless such a form of contract is used, the courts may be faced (as here) with what is essentially an artificial state of affairs.'

Chichester had been sent an order in two parts, one part of which they were requested to sign and return. The judge said, at page 107:

> 'The common wording at the bottom of each of the two fronting pages of the forms were:
> "THE TERMS AND CONDITIONS OF PURCHASE NOS 1–18 INCLUSIVE AS SET OUT ON THE REVERSE HEREOF ARE EXPRESSLY DECLARED TO APPLY TO THE PURCHASE ORDER.
> You are requested to sign the acceptance of order and return same within seven days. Any delivery made and accepted will constitute an acceptance of this order."
> This clearly was not an acceptance of the offer made by Chichester on 8 February and confirmed by them by their letter of 14 February. It killed Chichester's earlier offer. It was, I find, a counter-offer.
> It was argued that the conduct of Chichester thereafter in preparing for manufacture of the joinery constituted an acceptance. If this submission had been open to Mowlem on the pleadings, I would have had no hesitation in rejecting the argument on two quite separate grounds. First, the terms of their Purchase Order only contemplated two forms of

acceptance: (1) written acceptance; (2) delivery and acceptance of the goods. Further, the evidence falls far short of that necessary to establish acceptance by conduct. Chichester neither orally nor in writing either within seven days or at all accepted Mowlem's order prior to 30 April 1985.'

In the event, the form was not signed and returned but later Chichester used their acknowledgement of order letter:

'On 30 April 1985 – a substantial time before any deliveries were made – Chichester sent a printed form headed "ACKNOWLEDGEMENT OF ORDER". It read:

"Dear Sirs, We acknowledge with thanks receipt of your valued order of" – and there were then inserted the words 'recent date' – "which will receive our best attention. Your order is, unless expressly agreed otherwise in writing, accepted subject to the conditions overleaf.

Please check the following specification and inform us within two days in the event of any errors or omissions. If we are not advised of any errors we will assume all details are correctly shown and manufacture will proceed on that basis."

There then appeared in capitals the words:

"A COPY OF THIS ACKNOWLEDGEMENT IS ALSO A SPECIFICATION TO OUR WORKS SO PLEASE READ THIS CAREFULLY."

On the back were set the identical conditions to those on their original 1984 quotation.'

The judge concluded that Mowlem accepted Chichester's terms by acceptance of delivery. This was made more certain since Mowlem's own conditions had stated that an accepted delivery would be evidence of acceptance of the proffered conditions. As the judge stated:

'Although it is headed "Acknowledgement" and not "Acceptance", some of the sentences I have read are much more consistent with acceptance than acknowledgement. It fails to indicate the contrary to Chichester's condition 2 that discount was 5%. It does not deal at all with the non-adjustable 10% element unless condition 1(ii) is relevant.

But I am unable to spell out of the relevant facts any intention by Chichester to be bound by Mowlem's conditions.

A necessary conclusion therefore is that this acknowledgement constituted a counter-offer thus killing Mowlem's 14 March 1985 offer.

The question for determination is whether the subsequent acceptance by Mowlem of joinery when it came to be delivered on the site constituted an acceptance of Chichester's counter-offer of 30 April and therefore subject to their conditions, or whether the parties were never *ad idem*.

These cases are never easy. After some hesitation I have reached the conclusion that Mowlem did accept Chichester's terms. Mowlem at an earlier stage were specifically offering to treat a contract as complete if, on joinery being delivered to site, it was accepted by them. In all those circumstances, I find that this preliminary issue is decided in favour of Chichester.'

It appears from this judgment that it was not whose offer was last that was of major importance for the judge, but whether the parties by their action were *ad idem*. There was clearly good reason to doubt that they were. Also significant in this case is the fact that the contractor, in its counter-offer, stated clearly what action by the subcontractor it would consider to be an acceptance of that offer. In the absence of such clear provisions the decision of the judge might have been different. What is clear from this decision is that where there are defined requirements for acceptance which are not fulfilled, it will be more difficult to hold that the subcontractor has, by its subsequent performance, accepted those terms and conditions. In such a situation it is likely that a judge would conclude that the parties were not *ad idem* and that the terms could not be defined solely by reference to the contractor's or subcontractor's offer.

This conclusion creates a dilemma for contractors, who frequently seek acceptance to their letters of intent or subcontract orders by requiring the subcontractor to sign and return a copy of the letter or order. Should the subcontractor fail to return the copy letter or order there is unlikely to be a binding subcontract until the subcontractor commences performance. Provided work starts, the letter or order will become accepted, but where an immediate start is not required or the start becomes delayed, it is likely that no binding agreement will exist and the subcontractor, at a later date, will be able to escape performance or seek other terms or conditions more onerous to the contractor.

It has long been established law of contract that the offeror cannot impose a contract on another party by stating that, if no reply is received by a certain date the offer will be regarded as accepted. The classic case is *Felthouse* v. *Bindley* (1862) 11 CBNS 869, where in concluding the purchase of a horse, the would-be purchaser wrote to the seller, his nephew, as follows:

> 'Dear Nephew, Your price, I admit, was 30 guineas. I offered £30 – never offered more: and you said the horse was mine. However as there may be a mistake about him, I will split the difference – £30 15s . . . If I hear no more about him, I consider the horse mine at £30 15s.'

In his judgment Mr Justice Willes said:

> 'The uncle had no right to impose upon the nephew a sale of his horse for £30.15s unless he chose to comply with the condition of writing to repudiate the offer . . . It is clear . . . that the nephew in his own mind intended his uncle to have the horse at the price which he (the uncle) had named – £30.15s: but he had not communicated such his intention to his uncle or done anything to bind himself. Nothing therefore had been done to vest the property in the horse in the plaintiff.'

Letters of intent

For various reasons a contractor may require a subcontractor to commence work either to the works themselves or in preparation for work on site, prior to the parties finalising a subcontract. In many cases this will be because the contractor anticipates conducting further negotiations with regards to the terms and or requirements of the subcontract.

The phrase 'letter of intent' is used in differing situations and with different effect:

1. The contractor intends to prepare a formal agreement for both parties to sign, but for whatever reason such a draft agreement is not available at the time the

contractor requires commencement of the subcontract work. In this case the letter of intent will set out in detail all the relevant documents that will be incorporated into the agreement. The subcontractor may or may not be required to sign and return a copy of that letter to signify its agreement to the stated terms. Although such a document is intended to be temporary, it will have the same effect as a concluded agreement other than not being under seal until finalised as a deed, where relevant.

2. The letter of intent is a request to commence work or preparation for the work, with an undertaking to make payment – perhaps for the subcontractor's actual costs only if the works do not proceed or do not proceed with that subcontractor.
3. The letter of intent is exactly that – an advice to the subcontractor that the contractor is still interested in the subcontractor's offer, but that it cannot or does not wish, at that time, to proceed to a binding agreement.

These situations have very different results. In the first situation, as there are no new proposals or changed conditions the letter constitutes an acceptance, and a binding subcontract comes into being when the contractor sends that letter. In the second situation the contractor is making an offer, which the subcontractor can accept or not or may make a counter-offer. Once accepted by signature or other means, the parties are bound to that agreement and to the extent of that offer only. So if the letter mentions carrying out design work, then there is no obligation to do more than design work; if the letter mentions work up to a set value, then once that value of work has been completed the agreement may die unless continued in the full knowledge and agreement of both parties. The third situation binds neither party to anything and is no more than what it says: a statement of intent.

The current law with regards to letters of intent has been defined by Judge Humphrey Lloyd QC in the judgment on *Durabella Ltd* v. *J. Jarvis & Sons Ltd* (2001) 83 Con LR 145:

> '7. I do consider that the terms of the "letter of intent" indicated that Jarvis intended to create a legal relationship with Durabella. It is now well established where a "letter of intent" authorises work, materials or services to be provided pending the conclusion of some further agreement it will, if accepted, constitute a contract (or part of a contract) for what it requires. The "letter of intent" may never be supplanted by a further agreement. It will be the contract for the work etc. Many large projects have been carried out on such a basis. It is equally well established that in a commercial relationship the court will try to establish a contract or contracts since that would be consistent with the parties' presumed expectations. On the other hand a contract cannot exist unless it is clear that, viewed objectively, the parties were in fact agreed on all the matters which they considered necessary and which are necessary to form a contract.'

In the last sentence the judge makes the point that there cannot be a binding contract unless there is agreement between the parties, but that the express agreement need only extend to those matters which, at the time, the parties considered necessary. This is elaborated in his next paragraph:

> 'The programme had yet to be agreed but it was not considered to be a problem and, indeed, was not a problem. Jarvis' letter was treated by Durabella as having contractual

force – see its letters of 14 and 21 December 1995 and 9 January 1996. The first two letters are not simply administrative as there would be no point in them unless an order was being entered in Durabella's books. Even if that conclusion were incorrect the letter of 9 January was clearly an acceptance of Jarvis' letter. It did not constitute a counter-offer as there was nothing in it which was inconsistent with Jarvis' letter. It was not then or subsequently treated by Jarvis as containing anything at variance with Jarvis' requirements. The reference to Durabella's estimate, as such, is insufficient to displace the terms required by Jarvis. There was no reassertion of Durabella's terms. There was indeed no true battle of forms: Jarvis had made its terms clear when contracting for the initial three flats. Durabella's case correctly, in my judgment, presupposes that the Jarvis' letter was an offer. That offer was either accepted by Durabella's letters, in particular its letter of 9 January, or if programming were considered to be vital to a contract (which it normally is but in this instance appears not to have been seen as crucial to the creation of a binding agreement), then by its subsequent conduct in carrying out the work in accordance with arrangements agreed with Jarvis. It is in my judgment clear that Jarvis intended that its formal order was being postponed solely so that it could record the results of the survey and measurement, i.e. the quantity to be paid for at the agreed rate.'

The question then arises of what is crucial to the creation of a binding agreement. It seems likely from Judge Lloyd's judgment that what is necessary is limited to matters that the parties considered crucial at the time they agreed to the subcontractor commencing work. What that includes will differ from project to project but may exclude many matters often regarded as essential to a binding agreement, such as price, time, detailed terms and conditions. It is submitted that there is a difference between an express intention to agree a price or period for the works or other term, sometime in the future, and an ill-defined subcontract where terms may be implied, such as for the payment of a fair and reasonable price or for the work to be carried out in a reasonable time. In the first it may be considered that no binding subcontract exists until agreement is reached, while in the second there will be a binding subcontract with undefined terms.

In constructing the terms and conditions, arising out of an accepted letter of intent, the courts will look at the background to the letter of intent and consider the intentions of the parties at that time. This is in complete contrast to the usual approach with regards to a concluded agreement when the courts will generally only consider the actual words of the contract documents. In *Comyn Ching* v. *Radius plc* [1994] ORB 728, Judge Bowsher QC made the following comment:

'Very often, what is called a letter of intent is properly construed as a binding agreement or as an offer capable of acceptance. However, I do not so construe the letter of 25 November, 1988.

The correct approach when considering the effect of a letter of intent is "to look for the purpose of construing it at the document itself, at the surrounding circumstances, and at what happened when it was brought into existence. The fact that it has the particular label that it has does not brand it at the outset as a contractual document or as a 'non-contractual' document".'

In establishing the intentions of the parties the courts will look at what the subcontractor offered to do, rather than what the contractor sought to subcontract. This concept can be seen in the Court of Appeal judgment in *Monk Construction Ltd* v. *Norwich Union Life Assurance Society* (1992) 62 BLR 107, where Lord Justice Neill, quoting from the judgment of Mr Justice Goff in *British Steel Corporation* v. *Cleveland Bridge & Engineering Co Ltd* (1981) 24 BLR 94, said at page 123:

> 'It is clear that Mr Justice Goff regarded the nature of this dispute as a significant reason for holding that no 'if' contract had been concluded because "it would be an extraordinary result if, by acting on [the buyer's] request in such circumstances, [the seller] were to assume an unlimited liability for his contractual performance, when he would never assume such liability under any contract which he entered into".'

Letters of intent frequently contain words to the effect that they are 'subject to contract' or that 'there is no intention to create legal relations' or similar wording. The phrase 'subject to contract' was considered by the Court of Appeal in the case of *Fraser Williams* v. *Prudential Holborn* (1993) 64 BLR 1, when the use of the phrase 'subject to contract' was held to prevent the commencement of work as an acceptance of the offer contained within the letter of intent. Lord Justice Dillon put it in these words, at page 123:

> 'The words "subject to contract", which are very well known, cannot be construed as meaning merely "subject to written acceptance so as to form a contract".'

The conclusion is clear that the issuing of a letter of intent 'subject to contract' will not, without some further action or clarification of the intention, create a contract in writing. However, where the subcontractor acts upon the letter of intent, then it is submitted that the principle set out in the judgment in *Durabella* will create a contractual relationship, which may or may not be for the entire works.

The case of *Williams* v. *Prudential* can be distinguished from the acceptance of a letter of intent because the phrase 'subject to contract' was in the contractor's offer to the employer and not in a letter of intent requesting a limited commencement of the work under an agreement subject to contract. The effect of such words will depend on the context in which they are used. The law now regards letters of intent as generally being offers, where the offeror anticipates that the recipient will act on it. Offers have no legal effect since they require acceptance before the parties become legally bound. Unless there is a clear indication that the offer may not be accepted without some specific further agreement, or that some specific action is required of the subcontractor such as for it to confirm its acceptance in writing or to provide, for example, a method statement or tax certificate, then acceptance either in writing or by conduct will, it is suggested, create a legally binding contract for the work authorised by the letter of intent and no more.

Judge Thornton QC in his judgment in *Hall & Tawse* v. *Ivory Gate* defines the usual effect of a letter of intent as follows:

> 'A letter of intent is usually an unilateral assurance intended to have contractual effect if acted upon, whereby reasonable expenditure reasonably incurred in reliance upon such a letter will be reimbursed. Such a letter places no obligation on the recipient to act upon it and there is usually no obligation to continue with or to undertake any defined parcel of work, the recipient being free to stop work at any time. The effect of such a letter is to promise reasonable reimbursement if the recipient does act upon it.'

Having discussed the contractual effect of the letter of intent, Judge Thornton QC describes this arrangement as the provisional contract:

> 'I propose, therefore, to refer to this contract as "the provisional contract". It envisages that the parties will continue to negotiate about the form of the contract which would govern the work and which would supersede the provisional contract once all those terms have been agreed.'

Lord Justice Aldous, in the case of *Carillion Construction* v. *Ballast*, where the subcontractor was requested by the contractor to commence work under a letter of intent and did so on the basis that a formal contract was to be prepared by the contractor, referred to the letter of intent as an "interim contractual arrangement", at paragraph 33:

> 'The conclusion that the parties did not intend to enter into the contract as opposed to an interim contractual arrangement is demonstrated by their conduct. They continued to try to resolve the difficulties.'

Both the above cases make it clear that the law recognises that there can be a contractual situation on a provisional or interim basis while further terms and/or conditions are discussed and agreed. There does not appear to be any limit as to the matters that may be left to be agreed after the parties enter into the provisional or interim contract.

A further feature of many letters of intent is that the scope of the works is limited by either extent or value. Where this is the case, the subcontractor's obligations to the contractor, once he accepts the letter of intent either expressly or by commencing work, are limited but may be extended by subsequent agreement to extend the scope or value. One factor that may affect the subcontractor's obligations is the intended method of reimbursement for work carried out. Payment under letters of intent is often stated to be for reasonable costs, but may be in accordance with a priced bill or schedule of rates. Where payment is to be in accordance with the subcontractor's offer, it will be more likely that the subcontractor will be bound to complete the entire scope of the work as defined in the letter of intent, than where the request is simply to commence design or mobilisation and to be recompensed only for costs.

On this basis if, for example, the scope of the work is limited by value, once the subcontractor has completed work to that value it will have several choices: to either accept a further offer to do additional work, or withdraw from the project or negotiate an effective agreement for the completion of the subcontract works.

The letter of intent will frequently limit the amount of payment to be made to the subcontractor if the project does not proceed or no agreement is concluded. This is often on the basis of proven or reasonable expenditure only. The intention of such arrangements is to recompense the subcontractor if it does not carry out the anticipated subcontract work because either the work does not proceed or at an early stage the parties decide they cannot reach agreement on a satisfactory way forward.

However, should the parties fail to conclude an agreement but work proceeds to completion, then the subcontractor will probably not be limited to his proven costs, with no profit etc. This was the main thrust in the case of *Monk Construction* v. *Norwich Union*, where the work on a large new building was started on the basis of a letter of intent for the contractor to plan and mobilise for the project. In the event, the work proceeded and Monk completed the work tendered for. The employer claimed that its obligation was limited to paying only the contractor's reasonable costs with no profit allowance. Against this claim the House of Lords found that, once the work had proceeded beyond that defined in the letter of intent, the contractor was to be paid in accordance with its original offer as being the basis of a *quantum meruit*.

It is uncertain at what point the courts will decide that the subcontractor becomes entitled to more than the remuneration agreed in the letter of intent, and to be paid a reasonable return for its work. It will depend on the actual wording of the letter of intent, but it is suggested that this point will frequently be when the subcontractor commences the permanent works, but it will not be later than when the subcontractor reaches the limitation of the scope or expenditure defined in the letter of intent. Where the scope to be paid for on the basis of reasonable actual cost becomes exceeded, it is likely that the subcontractor's entitlement will then be a reasonable price for the entire work carried out to that time, as well as thereafter.

Where the value of the work to be executed under the letter of intent is defined, only in extreme circumstances will a subcontractor not be able to recover further sums for work carried out under the contractor's direction and/or with its knowledge and approval. There may be specific requirements to advise the contractor when the capped value is being approached, or there may be a clear condition within the letter of intent that funding does not allow for payments to be made in excess of the cap; in such circumstances the right of the subcontractor to further payment may be denied, at least until such funding becomes available.

As mentioned, the law regarding letters of intent seems have shifted towards the finding of a binding contract in the terms of such a letter. In the case of *Harvey Shopfitters Ltd* v. *ADI Ltd* (6 March 2003, unreported), Mr Recorder John Uff QC had to consider the effect of a letter of intent which had requested the recipient contractor to sign and return a copy of the letter by the words, 'If you are agreeable to the foregoing please sign the enclosed copy of this letter and return it to me'. The contractor did so, but both the date of the letter and its return were after the actual on-site start date.

Mr Uff considered a number of authorities and summed up his approach to the effect of the letter of intent in the following words, at paragraph 13:

> 'Without discourtesy to counsel and the detailed written submissions provided, the appropriate approach to the issues can be summarised in this way:
>
> (a) the courts now adopt a practical approach to whether and what agreement should be upheld;
> (b) niceties which might on a more traditional approach have been regarded as precluding agreement will not now be so regarded unless essential to the basis of the agreement;
> (c) this is the more so where the contract has been fully performed.'

Inevitably with letters of intent, which are normally written in haste without in-depth consideration, the words may be ambiguous and their meaning uncertain. In the *Harvey Shopfitters* case, the judge had to consider the effect of the words used, in particular the sentence, 'If, for any unforeseen reason, the contract should fail to proceed and be formalised, then any reasonable expenditure incurred by you in connection with the above will be reimbursed on a *quantum meruit* basis.' The judge refused to hold that the phrase 'fail to proceed and be formalised' meant that because the contract had not been formalised, the contractor was to be paid on the basis of its reasonable expenditure and on a *quantum meruit* basis; he held that, since the contract work had proceeded, the intended terms of that contract applied. He said, at paragraph 15:

> 'In my judgment the words "fail to proceed and be formalised" are not to read disjunctively. As a matter of ordinary construction "and" means "and" not "or". Furthermore, having regard to the consequences spelled out, it would be most surprising if the parties intended the mere failure to formalise the contract document to lead to the result that the careful process of tendering and pricing should be thrown over in favour of the uncertainty of *quantum meruit*. The second sentence of the fourth paragraph has ample scope when limited to failure to proceed with the works, which necessarily means that the contract will not be formalised. I therefore reject the claimant's primary contention. It follows that the agreement between the parties was a lump sum basis.'

And with regards to the incorporation of the IFC Conditions he said:

> 'In my judgment this affords no reason why the IFC Conditions should not have been imported by agreement. The argument against importing the IFC Conditions depends crucially on confining the agreement to the letter of 7 July. That letter, however, expressly contemplates the preparation of full contract documents. Furthermore, the claimant concedes that the scope of works is that set out in the tender documents. Such a concession was unavoidable, given the reference to the tender in the letter of 7 July 1998. Once the tender and the documents on which it is based are taken into account it becomes clear, in my judgment, that the parties intended to contract on the IFC Conditions, which accordingly formed part of the agreement made on the 7 July or 28 July 1998.'

This is an important decision since many subcontracts are formed in a similar manner, by a letter of intent, anticipating that a formal agreement will be concluded in standard terms as stated, at a later date. Whether the terms referred to will form the basis of the subcontract will largely depend on the subsequent actions of the parties. In the *Harvey Shopfitters* case the judge found that the parties conducted the contract as though the stated terms applied, saying at paragraph 22:

> 'What the parties in fact did, in my judgment, was to behave as though their relationship was governed by the IFC Conditions, particularly in relation to pricing and certification of the work and the certification of completion and extensions of time. Furthermore, at the end of the project the claimant and the architect sat down to seek to agree a final account using procedures which were referable to the IFC Conditions. Specifically, both parties recognised and accepted that provisional sums, of which a significant number existed in the contract documents, needed to be formally omitted and replaced with Architect's Instructions, all of which was referable specifically to the IFC Conditions. While the parties rarely, if ever, made actual reference to the terms of a standard form, I am in no doubt that, had they been asked which standard form applied, their answer without further consideration would have been IFC 84. I therefore find that there was indeed an agreed convention on the basis of which the parties performed the contract, for this purpose the defendant being represented by the architect as its agent. This convention was acted on without demur or exception by both parties and I find that it would not be just or equitable for the claimant now to be permitted to restyle from this common assumption. Accordingly, in addition to the findings set out in Part I of this judgment, I find that

the claimant would be estopped from denying that the IFC 84 Conditions of Contract applied to the works which are the subject matter of the contract.'

One problem the judge did not have to address in *Harvey Shopfitters* was the effect of amendments to the standard form. In many subcontracts, the subcontractor is advised either as part of the tender enquiry or more frequently in the letter of intent, that the formal agreement is to be a standard form as amended, or as amended by a document which the contractor has yet to provide to the subcontractor. It is suggested that such a reference to a deviation from the standard form will require clear acceptance by the subcontractor of the changes proposed, for such a reference to be effective. A general reference to standard terms as amended, or even to be amended in accordance with a stated document, without clear wording will at most be an agreement to agree changes and consequently may not be binding at law. In any event subcontractors are well advised not to proceed with any work authorised by reference to documents not in their possession, especially any document relating to the terms of the subcontract itself.

Most letters of intent make provision for the situation where the work does not proceed or the parties fail to reach an agreement, and the subcontractor as a consequence does not complete the Works. It is normal in such circumstances for the subcontractor's right to be paid to be limited to its proven actual costs, or reasonable costs or direct costs. Any entitlement to loss of profit is likely to be expressly excluded. It is often argued that phrases such as 'proven costs' will exclude any allowance for the subcontractor's overheads and profit. The courts, however, have been reluctant to accept that a subcontractor who proceeds to execute work on the instructions of a contractor will not be able to recover, in addition to the costs of labour, plant and materials, including costs of work done by sub-subcontractors, an allowance for its overhead costs, such as costs related to the provision of support staff not based on site or employed solely on the specific work, as well as contribution to the running of the subcontractor's business.

In *Monk Construction* v. *Norwich Union*, the Court of Appeal did not need to consider the meaning of the phrase 'proven costs' since they found that it was only to apply if the work did not proceed beyond mobilisation and the ordering of some materials. They did, however, find some evidence that a subsequent letter had indicated that 'proven costs' was to include an allowance for profit. Lord Justice Neill said, at page 124:

'Norwich placed considerable reliance on the letter from Mr Vest [Manager for Monk] dated 30 June 1989. In my judgment, however, this letter had no contractual force as it was written long after work began. In any event Mr Vest clearly had in mind that Monk were to be entitled to a reasonable profit in addition to the "proven costs".'

This suggests that the judge considered that had Monk been bound to the 'proven costs' provision, then they would have been entitled to an allowance for profit in addition.

In the Scottish case of *Robertson Group (Construction) Limited* v. *(First) Amey-Miller (Edinburgh) Joint Venture; (Second) Amey Programme Management Limited and (Third) Miller Construction (UK) Limited* [2005] BLR 491, Lord Drummond Young had to consider the meaning of the phrase 'all direct costs and directly incurred losses', in a situation where an initial letter of intent with a capped expenditure of £500,000 was subsequently replaced with one capped at £5,000,000. Work proceeded beyond the initial cap when the parties decided that they could not agree terms and it was decided that no further work should be carried out under the letter of intent.

Lord Drummond Young compared 'direct costs and directly incurred losses' with the phrase 'direct loss and/or expense' used in the JCT forms of contract, and concluded, at paragraph 6, that:

> 'the words "cost" and "expense" are regularly used as synonyms.'

He considered the meaning of the words 'direct' and 'directly', by reference to several authorities both Scottish and English, then, at paragraph 9, stated that in the case of a breach of contract:

> 'the general principle is that the innocent party is entitled to be put in the same position as he would have been if the contract had been performed;'

and that in such a situation:

> 'It is obvious that a loss of that nature may include both a loss of profit and a contribution to general overheads.'

He went on to say:

> 'The present case, of course, does not involve a breach of contract; it rather involves a specific contractual entitlement to remuneration. Nevertheless, it is of the nature of a contract that the party who supplies goods and services expects, or at least hopes, to make a profit. The intention to make a profit lies at the heart of all, or nearly all, commercial activity, and the law must recognise that elementary economic fact. I am accordingly of opinion that the failure of a contractor to make a profit should be accounted a "loss" not only in calculating damages for breach of contract but also in construing contractual terms relating to payment for goods and services. The same is true of earning a contribution to general corporate overheads; indeed, until such a contribution has been earned it probably cannot be said that any profit has been generated.
>
> 10. If the word "losses" did not cover general overheads and an element of profit, the pursuers would have to carry their corporate overheads from other sources of income, with no contribution from the works at the High School unless a formal contract were concluded. Likewise, the pursuers would have no possibility of profit from those works unless a formal contract were concluded. I do not think that a probable construction. Indeed, the head office overheads are likely to cover such matters as accounting and bookkeeping work carried out in respect of the present contract; they will also cover the salaries of draughtsmen and surveyors employed at head office who can be expected to perform work for the purposes of the present contract. Thus part of the general overheads will be referable to work carried out on the present contract. If the defenders' construction is correct, however, no contribution towards the cost of that work would be earned from the contract, unless the salaries of draughtsmen, surveyors, accounting staff and the like can be allocated on some sort of time basis. Such a procedure would be unusual, at least for accounting staff, and complicated. I do not believe that it would have been within the parties' contemplation when work started following the letter of 12 October 2001.'

In the *Robertson Group* case, since the contractor was to be paid both costs and losses it was not necessary to distinguish the head under which overheads and profit were to be paid. However, Lord Drummond Young made it clear that he considered that both could be included by the phrase 'direct costs', without the need to rely on the additional right to recover directly incurred losses, in paragraph 12:

> 'In addition, on the construction that I have adopted, the pursuers will be entitled to a contribution towards their general corporate overheads, and also to an element of profit, based on the content of the relevant instruction. It seems to me to be quite natural to describe such items as arising naturally, in the usual course of things, from the instructions to carry out work. Moreover, if no recovery were permitted, the failure to earn a contribution to overheads and profit would in my view properly be described as a "loss". The calculation of such a loss will normally be achieved by determining appropriate percentages to cover general overheads and profit, based on the actual costs of carrying out the instruction (that is, the cost of labour, plant, materials, subcontractors and site overheads).'

Many letters of intent expressly exclude the right to recover the loss of profit suffered by the subcontractor if the work does not proceed or a subcontract is not concluded. In the light of the judgment in *Robertson* v. *Amey-Miller* it is suggested that such loss of profit would be limited to the work not executed by the subcontractor, and that the subcontractor should not be denied profit on the work actually carried out under the letter of intent. For the subcontractor to be so denied, it is suggested, very precise wording would be required.

Subject to contract

It is generally held, where an offer is made 'subject to contract', that the offer cannot be accepted by the offeree and the obligations enforced without the execution of a formal agreement. This will be especially true in the case of contracts for sale but may not necessarily be the case where services are to be provided and in fact have been provided.

In the case of an offer to supply goods, 'subject to contract' will normally be an offer without terms, for example an offer to supply goods at a price but with no reference as to a date for performance or method of delivery. In the case of offers to supply services, the phrase 'subject to contract' may be intended, such as when used in a letter of intent, to indicate that acceptance by commencing work will only form a provisional contract, such as envisaged by Judge Thornton QC in his judgment in *Hall & Tawse* v. *Ivory Gate*.

It may be possible to make a distinction between an initial offer by a tenderer and a counter-offer by the contractee, especially when the counter-offer incorporates a request to commence work and sets out temporary payment provisions such as to be paid reasonable costs incurred.

It is suggested that it is possible to have an initial or provisional contract by way of acceptance of an offer which is 'subject to contract'. Thus a letter of intent requesting the subcontractor to commence work, if accepted, becomes a binding subcontract for the work requested and on the terms set out in the letter. At the same time the offer for the balance of the work remains subject to contract and at a subsequent date may be rejected by the offeree, when neither party will have any rights or obligations other than those contained within the provisional contract.

In the Court of Appeal case of *Fraser Williams* v. *Prudential*, the court had to decide whether Prudential had determined Fraser Williams' employment in breach of a contract between them. There had been an exchange of letters, the last of which had contained terms proposed by Fraser Williams. Fraser Williams had sought payment under those terms and had been paid accordingly. The judge at first instance had concluded:

> 'that neither party could contend that, the "subject to contract" provision in the proposal was still operative at the time of termination on 5 May 1989.'

The Court of Appeal rejected the judge's finding and found that there was no concluded contract.

Lord Justice Kennedy stated what Prudential's obligations to Fraser Williams were, at page 10:

> 'No doubt both Fraser Williams and Prudential then expected problems to be resolved as the work proceeded, including the problem of there being no written contract to which they both expressly agreed, so that ultimately Fraser Williams would be paid in accordance with the written contract rather than on a basis of *quantum merit*.'

Whether or not payment on a *quantum meruit* basis is contractual may depend on whether or not there is a formal request to commence work, such as by way of a letter of intent.

Lord Justice Dillon in a dissenting judgment, at page 18, refers to whether after the request for Fraser Williams to commence work the uncertainties were such that:

> 'in law there could not have been a binding long-term contract between the parties.'

The implication is that in the absence of a 'long-term' contract there might be a 'short-term' one. Since there does not appear to have been any dispute whether Fraser Williams were to be paid for the work they did, the court had no reason to consider the existence of such an arrangement.

A similar situation arose in the more recent case of *The Rugby Group Limited* v. *ProForce Recruit Limited* [2005] EWHC 70 (QB), which was an appeal from the decision of the Senior Master. The parties signed an agreement for the provision of services by ProForce to Rugby, which incorporated the phrase 'subject to contract' and concluded:

> 'This contract will be of a minimum two-year period and will be re-negotiable at the end of that period. During that period ProForce will hold preferred supplier status.'

A dispute between the parties arose as to the meaning and effect of the sentence, 'During that period ProForce will hold preferred supplier status'.

The judge, Mr Justice Field, first dealt with the 'subject to contract' issue and found that there was an enforceable contract:

> '16. I deal first with the contention that the agreement is not an enforceable contract because it contains the words "subject to contract". In general, except in a very strong and exceptional case, the effect of these words in an agreement is to prevent an executory contract from coming into existence because they are taken to mean that until a further contract has been executed neither party is to owe the other any contractual obligation. However, in this case, save for the alleged breach, the agreement cannot be regarded as being executory

> because after it was signed the parties did those things that the agreement contemplated that each should do for the benefit of the other. Thus after 31 July 1991 ProForce paid for the Vac Press and the Roadsweeper and supplied the personnel and equipment defined in the agreement and Rugby paid the stipulated monthly charges for personnel and equipment. This being the case it is my view that the parties are to be taken to have entered into an implied binding contract on the terms of the agreement.'

After reviewing a number of authorities as to the construction of contracts, he then concludes that there was no breach of contract by Rugby:

> '25. I reject Mr Sweeting's [Counsel for Proforce's] submissions. At the time the agreement was signed it was possible that Rugby would introduce a preferred supplier system in respect of the site over the next two years. If Rugby were to have introduced such a system, as they were planning to do at the end of 1999, the conferral of preferred supplier status on ProForce would have constituted a very real benefit because ProForce would then have been in the inner ring of suppliers and ahead of their competitors that were outside the ring. The price paid for the machinery by ProForce was £72,500 and ProForce were under an obligation to maintain it and provide it at the site. However over the term of the agreement ProForce were due to be paid £153,400 by way of equipment charges and at the end of the term if it was not purchased back by Rugby, it would have had some residual value. It is not possible to tell if ProForce's equipment obligations would be covered by the equipment charges but the figures in the agreement show in my view that the parties were not contracting on the basis that the conferral of [preferred] supplier status was intended materially to off-set ProForce's equipment obligations.
>
> 26. As it happened, Rugby did not adopt a preferred supplier system for the site during the contractual term. Instead they looked to a number of suppliers of labour who did not have the status of preferred supplier to satisfy their non-cleaning requirements. In my judgment, in acting in this way Rugby were not in breach of the agreement.'

This decision may also be analysed on the principle of a preliminary contract and with the intention of a future amended agreement, which was 'subject to contract'. In this case the obligations to provide labour and plant and to pay for these at the agreed rates were the subject of a binding agreement. It was the potential obligation with regards to 'preferred supplier' status that was subject to contract.

Capped price or expenditure

As already mentioned, a letter of intent, if accepted, will normally create a binding agreement between the parties, the extent and terms of which are limited to the obligations set out in the letter. For the subcontractor, these obligations may be to proceed with the entire subcontract work or may be limited to specific activities such as design, procurement of materials and/or mobilisation of labour and plant. In other cases, the extent of the subcontractor's work may be limited by date or value. For the contractor, its obligations may be limited to pay 'proven costs' or 'reasonable costs' with possibly no obligation to contribute to the subcontractor's overheads or to give it an allowance for profit. (See also the commentary on the case of *Robertson* v. *Amey-Miller*, in the previous section on letters of intent.)

Once the initial obligation has been fulfilled, then the parties will have no further obligations to each other and further performance will be subject to a new or extended agreement. In the Court of Appeal case of *Monk Construction* v. *Norwich*

Union, the letter of intent requested the mobilisation of the contractor and the ordering of materials and was expressed in the following words:

> 'Our client has instructed us to confirm that this letter is to be taken as authority for you to proceed with mobilisation and ordering of materials up to a maximum expenditure of £100,000.
>
> In the event that our client should not continue a contract with you, your entitlement will be limited to the proven costs incurred by you in accordance with the authority given by this letter.'

The works proceeded beyond the mobilisation stage and Monk completed the work it had tendered for. A dispute then arose as to the sum to be paid to Monk for that work. Norwich Union sought to infer from the letter of intent that, in the absence of a formal agreement, any work carried out by Monk was to be paid for on the basis set out in the letter of intent, i.e. limited to 'proven costs incurred'. At the Court of Appeal the judgment by Lord Justice Neill was that this provision only applied to the mobilisation and ordering of certain materials and only if the work to the substructure phase II did not proceed. Lord Justice Neill said, at page 124:

> 'For my part I do not find it necessary to come to a conclusion on all the various arguments addressed to us by Monk. In my view the provision for recovery of "proven costs" was intended to deal with the situation if no contract was signed and if no work on the sub-structure phase II project (apart from mobilisation and the ordering of some materials) was carried out by Monk. In that event Monk were not to be out of pocket because they had brought huts and stores etc. onto the site and had ordered materials for the contract in advance. I agree with the judge that the "proven costs" formula was never intended to apply to the execution of the main contract, nor was it intended to apply if work on the main contract began.'

Lord Justice Neill goes on to consider the possibility of there having been an 'if' contract:

> 'I recognise that there may be cases where a letter of intent provides a satisfactory basis for an "if" contract. It may sometimes be possible to imply terms which are missing from the letter of intent itself. But an "if" contract must contain the necessary terms. It must also be clear that the "if" contract is to apply to the main contract work if no formal agreement is ever reached. In the present case from an examination of the documents and from reading the evidence to which we were referred I am satisfied that the parties never intended that the third and fourth paragraphs of the letter of intent dated 11 July should govern the remuneration of Monk if they undertook the work on sub-structure phase II.'

Where the obligation is to carry out a specific operation of work up to a specified date, there will be little doubt when this obligation has been fulfilled; however, there may be less certainty when a capped price has been reached. Construction work is notoriously susceptible to change and variation, both in requirements and as a result of design development, and there is the possibility of claims for loss and/or expense, etc. Where the subcontractor is to be paid proven costs, it will be a relatively simple matter for the subcontractor to know when the cap has been reached. However, where the valuation of the subcontractor's work is to be carried out in accordance with the valuation procedures under the terms of the subcontract, three possibilities arise: first, the valuation by the subcontractor, which may be high; secondly, the contractor's valuation, which is likely to be low; and finally, the true valuation as decided by an independent third party, which will not be known at the time the cap is reached. Since the only valuation available to the subcontractor at the time is its own, it is considered that provided the valuation can be shown to have been a genuine valuation of its works, then the subcontractor will have fulfilled its obligations under such a provisional subcontract if its valuation approximates to the cap, which in any event can only be regarded as an approximate value.

Further considerations arise where the obligation relates to several matters, such as the work to be carried out, the period for the works and the value is capped. This was the situation in *AC Controls Limited* v. *British Broadcasting Corporation* [2002] 89 Con LR 52, where Judge Thornton QC had to consider the effect of a provision for a cap:

> 'Subject to the terms of this letter, ACC is authorised to proceed with the works up to a total value of £250,000 or any other sum which may subsequently be notified to you in writing by the BBC.'

However, earlier in the letter of intent it had stated:

> 'Forthwith following agreement as to the contents of this letter ACC shall commence and carry out survey work of all sites set out in the Invitation to Tender Document and shall prepare and submit for approval to the BBC a programme which shall identify the order and timing of all design and installation works and all other relevant matters. ACC shall use its best endeavours to accommodate any BBC observations on the programme submitted and to obtain the BBC's agreement to the programme within a reasonable period.'

While the original letter of intent was for survey work only, the BBC issued a further letter, which stated:

> 'Further to the above letter you are hereby authorised to proceed with the works to an additional value of £250,000 making a total authorised value of £500,000.'

The judge found that the acceptance of the second letter, far from putting a cap on the contractor's obligations, by its express words required the contractor to proceed with the entire works defined in the specification:

> '55. In essence, the letter of 7 July 1999 constituted a major and fundamental variation to the original letter of intent. It was, in reality, a second offer to enter into a second "if" contract. The terms of this second offer were that ACC was to embark on the access control system project and to start to carry out all the work involved that was defined in the specification and associated documents. Given the recent agreement as to the contents of the programme of work and the context in which this letter was sent, it was clearly envisaged by both parties that that work was to be carried out in accordance with that agreed programme. It follows that, in context, what was being offered by the BBC in the phrase: "you are hereby authorised to proceed with the works" was for ACC to start to undertake all remaining access control work in accordance with the agreed programme.'

He found that the effect of the cap was to restrict the right of the BBC to determine, in the event that it had not been possible to conclude a formal agreement. The judge said:

> '68. The BBC was also providing for the possibility of determining this interim arrangement and bringing the relationship to a premature end, but only if it: "determines that it is not possible to conclude a formal agreement". Thus, in the absence of such a determination, the BBC neither envisaged terminating the relationship prematurely before the works had been completed nor provided for that possibility in the contract made by ACC's acceptance of the offer contained in the letter of 7 July 1999.
>
> . . .
>
> 81. In construing the meaning and effect of this cap provision, the following factors must be taken into account:
>
> 1. The terms of the letter of 4 June 1999 were to govern the enlarged contractual arrangement created by the letter of 7 July 1999. Under that earlier letter, ACC was contractually committed to complete the survey work. Thus, the BBC was indicating in the letter of 7 July 1999 that a similar arrangement was to apply to the enlarged arrangement.
> 2. The value of the work at any one time could only be determined by the BBC and ACC would never know, with any certainty, when it had exceeded the cap on the authorised value of its work. Moreover, any valuation would inevitably fail to take into account a significant amount of work for which ACC would reasonably be entitled to payment if the arrangement was stopped at short notice
> 3. Thus, ACC could not terminate unilaterally the work it was carrying out and would never know the point in time when it had reached any cap or limitation of value.
>
> 82. With these considerations in mind, the meaning of this cap provision becomes clear. The words: "authorised to proceed up to a value of . . ." must be read with the whole of clause VII which follows them. Taken together, the words mean that the work to be carried out was subject to two limitations:
>
> 1. The BBC could terminate at any time but only if it had determined that a formal agreement could not be concluded.
> 2. Additionally, the BBC could terminate if and when ACC's value of work, as determined by Hansomb, reached the defined value or at any time thereafter.
>
> The onus on termination in either situation lay with the BBC. Unless and until the BBC terminated in one or other situation, ACC had to continue to work and would be entitled to payment for the reasonable value of work up to the point of termination.'

While this judgment considers only a specific letter of intent, it again demonstrates the courts' approach that such letters form a binding agreement, which will require express wording to allow either party to escape its natural obligations, on the one side to proceed with the work to specification and to time, and on the other side to pay as if a formal agreement had been concluded. The word preliminary is more likely to infer that the contract is preliminary in nature than in effect.

Different circumstances gave rise to a very different conclusion when Judge Richard Harvey QC gave judgment in the case of *Emcor Drake & Scull Ltd* v. *Sir Robert McAlpine Ltd* [2004] EWHC 1017 (TCC). The parties had carried out negotiations over a long period. There was general agreement as to the price for the work and that the terms of the subcontract would precisely reflect the main contract conditions or be 'synallagmatic'.

By August 2001 McAlpine were keen for Emcor to commence work despite no agreement on the wording of the subcontract terms. McAlpine, for internal procedural reasons, undertook to issue a short order limiting the value of work to be undertaken by Emcor to £1m. This was to be in exchange for a letter from Emcor as drafted by McAlpine setting out the state of the agreement as reached at that time. Judge Richard Harvey QC recorded:

> '6. On 21 June 2001 a conversation took place between Mr Wallace [Quantity Surveyor for SRM] and Mr Taylor [Director of EDS]. Both gentlemen were in cars and were using mobile phones. Mr Wallace's evidence relating to that call may be summarised as follows. Mr Wallace said to Mr Taylor that SRM could not authorise payment of one penny to EDS until complete agreement had been reached and recorded; that a letter was to be drafted to be sent back by EDS with a short order as a way of providing a temporary order number to allow EDS to be paid; that the letter would contain all the agreements that had been made and would identify all that would form the formal subcontract order; and that EDS would be obliged to execute the formal order as soon as it was prepared, to take over from the letter.'

In fact McAlpine issued several short orders. Successive orders increased the scope of the work to be provided as well as the capped sum, which by August 2002 had increased to £14m. Although McAlpine had prepared several draft agreements, none had strictly reflected the provisions of the main contract as agreed and consequently no concluded agreement was reached.

In October matters came to a head and Emcor notified McAlpine that they would not accept further purchase orders and gave notice that they would abandon the subcontract. McAlpine took Emcor's notice as being a repudiatory breach and forced Emcor to leave site on a few hours' notice. In his judgment of the preliminary issues before him, Judge Richard Harvey QC was required to decide whether Emcor was under an obligation to complete the whole of the mechanical and electrical works on the Dudley Hospital PFI project. His answer was no.

The judge's reason for this decision appears to have been because McAlpine had not given Emcor an order for the full subcontract sum but had chosen to release the work progressively, possibly because they believed they would have a hold on Emcor to conclude a subcontract in the agreed form. If such was their intent it failed, and Emcor having fulfilled their obligation under the short contract were not in repudiatory breach. In this instance the capped rate under the preliminary contract referred to the work to be carried out by Emcor and in no way extended to the complete works. It should also be noted that there was no question of Emcor's entitlement to payment being limited to the value of the cap despite having carried out work exceeding the order value. In addition, by having carried out work to a value exceeding that stated in the order, Emcor had fulfilled its obligation under the subcontract, which was discharged.

Where the subcontractor is acting on a letter of intent and the sum is capped, there is unlikely to be agreement on whether the letter of intent covers the entire subcontract work unless there is agreement on the value of the work to be carried out by the subcontractor, including the valuation of all variations and other matters such as loss and/or expense. If there is a dispute as to the value, the contractor may have valued the work to be below the cap while the subcontractor may have valued it higher than the cap.

In such circumstances the risk for the subcontractor is that if it withdraws when the cap is reached on its valuation, it is likely to be in repudiatory breach if the contractor's valuation is subsequently found to be correct. Conversely, should its valuation be found to be correct, then if it completes the work it may not be entitled to the additional payment because it will have exceeded the cap in breach of its subcontract.

It was such a situation that was considered by Judge Richard Seymour QC in the case of *Mowlem plc (trading as Mowlem Marine)* v. *Stena Line Ports Limited* [2004] EWHC 2206 (TCC), where Mowlem carried out major works to Holyhead harbour. The parties never concluded a formal agreement but operated on a series of 14 letters of intent, each of which capped the value of the work to be carried out. The relevant wording within the final letter of intent was:

> 'In the event that this letter of intent is rescinded and the Contract is not awarded to you, Stena Line Ports Ltd's total obligations to yourselves and any other parties will be limited to a maximum of £10,000,000.00.
>
> We confirm that Stena Line Ports Ltd's commitment to expenditure up to a maximum of £10,000,000.00 will enable you to proceed with the Works in accordance with your programme, until 18 July 2003.'

In June 2003 Mowlem had advised Stena that in their opinion the true value of the Works, when completed, would exceed £10m, and that there was a further issue, 'the rock issue', to be decided which they considered would further increase the final contract sum.

Mowlem argued that the final letter of intent had the effect of raising the cap on expenditure in the event that Mowlem were allowed to continue working after 18 July 2003. Under these circumstances should Mowlem exceed the cap, then Stena would pay a reasonable sum for these works. There was a separate issue with regard to instructions issued to Mowlem after the final letter of intent.

The judge rejected Mowlem's arguments and found that the cap held. In his conclusion he said, at paragraph 44:

> '... I am satisfied that the letter of intent dated 4 July 2003 did not have effect only until 18 July 2003 or only in relation to work done before that date or only in relation to work instructed before that date.... it would make no commercial sense to have a final limit on Stena's obligations to make payment which could be avoided by the simple expedient of continuing to carry out work after 18 July 2003. It would be even more bizarre commercially if financial limitation on Stena's obligations could be avoided simply by Mowlem exceeding that limit.'

A further difficulty for a subcontractor operating under a letter of intent for the entire works, where the contractor's payment obligation is capped, is that the closer the value of completed work comes to the stated cap, the more a refusal by the subcontractor to proceed further, especially if the contractor believes the capped value is sufficient for the completion of the subcontract work, may be regarded as the subcontractor subjecting the contractor to unlawful duress to conclude an agreement.

It is frequently the case, as in *Mowlem* v. *Stena*, that following the issue of a letter of intent the parties give little or no thought to the consequences of continuing working under such limitations, neither do they make any positive effort to conclude a binding agreement. In the *Mowlem* case the parties solved their immediate difficulties by Stena issuing a further letter of intent raising the value of the cap. Had Mowlem at an early stage refused to accept a letter of intent with a cap, then Mowlem's position would have been that which it considered it should have been, i.e. one where a true value for the contract works was to be calculated in line with the offered price and offered terms and conditions.

It is suggested therefore that a subcontractor accepting a letter of intent with a capped value should, a reasonable time before the cap is considered to have been reached, obtain either a concluded agreement or demand and an open-ended letter of intent. After all, there comes a time when the contractor has no intention that the subcontractor will not complete the works, and the letter of intent is no longer intended to be only a temporary contract.

Incorporation of terms – general principles

Where terms are not expressly stated in the articles of agreement or in an accepted offer, they may be incorporated by reference to other documents. In construction subcontracts, the most common documents implied into a subcontract by reference will be a standard form of subcontract terms and conditions, either an industry standard or a form bespoke to the contractor, and/or by reference to the terms and conditions of the main contract.

The law accepts in principle that the terms of a subcontract can be founded on a document referred to in an offer or signed agreement. Indeed, the HGCRA makes reference to terms in writing a basis for a subcontract in writing.

Where the parties enter into a preliminary subcontract, such as may result from a letter of intent, and it makes express reference to a standard form of subcontract, then so far as those standard terms are not inconsistent with the terms of the preliminary subcontract they will be the terms of the subcontract. In the case of *Hackwood Ltd* v. *Areen Design Services Ltd* [2005] EWHC 2322 (TCC), a letter of intent made reference to the terms being a JCT contract. At a later date, Areen Design sought to commence an arbitration under the provisions of the JCT contract. Hackwood resisted the arbitration on the grounds that since there was no concluded JCT agreement, there was no agreement to arbitrate. Mr Justice Field found that the letter of intent did incorporate the JCT conditions and said:

> '17. It is common ground that a valid and enforceable contract came into existence upon ADS's acceptance of the terms of the 4 June letter by starting work on 5 June 2001. The question I have to decide is whether under this interim contract the provisions of the JCT contract were to apply. In my judgment, this question must be answered by an objective assessment of the parties' intentions having regard to the terms of the letter and the matrix of facts in which it is set. Since the parties intended the letter to record the terms of their interim contract and it is not contended that that contract was varied by subsequent contract or that there is an estoppel, no regard can be had to the parties' conduct subsequent to the 5 June 2001: see *James Miller & Partners Ltd* v. *Whitworth Street Estates (Manchester) Ltd* [1970] AC 583; *Schuler (L) AG* v. *Wickman Machine Tool Sales Ltd* [1974] AC 235.

> '18. By the terms of the 4 June letter ADS agree to carry out the works specified in the Contractor's Proposal for a sum of £13.2 million and in accordance with a programme on

> site of 65 weeks completing on 2 September 2002. Although ADS are obliged to complete the works, Hackwood are free to terminate the contract at any time. ADS are only to have a contractual right to complete the works if a formal JCT contract is signed following agreement on all outstanding issues.
>
> '19. In my judgment, construed as a whole and in its context, the effect of the 4 June letter was to incorporate terms of the JCT contract into the interim contract save to the extent that those terms were inconsistent with the terms of the letter.'

SBCSub/C at clause 2.5 requires the subcontractor to observe, perform and comply with the provisions of the main contract so far as they relate and apply to the subcontract.

Practical difficulties can arise in the incorporation into the subcontract of terms by reference, due on the one hand to the contractor failing to make clear from the outset the terms it intends should apply and/or their details, and on the other hand the subcontractor failing to establish the terms to which it is agreeing. Many contractors will seek to subcontract either on their own bespoke terms or by incorporating significant amendments to the appropriate standard form. Neither is likely to be to the benefit of the subcontractor and in many instances will impose onerous obligations and/or restrictions on the subcontractor.

The more onerous the clause to be incorporated into the agreement the clearer must be the notice to the accepting party as to the nature of the clause. This point was strongly made by Lord Justice Denning in the case of *J. Spurling Ltd* v. *Bradshaw* [1956] 1 WLR 461, when he said, at page 466:

> 'I quite agree that the more unreasonable a clause is, the greater the notice which must be given of it. Some clauses which I have seen would need to be printed in red ink on the face of the document with a red hand pointing to it before the notice could be held to be sufficient.'

It will be good practice for a contractor intending to subcontract under its own terms or by amendments to a standard form of subcontract, to make its intention clear with its enquiry and at that time provide the tenderers with a copy that can be considered and where appropriate suitable provision made within the offer. It is not uncommon for contractors to seek tenders without any reference to the intended terms, and only at a late stage in the post-tender negotiations to introduce its requirements for the prospective subcontractor to agree. Such action may be regarded as sharp practice, and while probably not unlawful it may not in some circumstances be upheld as effectively incorporated into the subcontract.

It is of course open to the subcontractor not to accept such terms or at least to seek time to read and consider such a proposal. In practice this is rarely done, often to the subcontractor's detriment. A more extreme but by no means uncommon situation is that the contractor's terms are not introduced until the issue of the subcontractor's order or agreement for signature. Such action will generally be a counter-offer made by the contractor, requiring acceptance by the subcontractor. It will be open to the subcontractor to reject it up to the time for commencement of the subcontract works, when the subcontractor will be considered to have accepted the contractor's terms by performance. A contractor adopting this approach will be at risk until such time as its counter-offer is accepted, since rejection shortly before performance is due may leave the contractor with no alternative but to accept the subcontractor's terms.

Following the judgment of Mr Justice Jackson in the case of *Machenair Limited* v. *Gill & Wilkinson Limited* [2005] EWHC 445 (TCC), there must be doubt whether merely referring to bespoke terms is sufficient action by the contractor to ensure that those terms are effectively incorporated into the subcontract. The judge described the nature of the subcontract document and his analysis of its effect:

> '30. Each of the three purchase orders sent by Gill to Machenair had a note at the bottom stating, "This order is placed subject to our conditions of purchase, a copy of which is available upon request." Mr Christopher Dodd [Counsel] who appears for Gill, contends that this note had the effect of incorporating Gill's standard conditions into the contract. Mr Anthony Edwards [Counsel] who appears for Machenair, contends that Gill's standard conditions were not incorporated. He points out that on the first two purchase orders the crucial words were substantially obliterated by the fax header. In response to this last point Mr Dodd contends that a hard copy of the purchase orders was also sent to Machenair. Furthermore, the obliteration was only partial.
>
> '31. In relation to this issue my conclusions are as follows:
>
> (1) On balance of probabilities Gill did not send hard copies of their purchase orders to Machenair. Gill simply relied on the faxes. In relation to this issue I note that no hard copies of the purchase orders have been disclosed by Machenair.
> (2) The obliteration by the fax header was of such an extent that it was unreasonable to expect the recipient to decipher the words underneath.
> (3) In any event (even if I am wrong in my previous conclusions) the words at the bottom of the purchase orders were not sufficient to incorporate Gill's standard conditions into the sub-subcontract. Gill's printed conditions are extensive. They are not one of the standard forms used within the industry. They were never supplied to Machenair. The note on the purchase orders did not contain any form of résumé of those conditions. In my view, the present case should be distinguished from the authorities cited in footnote 64 on page 714 of the 29th edition of *Chitty on Contracts*.
>
> '32. For all these reasons I conclude that Gill's standard conditions were not incorporated into the sub-subcontract. Furthermore, I do not see anything in the previous course of dealings between the parties which had that effect.'

While this judgment was made on the specific facts, it does indicate that relying on terms that are 'available on request' is not sufficient to ensure their incorporation. The contractor will need to take positive steps, first to ensure the subcontractor knows what terms are to be incorporated and secondly to ensure that those conditions are in fact available to the subcontractor, such as enclosing a copy with the order.

The Court of Appeal case in *Poseidon Freight Forwarding Co Ltd* v. *Davies Turner Southern Ltd and Another* [1996] 2 Lloyds Rep 388, raises two further circumstances which not infrequently arise in building subcontracts. First, the situation where the relevant terms are said to be those detailed on the reverse of an order or letter of acceptance, but because it is faxed to the other party it does not contain the conditions as stated. If the faxing party sends the terms on a following sheet under a joint cover note, then it is probable that such action would be considered to be effective for the purpose of communicating those terms. However, should the faxing party fail to attempt to transmit those terms at that time, the courts may hold that this is evidence that those terms were not to be incorporated into the subcontract. In *Poseidon* v. *Davies Turner*, Lord Justice Leggatt reached this conclusion, at page 394:

'Although reference is sometimes made to a "battle of forms" as each side strives to render its own form applicable, this case is an example of a party whose terms are in issue doing nothing, probably because his representatives did not think about it. If they had, it is difficult to suppose that they would have regarded not sending their terms with the contractual letter of June 3, 1991 as being a propitious way of ensuring that they applied, nor does it look as though they hoped by keeping quiet about their terms to render them applicable as sometimes occurs. This is not a case where a party declares that the terms are available for inspection. It is a case where, on documents sent by fax, reference is made to terms stated on the back, which are, however, not stated or otherwise communicated. Since what was described as being on the back was not sent, it was a more cogent inference that the terms were not intended to apply.

This impression would have been confirmed by the subsequent course of dealing in which Davies Turner continued to refrain from sending the terms and from contending that any applied. If that was notice of anything, it was notice that, in the implementation of their arrangements for co-operation, Davies Turner were not seeking to insist on the terms that governed their other dealings.'

The second issue to arise from the *Poseidon* case is that where a party's standard terms are manifestly intended for the situation where there is a presumed relationship between the parties, and in the relevant situation the relationship between the parties is of a different kind, then the proffered terms may be found not to be relevant and therefore not to apply. In *Poseidon*, the terms were considered to be for the situation where Davies Turner shipped goods under their own bills of lading, which contrasted with the actual situation where they were acting as agents.

While in practice it will generally be the contractor's terms that are to be incorporated into the subcontract, the terms in question could be proffered by the subcontractor. While it may be relatively easy to demonstrate that a subcontractor has accepted the contractor's terms by conduct, by having commenced work, it will be more difficult to find that the contractor has accepted the terms offered by the subcontractor, without a written acceptance. Where there is a written acceptance such terms will be regarded as incorporated even where the contractor claims not to have read or considered those terms.

In the Court of Appeal decision in the case of *L'Estrange* v. *F. Graucob Limited* [1934] 2 KB 394, Lord Scrutton said, at page 403:

'When a document containing contractual terms is signed, then, in the absence of fraud, or, I will add, misrepresentation, the party signing it is bound, and it is wholly immaterial whether he has read the document or not.'

In this case the terms sought to be incorporated were written on an order form for the supply and installation of an automatic cigarette machine. Lord Scrutton went on to make the point that an order may either be an acceptance or a proposal which may be accepted.

Referring to the order, Lord Justice Maugham said, at page 405:

'The brown paper document is not a formal instrument of that character (a formal contract, such as a deed), yet, in my opinion, having been signed it may well constitute a contract in writing. A reference to any of the text-books dealing with the law of contract will provide many cases of verbal acceptance of a written offer, in which the courts have held that the written offer and the acceptance, even though only verbal, together constituted a contract in writing, which could not be altered by extraneous evidence. The rule may not operate equitably in all cases, but it is unquestionably binding in law.'

Later Lord Justice Maugham stated that it is always possible that there could be a pre-existing oral contract prior to the signed order, when it could be argued that the order did not represent the terms to be relied on. He said, at page 406:

'In a case of this nature it is possible that the document signed by a contracting party may not be the contract, but merely a memorandum in writing of a preceding verbal contract between the parties, and if in this case it appeared that the document in question was only a memorandum of a previous contract which had not contained the clause excluding all conditions and warranties, the plaintiff might have relied upon the case of *Roe* v. *Naylor (No. 2)* [(1919) 87 LJ (KB) 958] and contended successfully that, as the clause was not a part of the contract, she was not bound by it. In my judgment, however, such a view as that is excluded here, because on the facts there was no preceding verbal agreement between the parties.'

In construction projects it not infrequently happens that companies, whose primary business is the manufacture and supply of building components, agree to undertake the erection or installation of their components as a subcontractor, rather than as a supplier. In such circumstances the principle stated by Lord Justice Leggatt in *Poseidon*, at page 393, may apply and the proffered terms may be regarded as of no effect. Lord Justice Leggatt said:

'In this case, the principal dealings by Davies Turner were with cargo-owners in cases where Davies Turner shipped under their own bills of lading. That is to be contrasted with the arrangement under which Davies Turner and Poseidon were prepared to act as, what they might have called, "agents" for each other. That was a relationship of co-operation with regard to which the judge accepted that Mr Liu [witness for Poseidon] genuinely assumed that the standard terms of Davies Turner were not intended to apply. The judge was entitled to find that the assumption that these terms applied, as they did, and applied only to contracts of carriage where Davies Turner shipped goods under their own bills of lading, was a reasonable assumption in the circumstances.

In my judgment, the most natural construction that a foreign freight forwarder such as Poseidon might be expected to put upon the reference to "transacting business", would be that it was intended to apply to contracts between Davies Turner and their customers for whom they would be transporting or arranging for the transport of goods. That would explain the presence of the BIFA [British International Freight Association] terms on some Davies Turner documents. It is also consistent with the facts (a) that many of the documents sent by Davies Turner (to) Poseidon contained no reference at all to the BIFA terms; (b) that those terms were not shown on the back of most documents sent to Poseidon on which they were stated to be shown; and (c) that the terms were actually sent to Poseidon only in connection with eastbound shipments. Since the terms were not in fact drawn to Poseidon's attention, they assumed that the terms were not intended to apply to them. The notion of "transacting business" is hardly apt to a long term co-operation agency relationship under which Poseidon were entitled to expect equality and mutuality, not limitation of liability of Davies Turner alone.'

A further risk for a contractor is that it may fail to update its terms and/or procedures. Standard forms are continually being revised by drafting panels to incorporate changes in the law by way of both decided cases and new or amended statutes. It is not uncommon to find contractor's terms that are defective by not having been kept up to date or where the associated documentation does not reflect the revised terms. The more documents incorporated into the subcontract, the greater the risk of conflict and ambiguity.

A requirement that the provisions of the main contract apply so far as they relate to the subcontract, is a limited obligation which can be construed as consistent with the decision in *Brightside Kilpatrick Engineering Services* v. *Mitchell Construction (1973) Ltd* (1975) 1 BLR 62, referred to later in this chapter. It is incumbent on the subcontractor to become acquainted with the relevant provisions of the main contract. This is rarely done, and consequently subcontractors can discover that they have been acting in breach or have failed to make allowance for the requirements incorporated by such a clause.

There may also be an obligation owed by the contractor to the subcontractor; for example, SBCSub/C at clause 3.24 provides that the contractor may be required by the subcontractor to obtain for the subcontractor any right or benefit of the main contract. This is subject to a request by the subcontractor and limited to those provisions of the main contract that are applicable to the subcontract works and are not inconsistent with the express terms of the subcontract. Clause 3.24 goes on to make any action by the contractor under this clause to be at the subcontractor's cost, and the contractor is entitled to obtain both an indemnity and security for costs as is reasonable for the contractor to require.

It is not clear why SBCSub/C makes provision for the contractor's action to be at the subcontractor's cost, or indeed what costs could be passed to the subcontractor on these grounds. There could be an implied obligation for the contractor to obtain such rights or benefits and for the employer to provide them, and that failure to do so would be a breach of the subcontract conditions.

The problem for those seeking to determine the true construction of the resultant subcontract, comprising both the terms of the subcontract and by incorporation those of the main contract, is: which of several possibilities that may exist is to be regarded as the actual intent and agreement between the parties; and how are terms not directly applicable, such as those referring to a third party, to be interpreted.

Where reference is made for the terms and conditions of the subcontract to be those of a recognised standard form, it will be necessary to identify which revision is intended. In the event that no express mention is made, the presumption will be that the parties intended to use the revision current at the time of the placing of the subcontract. Where a revision is published between the date of the offer and the date of its acceptance, it may be presumed that the revision current at the time of offer applies, unless express reference has been made to a later revision in the post-tender pre-contract period. However, where reference is expressly made to a number of revisions, the subcontract will only include those revisions expressly included. In the case of *Aqua Design & Play International Ltd (in liquidation t/a Aqua Design) and Fenlock Hansen Ltd (t/a Fendor Hansen)* v. *Kier Regional Ltd (t/a French Kier Anglia)* (2002) 82 Con LR 117, the courts had to decide whether an amendment to the latest revision of DOM/1 had been incorporated into the subcontract. If it was not incorporated, then the contractor Kier could resist payment to the subcontractor on the basis that the employer was insolvent. The Court of Appeal reversed the decision at first instance with the following words of Lord Justice Peter Gibson:

> '23. The question of construction to which this appeal gives rise is a short one. In Article 1.3 of the Articles of Agreement is the reference to "the subcontract conditions for use with the Domestic Subcontract DOM/1, including Amendments 1 to 10 thereof inclusive and published by the Construction Confederation", a reference to the conditions as corrected by the Schedule of Corrections published by the Construction Confederation or a reference to the uncorrected conditions.
>
> 24. In determining that question of construction the court is concerned to ascertain the objective intention of the parties, and it does that by considering the contract as a whole against the relevant background of facts known to both parties. Where the judge went wrong, as I would respectfully suggest, is in looking too narrowly at the meaning of the word "published" such as one might obtain from a dictionary as distinct from looking to the meaning which the parties to the particular contract must be taken to have intended. That may or may not conform to the dictionary meaning. I do not say that the meaning held by the judge to be correct is not a possible meaning of Article 1.3 looked in isolation, but I think that it is tolerably clear that, in the context in which the words in question are used, that is not the right meaning. Had it been the correct meaning, it would have been unnecessary for 16 of the 21 corrections in the Correction Schedule to have been repeated, as they are in Kier's own form 8115a. Of the remaining five corrections, four relate to two clauses wholly replaced by Kier with different clauses of its own wording and the final correction, that relating to clause 32, is left in solitary state: not repeated, not deleted, not amended. Given the evident care with which the corrections have otherwise been dealt with in form 8115a, I find it impossible not to regard that treatment of clause 32 as significant.'

The deciding issue in this case is the fact that Kier had not just made reference to the form of subcontract or even to the form including amendments 1 to 10, but had transcribed those amendments into its own schedule of amendments. On this basis any amendment not set out in Kier's own correction schedule could not be regarded as incorporated into the subcontract.

Where an industry standard such as SBCSub/C is incorporated, there will normally be reference to an appendix or other document, which defines the specific requirements of the subcontract. Without the appendix, many of the terms can have only limited effect. For example, SBCSub/A defines the programme for the works in the Sub-Contract Particulars at item 5. Without this condition being defined, clause 2.3 of SBCSub/C is inoperable.

Reference to the incorporation of the terms of the main contract, especially where this is the head contract involving reference to a third party, such as employer's agent, may make interpretation particularly difficult. In the case of *Brightside Kilpatrick* v. *Mitchell*, Lord Justice Buckley found, at page 65, that the terms of the main contract were implied into the subcontract conditions only to the extent that they were 'matters relating to subcontractors' and only to the extent there was 'no conflict between the head contract and the subcontract':

> '. . . it is not easy to know precisely what these words were intended to mean, but I am much inclined to the view that upon their true construction what they say, and all that they say, is this: that the subcontractual relationship between the contractor and the subcontractor shall be such as to be consistent with all those terms in the head contract that

specifically deal with matters relating to subcontractors, so that there shall be no conflict between the head contract and the subcontract.'

A case that followed the *Brightside Kilpatrick* decision was *Comorex Ltd* v. *Costelloe Tunnelling (London) Ltd* (1995) SLT 1217, where Temporary Judge T. G. Coutts QC had to decide whether an obligation to indemnify the employer could be passed down the contractual chain by incorporation of the terms of the main contract by reference. In that case the words used in the subcontract were stated at page 1218D:

'. . . The contract between the pursuers and the defenders is contained in 14.4, 5 and 6 of process. In 14.4, a letter addressed to the defenders, it is said:

"Further to our recent discussion we would confirm that your conditions of contract will be as ours with our client, i.e. General Conditions of Contract for Mains and Service Laying Period Contracts for Minor Works April 1977 Edition.

Please find copy of same enclosed.

It is a condition of your contract that no variation will be allowed as we are tied to those conditions."

These contractual terms would appear to have been accepted by the defenders.'

In the event, it was the work of a sub-subcontractor that led to damage to a gas pipe and the contractor paid the employer £34,953 in settlement of the claim and sought to recover the same sum from the subcontractor under the incorporated indemnity term. The judge found that such a term was not to be incorporated, at page 1220D:

'In my judgment the purpose of the intimation in 14.4 of process is to put the defenders upon notice that the pursuers are governed by particular contract conditions and the intention of the parties was that the defenders should be bound to perform their subcontract for the pursuers in such a way as not to put the pursuers in breach of contract with the employer. To that extent the general conditions might be said to be imported by reference. The references to "the engineer" and "drawings" reinforce that view, as does the express provision in 14.4 that there would be no variation permitted. That reading gives context to some, at least, of the conditions when read as between contractor and subcontractor.

However, far from incorporating an indemnity clause, let alone an indemnity clause which would have the effect of making the defenders responsible for the negligence of their subcontractors and liable to relieve the pursuers from their contractual indemnity to the employer, the words used are inadequate to provide for such a result. Clause 19 is not apt as it stands to import an indemnity on a subcontractor. Nor is it apt to make a subcontractor liable to relieve a contractor of his contractual obligation to indemnify the employer. The pursuers are not the employers in terms of the general conditions and could not be made so without manipulation of the terms of the contract. I do not consider that a transportation of "contractor" (the pursuers) for "employer", and "subcontractor" (the defenders) for "contractor" can be sensibly made . . .'

Where the contractor refers to the main contract by the title of the standard form, and states that those will be the terms of the subcontract, the courts are likely to find that the actual terms of the subcontract are the terms of the standard form of subcontract intended for use with that form of main contract. For example, where the terms were stated to be JCT 2005 it is likely that this is to be interpreted as meaning SBCSub/C or one of its derivatives. Such a situation arose in the case of *Modern Building Wales Ltd* v. *Limmer & Trinidad Co Ltd* (1975) 14 BLR 101, where the judge, at page 108, accepted expert evidence that the words 'form of nominated subcontract (RIBA 1965 edition)' could be taken to mean the then standard Green Form published by BEC Publications:

> 'I have no doubt that anyone in the industry would understand, as I understand, the words "in full accordance with the appropriate form for nominated subcontractors (RIBA 1965 edition)" appearing on the plaintiffs' order of 18 December 1968 to refer to what is known as "The Green Form".'

When a contractor refers by name to a standard form of main contract it can be considered to mean the relevant standard form of subcontract, but it is far from clear what is meant when the terms of that form of main contract are to be implied into a defined form of subcontract. In the case of *Aughton Ltd (formally Aughton Group Ltd)* v. *M. F. Kent Services Ltd* (1991) 57 BLR 1, Lord Justice Ralph Gibson considered the effect of incorporation of terms, albeit in the context of whether the arbitration provisions could be incorporated by reference, and in the process made several comments of general relevance, at page 20:

> 'Lord Justice Oliver pointed out in *The Varenna*, in examining the effect of words of incorporation, there are likely to be two stages of that enquiry because incorporation in wide general terms will often incorporate into the first contract terms not strictly appropriate to it. It then becomes necessary to see whether those terms are so clearly inconsistent with the first contract that they must be rejected or whether the intention to incorporate a particular clause is so clearly expressed as to require, by necessary implication, modification of the incorporated clause so as to adapt it to the first contract into which it is to be incorporated.
>
> The distinction between conditions of a contract which define the rights and obligations of the parties with reference to the subject matter (e.g. the standard and extent of the work, the time for completion, and the power to vary the work by instructions, together with the terms for payment and the control of the relationship of the parties on the site), on the one hand, and on the other hand, those which control or affect the rights of the parties to enforce those rights and obligations by proceedings at law was, as stated above, emphasised in *Thomas* v. *Portsea* [1912] AC 1.'

Aughton establishes the principle that not every term of a printed form of contract can be simply incorporated by reference to that form of contract. In *John Martin Hoyes Directional Drilling Limited* v. *R. E. Docwra Limited* [2000] 1999-TCC-117, Judge Seymour QC had to consider the effect of the general incorporation of a heavily amended version of the GC/Works/1 form of contract into the subcontract. The words, 'The terms and conditions of this enquiry, will form part of the Sub-Contract Agreement, together with the terms and conditions of the Main Contract as stated in the attached appendix', were incorporated into the tender enquiry and in turn incorporated, as a listed document, into the subcontract. The judge first set out the usual effect of incorporating documents, at page 11:

> 'The usual effect of incorporating a document by reference into another document is that the second document is then read as if the first document was set out verbatim in it.'

Mr Rees (Counsel for Docwra) suggested that the GC/Works/1 form of contract should be read with the names of the parties changed round, such that the contractor became the employer and the subcontractor became the contractor. The judge's opinion of this suggestion was expressed in the following words, at page 11:

> 'Such verbal alteration and substitution cannot be justified by reference to any express provision of the Subcontract Form of Agreement in fact executed between Docwra and JMH. It can only be justified by some form of implication. The basis for any such implication can, I think, only be found if it could be said that the implication represents the actual but unexpressed agreement of the parties objectively determined.'

The judge then summed up the difficulties with regards to incorporation of terms, by the casual reference to documents including another contract between different parties. He stated that this was rarely done with any due consideration:

> 'The documents so listed included: "RE Docwra Limited Subcontract Enquiry dated 18 December 1997"; a letter of that date written by Docwra to JMH had enclosed with it an undated document entitled "Subcontract Enquiry" in which appeared the sentence: "The terms and conditions of this enquiry will form part of the Sub-Contract Agreement, together with the terms and conditions of the Main Contract as stated in the attached appendix" and the appendix referred to identified the form of main contract as GC/Works/1. It is a long chain.
>
> The length of the chain is, it seems to me, relevant to determining what, if any, implication is appropriate. Also relevant, in my judgment, is the fact that the sentence which I have just quoted followed immediately upon an obligation upon JMH to "enter into an ICE Blue Form of Subcontract if required".
>
> The problem of construction of the Subcontract Form of Agreement with which I am faced is not unique. It not infrequently happens, particularly in the context of subcontracts in the construction industry, that the provisions of another contract between different parties, such as a main contract, is casually incorporated by reference. The reality will often be that this happens mechanically, possibly without any real actual thought at all and certainly not with any serious attempt to address the implications of what is being done. However, once both parties have, on the face of it, agreed to the incorporation by reference of the other contract, the task of the court is, by construction or implication or a combination of the two, to divine the objective intention of parties who actually gave little or no thought to the matter.'

The judge then reviewed a number of cases and concluded, at page 14:

> 'These problems demonstrate, in my judgment, that it cannot have been the actual but unexpressed intention of Docwra and JMH, objectively determined, that a particular form of GC/Works/1, that is to say, amended rather than unamended, or vice versa, be incorporated into the Subcontract Form of Agreement with such verbal modifications as contended for by Mr Rees [Counsel for Docwra].'

From the Docwra case, it is clear that in the subcontract context the courts have little respect for the blanket incorporation of terms from one contract into another, and will be slow to give meaning to terms that require the implication of changes to the words before they can become operable. Such terms will include the substitution of the contractor for another party, such as the architect, as well as for the employer. It is unlikely that the courts will give effect to such terms unless it is clear that this was the objectively determined intention of both parties.

In trying to incorporate main contract forms into subcontracts and substituting the names of subcontractor and contractor for contractor and other parties to the main contract, the contractor will normally need to assume the role of more than one person within the main contract conditions. This frequently leads to 'absurd' or 'utterly fantastic' situations, as discussed by Judge Humphrey Lloyd QC in the case of *Cegelec Projects Limited* v. *Pirelli Construction Company Limited* [1997] No. 1997 ORB 646. Judge Lloyd pointed out, at paragraphs 29 and 30, the absurdities of such arrangements within the specific dispute resolution procedures of the subcontract between those parties:

> '29. Cegelec's case was however that the importation of clause 94 had the effect of requiring Pirelli to submit any dispute not to the engineer under the main contract but to Cegelec itself. In my judgment it is verging on the absurd to think that, as suggested by Cegelec, amicable settlement is represented by a requirement to put a dispute to one of the contracting parties for a decision which if not accepted will become final and binding. The only purpose of such a requirement would be to delay matters for up to three months for a decision need not be given earlier. Whilst there can be value in a "cooling-off" period I can see no reason why Pirelli should be taken to have agreed to a period the length of which will be determined by Cegelec. Such a procedure is at the least not conducive to a dispute being settled amicably and is arguably the antithesis of amicable settlement. In any event it conflicts with clause 19 which permits disputes to be referred "at any time" which is consistent with a party deciding when negotiations (the most usual form of amicable settlement) will not achieve their end. In addition there are the odd consequences demonstrated by Mr Elliot [Counsel for Pirelli], e.g. that Cegelec would have to give itself notices etc. Mr Ramsey [Counsel for Cegelec] suggested that in practice the decision would be taken by a director of Cegelec or someone else not connected with the dispute. There is nothing in the contract to that effect and such an unusual stipulation certainly could not be read into it.
>
> 30. Settling a dispute amicably may also mean a discussion as to how best to settle that particular dispute, perhaps with the assistance of a third party. It might therefore cover conciliation but it does not do so in this instance since conciliation is only available on Cegelec's case after it has taken a decision which either it does not accept (a further absurdity) or which Pirelli does not accept.'

Since most main contracts, unlike most subcontracts, make reference to third parties such as the architect, contract administrator or engineer, such difficulties will frequently occur.

In the Scottish case of *Wescol Structures Limited* v. *Miller Construction Limited* (8 May 1998, unreported), Lord Penrose was if anything more scathing as to the meaning of the phrase 'The conditions of contract to be as main contract' when he said, at paragraph 29:

> 'It must have been clear to Miller that simple incorporation of the conditions of the main contract was impossible as between them and a domestic subcontractor. The notion is totally without sense. Incorporation of main contract conditions is notoriously problematical in any event. Wholesale incorporation without modification would confuse lines of contractual obligation and right, and would be fundamentally at odds with any notion of privity of contract at different levels of the contractual hierarchy or of mutuality as

between Miller and Wescol. Mr Cormack [Counsel for Miller] contended that whatever the sentence meant it was inconsistent with the application of DOM/1 or 2 conditions of subcontract.'

It is of course possible to construct a form of subcontract based almost entirely on the incorporation of the terms of the main contract. One such form is the subcontract for use with the MF/1 form of contract, for use by services contractors. Under those subcontract arrangements a major section, section 3, deals with the main contract and the incorporation of its terms in the following way:

'Main Contract

3.1 The Conditions shall be deemed to be incorporated in this Agreement and as between the Contractor and the Subcontractor shall apply to the Sub-Contract Works as they apply to the Works. The Contractor shall provide the Subcontractor with a copy of the Main Contract other than the details of the Contractor's prices.
3.2 Unless the context otherwise requires, the provisions of the Main Contract shall apply to the Subcontract as if the Contractor were the Purchaser therein stated and the Subcontractor were the Contractor thereunder.
3.3 Where under the Main Contract and the Conditions any liability of the Contractor is to be determined or limited by reference to the Contract Price or Contract Value as defined therein, the liability of the Subcontractor hereunder shall be determined as if the expressions Subcontract Price and Subcontract Value were respectively substituted therefore.
3.4 Subject to Sub-Clause 3.3 and 3.5 the Subcontractor shall indemnify the Contractor against every liability which the Contractor may incur to any other person whatsoever and against all claims, demands, proceedings, damages, costs and expenses made against or incurred by the Contractor by reason of any breach by the Subcontractor of the Subcontract, but in no case whatsoever (except claims resulting from death or injury to any person caused by the negligence of the Subcontractor for which no limit applies) shall the Subcontractor's liability hereunder exceed the Subcontract Price.
3.5 The Contractor shall indemnify and hold harmless the Subcontractor from and against any and all claims made against the Contractor in respect of which the Contractor is entitled to an indemnity from the Subcontractor under Sub-Clause 3.4 to the extent that such claims exceed the Subcontract Price (except for claims to which no limit applies under clause 3.4).
3.6 The Contractor agrees and declares that every limitation and exclusion of liability of the Contractor contained in the Main Contract shall as between the Contractor and the Subcontractor extend to protect the Subcontractor, his servants or agents (with the exception of liability for death or personal injury caused by wilful or negligent acts or omissions).'

Where the parties have agreed or at least raised no objection the courts may find that the dispute resolution procedures in the main contract will be incorporated into the subcontract. In the Canadian case of *Altec Electric Ltd* v. *J. V. Driver Projects Ltd* (2002) 15 CLR (3rd) 199, it was held that not only were the terms and procedures of the main contract to apply to the subcontract but that the main contractor had in no way waived those terms by its actions. The judge, Master Funduk, found, at paragraph 5:

'The defendant subcontracted some of the work to the plaintiff and there is a written subcontract. Article 21.1 of the subcontract says that the general conditions of the prime contract are to be "read into and form part of" the subcontract. It is acknowledged by both counsel that the dispute resolution provisions in the main contract, above set out, apply to disputes between the plaintiff and the defendant.'

In the *Altec* v. *Driver* case, counsel for both parties appear to have accepted that the terms of the main contract were incorporated into the subcontract. It is far from certain that this is the situation in English law. The quotation from *Aughton* v. *Kent* noted earlier held that terms 'which control or affect the rights of the parties to enforce those rights and obligations by proceeding at law' must be distinguished from general conditions of a head contract loosely incorporated by reference into the subcontract. Such provisions, although they may be incorporated within the subcontract terms, in fact form a separate contract, which may be formed at the same time.

The incorporation of arbitration clauses has been the subject of several decisions of the courts, which have shown great reluctance to incorporate such clauses by reference without clear evidence of the parties' express agreement to such a provision, and consequently denying their normal right to the courts. In the Court of Appeal case of *Aughton* v. *Kent*, Lord Justice Ralph Gibson said, at page 20:

'In examining the effect of words of incorporation, there are likely to be two stages of that enquiry because incorporation in wide general terms will often incorporate into the first contract terms not strictly appropriate to it. It then becomes necessary to see whether those terms are so clearly inconsistent with the first contract that they must be rejected or whether the intention to incorporate a particular clause is so clearly expressed as to require, by necessary implication, modification of the incorporated clause so as to adapt it to the first contract into which it is incorporated. That approach, in particular, in my judgment, is not to be restricted to charterparty/bills of lading cases but is of general application.'

The judge went on to distinguish between conditions that control the work on site and those that affect the rights of the parties to enforce those rights at law:

'The distinction between conditions of a contract which define the rights and obligations of the parties with reference to the subject matter (e.g. the standard and extent of the work, the time for completion, and the power to vary the work by instructions, together with the terms for payment and the control of the relationship of the parties on the site), on the one hand, and on the other hand, those which control or affect the rights of the parties to enforce those rights and obligations by proceedings at law was, as stated above, emphasised in *Thomas* v. *Portsea*. That distinction is as relevant, in my judgment, in a case of this nature about engineering work as it is in a case of charterparty and bill of lading. It provides good reason for requiring that an alleged intention of the parties to exclude the ordinary right of access to the court by an arbitration agreement, which may well include special terms of limitation, be clearly demonstrated from the terms of the contract.'

It must be considered whether the same or similar level of certainty is required for other types of dispute resolution. From the above discussion in *Aughton*, it would appear to apply to any method of dispute resolution that prevents or restricts a party's right to refer the dispute to the courts. This would probably include such procedures as expert determination, but might also be extended to clauses making a decision of one party final and binding, such as referred to at paragraph 20(a) of the Scheme for Construction Contracts (England and Wales) Regulations (the Scheme).

In *Cegelec Projects* v. *Pirelli Construction*, Judge Humphrey Lloyd QC had to decide whether the dispute resolution process incorporated within the main contract, which required an extensive process to be completed before a dispute could be referred to arbitration, was by reference incorporated into the subcontract. He rejected the main contractor's claim that the main contract procedures were incorporated within the subcontract, on two grounds. First, they were contrary to the express provisions of the subcontract, which required the parties to seek an amicable settlement. He said, at paragraph 30:

> '30. Settling a dispute amicably may also mean a discussion as to how to settle that particular dispute, perhaps with the assistance of a third party. It might therefore cover conciliation but it does not do so in this instance since conciliation is only available on Cegelec's case after it has taken a decision which either it does not accept (a further absurdity) or which Pirelli does not accept. Moreover under the LUL [London Underground Ltd] Conciliation Procedure conciliation could be terminated at any time. To import a conciliation procedure such as the lengthy procedure devised by LUL for use between it and a contractor and to make it applicable lock stock and barrel by means of provisions such as those in clauses 3.1 and 3.2 seems to me to be artificial and so far as removed from reality that it cannot be treated as representing the intentions of Cegelec and Pirelli as what might constitute amicable settlement in clause 19. If that is what was intended it would have been signalled. Therefore I do not consider that the reference in clause 19 to amicable settlement imports the main contract procedures which are not voluntary (which is usually of the essence of amicable settlement) but obligatory.'

Secondly, the incorporated term relied on by Cegelec referred to the 'subcontract works'. The judge held that a dispute resolution process did not constitute 'subcontract works' but was related to 'subcontract rights and obligations':

> '31. Even on a semantic level I do not consider that clause 3.1 is effective to import clause 94. The main contract is to "apply to the SUBCONTRACT WORKS as it applies to the Works except where amended by the Subcontract". In my judgment this means that those provisions of the main contract which are referable to the execution and completion of the subcontract works are to be applied without the cumbersome need to repeat them. Cegelec is therefore to be placed in a "back-to-back" position as regards the work itself. It is not apt to transpose to the subcontract level a provision such as clause 94 which has no relevance to any of the subcontractor's obligations in relation to the Works. Judge Hicks QC helpfully described provisions such as those found in clause 94 as "second-order terms" as they do not regulate the parties' substantive rights and obligations. In addition as I have already stated clause 19 must be regarded as amending it so that it has no effect as clause 3.1 does not apply to provisions that are amended by the subcontract.'

Judge Hicks QC in *Roche Products Ltd and Celltech Therapeutics Ltd* v. *Freeman Process Systems Ltd and Haden MacLellan Holdings plc, Black Country Development Corporation* v. *Kier Construction Ltd* (1996) 80 BLR 102, again referring to the possible incorporation of an arbitration clause said, at page 114F:

> 'Although commonly called arbitration "clauses" they are not ordinary first-order terms of the contract, regulating the parties' substantive rights and obligations under it, but

> second-order terms about the contract and in particular about the resolution of disputes in relation to it. In that sense they are contracts in their own right with an independent existence. Thus they may survive the discharge of the underlying contract and indeed govern disputes as to whether it has been discharged and if so how and with what consequences. It is therefore right that clear evidence of the intention to incorporate them should be needed, and that in particular caution should be exercised where the incorporation relied upon is not specific to the subject contract but arises by reference to another contract, for then there may be good ground for questioning whether the incorporation of the substantive provisions of that other contract was intended to carry the separate contract between the parties to it as to how disputes between them under it should be resolved. Moreover there may be difficulty in "transplanting" the arbitration clause in such a way as to be workable in a different context.'

It appears from this judgment that general terms of incorporation relating to the subcontract works may not apply to any dispute resolution process including adjudication, without clear evidence that it was the intention of the parties to incorporate such terms. Thus conditions seeking to incorporate terms, such as the rules for conducting an adjudication, the naming or the selection of an adjudicator or of an adjudicator nominating body, are unlikely to be incorporated into a subcontract, without express conditions that they will be actually incorporated.

As stated above, many standard forms require the completion of an appendix to make certain of the terms operable. Such requirements include the period for carrying out the work, rates of retention, and appointment of arbitrators and/or adjudicators. Where the parties fail to complete such an appendix, it is not considered that the provisions of the appendix of the main contract can be incorporated just by reference to the main contract in total. It is possible that where there is a default provision in the appendix in the standard form, such as in SBCSub/A, for the arbitrator nominating body, that this default provision would apply just by reference to and incorporation of that standard form. However, the effect of such incorporation by reference is likely to be limited to this extent.

SBCSub/C requires the subcontractor to indemnify the contractor for specific matters, including specific provisions of the main contract. These are set out in clause 2.5, which has the following requirements:

1. Subject to the contractor's obligations under the main contract being identified under the 'Schedule of Information' and that they relate to the subcontract works then the subcontractor is to 'observe, perform and comply' with the obligations of the contractor under the main contract. Such obligations are without limitation but include reference to clauses:
   2.10 (*Levels and setting out*);
   2.21 (*Fees or charges legally demandable*);
   2.22 and 2.23 (*Royalties and patent rights*);
   3.22 and 3.23 (*Antiquities*).
   The subcontractor is required to indemnify the contractor against two events:
   (a) a breach, non-observance or non-performance of the provisions of the main contract;
   (b) any act or omission which gives rise to a liability of the main contractor to the employer under the main contract.
2. There is a general requirement for the subcontractor to 'indemnify and hold harmless' the contractor for any negligence or breach of duty by the

subcontractor. Specific reference is made to misuse of scaffolding or other property belonging to or provided by the contractor.

It will be noted that at clause 2.5 reference is made to specific conditions of the main contract. This sub-clause can only be effective where SBCSub/C is used in conjunction with a JCT 2005 Standard Building Contract. Without express alterations, this sub-clause will be ineffective with other forms of main contract.

The requirement of the subcontractor is limited to those matters included in the Schedule of Information and in so far as they relate to the Sub-Contract Works. This will exclude any requirement within the main contract where action is stated to be by the contractor.

The sub-sub-clause requires the subcontractor to indemnify the contractor against and from (1) a breach or non-performance of the provisions of the main contract and (2) an act or omission which leads to a liability of the contractor to the employer under the provisions of the main contract. Any provision requiring a subcontractor to know the provisions of another contract must in theory place a very substantial burden on that subcontractor. In practice, very few subcontractors will seek to see the provisions of the main contract, let alone retain an exact knowledge of those provisions. In general, the subcontractor will rely on there being nothing of an unusual nature contained in them. These are dangerous assumptions for a subcontractor to make and it should either obtain a clear statement of the provisions of the main contract to be observed by the subcontractor, or limit its liability to matters specifically advised to it by the contractor. While the main clause 2.5.1 refers to the main contract conditions, sub-clauses 2.5.1.1 and 2.5.1.2 refer to the provisions of the main contract, thereby seeking to incorporate any requirement of any document comprising the main contract.

Should the contractor seek to rely on some particularly unusual or onerous condition of the main contract, which it has not expressly drawn to the subcontractor's attention, then the provisions of clause 2.5.1.1 and 2.5.1.2 may not, by reference to the judgment in *Interfoto Picture Library Ltd* v. *Stiletto Visual Programmes Ltd* [1988] 1 All ER 348, be enforceable. In that Court of Appeal case, where the supplier of photographs on loan relied on a clause on the reverse of the delivery note for charging a high daily rate, it was held by both appeal judges that such a condition was inoperative where it had not been drawn to the hirer's notice specifically. Lord Justice Dillon said, at page 352:

> '. . . if one condition in a set of printed conditions is particularly onerous or unusual, the party seeking to enforce it must show that that particular condition was fairly brought to the attention of the other party.'

Lord Bingham used similar words, at page 357:

> 'The defendants are not to relieved of that liability because they did not read the condition, although doubtless they did not; but in my judgment they are to relieved because the plaintiffs did not do what was necessary to draw this unreasonable and extortionate clause fairly to their attention.'

The effect of incorporation of terms is mentioned by Judge Humphrey Lloyd QC in the case of *Cegelec Projects* v. *Pirelli Construction*, when at paragraph 14 he quotes from the judgment of Judge Hicks in *Roche Products and Celltech* v. *Freeman Process Systems and Haden Maclellan,* at page 114:

> 'Acceptance of a test of intention, and rejection of any requirement that the presence or absence of particular formulae be determinative, do not preclude a recognition that some types of provision may require clearer or stronger evidence than others before an intention to include them may be inferred. That may be because they are particularly onerous or unusual or for some other reason.'

He went on to say that he found difficulty in regarding arbitration provisions as onerous or unusual. However, it could be that in specific situations the express terms may be unusual for the type of provision which it is sought to incorporate.

From the decided cases there is little doubt that the courts are wary of conditions incorporated en masse by reference to a document in another contract, since such conditions will rarely have been considered by the parties and are consequently unlikely individually to have been the subject of agreement. A likely exception is where the terms relied on are well known and unamended so that there will be little excuse for the accepting party not being aware of the detailed terms it is accepting.

Further to the general clause 2.5 described above, SBCSub/C from time to time makes reference to specific clauses of the main contract. Provided the SBCSub/C form of subcontract is being used on a project where the main contract is a JCT 2005 Standard Form of Building Contract, there should be certainty as to the interpretation of the subcontract and the clause will be effective.

Difficulties in interpretation will be likely to occur when either the clauses referred to in the main contract are deleted or significantly amended, or the main contract is of a different form containing no such clause or condition having a similar effect, even with a different clause reference. Such a situation will arise, for example, when SBCSub/C is used with other JCT forms of main contract, such as the intermediate or minor works forms or forms published by other bodies such as ICE or IChemE.

One possible interpretation of such clauses will be to refer to the clause in the Standard Form of Building Contract and see if that interpretation will conform to the requirements and provisions of the main contract. A more extreme view would be to consider the effect of any clause in the actual main contract, with the number referred to, and in all probability find that as a result the subcontract provision was unintelligible and unworkable. In any event, this will involve an implied term which in accordance with the judgment of Judge Seymour QC in *Hoyes* v *Docwra*, referred to earlier, will need to be shown to represent the actual but unexpressed agreement of the parties objectively determined.

It is for the parties, in particular the contractor, to ensure that the meaning of each clause is certain. Where there is ambiguity, the courts are likely to give such clauses the meaning most favourable to the subcontractor.

Such an issue arose in the case of *R. G. Carter Limited* v. *Edmund Nuttall Limited* [2000] HT-00-230, when Judge Thornton QC had to consider the interpretation of the subcontract conditions. In that case, the subcontract was generally on the DOM/1 1998 conditions; however, the specific provisions provided in the Appendix and in particular Appendix Part 8, which specified the adjudicator as Mr Brewer, referred to clauses in the subcontract by number. With regards to Appendix Part 8 that reference was to clause 24, which under the earlier editions of

DOM/1 provided for limited adjudication where there was a disputed set-off, and not clause 38A in the 1998 edition providing for statutory adjudication as required by the HGCRA.

The contractor contended that the clause number reference in Appendix Part 8 should be read as a reference to clause 38A in the 1998 revision. Judge Thornton QC found, at page 20, that there were a number of possible interpretations of the subcontract as written and by applying the *contra proferentem* rule found Appendix Part 8 to be of no effect:

> 'Therefore, as I see it, the court is thrown, as an aid to construction on what, at any rate until recently, used to be called the "contra proferentem" rule – those who seek to purge the legal lexicon of the Latin language have yet to provide us with the modern translation of that verbal coupling. What that rule provides for, in summary, is that, where there is conflict or ambiguity in a clause that has been drafted by and proffered by one of the parties to the contract and which has been accepted by the other party to the contract, then, whichever of the constructions that is least favourable to the proffering party is the one that the court should adopt. It is clear to me that the least favourable construction of a clause drafted by and proffered by the claimant is that which would treat the reference to Appendix Part 8 and clause 24 as being references to the old DOM/1 conditions, which, since they form no part of the contractual relationship between the parties, is to be regarded as a reference that is of no effect. In other words, I should not give effect to these provisions, leaving it open, if the question ever arose in a different context between these parties, as to whether this is a reference to what I have referred to as "add-on" adjudication procedure provided for by clause 24 to set off disputes.
>
> However, so far as any of the disputes or possible disputes are concerned, on that approach to the construction of this appendix, the reference to Mr Brewer is of no effect at all.'

Where terms are to be incorporated into the subcontract, by statute, such as for payment and adjudication under the HGCRA, it is considered that the incorporation of the Scheme will have effect before incorporation from the main contract. This is simply because the incorporation of terms from another contract into the subcontract is not to override the terms of the subcontract. In the absence of express provisions such as those required by the HGCRA, the Scheme is incorporated and consequently becomes automatically part of the subcontract conditions. Thus if there is no or inadequate provision for adjudication in the subcontract, then the Scheme will be incorporated, which allows for use of a nominated adjudicator and/or specified nominating body if "named in the contract (subcontract)". Similar situations arise from the payment provisions, with regards to retention, for example. Since the Scheme makes no provision for retention, it follows that without express provision in the subcontract terms there will be no provision for retention and such provision will not be incorporated by reference to the terms of the main contract, since the subcontract will be complete without such provision.

Incorporation of terms – contractor's proposals

Either as an express requirement of the invitation to tender or as a matter raised during the pre-contract negotiations between the contractor and the employer, the contractor may undertake to limit its freedom, to comply with the specification and other requirements, to specific or limited proposals. Such limitations may include the source of materials, the methods of carrying out the works or, in the case of contractor-designed matters, the design solution.

The purpose of agreeing such matters in advance will be to guard the employer from the risk of having to issue a variation instruction, with the associated effects on both time and cost, should the contractor's proposals be unacceptable to the employer despite being contractually compliant.

Once such matters are agreed and incorporated within the contract conditions, it will be necessary for the contractor to impose the agreed restrictions on its subcontractors. This should be done by an express requirement of the subcontract but may be incorporated by direct incorporation of other documents or by reference to other documents. Provided the contractor effectively incorporates its accepted proposals into the subcontract, the subcontractor will be obliged to carry out its work in accordance with those proposals for the subcontract sum.

Problems may arise in the event that for some reason the contractor wishes or needs to change the proposals as accepted, or it becomes impossible to comply with those proposals. In these circumstances it will be for the contractor to obtain agreement from the employer to the alternative proposals, which the contractor may have discussed in advance with the subcontractor. In such circumstances it is possible that while the change will be a valid variation to the subcontractor, to be valued under the terms of the subcontract, the change may not be a variation under the contract and the contractor may not be entitled to any change in the valuation of the work.

In the Court of Appeal case of *Holland Dredging (UK) Ltd* v. *The Dredging & Construction Co Ltd and Imperial Chemical Industries Plc (Third Party)* (1987) 37 BLR 1, the court had to consider the effect on the subcontract of the agreed method statement incorporated within the contract. Holland was subcontractor to Dredging and Construction for the laying of a submarine pipe at Ardeer on the Clyde estuary. Dredging and Construction's contract was to dredge and lay the pipe and then to backfill the trench to the original seabed level. Dredging and Construction's accepted method statement required the dredging material to be stockpiled on the seabed a limited distance from the trench and then, after the installation of the pipe, to be redredged and deposited in the trench. It was anticipated that a limited amount of material might be lost in the process and that the shortfall was to be made up by additional dredging in the stockpile area. It was on these requirements that the work was subcontracted to Holland.

In the event, owing to a combination of the nature of the dredged material, the depth of the trench and the sea movements, it became impossible to fill the trench using that dredged material. As a consequence the method of backfill was changed to the use of land-quarried material, for which Holland sought a very significant additional sum. The court found that while the change of method was a change to Holland, for which they were entitled to payment, the situation was different under the main contract. Lord Justice Purchas found that, despite the contractor's method statement not being a subcontract document, since it was referred to in the bill of quantities and the conditions of the main contract, it was the basis of the subcontract. He said, at page 31:

> '"The Sub-Contract Works" are described in Part B of the Schedule and the relevant expression is "Backfilling . . . *with suitable dredged material*". The Second Schedule does not import

> liabilities arising under the main contract the particulars of which are defined for the purposes of the subcontract in the First Schedule, as already set out in this judgment, and are imported into the subcontract only for the purposes of clause 3 of the Blue Form. To this extent in the exercise of construing the subcontract reference may be had to the specification of works, the contract drawings, the bill of quantities and the ICI General Conditions. Neither the method statement submitted with the tender nor the September minutes are referred to in the First Schedule: but the former is referred to both in the BOQ (in several places) and in Condition 4 of the ICI General Conditions. It has not been suggested that the September minutes form part of or are in any way relevant to the construction of the subcontract and the judge rightly ignored them for this purpose. In my judgment the method statement was directly relevant to the liabilities under the subcontract.'

On this basis he found that there was no liability under the subcontract for Holland to obtain material from sources other than the dumping ground. He said, at page 33:

> 'I find it impossible to construe the liability under the subcontract to extend to obtaining material from any other source. Mr Williams [Counsel for ICI] conceded that there was nothing in the contractual documents to indicate that any party contemplated the winning of any material from any source than the dumping ground.'

Lord Justice Purchas then looked at the terms of the main contract and concluded that they were wider than those of the subcontract, so while Holland were able to recover costs from Dredging and Construction they could not in turn recover such costs from the employer.

Although the *Holland* case relates to the effect of a contractor's method statement, it is suggested that the same principle applies to any contractor's proposal accepted by the employer and imposed on the subcontractor.

Waiver

The terms of the subcontract may limit the subcontractor's entitlement to relief from sanction and/or benefit by way of additional payment or otherwise, by first requiring a defined action by the subcontractor. Such action may involve the giving of a notice within an express time period or by specified dates. Where these requirements are stated to be a condition precedent they may be rigorously enforced especially where such a failure, by the subcontractor, will prejudice the contractor. This will be especially true where the contractor is required to act under the main contract within an express period, and his failure to act in time might prevent the contractor obtaining, under the main contract, the same relief that the subcontractor seeks from the contractor. However, in many cases the issue of a timely notice or response will not be seen as more than a requirement for the smooth administration of the subcontract.

In many instances, the parties may not abide strictly to the precise terms of the subcontract, especially where these are imprecisely defined and/or uncertain. For example, the subcontract may require the submission of daywork sheets within a stated time of the work being carried out, or that an entitlement to be paid on a daywork basis is dependent on the prenotification that such a method of payment will be relied on, or where instructions are considered to have a cost implication that a notice, with anticipated cost, is to be given within a stated period of the instruction or prior to commencing work on that instruction. Similarly, the subcontractor may allow the contractor to make payments late and/or deduct discount or withhold retention when such is not strictly allowable under the subcontract arrangements.

Generally, it will not be possible, where either party has regularly waived their rights due from the other party under the strict provisions of the subcontract and payment has been made and accepted without question, for the aggrieved party to deny such relaxation retrospectively. In the Court of Appeal case of *Alfred C. Toepfer* v. *Peter Cremer* [1975] 2 Lloyds Rep 118, it was held by Lord Denning MR, at page 123, applying the principle established in an earlier case that:

> 'When one person has led another to believe that a particular transaction is valid and correct, he cannot thereafter be allowed to say that it is invalid or incorrect where it would be unfair or unjust to allow him to do so. It is a kind of estoppel. He cannot blow hot and cold according as it suits his book.'

It will however be possible, at any stage, to notify the other party that from that time on they will seek to uphold the strict terms and provisions of the subcontract. In the Court of Appeal case of *Charles Rickards Ltd* v. *Oppenheim* [1950] 1 All ER 420, where the delivery of a motor car went well beyond the contract date, it was held that the purchaser could set a revised reasonable date for performance and could refuse performance when such later date was not achieved. Lord Denning found that the original contract date had been waived, at page 423:

> 'I agree that that initial time was waived by reason of requests for delivery which the defendant made after March 1948, and that, if delivery had been tendered in compliance with those requests, the defendant could not have refused to accept. Supposing, for instance, delivery had been tendered in April, May or June 1948, the defendant would have had no answer.
>
> If the defendant, as he did, led the plaintiffs to believe that he would not insist on the stipulation as to time, and that, if they carried out the work, he would accept it, and they did it, he could not afterwards set up the stipulation in regard to time against them. Whether it be called waiver or forbearance on his part, or an agreed variation or substituted performance, does not matter. It is a kind of estoppel. By his conduct he made a promise not to insist on his strict legal rights. That promise was intended to be binding, intended to be acted on, and was, in fact, acted on. He cannot afterwards go back on it.'

Lord Denning then described the subsequent failure by the supplier to perform, and concluded, at page 423, that the purchaser, having been lenient, was entitled to give reasonable notice making time of the essence:

> 'It would be most unreasonable if, having been lenient and having waived the initial express time, he should thereby have prevented himself from ever thereafter insisting on reasonably quick delivery. In my judgment, he was entitled to give a reasonable notice making time of the essence of the matter.'

Lord Denning later summed up the requirements for the existence of a waiver, at page 425:

> 'On this point I would say that to constitute waiver there must be conduct which leads the other party reasonably to believe that the strict rights will not be insisted on. The whole essence of waiver is that there must be an intention to affect the legal relations of the parties. If that cannot properly be inferred from the conduct, there is no waiver.'

What is less certain is whether a subcontractor can be relieved of the strict requirements of the subcontract, to report all causes of delay or to prevalue changes and the like, if the number and extent of such events becomes unreasonably high. Such circumstances frequently arise towards the end of a project where there is an urgency to complete the Works, frequently by accelerating work to recover lost time. In the resultant hurly-burly neither party may strictly perform their obligations by way of the issue and/or confirmation of instructions.

Where the subcontract is placed by way of a letter of intent and a preliminary contract is formed, and the stated intention of the contractor is to formalise matters by preparing articles of agreement for signature, and after the letter of intent is accepted the contractor takes no further action to formalise matters, or does not do so until sometime after a dispute arises, then it may be considered that the contractor has waived its right to rely on the strict interpretation of the terms proposed.

In the case of *Harvey Shopfitters* v. *ADI*, referred to earlier in the section on letters of intent, such a situation arose where the employer sought to rely 'on the complete absence of any Notices of Delay' by way of a defence to the contractor's claim for an extension of time. It is a common situation that the parties to construction contracts, both at main contract and at subcontract level, proceed in an informal manner relying on trust and good faith. Such an approach has generally been encouraged throughout the industry by reference to 'partnering' and 'best practice' procedures, etc. In addition, many subcontractors are reluctant to be considered to be 'contractual' by the contractor, and often receive letters of objection and/or adverse remarks or comments when seeking to record matters of delay and other directions.

The fact remains that most standard forms of subcontract do require both parties to administer the contract in defined ways, for example that all instructions are to be in writing and notices of payment are to be issued within five days of the payment due date, both requirements being 'more honoured in the breach than the observance'. It is frequently the result that the contractor adopts a relaxed approach to the specific requirements of the subcontract, whether owing to its own inefficiencies or because it did not require the procedures of the written form to be closely followed. In *Harvey Shopfitters*, Mr Recorder John Uff QC found that evidence of such a relaxed approach to the strict terms and procedures of the contract was shown by the fact that, following the issue of the letter of intent, no formal agreement had been prepared for signature. The judge therefore rejected the employer's defence, at paragraph 38:

> 'As regards the contribution of the architect, Mr Evans, it is relevant to note that, while reliance was placed on the accuracy of his minutes of meetings, it was Mr Evans who sent the letter of 7 July 1998 and then took no further step to issue formal Contract Documents to be signed by the parties. One explanation might be that Mr Evans' attitude to the contract was sloppy verging on being negligent. But having seen Mr Evans in the witness box, I reject such an explanation. Mr Evans was a precise and careful architect when this

> was necessary and in my view the failure to draw up a formal contract is consistent with the degree of informality to which Mr Nolan [Director of Harvey Shopfitters] testified. While I have no doubt that all parties believed there was a binding contract in existence, no one showed any sign of wishing to stand by the letter of it nor to enforce the small print in the way now proposed. No formal waiver was alleged in this regard but it is worth noting that the defendant alleged and has now succeeded in demonstrating waiver as its alternative case in relation to the existence of a binding contract. That point further emphasises the degree of informality that I am satisfied existed, not only in the drawing up of the contract but, more importantly for present purposes, in its performance.'

The conclusion to be drawn from this judgment is that if contractors wish to rely on the full performance of the terms and procedures of the subcontract, they must be able to demonstrate that they have themselves adhered fully to the terms and procedures, which will rarely be the case.

Lord Denning in the case of *Hoenig* v. *Isaacs* [1952] 2 All ER 176 found that there was a waiver not to pay the contract price where the employer took possession of the work before it was entirely complete. He put it in these words, at page 180:

> 'Even if entire performance was a condition precedent, nevertheless the result would be the same, because I think the condition was waived. It is always open to a party to waive a condition which is inserted for his benefit. What amounts to a waiver depends on the circumstances. If this was an entire contract, then, when the plaintiff tendered the work to the defendant as being a fulfilment of the contract, the defendant could have refused to accept it until the defects were made good, in which case he would not have been liable for the balance of the price until they were made good. But he did not refuse to accept the work. On the contrary, he entered into possession of the flat and used the furniture as his own, including the defective items. That was a clear waiver of the condition precedent.'

In the subcontract situation, it is considered that this judgment will mean that where a contractor takes possession of the subcontractor's work and allows the work of following trades to proceed, even if the work of the first subcontractor is defective, the contractor will be deemed to have waived or at least limited its rights to recover the natural consequences of those defects which are known at the time of acceptance.

Conversely, where the contractor requires the subcontractor to commence work before the entire subcontract works are available, then it is suggested that the contractor will waive any right to require that the subcontract work be completed within the period set for it in the subcontract, except where the subcontract expressly makes provision for staged release to the subcontractor. In that situation the subcontractor's right to be relieved of the obligation to complete within the period set out in the subcontract will relate to the contractor's failure to release a stage on time as defined in the subcontract.

New terms

While, as discussed above, it is possible for one party to waive its rights under the subcontract, such a waiver does not change the terms of the subcontract. As stated by Lord Justice Ward in *Clarke & Sons* v. *ACT Construction* (2002) 85 Con LR 1, at paragraph 33, 'No term can be implied by a subsequent course of conduct':

> 'If the agreement was made in February 1992, then its terms would be fixed by reference to what was said at that time or by the course of dealing prior to that time. The claim is therefore misconceived in seeking to rely on the interim applications raised after the contract had been made. No term can be implied from a subsequent course of conduct.'

It follows that the terms of the subcontract are those established at the time of the conclusion of the subcontract and can only be changed or new terms added by express agreement between the parties.

This principle is of considerable relevance where, as so often happens, a subcontract is formed on a preliminary basis by a letter of intent or similar. A subcontract on the terms of that letter of intent will be concluded if accepted by the subcontractor, whether expressly in writing or by performance. Where, as frequently happens at a later date, the subcontractor is sent a formal subcontract order or a draft agreement for the subcontractor's signature, then should the terms be different from those of the preliminary subcontract, they will require formal acceptance by the subcontractor, which acceptance it will not be bound to give.

Terms arising in the course of dealing

Where a contractor and subcontractor frequently and/or regularly work together and there is a well-established pattern of operation or course of dealing such that the parties can say 'this is how it is always done', then the court may well imply into a new subcontract those terms and conditions, even if they have not been expressly used in that instance. A typical situation would be where a contractor always used its standard form by including reference to such terms in its subcontract order. If on a subsequent occasion it failed to make such reference, then unless it were to be repugnant to the express provisions of the new subcontract, it is likely that those terms would be included in that subcontract as well.

In the shipping case of *Circle Freight International Ltd (t/a Mogul Air)* v. *Medeast Gulf Exports Ltd (t/a Gulf Export)* [1988] 2 Lloyds Rep 427, the Court of Appeal was required to decide whether standard terms had been incorporated into a new contract. Goods had been stolen from a van and the forwarding agents denied liability under the standing trading conditions of the Institute of Freight Forwarders Ltd (IFF), 1981 edition. The court established that 'individual contracts were made orally by telephone', that 'an invoice would be sent to the defendant and this was the only document that passed' and that 'at least 11 such invoices were sent between March and August 1983'. At the bottom of each invoice in small type were the following words:

> 'All business is transacted by the company under the current trading conditions of the Institute of Freight Forwarders a copy of which is available on request.'

There was no evidence that the defendant ever asked for a copy of these conditions and none was ever sent.

Lord Justice Taylor found that the judge at first instance had found that the conditions referred to on the invoice were not effectively incorporated into the contract and could not be relied on. Lord Juctice Taylor said, at page 430:

'The learned judge referred to three cases on this first issue – *McCutcheon* v. *David MacBrayne Ltd* [1964] 1 Lloyds Rep 16; [1964] 1 WLR 125; *Hardwick Game Farm* v. *Suffolk Agricultural Poultry Producers Association* [1968] 1 Lloyds Rep 547; [1969] 2 AC 31, and *Keeton Sons & Co* v. *Carl Prior Ltd*, an unreported decision of this court given on 14 March 1985. He adopted the text proposed in the latter case by Lord Justice Ackner (as he was then known) at p. 4 of the transcript.

"The question in a case of this kind must always be, 'has reasonable notice of the terms been given?' This is essentially a question of fact depending on the circumstances of the case, and in particular on the nature of the business and position of the parties to the transaction."

The crucial part of the learned judge's reasoning on this issue is at page 9H of his judgment. He said:

"In my view, the plaintiffs have to show that they have taken reasonable steps to bring to the notice and the attention the term in question. As I have said, there is no contractual document in this case which refers to them, unlike the Keeton case, and unlike the Hardwick Game Farm case where the contract note contained the relevant clause – indeed, no contract document which even refers to the sets of conditions.

In my judgment, if they wish to rely upon a course of dealing as incorporating written terms, it is only reasonable that the specific term should be set out and not left to the defendants to chase around and find out for themselves. I am not satisfied in this case that the plaintiffs did take reasonable steps to draw the defendants' attention to the specific terms of exclusion of liability upon which they seek to rely. Therefore, in my judgment, the defence to the counter claim fails." '

Lord Justice Taylor's commentary on the above said, at page 430:

'Inherent in that reasoning are two propositions which are challenged by the appellants. First, that the specific condition relied upon must be drawn in terms to the customer's attention, and a reference to the conditions generally with the offer of a copy of them on request will not suffice. Secondly, that notice of the condition must be in a contractual document. Mr Malms [Counsel] for the appellants submits that the proper test here was as follows. Where there is a course of dealing, conditions will be incorporated into a contract where by words or conduct each party can be said to have led the other to believe the conditions were accepted.'

Lord Justice Taylor then considered a number of authorities and concluded, at page 433, that the standard terms relied on by the forwarding agents had in fact been incorporated by way of normal course of dealing:

'. . . it is not necessary to the incorporation of trading terms into a contract that they should be specifically set out provided that they are conditions in common form or usual terms in the relevant business. It is sufficient if adequate notice is given identifying and relying upon the conditions and they are available on request. Other considerations apply if the conditions or any of them are particularly onerous or unusual.

Again, it is not necessary that notice of the conditions should be contained in a contractual document where there has been a course of dealing.

Here, the parties were commercial companies. There had been a course of dealing in which at least 11 invoices had been sent giving notice that business was conducted on the IFF terms at a place on the document where it was plain to be seen. Mr Zacaria [Director of Medeast] knew that some terms applied. He knew that forwarding agents might impose terms which would frequently be standard terms and would sometimes or frequently deal with risk. He never sought to ask for or about the terms of business. The IFF conditions

> are not particularly onerous or unusual and indeed, are in common use. In these circumstances, despite Mr Gompertz's [Counsel for Circle Freight's] clear and succinct argument to the contrary, I consider that reasonable notice of the terms was given by the plaintiffs. Putting it another way, I consider that the defendants' conduct in continuing the course of business after at least 11 notices of the terms and omitting to request a sight of them would have led and did lead the plaintiffs reasonably to believe the defendants accepted their terms. In those circumstances it is irrelevant that in fact Mr Zacaria did not read the notices.
>
> Accordingly, I consider that the learned judge erred in his conclusion on the first issue and I would hold that the IFF conditions were incorporated in this contract.'

It must, however, be stressed that those terms must be shown to be the way those parties normally did business. It will not be sufficient to show they were frequently the terms used or had been used recently in such subcontracts. It is suggested that it is necessary that both parties would have no reason to anticipate any other terms being used.

In *Clarke* v. *ACT Construction*, Lord Justice Ward rejected the request to find that terms were incorporated by way of course of dealing, because various terms had been used between the parties over the years, and found on the facts that it was not true in that case that there was one clearly established procedure in use between the parties, at paragraph 34:

> 'It seems to me, therefore, that various means of charging were deployed and no sufficiently coherent and settled practice seems to have emerged to justify the allegation of a course of dealing.'

It is suggested that in building subcontracts there will be few situations where the contractor will be able to rely on terms being incorporated solely on the basis of their normal course of dealing, since in practice contractors rarely give a subcontractor sufficient repeat business to establish a course of dealing, and in any event each project is likely to produce necessary differences in the contractual provisions.

On the other hand, where a contractor refers to its standard terms in the concluding subcontract document, such as a subcontract order, and the subcontractor makes no formal objections, then such terms are likely to become the terms of the subcontract. Subcontractors must therefore be very much aware of the possibility of a contractor introducing its terms at a very late stage in the pre-contract negotiations. Such an order will normally not form a subcontract but will be a counter-offer, which the subcontractor need not accept. However, if the subcontractor does accept, whether by performance or otherwise, then it is likely to be bound by such terms unless it can demonstrate that such terms are so unusual and/or onerous as to have required the contractor to have brought these specifically to the subcontractor's notice. In such circumstances it is likely that only such unusual and/or onerous terms will be unenforceable, the remaining terms being effective.

Failure to make express provision within an agreement

Most standard forms of subcontract make wide provisions for the contractor to instruct change to the subcontractor's obligations under the subcontract and provide for suitable compensation to the subcontractor as a consequence. However, where there are no specific provisions within the subcontract to facilitate certain situations or requirements of the contractor, then there will be no express mechanism for compensating the subcontractor and/or relieving it from the consequences of those situations or requirements.

For example, if there were no provisions for extending the period for the subcontract works and the subcontract has an express requirement to complete by a certain date or within a certain time, then it is likely that terms as necessary will be implied into the subcontract to allow for the subcontractor to be compensated for the effect of the delaying event or events caused by the contractor. A further example might be the early completion and handover of parts of the subcontract works. Even where there is no express provision for such a situation, it is likely that a term would be implied into the subcontract that, for that part of the work, the provisions relating to completion of the subcontract as a whole would apply. These might include the release of retention and the taking over of responsibility for protection, insurance for the work, etc.

In such circumstances regard may be had to normal trade custom and practice as to the nature of the term that may be implied into the subcontract. Such implied term or terms may be found in relevant standard forms such as SBCSub/C. Reference to standard forms of subcontract may be specifically relevant where either a short form of subcontract, such as ShortSub or Sub/Sub, is used or where reference is made to a standard form of main contract, for which there exists a specific subcontract form for use in those circumstances.

Failure to conclude an agreement

Where the arrangements under which the subcontract work is to be carried out are uncertain or when the terms remain so far from agreement that it is impossible to find a subcontract in the normal form of offer and acceptance, the subcontractor will be entitled to be paid a reasonable sum for any work it carries out, in respect of a request from the contractor to do so.

It used to be considered that such situations were non-contractual and subject to the rules of Chancery; however, the Court of Appeal judgment in *Clarke* v. *ACT Construction* suggests that the law is changing towards adopting a contractual solution to such a situation. The judgment said that all that was necessary was an instruction to do work and an acceptance of that instruction. Lord Justice Ward said, at paragraph 27:

> 'In my view, the proper conclusion was to find that there was, as Mr Munro [Counsel] for Clarke submits, "a contractual *quantum meruit*". In focusing on the essential ingredients for "a building contract of some complexity" the judge may have lost sight of the fact that even if there is no entire contract, and especially even if there is no "formal" contract, there may still be an agreement to carry out work, the entire scope of which was not yet agreed, even if a price has not been agreed. Provided there is an instruction to do work and an acceptance of that instruction, then there is a contract and the law will imply into it an obligation to pay a reasonable sum for that work. That is what happened here.'

On the basis of this judgment, it would appear that were a contractor to request, in writing, that a subcontractor carry out work, provided that there is some certainty as to the nature and scope of what that work is and how the subcontractor is to be recompensed for it, even if this is only an implied obligation to pay a reasonable price, then upon the subcontractor responding to such a request, there comes into being a subcontract. If that subcontract is in writing, then the terms of the Scheme will be implied into it through the HGCRA.

In the case of *Hallamshire Ltd* v. *South Holland Council* (2004) 93 Con LR 103, the contractor had been required to carry out a second phase of a project for which it had contracted to do phase I. The employer wished work on site to commence although it found the contractor's offer too high. The parties therefore agreed that alterations would be made to the detailed design and specification, so as to make savings and thereby reduce the contract sum. In order to achieve an agreed contract sum for the phase II work, it would be necessary for the design team to amend the design and specification and for the quantity surveyor to prepare amended bills of quantities and for the contractor and quantity surveyor to agree new or revised rates as necessary.

Upon completion of phase II work, a dispute arose as to the price to be paid to the contractor. The employer sought to rely on the negotiated rates and the final certificate issued by the architect, while the contractor sought payment on a restitutory basis, claiming that there had been no concluded agreement and that the instruction to carry out the phase II works had only been an agreement to agree, which was not recognised as contractually enforceable in English law.

Judge Thornton QC upheld the arbitrator's decision, that the architect's instruction in question reflected the agreement to vary the existing contract and was not in fact a new contract at all. The judge said, at paragraph 31:

> 'The arbitrator made findings of fact to the effect that the parties negotiated a fresh contract whose intention was to vary the existing contract.'

While this effectively sidesteps the issue as to when an agreement to agree a price is not a binding contract, and when an agreed method of deciding the price is binding, the judge went on to consider the second question of law and stated, at paragraph 30:

> 'Did AI [Architect's Instruction] 51 require Jackson Coles [a firm of quantity surveyors] to agree the bills of quantities as a composite whole or could the necessary agreement of prices in the bills of quantities be achieved piecemeal?'

The judge considered the factual matrix behind the agreement and concluded by agreeing with the arbitrator's findings that there was no need for a formal offer of a total price and acceptance of such:

> '32. In answering the section, it is first necessary to set the relevant words of AI 51 into context. In considering these words, the entirety of the wording of AI 51 must be considered against the factual matrix common to both parties on the date that the variation agreement concerning AI 51 was reached. On that date, the defendant, as was known to the claimant, had defined the scope of phase 2 and had received a tender pricing all work items from the claimant which was unacceptable to it. However, work was to start prior to finalisation of the process of work scope and rate reduction which was intended to produce a price for phase 2 work acceptable to the defendant. This price reduction process was being carried out using the traditional process provided for defining and pricing vari-

ations within the JCT contract that the arbitrator had already defined. This included the process of omission, specification change and re-rating. Once that process had been concluded, as AI 51 stated, the bills of quantities that had been prepared in April 1997 would be revised to produce reduced costs and the drawings and specifications, also previously prepared, would be revised and reissued to reflect the amended items in the bill of quantities.

33. It followed that AI 51 defined with some precision the scope of phase 2 work. On the date it was issued on 16 July 1997, the scope of work was [not] that shown in the bills of quantities, drawings and specifications previously prepared but their revised state that existed at the date of AI 51. The costs were those previously tendered but as partially revised by the same date. Further revisions of work items, rates, prices, drawings and specification items were envisaged and these would be achieved by varying the scope of work and its cost as these stood on 16 July 1997. The final agreed package, arrived at by a complex piecemeal negotiation of individual work items, drawings, specification items and bill rates, would constitute the fair and reasonable costs that Jackson Coles would agree.

34. In the light of the wording of AI 51, taken against this factual background, it is clear that the arbitrator's understanding of the meaning of the relevant words of AI 51 is correct. The process of agreement required of Jackson Coles was one involving the agreement of individual work items and their cost, it was not one involving the negotiation and finalisation of an entire contract or a composite set of bills of quantities. It was not necessary, therefore, for the entirety of the bills of quantities, specification and drawings first to be negotiated and offered by one party as a composite package for acceptance by the other party.

35. It follows that, applying the arbitrator's findings of fact as to the agreed content of the bills of quantities and the mark up, there was agreement as to the cost of the work to be undertaken. No offer was needed and the absence of an acceptance, using the concepts of "offer and acceptance" as these are used in law relating to contract formation, was immaterial.'

The situation in both *Clarke* v. *ACT* and *Hallamshire* v. *South Holland* must be distinguished from the Court of Appeal case of *Courtney and Fairbairn Ltd* v. *Tolaini Brothers (Hotels) Ltd and The Thatched Barn Ltd* (1974) 2 BLR 97, by the degree of uncertainty and the state of the agreement reached between the parties. Also of significance is whether or not work was commenced. The courts appear to require a much greater level of certainty that an agreement was concluded, when the consequences of non-performance are the matter in dispute, rather than if the dispute is about work actually carried out, when the courts are more likely to find that a subcontract exists.

In the *Tolaini* case, Lord Denning MR, at page 100, found that there had been an exchange of letters between the parties, the first letter to Mr Tolaini, which said:

'In reply to your letter of the 10 April, I agree to the terms specified therein, and look forward to meeting the interested party regarding finance.'

In this situation Mr Courtney had offered to assist Mr Tolaini to obtain the money necessary to carry out the development in return for being given the construction work at rates to be agreed. In the event Mr Tolaini employed others to construct the Works. Lord Denning MR recorded these facts, at page 101:

> 'Those are the two letters on which the issue depends. But I will tell the subsequent events shortly. Mr Courtney did his best. He found a person interested who provided finance of £200,000 or more for the projects. Mr Tolaini on his side appointed his quantity surveyor with a view to negotiating with Mr Courtney the price for the construction work. But there were differences of opinion about price. And nothing was agreed. In the end Mr Tolaini did not employ Mr Courtney or his company to do the construction work. Mr Tolaini instructed other contractors and they completed the motel and other works. But then Mr Tolaini took advantage of the finance [which] Mr Courtney had made possible, but he did not employ Mr Courtney's company to do the work. Naturally enough, Mr Courtney was very upset.'

The judge at first instance found a building contract in the words recorded by Lord Denning:

> 'He said in his judgment that the letters:
>
> "gave rise to a binding and enforceable contract whereby the defendants undertook to employ the plaintiffs . . . to carry out the work referred to in [Mr Courtney's letter of 10 April 1969] at a price to be calculated by the addition of 5 per cent to the fair and reasonable cost of the work and the general overheads relating thereto".'

Lord Denning disagreed on the basis that there was no agreement as to price or means of calculating it:

> 'I am afraid that I have come to a different view from the judge. The reason is because I can find no agreement on price or any method by which the price was to be calculated. The agreement was only an agreement to "negotiate" fair and reasonable contract sums. The words of the letter are "your Quantity Surveyor to negotiate fair and reasonable contract sums in respect of each of the projects as they arise". Then there are words which show that estimates had not been agreed, but were yet to be agreed. The words are: "These [the contract sums] would incidentally be based upon agreed estimates of the net cost of work and general overheads with a margin for profit of 5 per cent." Those words show that there were no estimates agreed and no contract sums agreed. All was left to be agreed in the future.'

Thus there was no binding contract for Mr Courtney's company to carry out the building work. The state of agreement had not reached the stage of Mr Courtney being required to commence work in connection with the construction. This is the essential difference between *Tolaini* and *Clarke* v. *ACT*, where the construction work was carried out. Had Courtney and Tolaini reached an agreement as to the price or even a method of pricing, then a binding agreement might have existed, which would have been breached had Mr Tolaini then engaged another contractor to construct the work.

This is set out in Lord Denning's concluding paragraphs where he finds that there was no binding contract and that an agreement to negotiate does not give rise to an enforceable contract:

> 'In the ordinary course of things the architects and the quantity surveyors get out the specification and the bills of quantities. They are submitted to the contractors. They work out the figures and tender for the work at a named price; and there is a specified means of altering it up or down for extras or omissions and so forth, usually by means of an

architect's certificate. In the absence of some such machinery, the only contract which you might find is a contract to do the work for a reasonable sum or for a sum to be fixed by a third party. But here there is no such contract at all. There is no machinery for ascertaining the price except by negotiation. In other words, the price is still to be agreed. Seeing that there is no agreement on so fundamental a matter as the price, there is no contract.

But then this point was raised. Even if there was not a contract actually to build, was not there a contract to negotiate? In this case Mr Tolaini did instruct his quantity surveyor to negotiate, but the negotiations broke down. It may be suggested that the quantity surveyor was to blame for the failure of the negotiations. But does that give rise to a cause of action? There is very little guidance in the book about a contract to negotiate. It was touched on by Lord Wright in *Hillas & Co Ltd* v. *Arcos Ltd* [(1932) 38 Com Cas 23] where he said: "There is then no bargain except to negotiate, and negotiations may be fruitless and end without any contract ensuing".'

Then Lord Denning went on:

'. . . yet even then, in strict theory, there is a contract (if there is good consideration) to negotiate, though in the event of repudiation by one party the damages may be nominal, unless a jury think that the opportunity to negotiate was of some appreciable value to the injured party.'

Many commentators have regarded the Court of Appeal decision in *RJT Consulting Engineers Ltd* v. *DM Engineering (Northern Ireland) Ltd* [2002] BLR 217, as placing severe limitations on those subcontracts that can be referred to adjudication under the HGCRA, on the basis that the recorded terms are incomplete or deficient. It is submitted that this need not be the effect of that decision, which is that it is necessary to have all the agreed matters recorded in writing, it being for the parties to decide what needs to be agreed between them. Lord Justice Ward clearly defined the law:

'13. Section 107(2) gives three categories where the agreement is to be treated in writing. The first is where the agreement, whether or not it is signed by the parties, is made in writing. That must mean where the agreement is contained in a written document which stands as a record of the agreement and all that was contained in the agreement. The second category, an exchange of communications in writing, likewise is capable of containing all that needs to be known about the agreement. One is therefore led to believe by what used to be known as the *ejusdem generis* rule that the third category will be to the same effect, namely the evidence in writing of the whole agreement.

14. Sub-section (3) is consistent with that view. Where the parties agree by reference to terms which are in writing, the legislature is envisaging that all of the material terms are in writing and that the oral agreement refers to that written record.

15. Sub-section (4) allows an agreement to be evidenced in writing if it (the agreement) is recorded by one of the parties or by a third party with the authority of the parties to the agreement. What is there contemplated is, thus, a record (which by sub-section (6) can be in writing or a record by any means) of everything which has been said. Again it is a record of the whole agreement.'

In addition, where a dispute is to be referred to adjudication it is necessary for the matters relied on to be in writing. The judgment of Lord Justice Auld makes this very clear:

'21. I also agree that the appeal should be allowed. I do so, not because the whole agreement was not in writing in any of the forms for which section 107 of the 1996 Act makes provision, but because the material terms of the agreement were insufficiently recorded in writing in any of those forms.

22. Although clarity of agreement is a necessary adjunct of a statutory scheme for speedy interim adjudication, comprehensiveness for its own sake may not be. What is important is that the terms of the agreement material to the issue or issues giving rise to the reference should be clearly recorded in writing, not that every term, however trivial or unrelated to those issues, should be expressly recorded or incorporated by reference. For example, it would be absurd if a prolongation issue arising out of a written contract were to be denied a reference to adjudication for want of sufficient written specification or scheduling of matters wholly unrelated to the stage or nature of the work giving rise to the reference.'

In this case, the referring party in the adjudication had claimed that the other party had failed in its obligation and been in breach of its subcontract, but could not show that any of those obligations or duties had been expressed in writing, in any of the forms identified by the HGCRA or at all.

It is submitted that there is no conflict between the decisions in *Clarke* v. *ACT* and *RJT* v. *DM Engineering*. It is submitted that, if the dispute in *RJT* v. *DM Engineering* had concerned the right of RJT to be paid or had been limited to a general obligation to use reasonable skill and care, then such a contract might have been considered to have existed for the purposes of the HGCRA. It is the reliance on very detailed and explicit obligations, which were not in writing, that led the Court of Appeal to uphold the appeal against enforcement of the adjudicator's decision.

This interpretation of *RJT* is consistent with the judgment of Mr Justice Jackson in the case of *Trustees of the Stratfield Saye Estate* v. *AHL Construction Limited* [2004] EWHC 3286 (TCC), where, after quoting extensively from *RJT*, he said:

'46. . . . In my view, it is not possible to regard the reasoning of Lord Justice Auld as some kind of gloss upon or amplification of the reasoning of the majority. The reasoning of Lord Justice Auld, attractive though it is, does not form part of the ratio of *RJT*.

47. The principle of law which I derive from the majority judgments in *RJT* is this: an agreement is only evidenced in writing for the purposes of section 107, subsections (2), (3) and (4), if all the express terms of that agreement are recorded in writing. It is not sufficient to show that all terms material to the issues under adjudication have been recorded in writing.'

It is not necessary for the parties to consider and agree extensive terms and conditions but those they do agree must be recorded in writing. However, where they have agreed only a limited number of matters, it is possible that a dispute may arise which is so outside their agreement that it requires the incorporation of special terms, as was the case in *RJT*; then such a dispute is likely to fall outside the provisions of section 107 of the HGCRA.

When terms are left to be agreed

The process of reaching an agreement and concluding a subcontract will frequently be much more complex than a straightforward offer and acceptance. Negotiations as to the precise details of the agreement may continue over a long period and not infrequently after the commencement of the performance of the subcontract. The point at which the matters finally to be resolved cease to prevent there being an enforceable Sub-Contract Agreement, will under certain circumstances be finely balanced. For practical reasons the courts are more likely to find the existence of a binding subcontract once performance has commenced, and to find that any matters 'to be agreed' were not essential to the existence of an enforceable subcontract.

In the Scottish case of *Moyarget Developments Limited* v. *Mrs Rove Mathis and Others* [2005] CSOH 136, Lord Reed had to decide whether there had been a binding contract between the parties despite the fact that no formal agreement had been either prepared or executed. Even though the parties had never agreed the price to be paid for services to obtain planning permission for a housing development, Lord Reed after quoting from various authorities found that there was a binding contract. These authorities support the view that there is likely to be a presumption of a contract where work has been carried out. Lord Reed said, at page 29:

> 'Although the price to be paid for the services had not been agreed, and in the event was never agreed, that circumstance is not fatal to the existence of an enforceable contract, particularly where the services have in fact been provided. Thus, in *Avintour Ltd* v. *Ryder Airline Services Ltd* [1994] SC 270, at page 273, the court observed, in an opinion delivered by Lord President Hope:
>
> > "But there is an important difference between cases where nothing has been done by either party to implement the alleged contract and cases where a party to the alleged contract has already provided the goods or services for which he seeks payment. It is likely to be more difficult in the former case to enforce the contract if there is no agreement about the remuneration which is to be paid, because in the ordinary case the price is one of the essential matters upon which agreement is required. Where goods or services have been provided, however, the usual rule is that there is an obligation to pay for them unless they have been provided gratuitously. So it is easier in these cases, if there is no agreement about the price or remuneration, for an obligation to pay a reasonable sum to be implied."
>
> The court approved, at page 274, a passage in the judgment of Mr Justice Denning in *British Bank for Foreign Trade Ltd* v. *Novintex Ltd* [1994] 1 KB 623 (cited at pages 629–630):
>
> > "The principle to be deduced from the cases is that, if there is an essential term which has yet to be agreed and there is no express or implied provision for its solution, the result in point of law is that there is no binding contract. In seeing whether there is an implied provision for its solution, however, there is a difference between an arrangement which is wholly executory on both sides, and one which has been executed on one side or the other. In the ordinary way, if there is an arrangement to supply goods at a price 'to be agreed', or to perform services on terms 'to be agreed', then although, while the matter is still executory, there may be no binding contract, nevertheless, if it is executed on one side, that is, if the one does his part without having come to an agreement as to price or the terms, then the law will say that there is necessarily implied, from the conduct of the parties, a contract that, in default of agreement, a reasonable sum is to be paid."
>
> Reference might also be made to the judgment of Lord Denning MR in *F&G Sykes (Wessex) Ltd* v. *Fine Fare Ltd* [1967] 1 Lloyds Rep. 53 at pages 57–58.
>
> 75. In relation to the second matter – the intention to enter into a written contract – such an intention is not necessarily inconsistent with an intention to be bound by the existing

informal agreement with immediate effect. As was said by Viscount Haldane in *Gordon's Executors* [*Gordon's Executors* v. *Gordon* [1918] 1 SLT 407], at page 411:

"In a case such as the present it would of course have been open to those concerned to reach a definite and concluded agreement in conversation or by correspondence. Such an agreement is not the less a real one if the parties have, as part of its terms, stipulated that there is to be a further agreement embodying its substance and also other terms which they are subsequently to settle. In such a case the later agreement, when concluded and executed, will supersede the earlier one. But until then the earlier agreement stands and binds."

There may thus be a binding provisional agreement, as in the case of *Rossiter* v. *Miller* (1878) 3 App. Cas. 1124, discussed in *Gordon's Executors*. Whether the parties intended there to be *locus poenitentiae* pending the execution of a formal contract, or intended to be bound in the meantime by their informal agreement is, in the absence of any express provision, a matter of inference from the terms of the agreement and the surrounding circumstances.'

Deputy Judge Colin Reese QC in the case of *Alstom Signalling Limited (trading as Alstom Transport Information Solutions)* v. *Jarvis Facilities Limited (No 1)* (2004) 95 Con LR 55, had to decide whether or not a binding agreement existed or whether Jarvis was to be paid on a *quantum meruit* basis. The subcontract was on a target cost basis, where the contractor was to be paid their actual costs but was to share the savings or loss against an agreed target cost, referred to as pain/gain. While the principle of shared pain/gain was agreed between the parties to the subcontract, they had not agreed what was to be the method for calculating the amount of the pain/gain that would be borne by each party.

Alstom wished the allocation of the pain/gain to be directly proportional to each party's target cost, while Jarvis sought agreement that the pain/gain be on the basis of each party's achievement. During the post-tender period and during the course of the works there were a large number of meetings between the parties with the object of reaching agreement on the terms of the subcontract. These meetings are described in detail in the judgment and led the judge to the conclusion that there was clearly an agreement for the pain/gain to be shared between the parties; what had not been agreed was the mechanism by which this agreement was to be implemented.

The judge, in reaching his decision that there was sufficient agreement between the parties for there to be a binding agreement and that the differences of opinion between the parties were a dispute which might be decided by a tribunal, considered a number of authorities, quoting the relevant parts:

'53. Mr Bowdery [Council for Jarvis] developed his submissions in this way – this was, he said, a case where there was nothing more than an "agreement to agree" on something which was, contractually, an inessential subcontract term. No agreement was reached and accordingly, that was the end of the matter. He referred to the oft cited observations of Mr Justice Bingham (as he then was) and Lord Justice Lloyd in *Pagan SpA* v. *Feed Products Limited* [1987] 2 Lloyds Rep 601 at 610/611 and 619. In the judgment of Lord Justice Lloyd the principles to be derived from earlier authorities were conveniently re-stated. Those principles included –

"(2) Even if the parties have reached agreement on all the terms of the proposed contract, nevertheless they may intend that the contract shall not become binding until some further condition has been fulfilled. That is the ordinary 'subject to contract' case.

(3) Alternatively, they may intend that the contract shall not become binding until some further term or terms have been agreed; see *Love and Stewart* v. *Instone* [(1917) 33 TLR 475], where the parties failed to agree the intended strike clause, and *Hussey* v. *Horne-Payne* [(1879) 4 App Cas 311], where Lord Selborne said, at page 323:

'. . . The observation has often been made, that a contract established by letters may sometimes bind parties who, when they wrote those letters, did not imagine they were finally settling the terms of the agreement by which they were to be bound; and it appears to me that no such contract ought to be held established, even by letters which would otherwise be sufficient for the purpose, if it is clear, upon the facts, that there were other conditions of the intended contract, beyond and besides those expressed in the letters, which were still in a state of negotiation only, *and without the settlement of which the parties had no idea of concluding any agreement*. (Lord Justice Lloyd's emphasis.'

(4) Conversely, the parties may intend to be bound forthwith even though there are further terms still to be agreed or some further formality to be fulfilled (see *Love and Stewart* v. *Instone* per Lord Loreburn at page 476).

(5) If the parties fail to reach agreement on such further terms, the existing contract is not invalidated unless the failure to reach agreement on such further terms renders the contract as a whole unworkable or void for uncertainty.

(6) It is sometimes said that the parties must agree on essential terms and that it is only matters of detail which can be left over. This may be misleading, since the word 'essential' in that context is ambiguous. If by 'essential' one means a term without which the contract cannot be enforced then the statement is true; the law cannot enforce an incomplete contract. If by 'essential' one means a term which the parties have agreed to be essential for the formation of a binding contract, then the statement is tautologous. If by 'essential' one means only a term which the court regards as important as opposed to a term which the court regards as less important or a matter of detail, the statement is untrue. It is for the parties to decide whether they wish to be bound and, if so, by what terms, whether important or unimportant. It is the parties who are, in the memorable phrase coined by the judge: 'the masters of their contractual fate'. Of course the more important the term is, the less likely it is that the parties will have left it for future decision. But there is no legal obstacle which stands in the way of the parties agreeing to be bound now while deferring important matters to be agreed later. It happens every day when the parties enter into so-called 'heads of agreement' . . ." '

The above extract from the judgment in *Pagan* v. *Feed Products* sets out a range of situations which may exist and suggests the dividing point between a binding agreement with matters still unresolved and a situation where no subcontract has been concluded. Subject to the parties not having expressly stated that no subcontract exists or defining what has to be decided before a subcontract comes into being, then provided the agreed terms lead to a workable subcontract, whether the terms still to be decided are considered essential will be limited to those considered by the parties to be so.

The judge then considered an extract from *Walford* v. *Miles* [1992] 2 AC 12, at page 50:

'Mr Bowdery . . . submitted that when what the parties had left to be agreed was an appropriate pain/gain mechanism, it was unnecessary, unreasonable and beyond the court's powers to intervene by determining and imposing a pain/gain mechanism. In this context, he relied upon the observation of Lord Ackner in *Walford* v. *Miles* [1992] 2 AC 128. That was a case where plaintiffs had agreed, "subject to contract" to purchase a company and the premises from which it traded. The defendants decided to break off negotiations. The plaintiffs alleged that, so long as the defendants continued to desire to sell the business/property, they were legally bound to continue to negotiate with them in good faith. The House of Lords disagreed. The only speech was that of Lord Ackner who dealt with the general principles in this way –

"The reason why an agreement to negotiate, like an agreement to agree, is unenforceable, is simply because it lacks the necessary certainty. The same does not apply to an agreement to use best endeavours. This uncertainty is demonstrated in the instant case by the provision which it is said has to be implied in the agreement for the determination of the negotiations. How can a court be expected to decide whether, subjectively, a proper reason existed for the termination of negotiations? The answer suggested depends upon whether the negotiations have been determined 'in good faith'. However, the concept of a duty to carry on negotiations in good faith is inherently repugnant to the adversarial position of the parties when involved in negotiations. Each party to the negotiations is entitled to pursue his (or her) own interest, so long as he avoids making misrepresentations. To advance that interest he must be entitled, if he thinks it appropriate, to threaten to withdraw from further negotiations or to withdraw in fact, in the hope that the opposite party may seek to reopen the negotiations by offering him improved terms. Mr Naughton, of course, accepts that the agreement upon which he relies does not contain a duty to complete the negotiations. But that still leaves the vital question – how is a vendor ever to know that he is entitled to withdraw from further negotiations? How is the court to police such an 'agreement?' A duty to negotiate in good faith is as unworkable in practice as it is inherently inconsistent with the position of a negotiating party. It is here that the uncertainty lies. In my judgment, while negotiations are in existence either party is entitled to withdraw from those negotiations, at any time and for any reason. There can be thus no obligation to continue to negotiate until there is a 'proper reason' to withdraw. Accordingly a bare agreement to negotiate has no legal content." '

The significant difference between these two cases is that in one the parties were prepared to proceed with the execution of the construction work, while in the other they were not prepared to proceed with the sale of the business and the property. This was the point made by Alstom, who referred to the case of *Sykes* v. *Fine Fare* and was recorded by the judge at paragraph 54:

'Mr ter Haar [Counsel for Alstom] submitted that because the subcontract had been performed, this case was properly analogous with *F&G Sykes (Wessex) Limited* v. *Fine Fare Limited* [1967] 1 Lloyds Rep 53. In that case the court had declined to hold that the parties' failure to reach agreement on an important matter – the numbers of fowl to be supplied after the end of the first year of a long-term supply contract – had the result that there was no binding contract after the end of the first year. The contract contained an arbitration clause which provided in very wide terms for references to be made in respect of any "differences" which might arise. As Lord Denning MR succinctly summed the matter up –

"The provision that figures were to 'be agreed' does not nullify the contract. It can be made certain by reasonable figures being ascertained by an arbitrator."

And as Lord Justice Danckwerts put it (after he had read the arbitration clause) –

"The word 'differences' seems to me to be particularly apt for a case where the parties have not agreed: Viscount Dunedin said, in the case which was quoted to us of *May and Butcher, Ltd* v. *The King* [1934] 2 KB 17n: '. . . a failure to agree . . . is a very different thing from a dispute'. But it seems to me that the word 'difference' is particularly apt to describe that situation . . . the [terms] of this arbitration clause . . . seem to me to be sufficiently wide to cover a difference of the kind which has arisen in this case. The arbitrator has jurisdiction to decide the matter referred to him as to what is a reasonable amount for the defendants to have given notice of and which the suppliers should be in a position to supply . . ." '

Even here there is no agreement as to price, and where the contract has not been performed it is likely that, provided there is an agreement to sell and an agreed mechanism for deciding the price to be paid, then there will be an effective contract to sell. This was made clear in the House of Lords' decision quoted:

'55. Mr ter Haar [Counsel for Alstom] reminded me that in *Sudbrook Trading Estate Limited* v. *Eggleton* [1983] AC 444, the majority of the House of Lords had rejected the submission that option clauses in leases, which gave the tenant the right to purchase the reversion if notice was given within a specified period, constituted nothing more than unenforceable "agreements to agree". The purchase price in each of the option clauses was identically worded, save that minimum prices varied –

"... at such price not being less than [£12,000] as may be agreed upon by two valuers one to be nominated by the lessor and the other by the lessee and in default of such agreement by an umpire appointed by the . . . valuers."

The tenant gave the required notices and nominated its valuer. The landlords refused to nominate a valuer, with the result that the price fixing machinery could not operate. The majority of the House of Lords overruled earlier Court of Appeal authorities, which precluded the courts' stepping in where the parties had agreed on a particular price fixing machinery. The majority held that, in essence, the option contracts provided for the tenant to purchase the reversion at an objectively fair and reasonable price; that the machinery for fixing that price was a subsidiary and non-essential term of the contract; and, that on the breakdown of the agreed machinery for establishing the price, the court could undertake the necessary inquiries. Lord Fraser of Tullybelton explained this matter in this way –

"I recognise the logic of the reasoning which had led to the courts' refusing to substitute their own machinery for the machinery which has been agreed upon (by) the parties. But the result to which it leads is so remote from that which parties normally intend and expect, and is so inconvenient in practice, that there may in my opinion be some defect in the reasoning. I think the defect lies in construing the provisions for the mode of ascertaining the value as an essential part of the agreement. That may have been perfectly true early in the 19th century, when the valuer's profession and the rules of valuation were less well established than they are now. But at the present day, these provisions are only subsidiary to the main purposes of the agreement which is for sale and purchase of the property at a fair or reasonable value. In the ordinary case parties do not make any substantial distinction between an agreement to sell at a fair value, without specifying the mode of ascertaining the value, and an agreement to sell at a value to be ascertained by valuers appointed in the way provided in these leases. The true distinction is between those cases where the mode of ascertaining the price is an essential term of the contract, and those cases where the mode of ascertainment, though indicated in the contract, is subsidiary and non-essential: see *Fry on Specific Performance*, 6th edn (1921), pp. 167, 169, paragraphs 360, 364. The present case falls, in my opinion, into the latter category. Accordingly when the option was exercised there was constituted a complete contract for sale, and the clause should be construed as meaning that the price was to be a fair price . . ." '

Deputy Judge Colin Reese QC then reviewed several cases that had been referred to him, in detail, giving extracts from those judgments at paragraphs 56–59:

'56. *Sudbrook* was not a commercial contract but the basic principle could be seen to have been applied in a number of commercial cases to which I was referred. They were the decision of the Privy Council in *Queensland Electricity Generating Board* v. *New Hope Collieries Pty Ltd* [1988] 1 Lloyds Rep 205, the decision of the Court of Appeal in *Didymi Corp* v. *Atlantic Lines & Navigation Co Inc* [1989] 2 Lloyds Rep 108 and the decision of the Court of Appeal in *Mamidoil-Jetoil Greek Petroleum Co SA* v. *Okta Crude Oil Refinery* [2001] EWCA Civ 406 [2001] 2 Lloyds Rep 76.

57. The *Queensland Electricity* case concerned a 15-year coal supply contract between the Respondent Colliery and the Appellant Electricity Board. The price was agreed for the first 5 years with the agreement containing base prices which were adjustable by reference to "escalation" and "price variation" provisions. For sales/purchases after the first 5 years the general terms of the agreement were to continue but the base price and the price variation provisions were to be agreed. The agreement contained a comprehensive arbitration clause for the resolution of disputes or differences. It was argued by the Board that, after the first 5 years, the agreement constituted an unenforceable "agreement to agree". The argument was rejected. In giving the Advice of the Privy Council Sir Robin Cooke [judge] noted that the terms of the agreement indicated that it was intended by the parties to have legal effect for more than the first 5 years. He continued –

"... What other reasons could there be for making such elaborate provisions, emphasising its long-term nature? At the present day, *in cases where the parties have agreed on an arbitration or valuation clause in wide enough terms, the courts accord full weight to their manifest intention to create continuing legal relations. Arguments invoking alleged uncertainty, or alleged inadequacy in the machinery available to the courts for making contractual rights effective, exert minimal attraction ...*

... In accordance with the approach adopted in those cases, their Lordships have no doubt that here, by the agreement, *the parties undertook implied primary obligations to make reasonable endeavours to agree on terms of supply beyond the initial five-year period and, failing agreement and upon proper notice, to do everything reasonably necessary to procure the appointment of an arbitrator. Further, it is implicit in a commercial agreement of this kind that the terms of the new price structure are to be fair and reasonable as between the parties. That is the criterion or standard by which the arbitrator is to be guided.* If there are cases where the true meaning of the contract is that the arbitrator is to aim, not at objectively fair and reasonable terms, but merely at some result which appeals to him subjectively, they must be rare indeed and the present is certainly not one of them. The statements of basic intention in the recitals and in clause 9.1, together with the detailed pricing provisions for the first five years, supplement the ordinary implication of a fair and reasonable test. They lay down broad guidelines as to the object to be achieved; and how the system has worked during the first five years is likely to provide the arbitrator with much help in determining what is fair and reasonable for later periods."

58. In *Didymi*, the court was concerned with the enforceability of a clause in a charterparty which provided for the hire charges to be equitably increased or decreased in the event that the vessel's speed and/or fuel consumption varied from certain stipulated values. The owners claimed entitlement to increase hire charges. Although the charterparty contained an arbitration clause, High Court proceedings raising preliminary points of law were commenced. The charterers argued (inter alia) that the clause in question did not constitute a concluded agreement capable of enforcement. In the Commercial Court, Mr Justice Hobhouse rejected that argument saying –

"... [the provision] contains what are clearly intended to be words of obligation: 'Owners shall be indemnified'; 'such increase to be calculated'. This is not tentative: it is a liability which is capable of calculation. If the legal effectiveness of this provision is to be attacked it must be because, in truth, the liability is not capable of calculation or assessment and therefore is too vague or too uncertain to be the subject of a legally enforceable obligation ...

However ... I do not consider it right to categorise the provision as merely an agreement to agree. The words of a contract are used objectively to state the intention of the parties to the contract. They may do so skilfully or clumsily, but the function of the court is to extract from the words used their objective intention. The words of this paragraph do not disclose an intention merely to require an agreement. The words 'to be mutually agreed' are directory or mechanical and do not represent the substance of the provision ..."

and in dismissing the appeal, Lord Justice Bingham (as he then was) said –

"... I ask this question: Does the provision in issue in this case relate to an essential term of agreement or to the existence of any contract at all, or does it relate to a subsidiary and non-essential question of how a contractual liability to make payment according to a specified objective standard is to be quantified? I consider that this provision falls plainly in the second category.

The substantial provision in sub-clause (4) is that the –

... Owners shall be indemnified by way of increase of hire, such increase to be calculated ...

The procedure for calculation is in my judgment a matter of machinery, and I conclude that there was here a binding obligation to which effect can be given as a matter of contract."

He also accepted the owners' secondary submission, based on the *Sykes* case, that any defect in the provisions was cured by the arbitration clause which enabled any lack of agreement to be overcome (see again, page 115 of the report).

59. *Mamidoil* was the most recent of cases cited. A long-term (10 year) contract for the handling of crude oil was considered. The handling fee was fixed for the first 2 years of the contract. The parties then agreed fees for further periods up to the end of 1999. As a result of changed ownership/control the Refinery wished, if it could, to put an end to the contract when the period over which the agreed price applied expired. Although the contract contained an arbitration clause, a number of legal issues were raised in High Court proceedings. The Refinery argued that the 10 year term was a maximum term for which the contract might last but the continuation of the contract beyond the initial period for which the handling fee had been fixed was dependent on agreement(s) being reached as to the handling fee. The counter-argument was that this was a 10 year fixed term contract and, in the absence of agreement on the handling fee after the end of the initial period, a reasonable fee was to be fixed. The judgment of the Court of Appeal was given by Lord Justice Rix. Between paragraphs 50 and 68 of the judgment, he undertook a helpful review of many of the earlier authorities (but I note that neither *Queensland Electricity* nor the *Didymi* featured in this review) which I have read and noted. I too was specifically referred to the majority of those authorities in the course of argument but I see no need to cite them in the judgment. After completing his review Lord Justice Rix stated his conclusions in this way –

"69. In my judgment the following principles relevant to the present case can be deduced from these authorities, but this is intended to be in no way an exhaustive list: Each case must be decided on its own facts and on the construction of its own agreement. Subject to that:

Where no contract exists, the use of an expression such as 'to be agreed' in relation to an essential term is likely to prevent any contract coming into existence, on the ground of uncertainty. This may be summed up by the principle that 'you cannot agree to agree'.

Similarly, where no contract exists, the absence of agreement on essential terms of the agreement may prevent any contract coming into existence, again on the ground of uncertainty.

However, particularly in commercial dealings between parties who are familiar with the trade in question, and particularly where the parties have acted in the belief that they had a binding contract, the courts are willing to imply terms, where that is possible, to enable the contract to be carried out.

Where a contract has once come into existence, even the expression 'to be agreed' in relation to future executory obligations is not necessarily fatal to its continued existence.

Particularly in the case of contracts for future performance over a period, where the parties may desire or need to leave matters to be adjusted in the working out of their contract, the courts will assist the parties to do so, so as to preserve rather than destroy bargains, on the basis that what can be made certain is itself certain. Certum est quod certum reddi potest.

This is particularly the case where one party has either already had the advantage of some performance which reflects the parties' agreement on a long term relationship, or has had to make an investment premised on that agreement.

For these purposes an express stipulation for a reasonable or fair measure or price will be a sufficient criterion for the courts to act on. But even in the absence of express language the courts are prepared to imply an obligation in terms of what is reasonable.

Such implications are reflected but not exhausted by the statutory provision for the implication of a reasonable price now to be found in section 8(2) of the Sale of Goods Act, 1979 (and, in the case of services in section 15(1) of the Supply of Goods and Services Act, 1982).

The presence of an arbitration clause may assist the courts to hold a contract to be sufficiently certain or to be capable of being rendered so, presumably as indicating a commercial and contractual mechanism, which can be operated with the assistance of experts in the field, by which the parties, in the absence of agreement, may resolve their dispute."

When he turned to consider the facts of the case itself, he said –

"73. In my judgment, the 1993 contract should be viewed as a contract for a fixed period of 10 years and a term should be implied that in the absence of agreement reasonable fees should be determined for the period after 1994 . . .

(v) The contract does not expressly state that the fee after the end of 1994 is 'to be agreed'. It is simply silent as to what is to happen in that period. Therefore, this case is simply not presented with the difficulties which arise, in the face of 'to be agreed' language, where it is uncertain whether there is any contract at all. It cannot be said as was said in *May and Butcher* v. *The King*, that the statutory implication of a reasonable price, or an implication that the fee should be such reasonable fee as the arbitrator may decide, is excluded by express agreement that the parties were to agree the figure.

(vi) There is no evidence that the resolution of a reasonable fee would cause any difficulty at all. On the contrary, the evidence is the other way . . . In the absence of evidence to the contrary, I would infer that it was perfectly possible to derive from the agreements of price and price increases over the years objective criteria for working out a reasonable fee. *Thus, although it is true to say that the contract itself contained no mechanism or guidance (other than the arbitration clause) as to how a reasonable fee would be derived, I do not consider that the contract should fail on that ground.*

Contractually derived criteria or guidance may be of assistance in finding an implied term for a reasonable price: but the authorities indicate that the courts are well prepared to make the implication even in their absence.

(vii) I do not consider that the additional issues as to the length of any period of price revision or as to quantity discounts, raise any different problems. In practice agreement was made for one or two years. I see no difficulty in an arbitral tribunal or court finding that that was a reasonable period for which to fix the fee. The discount for throughput above a certain figure is simply another factor of price. Again, it gave the parties no difficulty over the years.

(viii) *The presence of an arbitration clause was not relied on as a particularly strong point by Mr Howard [Counsel for Mamidoil]. But I do not think that it is without its effect. It is a contractual mechanism for resolving 'any dispute . . . which cannot be amicably resolved'. It involves the parties' autonomous choice of their tribunal. It seems to me to be apposite to deal with a lacuna which, in common with other contracts which have to provide for future events, commercial practicalities lead parties to leave unresolved at the time of contract."'*

The judge concludes that the difference as to the method of determining the pain/gain to be allocated to each party is a dispute that may be determined by the court and states:

'60. Having considered the parties' respective submissions, in my judgment Mr ter Haar's submission that this is a case where there is a "difference" as to the pain/gain mechanism which the court is empowered to determine, is correct.'

In the case of *Alstom Signalling* v. *Jarvis Facilities*, discussed earlier, the question for the judge to decide was the nature of the pain/gain provision within the subcontract, which was therefore a matter of construction or interpretation of the Sub-Contract Agreement. This must be contrasted with the situation where there is to be implied into the subcontract a term, or at least a term fundamental to the dispute, when it is unlikely that a subcontract satisfying the requirements of section 107 of the HGCRA will be found. This point was made by Judge Raynor QC in the case of *Murray Building Services* v. *Spree Developments* [2004] TCC 4804:

'11. It is thus necessary (and, indeed, is conceded by Mr Jess [Counsel for Murray]) that in order to enforce this agreement by adjudication, the price, which is a vital term, must be recorded in writing within the meaning of the Act. That does not mean that the actual price must be stated. It would be sufficient if (as he contends) by a process of construction I were satisfied that the provision in the letter which says "forward your costs once finalised and agreed" means that the contract price will be that which is agreed with Brook [Consulting Engineer] subject to the $2^1/_2$% main contractor's discount. If that argument is right, then there will be a construction contract in writing within the meaning of the Act. What I have thus to determine is whether that argument is correct. I pause before considering that to note that it is agreed by both parties (in my view rightly) that if the matter is not one of construction, but falls to be determined by way of implication of a term, then that would not suffice to render the agreement an agreement in writing within the meaning of the Act.'

Thus if the case concerned the interpretation of the pricing agreement, that would be a matter of construction, which could be adjudicated, but if the dispute required the implication of a term for a specific pricing mechanism, then that would prevent the subcontract being considered a subcontract in writing within the definition of section 107 of the HGCRA.

In this case, Judge Raynor QC found that the express words of the subcontract order showed that there was no accepted price or effective pricing mechanism. This had been quoted by him:

> '6. The first was a fax dated 1 May 2002 (at page 321 of the bundle) on the headed notepaper of Spree Developments (the defendant) addressed to Mr Murray of Murray Building Services. It is a document that needs to be read in full and I will quote it in full:
>
> "Re: 3 The Parsonage, Manchester. Dear Andrew – Please find attached our order number 14362 relating to your quotation and specification issued by J. R. Book [Consulting Engineer]. Following my meeting with David Airey from J. R. Book and your subcontractor, Walsh Electrical Contractors, I have agreed with David Pender that they should commence on site on Monday 13 May (David Pender is of Walsh Electrical Contractors). This, I feel, will give you sufficient time to consider your programme of works for both the electrical and mechanical elements of the contract. Finally, I would be obliged if you would forward your costs once finalised and agreed with J. R. Book. Our order is given to enable you to order materials and attend site as referred to above."'

The judge did not find that there was an agreement between the parties to pay the price negotiated between the subcontractor and the employer's engineer, but that there was in fact no agreement on the fundamental issue of price. He said:

> '13. . . . what I do have regard to is the factual matrix, which is that Spree was the main contractor under the contract with Parsonage; that both parties knew that there was a scheme for both types of relevant work undertaken and revised by Book, and indeed there was express reference to that in the order of 1 May; that quotations had been tendered to Brook by the claimant, which quotations had not been seen by the defendant; that neither quotation had, at the time of the issuing of these documents, been accepted by Book, because that is, indeed, what is implicit in the fax of 1 May; that the defendant was to engage the claimant not as a nominated subcontractor but as a domestic subcontractor.
>
> 14. It is against that background that I now come back to the proper construction of these documents. The documents, to my mind, define sufficiently the scope of the work that was to be carried out; and indeed, it is not contended to the contrary before me. The order form specifically omits the price. The covering letter says, finally, (and I have quoted this before): "I would be obliged if you would forward your costs once finalised and agreed with J. R. Book."
>
> 15. It is said that, applying the factual matrix, those words, on their true construction mean that the claimant is to be entitled, by way of contract price, to the price agreed with Book. I am unable to accept that contention. It does not seem to me that that is a natural meaning of these words. It does not seem to me that they would be so understood by the hypothetical objective reasonable bystander. The fact is that the defendant chose to leave blank the contract sums. If the defendant had intended to say that the contract sums should be what was agreed with Book minus $2^1/_2$%, it would have been simplicity itself to say that. If the defendant had intended to say "we will agree to pay whatever you agree", it would have been simple to state that in the fax, if it was not in the order form. What Mr Fitton [Director of Spree Developments] in his letter stated was "I would be obliged if you would forward your costs once finalised and agreed with J. R. Book." It does not seem to me that that carries with it any implication that those will be the sums that will constitute the contract price.'

Finally the judge stressed that his decision related only to the enforcement of the adjudicator's decision and the question of whether there was a contract in writing for the purposes of the HGCRA:

> '17. I stress a number of things. First, although this means that the claimant will fail in an attempt to enforce the award, it does not mean that the claimant is without remedy. It will be open to the claimant to claim, on the basis of what it says in the contract, the sums due thereunder. What it cannot do is to enforce the adjudication award because the construction contract was not in writing within the meaning of the Act since the contract price (I find) was not stated in the documents relied on.'

What is not entirely clear from this judgment is whether there will never be a subcontract in writing if 'the contract price is not stated in the documents relied on', or only where this is material to the dispute as in that case. The judgment of Lord Justice Auld in *RJT Consulting* suggests that it might not be fatal for a subcontract, to be in writing for the purposes of the HGCRA, that the documents did not state the price if the dispute concerned some other matter such as the quality or scope of the work.

Unconscionable behaviour

When a party to an agreement signs a document setting out that agreement, that act of signing will normally be sufficient evidence that the nature of their agreement is as set out in that document. This will generally be the case irrespective of which party prepared the document setting out and recording the agreement.

In drawing up a written agreement it will be the intention of the draftsman to use clear and precise words in the hope that there is no ambiguity or other cause for uncertainty or confusion as to its meaning and the intentions of the parties. It will then be for the other party to consider carefully the draft and to be satisfied that it truly reflects the agreement between the parties, before signing the written agreement as presented to it.

It may happen that in writing up the agreement further issues may occur to the draftsman, which he may incorporate either to add clarification to the agreement or to strengthen its position. In the latter case the draftsman is no longer preparing the written agreement but is in fact making a renewed offer and as such should make this clear to the other party.

Where the drafting party behaves in this way, intending the other party to sign up to the agreement as drafted without being aware of the changes made to the agreement or the implication of those changes, then the agreement as signed may not be upheld.

The Court of Appeal considered this situation in the case of *Hurst Stores & Interiors Ltd* v. *M. L. Europe Property Ltd* (2004) 94 Con LR 66. In that case Hurst Stores were a works contractor on a project where Mace were construction managers for the employer M. L. Europe Property. At a late stage in the work but prior to completion, discussions took place as to the value of the works, including the valuation of the significant number of variation instructions to that date. The result of those discussions was to agree the value of the works as varied.

Mace prepared a statement of account generally in the format of their interim accounts and, in accordance with their normal practice, obtained a signature from Hurst Stores' site manager to signify agreement to that interim payment. Mace had used the phrase 'final account' and claimed that the agreement was to full and final settlement for all claims up to that date. At trial it was found that no such agreement existed, and Mace sought to have the decision at first instance overturned and for Hurst Stores to be denied any right to be paid for disruption costs in addition to the agreed and recorded valuation of the work.

Lord Justice Buxton referred to the case of *Commission for New Towns* v. *Cooper* and defined a three-step test to be met:

> '19. Against that background I turn to the two issues in the appeal and to the two declarations sought. First, rectification on grounds of unilateral mistake. It was agreed below and before us that the judge and ourselves should follow the admittedly obiter guidance given in this court in *Commission for New Towns* v. *Cooper* [1995] Ch 259. First, what was said by Lord Justice Stuart-Smith at page 280B of the report, where the Lord Justice said this:
>
> "... were it necessary to do so in this case, I would hold that where A intends B to be mistaken as to the construction of the agreement, so conducts himself that he diverts B's attention from discovering the mistake by making false and misleading statements, and B in fact makes the very mistake that A intends, then notwithstanding that A does not actually know, but merely suspects, that B is mistaken, and it cannot be shown that the mistake was induced by any misrepresentation, rectification may be granted. A's conduct is unconscionable and he cannot insist on performance in accordance to the strict letter of the contract; that is sufficient for rescission."
>
> Lord Justice Evans said, at page 292E:
>
> "... there is nothing unfair in holding them to the agreement which, to its knowledge, was the only one which MK's [Milton Keynes Development Corporation's] representatives intended to make. I would have no hesitation, if necessary, in holding that 'knowledge' in this context includes 'shut-eye' knowledge."
>
> The Lord Justice then went on to refer to the analysis of types of knowledge adopted by Lord Justice Peter Gibson (as he then was) in the well-known case of *Baden* v. *Société Générale* [1993] 1 WLR 509. In my judgment, it is not necessary to pursue the details of that analysis because it was sufficiently expressed as relevant to this chapter of the law in the terms adopted by Lord Justice Evans, already set out, with whose judgment Lord Justice Farquharson agreed.
>
> 20. What it therefore comes to is that there had to be established (i) Mr Mell [Project Manager of Hurst Stores] was mistaken as to the content of the document; (ii) Mace, through Mr Rumsey [negotiator for Mace], had actual, or what Lord Justice Evans called "shut-eye", knowledge of that mistake; and (iii) overall, the conduct of Mace was unconscionable.'

Lord Justice Buxton considered the facts of the case as found by the judge at first instance and found that his conclusion was one he was empowered to make:

> '24. ... The relevant question in that regard is agreed in the following terms and set out in paragraph 47 of the judgment:
>
> "... had it been proved that Mr Rumsey had actual knowledge that Mr Mell was mistaken as to the contents of the 27 April 2001 document or that, in this regard, Mr Rumsey wilfully shut his eyes to the obvious or that he wilfully and recklessly failed to make such enquiries as an honest and reasonable man would make?"

There was no evidence from Mr Rumsey. There was no clear evidence as to how the document reached Mr Mell from Mr Rumsey. The judge assumed that either it was handed to Mr Mell, or was passed to him through the normal documentary channels, in either case without any indication that it represented a departure from previous formulations. The judge set out those findings in paragraph 49 of his judgment:

"49. The possible factual situations which might have happened are these –

(1) Mr Rumsey, as he had said in the witness statement upon which Mr Mell commented in his evidence, personally handed the 27 April 2001 document to Mr Mell and invited him to sign it (without explaining to him that Mace had included in the document a previously unheralded invitation to Hurst to forego all as yet unrecognised contractual claims up to the date of signature). If he did so and saw that Mr Mell signed it without first taking time to read carefully through it, the probability is that he had actual knowledge of Mr Mell's mistake as to the effects of the document; alternatively,

(2) Mr Rumsey, as he had said, personally handed the 27 April 2001 document to Mr Mell (without explaining to him that Mace had included in the document a previously unheralded invitation to Hurst to forego all as yet unrecognised contractual claims up to the date of signature) but Mr Rumsey did not invite Mr Mell to sign it then and there. In the alternative, in line with the established site practice, Mr Rumsey forwarded the document to Mr Mell for him to collect from the Hurst site correspondence tray, (without including a covering note or in any other way drawing Mr Mell's attention to the fact that Mace had included in the document a previously unheralded invitation to Hurst to forego all as yet unrecognised contractual claims up to the date of signature). On either of these possible bases, the probability is that Mr Rumsey was wilfully shutting his eyes to the risk that Mr Mell would not notice the newly introduced, potentially prejudicial, words which he had no reason to suspect might be there. Alternatively, at the very least, on either of these possible bases, Mr Rumsey would have been a person wilfully and/or recklessly failing to take such steps as an honest and reasonable man would take if he knew (as on my findings Mr Rumsey did know) that the document he had prepared did not simply reflect the agreement(s) on CMI values which had been reached earlier that month after protracted negotiations."

The judge was dealing, as he had to, with the evidence available to him. He concluded that it did not matter which of the two explanations as to how the document had been transferred was correct. He still found, on the hypothesis that Mr Rumsey had not been present when Mr Mell signed the document (which was the hypothesis favourable to MLEP), that Mr Rumsey had had at least "shut-eye" knowledge of the mistake. There is no way it can be said that that conclusion was not open to him.'

Similar situations to that described above in *Hurst Stores* can arise when a contract agreement is being concluded. Where, for example, there have been extended negotiations as to the detail of specific provisions of a contract, the late insertion or omission by the drafting party to the disadvantage of the other may make such change unenforceable. In exceptional circumstances the courts may be persuaded to rectify the contract for unilateral mistake.

The law in this regard was reviewed by Mr Jules Sher QC in the case of *George Wimpey UK Limited (formerly Wimpey Homes Holdings Limited)* v. *V. I. Components Limited* [2004] EWHC 1374 (ch). In this case Wimpey were negotiating with V. I. Components, through an agent, to purchase a building plot in order to construct an estate of flats for sale. The intention was that the purchase price would be adjusted should, as was expected, the actual constructed building vary from that for which planning permission existed. In essence the value of each flat was considered to have two elements, one relating to the flat's location etc. within the development, the 'E' factor, and the second being proportional to the floor area. Thus to adjust the flat value, the E element was deducted from the total value and the remainder adjusted by proportion to the changed floor area, and finally the E factor added back to arrive at the new or adjusted value for that flat.

The formula for adjusting the second element was the subject of much negotiation and amendment. Towards the end of negotiations both parties became keen to conclude the agreement quickly. After final agreement was reached, the agent for V. I. Components, Mr Youens, sent Wimpey's representative, Mr Ketteridge, details of the final agreed formula for agreement. This formula failed to make provision for adding back the E factor. The judge found as a matter of fact that both V. I. Components' director, Mr Daykin, and Mr Youens were aware that this had been omitted and that the omission had never been agreed by Wimpey and might be overlooked in the hurry to finalise a formal agreement, which is what happened. The question for the decision of Mr Jules Sher QC was whether this constituted a unilateral mistake at law which allowed for rectification of the contract.

The judge reviewed the law in detail, by reference to the decided cases. First, he made the point that the courts will be slow to intervene to rewrite a contract:

> '65. This case concerns the boundary between legitimate negotiation and unfair dealing. In the cut and thrust of negotiations many things may happen that may lead to unhappiness with the end result on one side or the other. The court should be slow to intervene and rewrite a contract. It is not for the court to mend any side's bargain . . . What this case has now come down to is this: should the court intervene and rectify the formula where only one side, Wimpey, has made a mistake?'

The judge then referred to the judgment of Lord Justice Buckley in *Thomas Bates & Son* v. *Wyndam's (Lingerie) Ltd* [1981] 1 WLR 505 at pages 515 and 516, which set out a four-point test for rectification:

> '. . . first, that one party A erroneously believed that the document sought to be rectified contained a particular term or provision, or possibly did not contain a particular term or provision which, mistakenly, it did contain; secondly, that the other party B was aware of the omission or the inclusion and that it was due to a mistake on the part of A; thirdly, that B has omitted to draw the mistake to the notice of A. And I think there must be a fourth element involved, namely, that the mistake must be one calculated to benefit B.'

Mr Jules Sher QC decided that the only test in doubt in the Wimpey case was the second: was V. I. Components aware of the mistake on the part of Wimpey? In the *Bates* v. *Wyndam's* case Lord Justice Buckley had put the jurisdiction on the basis of "the equity of the position", while Lord Justice Eveleigh had put it on the basis of estoppel, at page 520:

> 'In a case like the present if one party alone knows that the instrument does not give effect to common intention and changes his mind without telling the other party, then he will be estopped from alleging that the common intention did not continue right up to the moment of the execution of the clause. There is no need to decide whether his conduct amounted to sharp practice. I think he might at that time have had no intention of taking

advantage of the mistake of the other party. I do not think that it is necessary to show that the mistake would benefit the party who is aware of it. It is enough that the inaccuracy of the instrument as drafted would be detrimental to the other party, and this may not always mean that it is beneficial to the one who knew of the mistake.'

Mr Jules Sher QC, at paragraph 68, refers to the case of *Agip SpA* v. *Navigazione Alta Italia SpA (The Nai Genova and Nai Superba)* [1984] 1 Lloyds Rep 353, where a claim for rectification for mistake failed because the judge had found as a matter of fact that the defendants were unaware of the plaintiffs' mistake:

'The judge's very positive finding of fact that the defendants were actually unaware of the plaintiffs' mistake is, of course, entirely consistent with this conclusion. The greater the degree of the defendants' carelessness in not detecting the relevant error – and Mr Romano [witness] frankly acknowledged in the course of his oral evidence that there had been a "great mistake" on his side – the more unrealistic it becomes for the plaintiffs to assert it was reasonably foreseeable by the defendants. In my judgment, on the evidence, the mistake which occurred was neither intended nor reasonably foreseeable.

. . . I might perhaps add that I strongly incline to the view that in the absence of estoppel, fraud, undue influence or a fiduciary relationship between the parties, the authorities do not in any circumstances permit the rectification of a contract on the grounds of unilateral mistake, unless the defendant had actual knowledge of the existence of the relevant mistaken belief at the time when the mistaken plaintiff signed the contract.'

Mr Jules Sher QC then, at paragraph 69, refers to the case of *Commission for the New Towns* v. *Cooper (Great Britain) Ltd*, where Lord Stuart-Smith is quoted as follows:

'I would hold that where A intends B to be mistaken as to the construction of the agreement, so conducts himself that he diverts B's attention from discovering the mistake by making false and misleading statements, and B in fact makes the very mistake that A intends, then notwithstanding that A does not actually know, but merely suspects, that B is mistaken, and it cannot be shown that the mistake was induced by any misrepresentation, rectification may be granted. A's conduct is unconscionable and he cannot insist on performance in accordance to the strict letter of the contract; that is sufficient for recission.'

Mr Joules Sher QC summarises the law in these words:

'70. The *Commission for New Towns* case was however decided (amongst other grounds) on the question whether the non-mistaken party had actual knowledge of the mistake made by the mistaken party. The judge below had found that there was no such actual knowledge. The Court of Appeal, however, held that wilfully shutting one's eyes to the obvious or wilfully and recklessly failing to make such inquiries as an honest and reasonable man would make constituted actual knowledge in law and that there was actual knowledge in this sense.'

The Court of Appeal subsequently overturned this judgment but not by way of a differing definition of the law but for procedural reasons. The main one was that Wimpey had not pleaded dishonesty by the directors of VI Components, but the judge had inferred it by reason of their conduct. The Court of Appeal were also discontented because, although it was for the main board directors of Wimpey and they alone to accept the terms, there was no evidence that they had ever intended to do otherwise, even if their agent had not anticipated such an agreement.

It is suggested that the effect of these decisions must be regarded more as a risk warning against those who would seek to trick another party into signing up to a document blatantly different in detail from an agreement actually reached. It must not be considered in any way as opening the door for a party having concluded a bad bargain, to have such a bargain changed or quashed. However, where a party with strong bargaining power at a late date issues a document for acceptance, which on its face reflects the concluded negotiations of the parties but in fact introduces new or revised terms to the accepting party's disadvantage, without drawing such changes to the accepting party's notice, then it is likely such action may be regarded as dishonest or fraudulent.

Chapter 3
Subcontractors Selected by a Third Party

Introduction

It is normally for the contractor to decide what, if any, elements of the work it will subcontract to others and which companies it chooses to have tender for the work. In many instances the contractor will be required to obtain prior agreement from the architect or contract administrator to subcontract any element of the works, but the choice both as to the subcontractors used and the work sublet will be the contractor's. It is many decades since an employer seeking an offer from a major building contractor could reasonably have expected that all the work would be carried out by employees of that contractor. In the current market, few large general contractors directly employ much if any of the workforce. Strangely, the provisions of the main contracts have not changed from requiring the contractor to obtain prior approval to sublet work, to a requirement for the proposed subcontractor to be approved.

Apart from vetting the source of labour to be employed on the project, either or both of the employer and design consultants may wish to select or limit the subcontractors employed on the works. This may either be generally for all trades or only for specified ones. There are several reasons for this, which include:

1. The employer may wish to use one company throughout its estate, works, etc. and thereby limit the problems for ongoing servicing, maintenance and spares. Such trades include the suppliers of lifts, escalators and mechanical shutters.
2. An employer may have an ongoing relationship with particular specialists, having developed component details that the employer wishes to maintain as part of its company profile. This is particularly true of the retail industry, which may have a number of preferred contractors for high profile elements such as counters, signs, etc.
3. An employer may also negotiate deals with product manufacturers and/or installers to ensure a readily available supply of components, where the market demand is very variable. By offering an exclusive contract in exchange for a minimum supply period and agreed pricing arrangements, there will be benefits for both employer and specialist contractor. Such agreements were common in government contracts where components might be required at short notice.
4. The increasing reliance on design development by specialist subcontractors may make it desirable for the specialist subcontractor to be appointed early and ahead of the appointment of the contractor, to ensure that sufficient time is available for the development of that part of the design.
5. Similarly, for items where the lead time for procurement and/or manufacture is long or where plant requires to be reserved well in advance of work on

site, it may be desirable to place a contract ahead of the placing of the main contract.

6. Just as the employer may have preferred specialists it requires to be employed, so the design consultants may have preferred and trusted specialists. This will be true especially where a high level of craftsmanship or artistic ability is required.

Whatever the reason for a third party seeking to decide who should be the specialist subcontractor, a variety of procedures are available. A prerequisite however is that such an intention is provided for within the main contract. Arrangements for some of these procedures have been set out in standard forms of main contract and the compilers of such forms have provided special, compatible forms of subcontract, while other arrangements may be agreed on an ad hoc basis. Common arrangements include:

1. specification;
2. novation;
3. naming;
4. nomination;
5. works contract under a management or construction management contract.

While these phrases commonly have the specific meaning given to them below, it must be acknowledged that such definitions are not universal and especially with bespoke contracts such words will have the meaning given to them by that contract and no other. Conversely where standard forms define precisely what is meant by a certain word or phrase then if something else is required by way of a procedure then the necessary changes to the standard form must be incorporated into the agreement. Otherwise it will be necessary to have a post-contract agreement or waiver.

Pre-tender discussions

During the early stages of the design development of a project the design consultants may need to obtain information and/or design details from specialist suppliers and installers. Generally, the specialist will regard approaches from consultants as marketing opportunities and will be very willing to give details of their product range and/or services.

The outcome of an initial meeting may be that the consultant is able to incorporate the information in the specialist's trade literature into the design without any further advice or assistance from the specialist. However, it is frequently the case that the consultant requires a deviation from the standard arrangement, either to the component or to the method of its installation. The specialist may promise to review the consultant's requirements and to provide further information.

Such undertakings are rarely done on any formal contractual basis. However, while the consultant will probably have an urgent need for the required information, the specialist may have many pressing contractual obligations to satisfy before being able to give due consideration to a service being provided free and without any certainty that it will eventually lead to a contract for paid work. A specialist offering any design development assistance must make clear to the party being assisted exactly what it is prepared to do and when, without any firm commitment from the other party. Unless this is done and that offer held to, the consultant may become frustrated with the specialist's performance and turn against it as the potential subcontractor for the work.

There will be no contract between the specialist and the consultant at this stage, and it is unlikely that there ever will be since the intention will be for the specialist to be a subcontractor to the contractor with possibly, in addition, a collateral contract or warranty agreement with the employer. However, the specialist will owe the consultant a duty of care in tort since it will know, or have reasonable grounds for believing, that the consultant is relying on the specialist's advice in the development of the consultant's design.

Thus where a specialist provides technical advice and/or drawings and/or specification by way of possible design solutions, it must make very clear the terms on which those are being provided. Such terms should exclude liability for the design, and should set out the limited basis against which such advice etc. has been given.

Specified subcontractor

The employer or design consultant can restrict the contractor's choice of subcontractor by specifying those with which it may subcontract the work. Although there is no reason why this should not be limited to one subcontractor, the SBC standard forms require the contractor to be able to select from a list of no less than three. Clause 3.8 at sub-clauses 1 and 2 requires:

'1. That where the contract bills require that measured work is described as to be carried out by a named person, then the contractor is to select one of those named in the contract documents. The selection from the list is to be at the contractor's sole discretion.
2. The list is to comprise not less than three persons and any of the parties may seek to have other persons added to the list, and such additions shall not be unreasonably denied, subject to such proposal being made before the execution of a binding subcontract.'

The requirement that there be no less than three subcontractors for the contractor to select from does not appear to have any reason behind it. If the employer or design consultant wishes to restrict the choice to one, for reasons only they know, why should the form of contract without amendment require the employer to select another two as well as the one required?

Similarly there is no reason why the contractor should be able to add others to the list. This provision is much more forceful than merely providing for the contractor to propose or suggest additions to the list. The fact is that by the words of the contract the contractor's proposed additions must be consented to unless there are reasonable grounds for withholding consent.

It is perhaps even more surprising that clause 3.8.3.2 allows the contractor, in the event that the selected list shows less than three persons, to subcontract the work to any person. Sub-clause 3 provides that:

'Should the list from which the contractor may make a selection be less than three persons, who are able and willing to do the work, then either

1. The parties will agree other names to make the list up to three, or
2. The contractor is free to subcontract in accordance with clause 3.7.'

Under these arrangements it is for the contractor to seek an offer from the specified subcontractor or, where more than one is specified, from any of the specified companies. It is not required that the contractor consider all of the proposed subcontractors; the contractor is free to eliminate any at the tender stage. The contractor can seek to impose any terms it wishes on the subcontractor, and it will be for the parties to reach agreement. Neither the employer nor the design consultant will normally have any involvement with the relations between the selected subcontractor and the contractor. It will be for the contractor to obtain performance from the subcontractor as necessary to enable the contractor to integrate the subcontractor's work into the Works as a whole. It will also be for the contractor to value and pay the subcontractor in accordance with the provisions of the agreement between them.

The contractor will be especially vulnerable during the period from when it receives the subcontractor's offer, which will form a basis for the contractor's offer to the employer, until the conclusion of the main contract, when the contractor will become in a position to conclude the subcontract. It is considered that there is a greater risk, with contracts which incorporate named subcontractors, of the selected subcontractor not having the necessary capacity or not being able to provide the required service at the tendered sum at the time the contractor seeks to conclude the subcontract, than will exist in the general situation, because of the restricted tender list. A tendering contractor would be advised to enter into a binding agreement with the prospective subcontractor, prior to the contractor making an offer for the main contract work, to the effect that its offer is binding in the event of the contractor being successful.

Any such pre-contract agreement requires that both the contractor and subcontractor are in agreement as to the terms and requirements before the contractor makes its offer to the employer, which is a circumstance that does not generally apply to subcontractor's offers at the tender stage of main contracts which are open to further negotiation either way.

In accepting the obligation to use a specified subcontractor, the contractor takes on risks that may not exist where either it carries out the work with its own labour or is free to sublet to any subcontractor of its choice. The specified subcontractor may be unknown to the contractor, who may have little idea as to the subcontractor's reliability and/or viability or the quality of its management and/or operatives. The contractor may therefore wish to seek some form of warranty from the employer that the specified subcontractor is capable and/or will perform as necessary, or the contractor may seek to be relieved of liability for non-performance in the event of a failure to perform as necessary by the specified subcontractor. Such provisions may be restricted to the contractor being relieved of a liability to damages in the event of late completion, or may include compensation to the contractor for its additional costs as well. Alternatively the contractor may seek a form of bond or warranty from the subcontractor, but this will generally increase the subcontractor's price and so make the contractor's bid less competitive.

Novation

Novation is the opposite of specification in that the subcontractor tenders to, and the subcontract is placed by, the employer, often on the advice of its consultant, and both the contractor and subcontractor are required to allow the employer to hand over its obligations and benefits under that contract to the contractor. There is not normally any opportunity, before the novation, for the contractor's terms and requirements to be notified to, let alone agreed by, the subcontractor. The main contract may or may not make suitable provisions for resolving any difficulties at the transfer stage. One of the reasons for a novation agreement may be to advance the procurement process because insufficient time has been allowed for such procurement after the appointment of the main contract. Any difficulties in completing the novation process will either leave the subcontractor as a directly employed contractor or involve some renegotiation of the main contract terms.

Once the novation has been successfully completed, the subcontractor becomes a domestic subcontractor in a similar manner to when the whole procurement process has been carried out by the contractor.

One special class of subcontractor is the design consultant, where the contractor is responsible for the design as well as the construction of the works. Such consultants may or may not include the architect but frequently include the design engineers, both for the structure and for the services. Considerable care is needed in selecting suitable terms and conditions since neither the normal terms for design consultants engaged by a building owner nor the standard terms of subcontract are suitable.

CIC/Nov Agr form of novation agreement, prepared by the Construction Industry Council, is a document to be used where the work of consultants is to be novated to a contractor. It is a very simple document limited to five terms.

The first mutually discharges the consultant and employer from further performance to each other of the obligations of the appointment:

> '1. As from the date of this Agreement the Employer releases and discharges the Consultant from the further performance of the Consultant's services and other obligations under the Appointment save for:
> (a) any obligations in the Appointment to provide warranties in favour of third parties at the request of the Employer; and
> (b) any obligations in the Appointment as to confidentiality to the extent that these do not conflict with the duties owed by the Consultant to the Contractor arising under this Agreement
>
> and, subject to all fees properly due and owing under the Appointment as at the date of this Agreement having been paid to the Consultant, the Consultant releases and discharges the Employer from further performance of the Employer's obligations under the Appointment.'

Under the second term, the consultant undertakes to carry out the remaining obligations for the contractor:

> '2. In respect of those services and any other obligations to be performed under the Appointment that have yet to be performed by the Consultant at the date of this Agreement, the Consultant agrees to perform those services and obligations (subject to the

variations if any set out in Schedule 1 to this Agreement) for the Contractor as if the Contractor were a party to the Appointment in place of the Employer and in accordance with and subject to the terms of the Appointment (including without restriction any limitation or exclusion of liability therein). The Consultant shall have no liability to the Employer arising out of the performance or non-performance of such services and obligations.'

Under the third term the contractor undertakes the balance of any obligation originally to be performed by the employer:

'3. In respect of those obligations to be performed under the Appointment that have yet to be performed by the Employer at the date of this Agreement (with the exception of any outstanding obligation to pay any fees properly due and owing under the Appointment as at the date of this Agreement), the Contractor agrees to perform those obligations (subject to the variations if any set out in Schedule 2 to this Agreement) as if the Contractor were a party to the Appointment in place of the Employer and in accordance with and subject to the terms of the Appointment (including without restriction any limitation or exclusion of liability therein). The Employer shall have no liability to the Consultant arising out of the performance or non-performance of such obligations, subject to all fees properly due and owing under the Appointment as at the date of this Agreement having been paid to the Consultant.'

The fourth term deals with any work the consultant may have carried out while contracted to the employer for which the contractor is liable to the employer under its contract:

'4. (a) Insofar as the Contractor is responsible under the Main Contract for services that have been performed by the Consultant for the Employer prior to the date of this Agreement, the Consultant warrants to the Contractor that all such services have been performed for the Employer in accordance with the terms of the Appointment.

(b) In any claim for loss suffered by the Contractor that is alleged to have arisen as a result of a breach by the Consultant of the warranty in this clause, the Consultant shall be entitled to rely on any limitation in the Appointment (including without restriction any limitation or exclusion of liability therein) and to raise the equivalent rights in defence of liability for such loss as if the claim were being brought by the Employer rather than the Contractor, save that the Consultant shall not be absolved from liability to the Contractor for such loss merely by virtue of the fact that the loss has not been suffered by the Employer.'

Finally, term five makes the agreement subject to the laws of England and Wales or Northern Ireland and denies rights under the Contracts (Rights of Third Parties) Act 1999.

CIC/Nov Agr is simply what it says – a novation agreement. It is of prime importance that the agreement to be novated is one that is suitable for novation. It is suggested that the standard forms of appointment for professional design services are not suitable. The guidance notes to CIC/Nov Agr makes this point within note 4:

'4. Another important point about this novation is that the interests and concerns of the employer and the contractor are not the same. Under the new contract, the consultant

> may perform the same services for the contractor as, under the old contract, he would have performed for the employer. However, he may not do so in the same way, as he is now under a duty to take the specific interests and concerns of the contractor into account. Likewise, if he had been engaged by the contractor from the outset (or at any other time prior to novation), he might have performed his services differently from the way he actually did for the employer.'

An alternative to CIC/Nov Agr is the CLL form. As with CIC/Nov Agr it is a three party agreement in which the first clauses deal with the transfer of obligations between the parties. Clause 1.4 seeks to give the contractor a right to recover any loss it suffers as a result of any negligent act, default or breach by the consultant prior to the novation:

> '1.4 Without prejudice to Clause 1.2, the Consultant warrants to the Contractor that it shall be liable for any loss or damage suffered or incurred by the Contractor arising out of any negligent act, default or breach by the Consultant in the performance of its obligations under the Appointment prior to the date of this Agreement. Subject to any limitations of liability in the Appointment, the Consultant shall be liable for such loss or damage notwithstanding that such loss or damage would not have been suffered or incurred by the Employer (or suffered or incurred to the same extent by the Employer).'

Clause 1.5 is an acknowledgement by the consultant that he has been paid all fees and expenses properly due to him up to the novation:

> '1.5 The Consultant acknowledges that all fees and expenses properly due to the Consultant under the Appointment up to the date of this Agreement have been paid by the Employer.'

This clause means that in the event of there being a dispute over fees and/or expenses between the consultant and the employer, a novation under these terms cannot take place.

In simple terms, a consultant working for the employer aims to satisfy to his best ability, or more probably has a duty to use reasonable skill and care to achieve, the objectives set by the employer by way of a commission brief. Depending on that brief, the priorities may be for quality, cost or time; when quality is the factor it may involve appearance, durability, flexibility or other design criteria. The contractor's requirements will generally be for the consultant to satisfy the contract obligations for the minimum cost. The contractor may in addition require the working of the design to be compatible with the planned method of construction. For example, components may be limited in size due to the selected hoisting and handling equipment, or the contractor may require components to be delivered in large units to minimise the number of crane operations. The consultant's design development must comply with such requirements.

Where the contractor is not required to accept the novation of existing consultants it will be able to employ or work with consultants of its choice to develop, at least in outline, its proposals at tender stage. This will enable a quick and effective start on the detailing of the work when the contractor is appointed. When the consultants are to be novated, not only will there be a period for the novation process but the consultants will not, before that time, be aware of the contractor's requirements and may in fact have carried out abortive work for which there will be no provision for a fee to the consultant.

While on the face of it there is an apparent benefit to the employer for the consulting team to be novated, in practice it is probably more satisfactory for the employer's consultants to remain bound to the employer, when they will be able to oversee the design development to safeguard the employer's interests, instead of being required to become 'gamekeeper turned poacher' and pursue the contractor's interests while ignoring those of the employer, for whom the consultant may wish to work again in the future.

In addition, the provision for the contractor and consultant to seek variations to the agreement by way of schedules 1 and 2 provides a potential delay to the conclusion of the novation agreement, and so the implementation of the contract. It will be necessary for the invitation to tender to have advised the potential contractor of the intention to novate the consultant agreement, and also, at tender stage, to advise the contractor of the details of such agreement so that it can make due allowance either by way of its tendered price or by proposed variations to agreements for the contractor's requirements of the consultant. In any event there can be no way in which the contractor can advise the consultant pre-tender of its requirements. Similarly, there is no provision for the consultant to have priced within, or to alter its fee arrangements to allow compliance with, the contractor's specific proposals or requirements.

A major concern for the contractor will be to obtain from the novated engineer, pre-novation, a warranty that his outline design satisfies all the matters specified in the employer's requirements and that the contractor may rely on that design for the pricing of the work. It will not be certain, should the consultant's work, pre-novation, prove to be deficient, that the contractor will have any claim, either against the employer for those deficiencies or against the consultant. Such a discrepancy was the issue in the case of *Blyth & Blyth Limited* v. *Carillion Construction Limited* (2001) 79 Con LR 142, where Lord Eassie had to consider a counter-claim by the contractor that the engineer was liable for the increased costs resulting from the finalised design incorporating additional reinforcement over and above that indicated in the tender enquiry for which the engineer had been responsible on behalf of the employer. Although the judge had to consider the express terms of the actual novation agreement, it is unlikely that without express terms a novated consultant would be liable to the contractor for any deficiencies in the tender or pre-contract documentation.

Lord Eassie summarised the contractor's case:

> '35. One ought, accordingly, to examine the legally logical position of the defenders' analysis to the effect that the Novation Agreement produced an essentially three-sided relationship whereby A (contractor) engages B (consultant) to perform services for, and give advice to, C (employer). Accepting for the moment that analysis to be correct, the questions which arise are whether the sufficiency of the performance by B (consultant) is to be judged by what is required by the destinee of the services C (the employer) and whether in the event of defective performance the losses or costs for which B (consultant) may be liable in damages are those reflecting the need to put C (employer) in the position of having received satisfactory service or, as the defenders contend, those said to have been suffered by A (contractor).

36. Counsel for the pursuers submitted that on the analysis and argument ultimately advanced by the defenders, the natural measure of loss was that suffered by the recipient of services said to be defective. It appears to me that such will normally be the case. If a husband contract with a surgeon for the treatment of his wife, the liability of the surgeon for negligently treating the wife will be the loss sustained by her. If a father engage a garage to repair his son's small, elderly car by a particular date and the garage fails timeously to repair it, the liability of the garage would be measured by the cost of hiring a replacement small car and not, say, the cost of the father's hiring a large limousine because the son has borrowed the family saloon.'

Lord Eassie discussed the possibility of the engineer having a liability to both the employer and the contractor concurrently, at paragraph 41:

'No authorities on this matter were cited to me, but in principle it appears to me that a valid claim by the employer against the engineer on the basis of defective design is not defeated by the employer's having the possibility of making a valid claim on a similar basis against the contractor by virtue of the design obligations undertaken by the contractor in the building contract. Thus, if defects were to emerge in the leisure centre requiring the execution of repairs and its closing down, and if the defects arose from defective design the genesis of which lay in what was done for the employer before the date of the tender, one would readily think the employer to have a perfectly presentable claim against the designer. The fact that the terms of the building contract might also give the employer a claim against the contractor by reason of his having effectively underwritten the preconstruction design, would not discharge or destroy the liability of the designer to the employer. Further, applying the litmus test of insolvency of the contractor, the employers' claim against the consultant must persist. There being such liability *in solidum* it may well be of course that questions of relief or apportionment between the two obligations would arise.'

Lord Eassie concluded by finding that there was no liability to the contractor by the novated engineer for any actions carried out by the engineer pre-novation:

'52. In these circumstances I have come to the conclusion that the pursuers are correct in their submission that the defenders cannot claim for their own losses, not losses conceived as having been suffered by the employer, based on alleged breach of duty by the pursuers committed prior to the date of novation and in relation to the duties then owned to the employer.'

The difficulty that is apparent in the *Blyth & Blyth* case is that the engineer without doubt made a very significant mistake for which he got away without sanction, leaving the contractor to pay the price of the engineer's defective work. Either the engineer was wrong in his pre-tender enquiry estimate of the quantity of the steel reinforcement necessary or, which is unlikely, he overdesigned at the detailing stage, causing the contractor to incur the cost of the unnecessary reinforcement. It is to address this potential difficulty that CIC/Nov Agr has introduced clause 4 and CIC clause 1.4.

While it is acknowledged that such an engineer will have a duty to the employer not to make an excessive allowance in its pre-tender enquiry estimate, it must be anticipated that the successful contractor will have tendered relying on the engineer's estimate as stated in the tender documents. No benefit can be found for an employer to require its consultant to incur fees in preparing estimates for reinforcement quantities, and then to state that such quantities are not to be relied on by those tendering for the work. Such an estimate can be distinguished from random figures inserted into a bill of quantities for the purpose of obtaining rates for the pricing of variations.

The consultant engineer, once novated to the contractor, owes a duty to the contractor to design the structure as economically as possible. There is no place for making provision for possible future changes, without an express requirement within the employer's requirements, as there might be where the commission is direct to the building owner. Should the engineer make excessive or unnecessary allowances for possible changes to the building or its use, where not specifically required to do so, then the engineer may well be liable to the contractor for any excess expense to the contractor as a consequence, which the contractor can demonstrate to have been incorporated into the detailed design. It is for all these reasons that a fair novation should contain a warranty from the engineer to the contractor as to the sufficiency of any pre-novation design or estimate.

Irrespective of the attempts made to redress the difficulties resulting from the judgment in *Blyth & Blyth*, it is suggested that in the case of design consultants, unlike a contract for the supply of goods, the service to be provided to a design and build contractor is so different from the service to be provided to an employer as not in any event to be capable of a straight novation, where all the obligations and rights of the novated party remain unaltered and all that changes is the party to which they are due.

Naming and nomination

The phrases 'named subcontractor' and 'nominated subcontractor' have been incorporated into the wording of main contracts in the past. The generally accepted difference is that, with regard to a named subcontractor the invitation to tender and selection of the successful tenderer rests with the employer's agent, while the formal tender acceptance is by the contractor and the resultant subcontract is to all intents a domestic subcontract; but with a nominated subcontractor the employer's agent retains some control for matters such as valuation of the work and the granting of extensions of time.

Thus in either case the employer or its consultants prepare the subcontract enquiry documents and select the tender list, and on receipt of the offers make the selection of the subcontractor to be employed. There will normally be a provision within the subcontractor's offer that it will enter into a subcontract with the appointed contractor. There will be a reciprocal obligation within the main contract to conclude an agreement with the selected subcontractor.

Naming and nomination can be distinguished from novation in that novation is the transfer of an existing contract to a new party, while in naming or nomination there will be no contractual obligations between the eventual parties to the subcontract that result from the acceptance of the request to tender, and there is no formal contract to carry out the work until that offer is accepted by the contractor or a signed agreement is concluded between the contractor and subcontractor.

However, it is probable that at the time the subcontractor's offer is accepted on behalf of the employer, there will come into existence a contract between the subcontractor and the employer which goes beyond the obligation to enter into a subcontract with the contractor but also forms an agreement to carry out the subcontract work itself. If that were not so, there could be no obligation to commence design and/or coordination work nor to procure materials and/or carry out off-site manufacture to meet the anticipated programme of works on site. The terms of such a contract will in principle be those of the specified form of named or nominated subcontract. The employer's obligation is to contract with a main contractor who will enter into a subcontract with the selected subcontractor to enable it to perform the work as anticipated in the invitation to tender. If for any reason the employer is either unable to achieve this situation on the agreed terms or subsequently decides not to proceed with the named or nominated subcontract work, then the employer will be in breach of the contract between it and the subcontractor and the subcontractor should be entitled to recover its losses as damages from the employer.

ICSub/NAM naming procedures

The ICSub/NAM form for naming under the IC form of main contract is in three parts: ICSub/NAM/IT – Invitation to tender, ICSub/NAM/T – Tender by subcontractor and ICSub/NAM/A – Articles of Agreement.

The form ICSub/NAM/IT is addressed to the prospective subcontractor and gives details of the project title and nature of the main contract works, and then states the nature of the subcontract work. The tenderer is then advised that:

> 'It is intended that the Named Subcontractor will be employed by the Contractor under a subcontract in the form of the Sub-Contract Agreement included in this document ICSub/NAM.'

Three possible procedures are set out which allow for either the provision of documents to contractors pricing the works or for the issue of an instruction to the contractor post-contract for expenditure of a provisional sum. The third option provides for arranging a replacement subcontractor, if such were required.

ICSub/NAM/IT then sets out the tender documents and provides for a choice of pricing documents, from the following list:

- a priced copy of the bills of quantities;
- a priced copy of the specification;
- a priced copy of the work schedules;
- a subcontract sum analysis;
- a schedule of rates.

together with an activity schedule.

The tender information is split into three sections: general information, main contract information and subcontract information. Each of these sections is to be filled in by the person seeking the tender and set out in the particulars of the intended subcontract.

Under ICSub/NAM/T, tender by subcontractor is addressed to the employer and the contractor, if already selected. After a statement to the effect that the tenderer has noted the tender information and has based its offer on that information, the form states the subcontractor's price for the work. There is then an undertaking to supply the priced document(s) and to enter into the completed form of Sub-Contract Agreement.

The subcontractor's offer is based not only on the documents supplied to it with the tender enquiry but also on the entries the subcontractor makes under items T1 to T5 of ICSub/NAM/T. These items relate to programme, attendances, fluctuations, dayworks and incorporation of the subcontract works into the main contract works.

Section T1 programme subsections 1–3 deal with drawing preparation, approval and off-site works; a period of time is inserted in each section giving a total for all pre-site work. The three stages set out are:

1. The period required by the subcontractor for the preparation of all the drawings etc. that it is necessary for the subcontractor to prepare including co-ordination, installation, shop or builders' work and any other details. The period is to be that required from receipt of the necessary drawings and specifications from the architect to the submission of the subcontract information to the architect for comment.
2. The period for the architect to make his initial comments on the subcontractor's drawings and other information. This is stated to be the period from receipt to return to the contractor.
3. The period for procurement of materials, fabrication and delivery to site.

It is to be noted that there is no provision within the stated period for the resubmission and/or amendment of the drawings, nor for the taking of site measurements, where these are required. As discussed elsewhere, a major cause of delay to subcontractors who are required to produce design development or shop drawings for approval or comment, is that these drawings are frequently returned with comments and requirements to change that do not relate to a failure to comply with the requirements of the specification or design intent drawings, but incorporate observations and/or wishes of the consultant designer that have no basis in the Sub-Contract Documents.

The period for approval of drawings at IT10 is to be stated by a single period and does not identify a period or date by which the architect may require such drawings. Since one of the major reasons for naming may be to obtain the specialist contractor's details at an early stage for incorporation in and/or coordination with the design of the works as a whole, this appears to be a curious omission by the drafters of this form.

It is to be noted that T1.1 refers to submission to the architect/contract administrator and it must be assumed that the intention is for submission to be made directly to the architect/contract administrator; certainly there is no time allowed for transmittal of drawings from subcontractor to contractor and from contractor to the architect/contract administrator and back, for which a period of two weeks is considered normal. Such direct contact between subcontractor and consultant designer has potential benefits with regards to ease of communication, but has serious difficulties with regards to the issue of change instructions and notification of delays and other contractual matters, which rest between the subcontractor and the contractor. Certainly within the terms of ICSub/NAM/C there is no provision for such direct contact.

At T1.4 the subcontractor is required to state the period of notice it requires to commence work on site and at T1.5 the period required to carry out the work on site after the period of notice. Strangely T1.4 expressly refers to the period of notice to commence work on site, whereas the subcontractor will require notice to commence its work whether on or off site. The period in T1.5 is that required by the subcontractor to carry out the subcontract works on site and is stated to be after the delivery period and notice to commence, and not as might be expected the period for delivery following the period of notice to commence.

The reference to sections in these items is no doubt intended to refer to where the Works under the main contract are to be carried out in sections. However, there is no reason why the subcontractor should not identify sections in its work for which it requires a separate notice to commence and/or period of time to carry out the work.

Finally, at T1.6 there is provision for the subcontractor to provide further details that it considers may quantify or clarify or are otherwise relevant to the carrying out of the Sub-Contract Works. It is to be anticipated that where the named subcontractor procedures are being used, this will relate to a trade or work which requires some special arrangements or provisions. It will be most important that the prospective subcontractor identifies any special needs of its work or method of working with regard to time in this section of its tender. It cannot be assumed that either the employer's agent or the contractor will be conversant with the procedures and/or needs of a specialist subcontractor's work or business.

At item T2 the subcontractor has the opportunity to provide details of any special attendances or other special requirements that the subcontractor will require to be provided by the contractor, in addition to those specified in item IT11 free of charge to the subcontractor. Again it is for the prospective subcontractor to identify fully its requirements. Under no circumstances should it assume that any attendance other than those clearly set out in ICSub/NAM/IT at item IT11 will be provided free of charge by the contractor. Therefore the subcontractor will be well advised to spell out in very precise terms the attendances it will require. It should not rely on phrases such as 'reasonable hoisting' or 'provision of water and electricity', which may be considered not to include for any connection or distribution and may exclude the provision to site accommodation. It will also be well advised to fill in this section in considerable detail and include items normally expected such as safety lighting and distribution of power and water and basic setting out by way of lines and levels as well as items specific to its work.

At item T5 the prospective subcontractor has the opportunity to identify when its work or parts of its work are to be regarded as incorporated into the main contract works for the purpose of relieving the subcontractor for liabilities for loss or damage to its work. While this item expressly relates to clause 6.7.8 and liability for the cost of rectification should the subcontractor's work be damaged after incorporation into the main contract works, in completing this section a prospective subcontractor may wish to consider other matters such as discussed elsewhere, including the sections on completion and temporary disconformity.

The form ICSub/NAM/A, Sub-Contract Agreement, is a simple form of agreement that can be either signed under hand or executed as a deed. The recitals relate directly to the enquiry and tender forms and the subcontract conditions ICSub/NAM/C.

The conditions set out in ICSub/NAM/C are generally those set out in the ICSub/C form. It should be noted that the conditions for use where the subcontractor is named provide no additional protection or benefits to or for the subcontractor over those in the standard form of subcontract ICSub/C.

Subcontractors tendering under the ICSub/NAM procedures must understand that their offer is more than just a simple offer open for acceptance. It is in addition a binding agreement to the effect that if selected they will contract with the contractor; the contractor may or may not have been identified at the time of the subcontractor's offer and so this may be a 'blind date'. It may be that, before committing itself to such an obligation, the subcontractor will wish to obtain details and/or assurances as to the prospective contractor tender list. In the event of either a major breach of the agreement between the subcontractor and the contractor or of the contractor becoming insolvent, the subcontractor will have no rights other than would exist under any other subcontract arrangement.

Notwithstanding the undertaking to enter into a subcontract within the timescales set out in the tender by the subcontractor, there is no provision for agreement as to the timing and sequencing of the subcontractor's works. Whether the very simplistic dates within which the work is expected to be commenced, set out at item IT9 of the invitation to tender, coupled with the unilateral requirements of the subcontractor set out at item T1 of the tender, will be sufficient and/or satisfactory, will to an extent depend on the nature and extent of the subcontractor's work. The use of named subcontractors will frequently be restricted to the provision and installation of feature components, such as gates, clocks, fountains, etc., where the on-site period may be relatively short and the interface with others limited.

However, many such feature items will require specialist builder's work and/or integration with other finishing activities, which are discussed more fully in the section 'Programming differing trades' in Chapter 4. There is no express provision for the integration of the subcontract work with that of the main contractor, either under ICSub/NAM/C or for the coordination of design detailing by the design consultant.

Nomination

The standard form of building contract SBC/Q, unlike its predecessors, has no provision for either the naming or nomination of subcontractors but is limited to specifying a list of three for selection by the contractor as described in the section 'Specified subcontractor' earlier in this chapter. It is of course open to the parties to a construction contract to agree procedures for nomination, which may be in the form of a three-way agreement in a similar manner to that of a works contractor under the management contract arrangements. However, without such express arrangements and agreement, the additional benefits, rights and obligations traditionally associated with nominated subcontracts will not apply either under the subcontract or under the main contract. Just attaching a specific name to a subcontract or making the selection and/or requesting the appointment in a certain way, does not imply in terms not expressly stated in the relevant Sub-Contract Documents.

Even where the main contract provides for both the specification and nomination of subcontractors, such as was the case in the 1983 JCT forms, the change to a single tender sought and obtained by the employer's agent will not make the selected subcontractor a nominated subcontractor without express agreement between the parties. This was the situation in the case of *Mowlem (Scotland) Limited* v. *Inverclyde Council* [2003] XA 29/02, where the Lord President said:

> '49. We are not persuaded that at the outset Structal was "named" within the meaning of clause 35.1.4 with a view to its being a nominated subcontractor. In regard to the curtain walling work, for which it was one of the persons listed in Bill No. 1, it was plainly not treated in the contract documents or in the communications between Mowlem and Inverclyde as a prospective nominated subcontractor. The fact that it was the only subcontractor which responded to the invitation to tender, and that it was the only subcontractor which could have offered to carry out the design and manufacture of curtain walling according to the Structal system does not, in our view, alter the position. Inverclyde could have required that in tendering Mowlem should treat Structal as a prospective nominated subcontractor but did not do so either expressly or by implication.'

It is believed that the JCT have dropped the provision for nominated subcontractors as a result of strong and prolonged campaigning by a certain senior member of the bar, on the grounds that such procedures tip the balance of contractual risk too far in favour of the contractor. However, there are many reasons why the employer or its professional team may wish to select a specific company or individual to carry out work on its project. Where this is done, either out of necessity or desire, it is difficult to understand why the risk of that selected company or individual who is perhaps unknown to the contractor, should, in the event of failure to perform, be regarded as the failure of the contractor and that the contractor should bear the loss both to itself and/or to the employer.

Such arguments are less persuasive where the employer's selection is made pre-tender, when the contractor has the opportunity to assess the risks of employing the selected subcontractor and either can make due allowance in its offer or decline to tender. As mentioned in the section 'Specified subcontractor' earlier in this chapter, SBC standard forms such as SBC/Q require that the specified list contains at least three names. This appears to be unsatisfactory for both the employer and the contractor, by denying on the one hand the employer the right to select its chosen specialist and on the other, the contractor freedom of choice.

Difficulties in third party selection

In all situations where enquiry documents are prepared by a party other than the contractor, there will be a real risk that the interfaces as to responsibility may become uncertain. Typically, a specification for one trade may define work to be carried out by another; for example, the fire stopping around service pipes and other builders' work or building work, in connection with services work.

As mentioned in the section 'Interface of design responsibilities' in Chapter 6, the problems of interface detailing and the associated responsibility for the actual work at the interface between trades becomes much more uncertain where the specialist contractor has a design development involvement rather than where it is contracting to construct work fully detailed and specified at the time of tender. There may be confusion not only as to the actual work to be done, but also as to the provision of attendances. Statements to the effect that the subcontractor will have free use of scaffolding may be taken to mean that it can use scaffolding that already exists or that it will be provided with all necessary scaffolding.

An inherent problem for third parties seeking to obtain offers from specialised subcontractors to be novated, named or nominated under the terms of a main contract will result should the offers received not be fully compliant. It is common practice in all construction contracts to seek to pass the risks associated with weather, changing markets for materials and labour, breakdowns of plant and/or equipment and other uncertain matters, down to the contractor and/or the subcontractor. It is equally common for a subcontractor at least in its initial offer to exclude most indefinable risks, thereby enabling it to make its most competitive bid, which can be adjusted by post-tender negotiation to reach a balance of risk most economical to both parties. It is this option to reach a post-tender compromise which does not readily exist where a third party is making the selection, since any compromise with the subcontractor must in turn lead to an adjustment to the main contract by either arranging for the contractor to assume the risks or for the employer to accept them and make any appropriate compensation arrangements with the contractor.

In addition to deciding the risks to be adopted by the novated, named or nominated subcontractor, it is necessary to consider the extent of the work to be undertaken. Many specifications prepared for incorporation in main contracts incorporate associated works under the specification for specific work; for example, the requirement to provide a conduit for electrical cable may include the cutting of the chase and the making good of the structure. A specialist excluding from its bid, for example, 'all building work' or 'all builders' work' may fail to define exactly what is and what is not incorporated in its offer. Another way to express excluded work is to relate the work to that achievable using the normal tools of the trade. Thus work requiring the use of a bricklayer's or pointing trowel would not be work attributable to an electrician.

It is true that when tendering against an enquiry as a novated, named or nominated subcontractor, the subcontractor will know that it is tendering either as the sole tenderer or against limited competition and should therefore be more willing to give a fully compliant bid. However, there may be a range of matters where either because of the uncertainty as to how the design may develop or for other issues which have yet to be resolved, such as the timing of the work or the means of access and/or hoisting, the subcontractor may require its price to be adjusted if necessary. Further, the enquiry may seek that the subcontractor include for matters for which it does not have the facilities and/or expertise to provide, when it will either require such matters to be carried out by others or to be paid an uncompetitive price for such work.

A common feature that may be incorporated into subcontracts with subcontractors selected by third parties is that the employer's consultants retain responsibility for the valuation and certification of the subcontractor's work. In addition, under certain circumstances the subcontractor may have the right to be paid directly by the employer. Such a right may create difficulties when the contractor considers it has a claim against the subcontractor by way of set-off.

Chapter 4
Programming of the Subcontract Work

Programme for work

It will normally be for the contractor to decide the nature of the subcontractor's obligation with regard to the time for and timing of the works. The most common method is to define a period for the carrying out of the subcontract works combined with a period of notice for the commencement of work on site. An alternative method is to require the subcontractor to carry out the works as and when available or as and when instructed by the contractor's project manager.

Neither of these methods is fully satisfactory to either party. Where a subcontractor is given a period of time to carry out the work, then subject to any express limitations in the subcontract, the subcontractor is at liberty 'to plan and perform the work as he pleases, provided that he finishes it by the time fixed in the subcontract'. This principle was restated in the judgment of Judge Gilliland QC in *Pigott Foundations Ltd* v. *Shepherd Construction Ltd* (1993) 67 BLR 49.

The contractor is likely to qualify the obligation of the subcontractor by a requirement to work to a project programme or to coordinate the subcontract works with others. SBCSub/C at clause 2.3 defines the time for commencement and completion as being to comply with the programme details and reasonably in accordance with the progress of the main contract Works. The programme is expressly stated to be that detailed in the Sub-Contract Particulars (item 5) and is subject to the subcontractor receiving a notice to commence work in accordance with those particulars and subject to clauses 2.16 to 2.19.

The reference to the Sub-Contract Particulars (item 5) is an instance where there is a need for the parties to complete and execute the article of agreement SBCSub/A, for this clause to have relevance. By specifically referring to the Sub-Contract Particulars (item 5) it is doubtful that any other arrangements will be effectively incorporated into the subcontract. Further, clause 2.3 makes the subcontractor's obligation to complete in accordance with the programme details subject to it having received a notice to commence in accordance with the particulars. The inference is that a failure by the contractor to issue a notice to commence in accordance with the Sub-Contract Particulars (item 5), will mean that the subcontractor is left with the obligation to carry out its work reasonably in accordance with the progress of the main contract Works, and not to complete in a stated time or to a pre-determined programme.

ICSub/C at clause 2.2 has an identical clause to SBCSub/C clause 2.3. Sub/MPF04 at clause 14.2 uses different words to much the same effect, except to incorporate a requirement to proceed regularly and diligently. In that clause the subcontractor is required, on being given access to the site to commence the subcontract works, to proceed with them as follows:

1. regularly and diligently;
2. in accordance with any specific dates or programme requirements identified in Appendix A; and
3. reasonably in accordance with the progress of the project.

so as to achieve practical completion within the period for completion.

The meaning of 'regularly and diligently' is discussed in the section 'Determination by the contractor' in Chapter 12. However, it is considered that a requirement to proceed 'regularly and diligently' is inconsistent with a requirement to proceed 'reasonably in accordance with the progress of the project'. The contractor may require one or the other, but not both. ShortSub has no express term relating to the time for commencement and/or completion, but just a general term at 5.1 to carry out the Works in accordance with the Sub-Contract Documents and with due diligence.

In *Pigott* v. *Shepherd*, Judge Gilliland QC stated, at page 61, the effect of a similar clause 11.1 in DOM/1:

> 'In my judgment the obligation of the subcontractor under clause 11.1 of DOM/1 to carry out and complete the subcontract works "reasonably in accordance with the progress of the Works" does not upon its true construction require the subcontractor to comply with the main contractor's programme of works nor does it entitle the main contractor to claim that the subcontractor must finish or complete a particular part of the subcontract works by a particular date in order to enable the main contractor to proceed with other parts of the works. The words "the progress of the Works" are in my judgment directed to requiring the subcontractor to carry out his subcontract works in such a manner as would not unreasonably interfere with the actual carrying out of any other works which can conveniently be carried out at the same time. The words do not however in my judgment require the subcontractor to plan his subcontract work so as to fit in with either any scheme of work of the main contractor or to finish any part of the subcontract works by a particular date so as to enable the main contractor to proceed with other parts of the work.'

One problem for the contractor is that the more he attempts to tie down the obligation of the subcontractor to carry out its works in an agreed sequence and/or to commence and complete specific areas to express dates, the more likely the contractor will itself be in breach of contract for failing to meet its own obligations to make the works available at a predetermined time. In addition, the contractor will also limit its right to advance the subcontract works if the general progress of the project is better than anticipated at the time of the conclusion of the subcontract.

Further, without a stated intention as to how the work is to be made available to the subcontractor, the subcontractor probably has a right to regard the contractor's obligation as being to make available the whole of the works for the whole of the subcontract period. While for most services and finishing trades this is rarely the intention or the anticipation of the parties, without some stated intention to the contrary there will be uncertainty and therefore unenforceability of any requirement by the contractor of the subcontractor which is affected by its access to the work.

If integrated programming is difficult for the on-site works, the potential problems are significantly larger when considering off-site or pre-site activities. There may be an express requirement for a subcontractor pre-fabricating components off-site, to visit site and take 'site dimensions'. Unless the date for taking such dimensions is agreed and established in the subcontract, there will be every possibility of a claim that the work was not available in time for the fulfilment of this obligation. Unless specifically stated in the Sub-Contract Particulars, neither drawing nor fabrication work need be carried out in any particular period in relation to work on site. It is for the subcontractor, in the absence of an express requirement in the subcontract, to balance its resources and even out its workload to its best advantage, so as to satisfy its contractual obligations to all its clients.

Where the subcontractor has to prepare and submit for approval shop drawings, these may be reliant on drawings by, and/or necessary for the drawing work of, others, including in some situations the obtaining of statutory approval by the employer's own consultants. Unless this has been fully thought through and an appropriate and integrated programme agreed and incorporated as a subcontract requirement, the on-site works are very likely to be delayed and/or disrupted.

Where the subcontractor is not to be paid for work carried out ahead of on-site installation, there will be a financial disincentive to carry out any off-site activities any earlier than is necessary to satisfy the strict requirements of the subcontract. Further, the 'just in time' philosophy, currently being advocated as good management practice, encourages the subcontractor in this approach to its work. The problem is that there will consequently be a very restricted and limited time and opportunity to absorb the effect of any change, such as that which frequently follows the submission of the specialist subcontractor's drawings for approval.

For many subcontractors with off-site works there will be the further complication of integrating the works of their own sub-subcontractors and suppliers. Unless and until each of these sub-suppliers has been contracted with, the subcontractor will be at risk. For this reason the potential subcontractor may qualify its offer as subject to availability at the time of order. Availability of subcomponents may not only be subject to market forces operating at the time of order, which may differ to those at the time of original enquiry, but in addition many materials and or processes are subject to seasonal changes. Materials from northern latitudes may be subject to the effects of winter weather, including being ice-bound in harbours. At other times of the year industries are subjected to annual shut-downs for seasonal holidays.

Programming differing trades

As mentioned above, the standard forms of subcontract have just one arrangement for establishing the subcontractor's obligations as to time, which follows the same basic concept as contracts in general, that there is a defined date by which the subcontract works must be completed. To allow for the contractor's potential for changed requirements and/or the uncertainty of its programme predictions, there is in SBCSub/C provision, by reference to item 5 of the Sub-Contract Particulars in Agreement, SBCSub/A, for the start date to float, with the subcontractor's period for the carrying out of its work being a stated period from the expiry of the notice to commence work on site.

The provisions of the Sub-Contract Particulars in Agreement, SBCSub/A, allow periods for off-site procurement and fabrication, but not for the preparation of shop drawings and approval, for a period of notice to commence work and a period for carrying out the work. In addition, subsection 4 allows for further details 'that may qualify or clarify the above or are otherwise relevant to the carrying out of the Sub-contract Work'. These are defined as follows:

1. The period required for the procurement of materials, fabrication and delivery to site prior to commencing work on site/work.
2. The required period of notice to commence work on site to enable a start to be made to the Sub-Contract Works.
3. The period required for the carrying out of the Sub-Contract Works on site after delivery and after the expiry of the period to commence work.
4. Further details or arrangements that may qualify or clarify the above or are otherwise relevant to the carrying out of the Sub-Contract Works.

There is also provision for several periods where the work is to be carried out in sections.

Sub/MPF04, in appendix A, states a period for completion which is to commence 'on the later of the expiry of the period (if any) for pre-site activities and the period of notice to commence, or such other period as may be established by the operation of clause 16 (extension of time)'. In addition, item 14 of the appendix gives further details as follows:

1. period (if any) required by the subcontractor for pre-site activities;
2. period of notice to be given to the subcontractor of the date when access to the site will be given;
3. specific dates or programme requirements.

It is suggested that except in relation to subcontracts where the obligations have little or no interdependence with other trades, then unless the parties agree, at least by way of intention, the major dependencies and requirements for the successful carrying out of both the subcontract works and the Works as a whole, there will be every possibility that the project, as a whole, will not progress successfully.

The arrangements in SBCSub/A (item 5), referred to above, are reasonably satisfactory where the subcontractor is one using materials readily available ex-stock and whose labour requirements can be drawn from the general skill base for that trade. Such trades will include plastering and painting and generally trades such as earthworks, brick and block works, carpentry, drylining and ceilings and much services work, provided appropriate arrangements are set out at subsection 4. However, even with these straightforward trades it will be important to the progress of the project that work is made available for the follow-on trade within the period set out in the project programme. In fact, it is the delaying of the follow-on subcontractor that will cause at least as much and probably more difficulty and probable loss to the contractor, than a failure by the subcontractor to complete within the period stated within the subcontract. In other words, it is the rate of commencement of the works or more particularly rate of release of work to others, that is generally a critical factor.

For most trades, the contractor's programme will only plan the works of the subcontractor as a single activity or as a single activity for each major phase of installation and/or different sections, areas or floors. In practice, while the work will generally be available, there may be a requirement to return at a later date to complete, after some further activity or event. Typical activities of this nature include infilling holes left in floors after the removal of cranes or completing external cladding after the removal of hoists or chutes. Making good around succeeding trades can range from brick or blockwork around service ducts and pipes, or plaster around switch and socket boxes, to roofing around rainwater outlets and/or other service penetrations.

The probable intention will be that the subcontract completion period will be for the bulk of the subcontract works, and since no consideration will have been given to the completion around the works of others, a lack of programme provision will make the subcontract one without effect as to time, which may become at large.

Again, much roofing work can be carried out in two distinct phases: a first stage which makes the building largely weathertight, such as the felt and batten stage in a traditional slate or tile roof. For any one roof or sub-roof this is likely to be of more critical effect than the actual completion of the entire subcontract works. The late delivery of, for example, a feature ridge termination finial may significantly delay final completion of the roofing works but is most unlikely to have any or any significant effect on the progress of the Works in general.

Trades involved in the construction of reinforced concrete work, by convention, consider completion of their subcontract works to be the date when concrete is placed to the last major structural section. In fact, there will at minimum be the further activities of curing and striking shutters, including any necessary back propping, followed by any making good or 'rubbing up' to the as-struck surface, and finally the clearing of materials and plant from the site. The reason for this assumption is that while contractor's project programmes will generally define their programme activities as 'shutter, reinforce and concrete' or just 'construct' there will be no separate activities for cure, strike and make good, which will be executed progressively for several weeks after the final concrete placing.

While some of the main building trades may be required to return to an area to complete around the work of subsequent trades, the reverse situation may exist with some finishing trades who may need to set out and install their fixings and connections prior to intermediate finishing trades carrying out their work. For the contractor, the extent or indeed the existence of such a requirement may not be known until the subcontractor's own detailing has been carried out and approved, possibly by a third party, such as the employer's consultants or other subcontractors. It is therefore prudent for the contractor to assume, when in doubt, that such an activity will be needed and to provide for it at item 5 of the Sub-Contract Particulars.

The contractor may require certain windows and other elements of external cladding to be left temporally uninstalled, so as to provide, for example, for temporary access from external contractor's hoists and/or scaffolds or for the disposal of rubbish and surplus materials via rubbish chutes. In other places it may be necessary to make provision for the tying in of scaffolds, cranes or hoists. In such situations the contractor will be preventing the regular progress of the subcontractor's work and may, in addition, prevent the subcontractor from completing its work by the due date.

A subcontract to provide temporary works, such as scaffolding, will be required to progress at the rate of those works that it serves, in the case of scaffolding, the work for which it provides access. In certain circumstances, such as scaffolds for maintenance or demolition work, there will be a requirement to provide a fully erected scaffold by a certain date. This should be defined as the subcontract completion date, with the provision for dismantling and removal from site subject to separate provisions.

Most building finishing trades will be required to visit the same unit or area of the works a number of times, as the work of other trades proceeds. Building services and joinery operations will generally be required to be carried out as first fix, second fix and finals operations. Ceiling subcontractors may be required to fix the ceiling grid as a separate operation to their tiling. Partition wall contractors will normally as a first stage install the frame and boarding to one side only, returning once the services have been installed in the void between surfaces to fix the boarding to the other face. None of these trades fit comfortably into a requirement for the subcontractor to carry out its work in a fixed period of time after a stated period of notice.

From the above it can be appreciated that the limited provisions of the Sub-Contract Particulars as set out in SBCSub/A are inadequate unless good use is made of subsection 4. The example form below is more likely to act as an effective skeleton agreement for on-site works.

Execution of the work on site

1. The contractor shall give to the subcontractor ______________ days' notice in writing of the availability of each section of the subcontract works.
2. The contractor shall select either:
   (a) That the subcontractor shall carry out the work on the basis of an agreed subcontract period.
   (b) That the subcontractor shall carry out the work within a reasonable period upon the work of each section becoming available.
3. Where 2a is selected
   3.1 The contractor's detailed objectives and requirements of the subcontractor are: [The contractor will set out its specific requirements relating to the project or trade, such as early completion of stages of the work or any areas that are to be completed after the general completion of the subcontract work.]
   3.2 The contractor is to release to the subcontractor the work at the end of the period of notice and the subcontractor is to complete within ______________ weeks of that date.
   3.3 Where due to the changed requirements of the contractor small sections of the subcontractor's work on site extend beyond the subcontract completion date the subcontractor will complete such elements within ______________ days of such elements being released to it.
   3.4 Where, after the issue of the notice to commence there is a requirement for further drawings and/or procurement or manufacture the required periods for design and/or procurement or manufacture are ______________ weeks.
4. Where 2b is selected
   4.1 The anticipated period of time for the subcontractor's work on site is ______________ weeks from the end of the period notice.
   4.2 The work is generally to be carried out in conformity with other trades and as set out on the contract programme ______________ or any revision thereof.
   4.3 The contractor will notify the subcontractor ______________ days in advance of each area of work or activity to be made available to the subcontractor and the target period for the handing back of that area of work or activity to the contractor.

4.4 Should the subcontractor consider that the target period set by the contractor under clause 4.3 above is unreasonable or unachievable, then the subcontractor shall so advise the contractor, giving its reasons for considering such target to be unreasonable or unachievable.
4.5 The subcontractor's obligation to progress the works as expeditiously as possible, having regard to the progress of the other trades on site, is not subject to the implementation of clauses 4.3 and 4.4 above.

Programming off-site or pre-site works

A large number of specialised trades may have an obligation to prepare and obtain approval for design and/or shop drawings. There will also, generally, be a need for a fabrication or manufacturing period and, in some instances, an earlier event of site survey. Despite these being obligations under the subcontract, there is frequently no requirement within the Sub-Contract Conditions defining the times and/ or availability of the site and/or information, for the execution of these activities.

At clause 2.7.4, SBCSub/C sets out the obligations of the parties to notify each other of their requirements for information. This clause requires either party, who has reason to believe that the other is not aware of the time by which it needs to provide such further drawings, details, information or directions, to advise the other party sufficiently in advance of the date such is required so that the other party can comply with clause 2.7. This clause is qualified by the phrase 'so far as reasonably practicable'. It must be sensible for both parties to issue a schedule of its requirements from the other party, giving the date by which each item is required, notwithstanding there being no express requirement to do so.

SBCSub/C fails to make a requirement for preparation of design information to be either a matter of pre-subcontract planning or a scheduled time period within the Sub-Contract Particulars of SBCSub/A. This, it is considered, is a major failing of this subcontract form and in practice of subcontracts in general. Even in subcontracts, where there is no requirement for the preparation and approval of shop details or significant periods of off-site manufacture and/or procurement of materials, on a well-managed project there should be programme time for the receipt and consideration and if necessary the reworking of job-specific method statements and works samples. Further, it must not be overlooked that many such pre-site activities will involve input and/or agreement from third parties who will also be working to tight schedules and will be unable to provide instant responses.

The need to prepare fabrication drawings may be the duty of the subcontractor, but if these are to be approved by the employer's consultants and/or integrated with the design work of others, it will be necessary to have the early and timely production of each of the drawings as an obligation under the subcontract. Not infrequently, detailing by specialist sub-subcontractors will involve more integration with the works of others than the work itself.

A good example of such a specialist item may be control gear and/or automatic operating equipment for doors and gates. While the doors or gates may be both an item to be installed late in the works programme and a locally fixed one, the control gear will need a separate space for its location. There will inevitably be a connection between the equipment, the control gear itself and perhaps, in addition, an operating point, all of which is likely to need advance planning and some early installation works such as conduit for connecting cables and wires.

In addition to the need to provide fabrication drawings, there may also be a requirement to provide material and/or works samples for approval/acceptance. There will be an implied suggestion and a distinct possibility that there may be a rejection, when time will be needed for a reselection and/or working of a further sample. Whether such rejection is as a result of a failure by the subcontractor or a changed requirement by the employer, it will give rise to an inevitable delay to the subcontract works, unless sufficient programme time for it has been incorporated. When planning periods for the approval of drawings, it is necessary to consider the period of time required to transmit the drawings along the contractual chain. Drawings may pass through as many as five sets of hands between a sub-subcontractor and a specialist consultant and back. This process of transmittal may well take much longer than the approval process itself.

Where there is an obligation for the specialist trade subcontract to take site dimensions, prior to manufacture, then there will be an obligation for the contractor to have constructed the adjoining work in due time, to allow for the taking of dimensions and subsequent fabrication. In the absence of an express provision within the subcontract as to the date of availability for the taking of dimensions, the subcontractor will be entitled to consider that the works will be available so as reasonably to suit its programme, which as mentioned above may have no direct relationship to either planned or actual installation on site.

The programme for the off-site fabrication, subject only to the obtaining of approvals and/or dimensions as discussed above, will be for the subcontractor to decide. It will be for the subcontractor to balance the work level within its production facility and those of its sub-suppliers with the cost of the advance financing of the work resulting from the manufacture ahead of the date planned for installation.

Also of significance when planning the subcontractor's fabrication process will be the changing requirements as work proceeds, both as to volume and condition for the storage requirements. Many fabricated units will be much more bulky than the materials from which they are constructed. Generally, finished products will also be more susceptible to damage than when in unfinished condition. The subcontractor may therefore manufacture its components to the unfinished state, at a time to integrate with its other work, but delay the finishing work until immediately prior to delivery to site.

This principle was discussed in the Court of Appeal decision in the case of *Greater London Council* v. *Cleveland Bridge & Engineering Co Ltd* (1986) 8 Con LR 30, which concerned the right of the trade contractor, Cleveland Bridge, to fluctuation costs arising out of the delay to the project. Although it had manufactured the steel components in good time, it had delayed the subsequent painting activity until just prior to delivery and installation. The GLC claimed that if the painting had been carried out soon after fabrication, which it could have been and which they claimed it should have been, then there would have been no entitlement to the increase in costs as claimed. On this basis it was claimed that the GLC should not be liable for additional costs arising from the contractor's decision to postpone, to the last moment, the painting work, which could have been done earlier at less cost to the GLC. Lord Justice Parker dismissed this suggestion, at page 48:

> 'It is contended, however, that if a manufacturer who has four years in which to complete a piece of equipment which could take as little as ten months to complete begins to do

the job at the beginning of the period, he is obliged, nevertheless, still to complete within the period in which it could have been completed. I ask myself why, and indeed Mr Lloyd [Counsel for GLC] was asked why. He conceded that, in taking a longer period, the contractor would not be in any way be in breach of his operation obligations, but it was said that he would be in breach of some financial obligation. Then one looks to see what this financial obligation could be. If it was said it was the obligation so to programme his work as to ensure that clause 51 provided the least benefit to himself and the maximum benefit to the employer, he would be quite unable to know at any one time, for the reasons which I have already explained, whether it would suit the employer better for him to start late, or whether it would suit the employer better for him to start early, or whether it would suit the employer that he should take a long period or a short period. The suggestion in my view is unworkable.'

The failure to plan and control effectively the pre- and off-site works is potentially far more likely to delay the Works, as a whole, than failing to plan the on-site work. The subcontractor's ability to respond to delay and/or change is, to a significant extent, limited by both drawing office and manufacturing capability. A major reason for this is the limited nature and amount of equipment owned by the subcontractor and/or space at its works. In addition, for some materials the actual process time, such as in the manufacture of bricks, will be finite, while for other materials the transport time from a distant source may be the limiting factor. The same changes are less likely to be crucial to on-site operations, where generally additional equipment and/or labour can be obtained at short notice, which is not the case for work in the drawing office and/or in the factory or fabrication shop, even at additional cost.

The provisions of SBCSub/A at item 5 fall far short of the requirements outlined above. Below are some outline provisions which seek to address these deficiencies:

Preparation of the subcontractor's design and/or shop drawings

1. The subcontractor undertakes to prepare and submit to the contractor its design and/or shop drawings on the basis of the contract drawings and other documents within ______________ weeks of appointment. Such drawings will indicate where the design information is inadequate and/or dimensions are to be finalised by site measurement.
2. The subcontractor undertakes to review and revise all design and/or shop drawings returned to it with comments within ______________ working days of receipt by it of those drawings.
3. The contractor requires the subcontractor to provide the following specific information for coordination with the design of others by the dates stated:

Procurement of materials and manufacture of the components for the subcontract works

1. The subcontractor requires ______________ weeks as a minimum procurement/manufacturing period from receipt of approval to manufacture to the delivery of components to site.
2. The subcontractor's anticipated manufacturing period for components is from ______________ (date) to ______________ (date) and upon appointment will reserve sufficient production time within the stated period for the manufacture of components comprising the subcontract works.
3. The subcontractor lists below the minimum production periods for additional or replacement components ______________ days/weeks.

Programme changes and amendments

The subcontractor's programme for on-site works may be subject to change either because of a changed requirement by the contractor, generally related to progress of other works on site, or because of problems in the subcontractor's own supply chain and/or labour availability. In the absence of clear provisions to the contrary, the subcontractor will generally have a contractual right to adjust its installation programme, if necessary, but only to the extent that it still complies with the specific requirements of the subcontract. Where the contractor requires to adjust the subcontractor's programme, to conform to its detailed programming and/or integration of other trades or to accommodate delays or better than anticipated progress or to accelerate other elements of the Works, then its changed requirements will necessarily require a clear instruction from the contractor of its new requirements which may lead to the subcontractor incurring additional costs.

It is probable that the subcontractor's right to compensation and/or payment where the programme is changed by an instruction or direction of the contractor, under the provisions of the subcontract, will differ from its rights where it is prevented from performing as intended under the subcontract by default of the contractor and/or its other subcontractors. As mentioned elsewhere, in general terms compensation for an instructed variation will be a valuation based pro-rata to the contract sum, while payment for either loss and/or expense under express provisions of the subcontract or general damages for other breaches will be on a proven loss basis.

It is perhaps one of the anomalies of the SBCSub/C form of subcontract that in the event of the subcontractor being delayed or being likely to be delayed by the contractor and/or his other subcontractors, the express conditions require the subcontractor to notify the contractor. It might have been expected that it would be for the contractor to advise the subcontractor of any anticipated delay to the subcontractor's work by others, and to instruct the subcontractor as to the contractor's new requirements, giving the subcontractor the maximum of notice and thereby reducing to a minimum the loss and/or expense likely to be incurred by the subcontractor and chargeable to the contractor. These requirements are set out in section 2 of SBCSub/C. Specifically, clause 2.18 limits the contractor's obligations to the giving of an extension of time. It is likely that any delay in giving clear instructions can be regarded as an act, omission or default by the contractor or other subcontractor. The provisions of clause 14.4 of Sub/MPF04 are in direct contrast to the provisions of SBCSub/C and require the contractor at all times to ensure that the subcontractor is aware of the actual and projected progress of the project.

Clause 2.18 requires the contractor to consider on receipt a notice from the subcontractor given under clause 2.17. If the contractor considers that both of the following requirements exist:

1. any of the events which are stated to be a cause of delay is a relevant subcontract event;
2. completion of the Sub-Contract Works or such works in any section is likely to be delayed thereby beyond the period or periods stated in the Sub-Contract Particulars (item 5) or any revised period or periods;

then the contractor is to give to the subcontractor an extension of time, which the contractor considers to be fair and reasonable, by revising the period for the completion of the subcontract work.

SBCSub/C at clause 2.19 gives a list of the relevant events to which clause 2.18 refers. The contractor's obligation to respond to a notice under clause 2.18 is to notify the subcontractor as to its granting of an extension of time, which is to be as soon as reasonably practicable but in any event within 16 weeks of the receipt of the notice from the subcontractor. There is no express provision or requirement for the contractor to issue any instructions to the subcontractor, other than for details of such loss and/or expense as the contractor requests, in order reasonably to enable the direct loss and/or expense to be agreed.

However, it is suggested that once a cause of loss and/or expense has been identified and if it is ongoing, there is no obligation for the subcontractor to continue to expend such loss without an express instruction to that effect from the contractor. Indeed, if the subcontractor does continue to expend such sums, without an instruction, it may be barred from recovery, as having taken accelerative measures on its own initiative. This restricted obligation of the subcontractor has been discussed in relation to the case of *Ascon Contracting Limited* v. *Alfred McAlpine Construction Isle of Man Limited* (1999) 66 Con LR 119, in the section 'Acceleration' in Chapter 8.

ICSub/C has at clause 2.12 similar provisions to SBCSub/C requiring the subcontractor to notify the contractor when it becomes reasonably apparent that 'the commencement, progress or completion of the Sub-Contract Works is likely to be delayed'. The contractor is to consider the notice and if it believes completion is likely to be delayed, the contractor shall as soon as possible establish the length of delay and make in writing a fair and reasonable extension to the period for completion.

Sub/MPF04, in contrast to both SBCSub/C and ICSub/C, acknowledges the principle that notification of delays is a duty of the contractor. Clause 14.4 requires that the contractor ensures that the subcontractor is aware of the progress of the project. This is to include both actual and projected progress and the date the contractor anticipates achieving completion.

In addition, at clauses 16.2 and 16.3, the subcontractor has similar duties to those in SBCSub/C and ICSub/C to notify the contractor when it becomes aware of delays to the progress of the subcontract works and to provide the necessary documentation to demonstrate the effect of any relevant event on that progress. Thus when the subcontractor is aware that the progress of the subcontract works is or is likely to be delayed, it is to notify the contractor of the cause of the delay and the anticipated effect on the progress and completion of the subcontract works. If the subcontractor considers the cause of the delay is one of those for which it is entitled to an adjustment to the period for completion of the subcontract works, then it must:

1. provide supporting documentation to demonstrate to the contractor the effect on the progress and completion of the Sub-Contract Works;
2. revise any documentation provided so that the contractor is at all times aware of the anticipated or actual effect of the cause of delay on the progress and completion of the Sub-Contract Works.

Where the subcontractor gives a notice under clause 16.3, the contractor has just 56 days, in contrast to 16 weeks under SBCSub/C, in which to advise the subcontractor of its decision as to the subcontractor's entitlement to a revised period for completion.

Alternative arrangements

Although the most common arrangements under construction subcontracts for the limiting and controlling of the timing of the subcontract work are as defined above, and comprise a defined subcontract period, a period of notice and perhaps an outline programme, there is no reason why this should be so. The parties are free to make whatever arrangements they wish. It may be that the inflexible nature of the standard subcontract conditions does not allow for the accommodation of the complexities and vagaries of site conditions and/or project design changes and development.

Where the subcontract arrangements provide for the subcontract works to be carried out in a fixed period after a stated period of notice, the subcontractor remains free to plan and organise its work as it wishes, provided it completes within the period for its work. This creates two specific problems:

First, since there is no provision for the progressive release of the work by the contractor to the subcontractor, the contractor can be held to be in default if the entire subcontracted work is not available at the end of the period of notice. If the entire work is not available at the date for commencement of the subcontractor's work on site and the contractor gives successive notices to the subcontractor, as further areas of its work become available there is no provision for reducing the period for the completion of those further areas to less than the subcontract period.

Secondly, there is no requirement under SBCSub/C or ICSub/C for the subcontractor to commence work at the end of the period of notice, unlike that required by appendix A in Sub/MPF04, or to proceed in a specific order or at a specific rate. A slow start by the subcontractor will not be a breach of its obligations other than by way of a vague requirement to proceed regularly and diligently, where such a provision exists. For many trades the effect of a slow or delayed start to the work of others may be very significant and lead to delay and/or loss to the contractor, even if the first subcontractor completes its work within its subcontract period.

One alternative arrangement, for example, is for the subcontractor to be required to provide labour and materials on demand, subject only to a specific period of notice. The benefits and disadvantages of such an arrangement were discussed in 'An Alternative Approach to Subcontract Arrangements Following Pigott' published in *Arbitration* Vol. 62, no. 4, November 1996. To be effective, there may need to be provisions for the situation where work becomes excessively delayed and/or there is a requirement for the rate of production to be very much increased.

Unless there are express requirements within the Sub-Contract Agreement for sectional completions and/or achievement of express stages in the work by agreed dates or within stated time-scales, a failure to achieve interim targets by the subcontractor will not be a breach of the subcontract. This is so even where this is a direct result of the subcontractor's failure, such as constructing defective work requiring removal and rebuilding. Only where achievement of such a stage by such a date is a condition of the subcontract will there be any liability on the subcontractor. Any breach of the subcontract arises not from the subcontractor having difficulties for which it is liable, but from its not recovering from them within the subcontract period. To this end it may be argued that the subcontractor must be allowed float, where it so plans, within its subcontract period for its recovery from those difficulties that it can reasonably anticipate could occur.

If a delay to the completion of the subcontract works occurs, then liability will ultimately depend on all the circumstances. To recover damages from the subcontractor the contractor will need to demonstrate a breach of an express condition of the subcontract by the subcontractor or alternatively a breach of a matter which flows naturally from the subcontract or was in the contemplation of the parties at the time of the subcontract and which has led to the loss or losses claimed. Just to be able to show that the subcontractor had problems of its own making or at its own risk, such as a shortage of resources at one time and/or that some defective work had to be rectified, does not constitute a breach of the subcontract or demonstrate the cause of the delayed completion or that the delay led to the loss being claimed by the contractor. The subcontractor's obligation is to take such steps as are necessary to recover the consequences of its own problems. It is suggested that where delays are due to differing factors or events, some due to the subcontractor and some to others, then any recovery action implemented by the subcontractor at its own cost to restore the situation will, where the lost time is regained, in the first instance, be regarded as recovering the losses for which the subcontractor is liable.

In considering possible alternative obligations under the subcontract, it will be essential to have regard to the nature of the work and the interface with others. It will often be more important for the subcontract works to progress so as to enable the next trade to be able to commence, a specified period after the release of a unit of production to the subcontractor, than for the subcontractor to have an obligation to complete the entire works in an express period. It is suggested that in many, if not most, subcontract arrangements, the contractor is more likely to suffer loss due to unsatisfactory progress during the duration of the subcontract works, than because the works are not entirely complete on the due date. However, unless there is an express requirement for the subcontractor to progress at a certain rate or achieve agreed milestones by specified dates, there will be no breach of the subcontract on which to claim, as damages, the loss suffered.

Programme where the subcontract period is not defined

Where no express agreement exists for the subcontract works to be carried out by a certain time or within a finite period, then time is said to be at large. This is taken to mean within a reasonable time, and a reasonable time is stated to mean reasonable in all the circumstances, and certainly will not be held to mean that the subcontractor is to work particularly fast or that he will warrant against any delays that occur in construction work, such as resources not being available on specific days, delays due to inclement weather, work being constructed which requires rectification and in circumstances where there has been delay by the contractor or the deployment of resources to work for which there is a pre-commitment elsewhere. However, the subcontractor may be liable to the contractor if the subcontractor fails to proceed at a reasonable rate. In the event of there being exceptional problems arising when there is an undefined period for execution of the subcontract works, there may be an implied term for the subcontractor to notify the contractor and seek his instructions as to whether to take accelerative and/or other measures.

It is of course possible for the parties, at any time, to agree a programme for the subcontract works. However, unless there is some consideration for that agreement, then it is unlikely to have contractual effect. This was the effect in the case of *Lester Williams* v. *Roffey Brothers & Nicholls (Contractors) Ltd* (1989) 48 BLR 69, referred to in the section 'Sectional completion and partial possession' in Chapter 10, where an agreed addition to the subcontract sum was held to be in consideration for the subcontractor agreeing to work in a defined sequence.

Programming following delay

Particular problems arise where the contractor's programme is running late and where it is not recognised that the Works programme is to be extended proportionally. This will be the situation where the delay is due to a default by the contractor and for a reason or event not leading to an extension of time under the main contract. Under such circumstances, the subcontractor's obligations to accelerate must be considered. These may include requirements for the subcontractor to reprogramme so as to recover the time lost and/or to work in changed circumstances, such as in more restricted and limited areas, overlapping its work more with the work of others and/or working extended hours. All such actions will be likely to lead to a reduction in productivity and consequently increased costs to the subcontractor. Rarely will there be any pre-arranged obligation or agreement for the subcontractor to spend money to recover losses due to reasons not of its own making or at its own risk. Without an express instruction and/or instructions from the contractor, it is unlikely that the subcontractor will have any right to recover such additional costs. It is considered that the issue of short-term programmes or revised programmes to completion could be held to be instructions, by the contractor, to vary the rate and/or sequence of construction from that required by the programme as previously agreed, issued or instructed. If in doubt, the subcontractor should confirm receipt of such changed requirements as an instruction and/or as a relevant event where there is a loss and/or expense provision within the subcontract. As mentioned earlier, it is doubtful if there is any right to continue to incur loss and/or expense for a continuing situation or circumstance without an express instruction from the contractor and consequently no right to recover such extra expenditure.

As mentioned above, the only express action available to the contractor under the SBCSub/C form of subcontract following delay to the regular progress of the subcontract works is to extend the period for the subcontract works. This is rarely the action that the contractor will wish or require. Frequently, even if not as a normal requirement, the contractor may require the subcontractor to compress and reduce the period for carrying out of its work. Other than the general right to issue directions, under clause 3.4 of SBCSub/C, there is no express provision for the contractor to instruct or require the acceleration of the subcontract work. Indeed, not only is there no express provision for the contractor to instruct such action, clause 2.18 makes the granting of an extension of time mandatory by the words 'the Contractor shall give an extension of time'. It seems, therefore, that whatever other directions the contractor may wish to give or in fact gives to the subcontractor, it must in addition grant an extension of time. The requirement for the contractor to extend the subcontract period is stated in both clause 2.18.2 and clause 2.18.4 to be no later than 16 weeks from receipt of a notice from the subcontractor. This is a wholly impractical period, with the possible exception of projects of very long duration, for the notification to the subcontractor of an extension of time, if the contractor is to keep control of the work. Indeed, on projects of whatever duration the contractor will need to instruct all its subcontractors, as soon as a delay to the works becomes apparent, what its changed requirements are.

Sub/MPF04 goes some way to assist the contractor to recover lost time under section 17, acceleration, where there is provision for the contractor to invite the subcontractor to make proposals for accelerating its work. Clause 17.1 provides that the contractor may, if it wishes to accelerate the works, request proposals from the subcontractor as to the possibility of achieving practical completion before the expiry of the period for completion. The subcontractor is to either:

1. make such proposals accordingly, identifying the time that will be saved and any additional costs that would be incurred; or
2. explain why it is impracticable to achieve practical completion at an earlier date.

Clause 17.2 allows for the contractor either to accept the Sub-Contractor's Proposals or to seek revised proposals. If the contractor wishes to accept the Sub-Contractor's Proposals, it must issue a change instruction identifying the agreed change to the period for completion and the additional cost.

Clause 17.3 restricts the contractor's right to reduce the period for completion of the subcontract work to two express provisions under clause 17 (Acceleration) and 22 (Cost savings and value improvements).

It is surprising that there is no express provision for the contractor to take the initiative and a more active role in the recovery of lost time, especially having regard to the contractor's responsibility for the overall coordination of the various trades. A more positive provision might have been anticipated, enabling the contractor to issue any instructions the contractor regarded as necessary to recover lost time, with provision for the subcontractor to indicate both the consequences of such instructions and its ability to comply.

In practice, it will frequently be the case that the full details of the subcontractor's work will not be known at the time the subcontract is agreed. Further, the contractor may not have had any or sufficient opportunity to plan in detail all the inter-trade interfaces. By placing an obligation on the subcontractor to commence at a stated time subject to notice and to complete the subcontract works within an express period, it is likely that without express words to the contrary, the subcontract will have implied into it an obligation that the work would be made available in sufficient time for the subcontractor to carry out its obligation in an ordered manner and at a reasonably steady rate. Further, unless the subcontract expressly indicates to the contrary, the subcontractor will be able to claim that its expectation was to have the whole of its work available on the commencement of its work.

Where the contractor fails to advise in reasonable time its inability to release elements of the subcontract works on time, it is likely to forfeit its right to have the subcontract works completed in the agreed period or by a specific date. Further, the contractor may itself be liable to charges from the subcontractor whether by way of financing charges on material obtained and/or manufactured ahead of being able to install them, or for labour allocated for the project that cannot at a late date be redeployed.

Delays and prolongation events occur in a number of forms and, subject to the express provisions of the subcontract, may have different effects:

1. The subcontracted works as a whole may be deferred but without affecting the period within which the works are to be completed. This will most commonly be the case with specialist components with a relatively short on-site duration, such as entrance doors or gates.
2. The subcontract works may proceed generally as intended but where no allowance has been made to leave out elements for completion later in the programme, there will be a delay to the subcontract works as a whole. Elements of external cladding left to provide hoist access or rubbish disposal, or glazing left to allow for scaffolding ties, are common examples.
3. The subcontractor may be required to commence on or about the anticipated date but work is only made available to it at a much slower rate than necessary.
4. For reasons not anticipated at the time the subcontract was concluded, the contractor may require the subcontractor to carry out its work in a radically changed manner, such as when work originally to have been completed as it progressed is to be carried out in several stages involving repeated visits to the same area and/or a considerably extended subcontract period, or where design development has incorporated more or changed interfaces between trades.

Subcontractor's programme

At any time, it is the right and in many cases the obligation of the subcontractor to plan its work and to submit such programme or programmes to the contractor for his information, advice or approval.

While many subcontracts require the preparation of programmes by the subcontractor, few subcontracts consider the effect of such programmes. The likely intention of a requirement for a subcontractor to submit its detailed programme for its work, is to allow the contractor to coordinate the requirements of its various subcontractors with each other and/or integrate their work with that of the contractor. In practice, it will be extremely difficult, if not impossible, for the contractor to achieve this aim. The contractor will almost certainly appoint its various trade subcontractors progressively and must either postpone accepting any programme until all are available or it will risk finding that the time available for the later appointed trades is unacceptable to them.

There is no right, without express provision within the subcontract, for the contractor to reject a subcontractor's programme that is compliant with the requirements of that subcontract. Where those requirements are limited to completing the subcontract works within a specified time from commencement on site, there can be no grounds for rejecting the proposed programme unless it overruns that period. Thus any changes the contractor wishes to make will require a variation instruction. Alternatively, the contractor will risk being in breach of the subcontract by failing to give access to the various sections of the work in due time.

If and when the contract works become delayed, the contractor will frequently seek the Sub-Contractor's Proposals to complete on time. It is generally the case that at least some of the subcontractor's work will be dependent on work by others. The subcontractor's response is often to request dates from the contractor as to when the work by others will become available. It is submitted that the subcontractor should, as requested, plan its work to completion, identifying the dates for release of work by others necessary to achieve that programme. It will then be for the contractor to obtain the necessary performance from others or issue to the subcontractor instructions as to the contractor's requirements, if necessary, by adjusting the date for completion of the subcontract works.

Extensions of time

The need for extensions of time arises from the use of pre-ascertained and liquidated damages (LADs). The granting of such extensions of time by the employer will relieve the contractor from an automatic liability, to a fixed rate of damages, for failure to complete the works by the date fixed in the contract. Such arrangements are appropriate to and generally incorporated into main contracts. In those circumstances, the contractor is responsible for the entire work and may, within certain stated restrictions, carry out the work to his own programme at his chosen rate and sequence, his only obligation being to complete by the due date.

The work of subcontractors, especially those for building services and finishing trades, will interface with the work of others. Attempts are made by contractors to incorporate within their subcontract terms limits as to the time within which the subcontractor is to complete, while at the same time reserving to the contractor flexibility to manage and coordinate its various sub-trades and activities. The extent to which this can be successfully achieved is doubtful, as was clearly decided in the judgment in *Pigott* v. *Shepherd*.

It will rarely be failure by a subcontractor to complete by the date for completion, stated in the Sub-Contract Documents, that will lead to loss to the contractor. Such loss will arise from the contractor's inability to proceed in a timely and effective manner with the work of the following trades. This was the case in *Pigott* v. *Shepherd*, where because piling works were carried out in a dispersed sequence, the construction of pile caps could not proceed until the subcontractor's work was almost complete. The losses suffered by the contractor were in the first instance those that arose from its inability to commence pile caps at the programmed date, and not because the subcontractor failed to complete its work within the subcontract period. It is for reasons such as this that subcontracts rarely incorporate an LAD clause, but rely on general or common law damages, which may be capped by an express term in the subcontract. Such damages will require the contractor to demonstrate causation and to prove actual loss, in addition to a breach of the subcontract terms. Where the subcontractor only has an obligation to complete in a fixed time, damages for the subcontractor's delay can only arise as a direct consequence of such failure.

The SBCSub/C form of subcontract at clause 2.17 sets out the requirements for the subcontractor to notify the contractor of delays and/or likely delays in three sub-clauses:

1. Should the subcontractor consider that the completion of its work is likely to be delayed, then the subcontractor is to give the contractor written notice setting

out the circumstances and so far as it is able the cause of the delay and where applicable any event which the subcontractor considers to be a 'Relevant Subcontract Event'.
2. The subcontractor is also in writing to give details of the effect of each event, including the estimated delay to the completion of its work.
3. The subcontractor is also to notify the contractor, again in writing, of any material change to its estimate of delay or any other information the contractor may reasonably require.

The requirements of ICSub/C are defined within clause 2.12.1: 'If and whenever it becomes reasonably apparent that the commencement, progress or completion of the Sub-Contract Works or of such works in any Section is being or is likely to be delayed, the subcontractor shall forthwith give written notice of the cause of the delay to the Contractor'. The provisions of Sub/MPF04 at clauses 16.2 and 16.3 have been set out in the section 'Programme changes and amendments' earlier in this chapter.

In practice, the main purpose of a clause 2.17 notification is to draw the contractor's attention to actions and/or inactions that may lead to a delay to the subcontractor's work. This will enable the contractor to take action, as appropriate, including a similar notification under the main contract, with the intention of his obtaining an extension of time and receiving from the contract administrator relief from the liability for LADs to the employer.

The requirement for notices under SBCSub/C clause 2.17 is complex and must be considered in conjunction with item 5 of the Sub-Contract Particulars set out in SBCSub/A. For the reasons set out in Chapter 10, Completion, it is for the contractor to issue a notice or notices for the commencement by the subcontractor of the work on site. Until the subcontractor receives that notice or notices, the subcontract commencement date will not have been established. Condition 2.3 makes the subcontractor's obligation to carry out and complete the works 'in accordance with the programme details stated in the Sub-Contract Particulars (item 5) and reasonably in accordance with the progress of the main contract Works or each relevant Section of them, subject to receipt by the subcontractor of notice to commence work in accordance with those particulars and subject to clauses 2.16 to 2.19'. Further, it is considered that such notice to commence will only be effective where the work to be available on a specified date is clearly defined. A general indication that a start may be made on site is not a notice that is compatible with the programme provisions of item 5.2 of the Sub-Contract Particulars, unless accompanied by a detailed schedule of release dates, either under item 5.4 of the Sub-Contract Particulars or as part of the notice to start on site.

The express provisions of SBCSub/C only require the giving of notices of delay by the subcontractor to the contractor, and by so doing reflect the requirements for notices under the main contract from the contractor to the employer. In practice, for many of the events listed in clause 2.19 of SBCSub/C, relevant subcontract events, the contractor will or should be aware of such events before the subcontractor. Matters where the subcontractor will have prior knowledge to the contractor will be those concerning the subcontractor's off-site and/or pre-site works, including matters delaying the provision of drawings and/or materials for the subcontract works.

Few, if any, subcontracts, and SBCSub/C is no exception, place any obligation on the contractor to keep the subcontractor notified of delays likely to affect the subcontractor's commencement date and/or progress on site; such a requirement is provided for by clause 14.3 of Sub/MPF04. The absence of such express requirements does not reduce the importance of subcontractor's being kept informed of all matters likely to affect or change either the subcontractor's ability to perform as planned or the contractor's actual requirements by way of performance by the subcontractor. It is inevitable that the subcontractor's ability to speed up, delay or change the sequence of its work will be enhanced if it is given the maximum advance notice. In addition, the resultant costs and/or loss in complying with such changed requirements will be more easily controlled if advance notice has been given.

Since the actual programme for the subcontractor's on-site work will only be defined and established following the issue of the notice or notices to commence work under clause 2.3, any requirement for the subcontractor to give notices under 2.17 will only relate to commencement, progress or completion of the subcontract works following the receipt of such notice or notices under clause 2.3, by the subcontractor.

It is suggested that the provisions of SBCSub/C will only be effective where the sections of item 5 of the Sub-Contract Particulars in SBCSub/A are completed. Further, where the contractor does not intend to release the whole of the subcontractor's work at the end of the period of notice to commence work on site, then the contractor will need to identify, under item 5.2, the time intervals at which the balance of the work will be released to the subcontractor. In the event that items 5.1, 5.2 and 5.3 are completed and no further detail is incorporated, there will be a presumption that the entire subcontract works will be available at the end of the period of notice. Should that not be achieved, then time may become at large, or the subcontractor have the subcontract period from the release of the last part of its work, in which to complete the work.

The subcontract clauses 2.16 to 2.19 all reflect the provisions of the JCT forms of main contract and seek to make the subcontract 'back to back' with the main contract. The effect of these provisions is discussed here.

Clause 2.18.2 requires the contractor to notify to the subcontractor within 16 weeks after receiving a notice under clause 2.17, details of the revised period for the subcontract works. In practice, delays on site are much more complex than can be resolved just by the granting of additional time to the subcontractor. They will generally require a rescheduling of the works, which must be coordinated with the work of others. Where the work schedule slips, for whatever reason, the contractor will need to keep the subcontractor advised of its revised and/or updated requirements, otherwise there is a probability that time may become at large and the subcontractor's obligation will become to complete in a reasonable time only. No directly relevant law is known, but the case of *Walker and Another* v. *The London & North Western Railway Company* [1876] Common Pleas 518, which was a main contract with no provision for an extension of time, gives some guidance as to the consequence of not maintaining an effective completion date. It is suggested, in the case of a subcontract, that there is also the need for an effective programme defining the contractor's requirements, following any delay to the progress of the subcontractor's work. In that case the court found at page 531:

'Delay was in part occasioned by the act of the board in ordering extra works and otherwise, and it was held that the board were, notwithstanding, entitled to determine the contract and take possession of the works, but there it was assumed that the clause had been put in force before the time originally specified for completion had arrived and before any extension of time had been given.

The clause in our opinion can only be acted on and enforced within the time fixed for the completion of the works, for time is clearly of the essence of the contract, and it is only with reference to the time so agreed that the rate of progress can be determined. If, as has happened, the time has been exceeded, there may be a new contract to complete in a reasonable time; but to give the clause in question any application to a reasonable time after the time originally fixed has expired, would be, without any express provision, to make the company judge in their own case of what was a reasonable time, and to enable them in their own favour to avail themselves of a most stringent and penal clause.'

The relevance of this to subcontract conditions is in regard to changes in the contract programme/schedule. Once the situation on site becomes different from that envisaged, either by the contract or subcontract programme or further later instructions as to programme, then the contractor will not be able to claim that the work should have been carried out to certain dates unless, it is suggested, it has reasonably adjusted the programme/schedule and kept the subcontractor advised of the contractor's current requirements.

Unless a programme of works is incorporated within the subcontract, as a subcontract document, clause 2.3 of SBCSub/C will not by itself place on the subcontractor the obligation to carry out its work to a sequence or schedule subsequently decided on by the contractor. This was the effect of the judgment in *Pigott* v. *Shepherd*, quoted above. Where no such programme is incorporated within the subcontract, or where there is such a programme but delays, for whatever reason, make it no longer relevant, then the contractor must issue its instructions or risk the time for the subcontract works becoming at large. Such instructions will be subject to performance and valuation under the appropriate provisions of the subcontract. The limited obligation of the contractor to grant an extension of time within 16 weeks will be of no benefit to it if it has failed to issue necessary instructions as described above. In this respect, it must be acknowledged that only on the largest projects or for subcontracts of extensive or wide-ranging works will a 16-week period for the granting of an extension of time fall before the expiry of the subcontract period. Further, the issue of instructions to work to a revised programme will themselves, in all probability, for the purposes of clauses 2.18 be the giving of an extension of time.

Clause 2.21 requires the subcontractor to allow to the contractor 'the amount of any direct loss and/or expense suffered or incurred by the Contractor and caused by that failure'. That failure is the failure to complete within the contract period. This may be taken as limiting the contractor's damages to those arising from failure to complete by the date for final completion. The subcontract makes no express provision for recovery of loss and/or expense resulting from breaches of subcontract by the subcontractor during the progress of the subcontract works, except for failure to complete any section, even if the subcontract contains appropriate provisions that may lead to such breaches being identifiable. By including and limiting the provision in clause 2.21 to a failure to complete, it is at least arguable that the parties are agreeing to limit liability to this express circumstance.

Since, as mentioned above, it is likely to be the release of areas of the work to other trades which will be of real significance to the contractor, such a limited right to loss and/or expense will be of very limited benefit to the contractor and would appear to deny the contractor any ability to recover the losses most likely to arise from slow progress by a subcontractor. Such loss to the contractor may be recoverable by agreement under clause 4.21, which relates to the regular progress of the main contract being affected by any act, omission or default by the subcontractor. Of course it will be necessary for the contractor to show the provision within the Sub-Contract Particulars on which it seeks to attribute such act, omission or default.

Financial planning

At all levels in the construction chain there is a need to provide finance for the project. For the subcontractor this will include any pre-site planning and procurement work, the cost of design and/or component manufacture, the purchase of materials, and hire of labour and plant, together with the costs of managing and running the business. Some of these costs will be financed by the subcontractor's credit facilities with their suppliers, and the payment of the workforce and staff in arrears. Further finance will be available from the subcontractor's own capital and from its bank by way of overdraft facilities.

The subcontractor will be able to anticipate the expenditure necessary to comply with the subcontract and compare this with its expected recovery by way of the valuation of the work and payments from the contractor, provided there is a defined programme for the work and an agreed payment mechanism. Where these are not fully defined it will be for the subcontractor to define them in its offer or clarify these matters during the pre-contract negotiations.

In some subcontract arrangements there is an express requirement for the subcontractor to provide an 'expenditure chart', which will define for each payment period the subcontractor's anticipated valuation. Where this is not an express requirement the subcontractor may consider that such a schedule should be prepared and be included as a subcontract document. In any event, for most building trades expenditure will be reasonably proportional to the amount of work carried out and therefore it will not be difficult to make an assessment of the likely valuations.

Finance must be considered a resource that may affect the regular progress of the work, just as much as the supply of labour or materials. Should the contractor, by failing to pay in accordance with the subcontract provisions, deny the subcontractor the resource necessary to proceed regularly with the work, this will be an act of prevention by the contractor, which may relieve the subcontractor of its obligation as to progress under the Sub-Contract Agreement and/or payment of its loss and/or expense flowing from such a breach by the contractor.

Where the contractor issues variation instructions or the subcontract value becomes increased for any other reason, then the express provision of SBCSub/C is for the contract period to be extended. Should the contractor wish to issue directions for the subcontracted work to be carried out in a reduced period, the subcontractor's ability to comply with such a direction may depend on obtaining the necessary financing required. In extreme circumstances such a direction may give the subcontractor reasonable grounds to consider that the direction is unreasonable, and refuse to comply. The provisions of SBCSub/C clause 3.5.1 strictly limit the subcontractor's right to object to a variation as defined at clause 5.1.2, which is limited to change to the kind or standard of materials or goods. It will then be for the contractor to decide what action it wishes to take, which could include relieving the problem by way of advanced or more frequent payments to the subcontractor.

Extensions of time under other subcontract arrangements

Some contractors seek to limit the subcontractor's right to an extension of time to that granted under the main contract, by use of clauses such as the following sample:

> The Sub-Contract Works are to be commenced on site within seven days of a written instruction to proceed and are to be completed within the Subcontract period subject only to such extension of time as the Main Contractor may allow where the Sub-Contract Works are delayed by causes which result in the granting of an extension of time under the Main Contract.

Such clauses only partially respond to the purpose and nature of an extension of time provision, which is to keep alive the provisions of the subcontract and the contractor's right to damages in the event of the subcontractor's breach. Since the subcontractor may also be delayed by matters for which the contractor is responsible, by the incorporation of such a clause the contractor is prevented from maintaining an effective subcontract programme in the event of the contractor's own default, which includes defaults of other subcontractors.

However, in the subcontract situation it is not so much the granting of an extension of time that will preserve the contractor's rights, but the contractor ensuring that at all times the subcontractor knows in detail what is required of it. It is for the subcontractor to be managed by the contractor under the provisions of the subcontract, not for the subcontractor to second guess the contractor's requirements.

Coordination

Many subcontracts incorporate a requirement for the subcontractor to coordinate its work with that of other subcontractors. There must be considerable uncertainty as to the meaning and effect of such a clause. What it cannot mean is that a subcontractor is empowered to instruct another to carry out its work in a certain manner or by a certain time, or any other matter.

At most, it is suggested, it is a requirement for the subcontractor to notify or advise other subcontractors of its intentions and to request appropriate action and/or advice where such is not possible. Only the contractor has a contractual link to each subcontractor and therefore the contractor is the only party able to manage interface problems and instruct as necessary.

Chapter 5
Organisation and Management of the Subcontract

Introduction

The organisation and management of the subcontract works for both contractual and practical reasons require the positive involvement of both contractor and subcontractor and are certainly not to be regarded as matters entirely handed down to the subcontractor. This statement is made despite many formal subcontracts being written as though the contractor had no further involvement in the construction process after the subcontract was concluded, other than to consider requests, notices and applications from the subcontractor.

Contrary to much perceived opinion, organisation and management of the subcontracted work will differ considerably from trade to trade and indeed from project to project. Both parties to the subcontract must identify the express aims and objectives necessary for individual subcontracts, and should not be restricted in their organisation and management either by the strict requirements of the form of agreement or by experience derived from previous projects.

However much of the total project is sublet by the contractor or retained by it to be carried out by its directly employed operatives, the contractor's main involvement is to procure the work from subcontractors who have the necessary facilities and/or resources and expertise to carry out the work required, and then to integrate that work into the project as a whole. The subcontractor's obligation is to provide the necessary labour, plant and materials to carry out the subcontracted work in the time and to the quality to satisfy the subcontract conditions and/or the contractor's requirements. This will require the integration of work on a number of projects so as to fulfil the subcontractor's obligations to all its clients.

These primary objectives will exist whether the subcontract is formalised in extensive documentation or subcontracted by way of a simple acceptance of an offer or in response to a request to execute work. It will not be the intention of the parties, at the time the subcontract is placed, that the work will be significantly changed or severely delayed, but such situations do frequently arise and it is right that the parties consider what is to be done in those circumstances.

It is the attempt to make provision for such changes that has led to the development not only of industry standard forms of subcontract, such as SBCSub/C, but to the extensive range of main contractor's own forms. Experience shows that the application of standard forms, whether an industry standard or otherwise, without detailed consideration of the specific trade and project circumstances may not lead to satisfactory arrangements for the contractor to be able to manage the subcontractor as the contractor requires, should difficulties arise.

Enquiry and tender

It is common practice in the UK for enquiries to subcontractors to be made at the time the contractor is tendering for the main contract. Because of the limited time available for the tendering process as a whole, enquiries to specialist subcontractors will be very basic, often little more than an extract from a bill of quantities, the appropriate specification pages and relevant drawings together with an enquiry letter, which may indicate the likely timing for the subcontract work, intended form of subcontract and other relevant data.

It will rarely, if ever, be the prospective contractor's intention to enter into any agreement or provisional acceptance of the subcontractor's priced quotation prior to the contractor being awarded the main contract. For certain trades, in particular piling where specialised equipment will be required on site shortly after appointment, the prospective contractor may request details of lead times so as to be able to qualify its offer as necessary.

A prospective subcontractor receiving an enquiry will know that the chance of its tender becoming a live subcontract will not be very high, except where the specification is exclusive to the work being done by others. Should the subcontractor receive requests for prices from more than one contractor, it may rate its chances proportionally better. Its objective must be to give an attractive price so that the contractor is likely, in the event of its being successful under the main contract enquiry, to come back to the subcontractor with the intent of finalising an agreement. However, the prospects of such an outcome will vary considerably from trade to trade and according to the anticipated time lag between the appointment of the contractor and the commencement of the subcontractor's work. Early trades and those with significant lead times will be most likely to be seriously considered upon the appointment of the contractor, together with those providing particularly specialised services. Trades such as plasterers and painters will have low priority for early appointment, and considerable scope for the contractor to obtain more competitive quotations.

The prospective contractor, having received quotations in response to its enquiries, must decide how to incorporate these into its own tender for the main contract. It will hope to be able to negotiate with the subcontractor whose quotation is lowest, once the contractor has a firm contract, and to conclude with the subcontractor an agreement for a lower price or rates and/or for a discount. In making any such allowance in its tender the prospective contractor takes the risk that by the time of its appointment the subcontractor may have filled its order book or have reviewed its price and found that it has underpriced the work, and withdraws its offer, or that the market place has changed and prices have risen.

Alternatively, on being appointed the contractor may re-tender the work of its trade contractors knowing that it is more attractive for prospective subcontractors to respond to an enquiry related to an actual appointment, rather than to a tendering contractor.

In any event, the appointed contractor will generally wish to review the offers received from specialist contractors, both with regard to the rates and price and in respect of more certainty as to the timing of the subcontract work and the terms of the subcontract. The technique used is often that of the carrot and the stick, by suggesting that, while the subcontractor's offer is attractive, the contractor is considering other offers, and hinting that a reduction in price or the acceptance of an additional obligation would probably make the subcontractor's offer successful.

A subcontractor's offer described by a different term, such as quotation, estimate or even budget price, unless clearly stated that it is not intended to be an offer open to acceptance, will be considered an offer in the terms as stated. If for any reason a potential subcontractor requires to review its offer before acceptance, then it must make this very clear. If, for example, the price is only to be regarded as 'provisional', then this must be clearly stated, and the prospective subcontractor should state what is necessary or the procedures to be adopted so as to finalise the price in the event that it is required to commence work before full agreement is reached. Typical matters that may be expressly excluded from a subcontractor's offer may include:

1. a knowledge of the site, where the subcontractor has not, as requested, visited the site, prior to submitting its tender;
2. similarly, a knowledge of the main contract conditions;
3. availability of labour, plant and/or materials at the time of acceptance.

On the other hand, the subcontractor's offer may be based on express conditions of its own. For example, a subcontractor whose production facilities have a shortage of work during a specified period, may make its offer expressly on the basis that production will be carried out at that time. The subcontractor may also make its offer on the basis of stated general conditions of subcontract, which may be either its own in-house conditions or an industry standard such as SBCSub/C. It is suggested that while specific conditions to the subcontractor's work or method of working should be seriously considered, the proposed use of general conditions biased towards the subcontractor will rarely be entertained by a contractor.

It will be normal good practice for a tendering company to review their quotation immediately before finalising the offer, to ensure that nothing has been overlooked with regard to underpricing the work and possible cost savings. In addition, there may be commercial adjustments to the estimate to make the offer more competitive or because of anticipated difficulties specific to the project. It is again good practice for a detailed record to be kept of the reasons for any such adjustments, especially if these involve a lump sum or percentage adjustment to the original sum. This is important since at a later date it may be necessary to demonstrate the basis of such adjustment when it comes to the valuation of variations. Such adjustments may be regarded as a one-off commercial settlement or an adjustment to the entire tender, or may be intended to relate to a certain aspect of the work only.

Post-tender, pre-subcontract

As mentioned earlier, a subcontractor's tender may not be fully compliant with the contractor's requirements, either because those were not fully defined in the tender enquiry or the subcontractor, for whatever reason, was unable or unwilling to consider all the matters requested. This will be especially true where the contractor's enquiry has been made prior to it being awarded the main contract, or where the subcontractor's offer was initially prepared in response to an enquiry from another contractor.

For the contractor it will be important to assess the balance between obtaining the most advantageous Sub-Contract Agreement as to price and terms, and the need to place the subcontract in time for the subcontractor to be able to obtain the necessary resources and mobilise labour, plant and materials so as to meet the requirements of the project programme. Where the subcontract includes design development to be integrated with the work of others and/or there is significant off-site manufacturing work, then the final date necessary for such work to be commenced, so as not to delay work on site, may be difficult for the contractor to assess and is frequently underestimated, with resultant delays to the project or the need to instruct accelerative measures at additional costs.

What is certain is that as the date for commencement of the subcontract work, whether on or off site, approaches so the strength of bargaining power moves from the contractor to the subcontractor. Should the contractor instruct or request a commencement of the subcontract work under a provisional subcontract, such as a letter of intent, there will be little incentive for a subcontractor to accept any amendment to the terms of that provisional subcontract that is to its disadvantage. It is unrealistic for a contractor to believe that, after the issuing of an instruction for the subcontractor to commence work under a provisional subcontract, because the contractor wishes to negotiate better terms at a later date it will be able to obtain from the subcontractor agreement to terms that the subcontractor was not prepared to accept prior to an instruction to commence.

It not infrequently happens that a contractor seeks to conclude a subcontract on its terms by issuing a subcontract order incorporating such terms. Such orders will, in many if not most cases, be a counter-offer and if not acted on by the subcontractor will be without any contractual effect. The contractor may consider it has a valid subcontract for the subcontractor to perform in a certain manner and to a certain programme, whereas in fact there is no subcontract at all. In such circumstances, if after a significant period of time and before commencing work, the subcontractor seeks to return to the negotiating table, the contractor may find itself in a cleft stick of having to accept the subcontractor's terms or seek too late an alternative subcontractor.

It cannot be emphasised too strongly that it is in a contractor's best interests to concede to terms in order to reach an agreement, even on less than ideal terms, while it still has the upper hand and still has other options available to it, rather than to delay the conclusion of a subcontract and place the contractor in a position of 'take it or leave it' from the subcontractor. If a concluded agreement is not reached in time, then the contractor will be at the mercy of a subcontractor who has the upper hand, whether or not the subcontractor's circumstances have changed during the period of negotiation. Such changes may include having obtained work from other contractors or now having a limited period for its own negotiations with its suppliers and sub-subcontractors.

From the subcontractor's position, once a contractor has concluded an agreement for the Works by way of the main contract, there becomes a greater level of possibility of the subcontractor concluding a subcontract. The subcontractor will generally be prepared to review and reconsider its pricing to make savings and thus improve its offer to the contractor, but at the same time the subcontractor must take whatever steps it needs to ascertain and/or agree details of all matters left outstanding at the time of its original tender. Such matters include:

1. the terms of the main contract;
2. the proposed terms of the subcontract especially any amendments to a standard form prepared by the contractor;
3. site arrangements;
4. programme for the subcontract work;
5. facilities to be provided by the contractor.

Most formal subcontract terms incorporate a clause to the effect that the subcontractor is deemed to know the requirements of the main contract. In practice, the idea that the subcontractor has in fact visited the contractor's office and made itself conversant with all relevant terms of the main contract just does not happen. The subcontractor should make it clear that it is not conversant with the terms of the main contract except as expressly stated in the subcontract document.

Many enquiry documents make reference to a standard form of subcontract such as SBCSub/C, but in addition to the contractor's schedule of amendments. Frequently such amendments are only supplied, if at all, at the time the subcontract is concluded or offered for signature. It is important for the subcontractor either to obtain details of such amendments and to assess the effect of the incorporation of such changes before finally concluding an agreement, or to reject their incorporation into the subcontract by deleting reference to such amendments. It must be remembered that the main purpose of making changes to a standard form is for the contractor to move the balance of the subcontract in favour of the contractor and to the subcontractor's disadvantage. By presenting the amendments as word changes the true effect is concealed and only by working through the whole subcontract will the effect be understood.

Many of the changes most commonly made by contractors will reduce the subcontractor's rights and compensation in the event of the contractor's failure to carry out its obligations under the subcontract; for example, reducing the rate of interest for late payment or the need to agree loss and/or expense before deduction, or a requirement for the referring party in an adjudication to pay all the costs. It might be suggested that the only conclusion to be drawn from the contractor's requirement for such changes to the standard form is that the contractor has no intention of fulfilling its obligations as set out, since if its intention was to perform the subcontract strictly to those terms, such amendments would be unnecessary.

There may be very specific conditions in the main contract which the subcontractor needs to be aware of and to have fully considered as part of its offer. The contractor may have good reason for wishing to deviate from the standard form, either because of the particular circumstances of the project or to be compatible with its in-house procedures and systems. Where such circumstances exist, the contractor could reasonably have been expected to set these out clearly at the time of the original enquiry.

The problems relating to the programming of the subcontract works and the differences between individual trades have been discussed in Chapter 4. Even if it is not possible to set out in advance dates for the various activities forming part of the subcontract, it is important that each party fully sets out its own requirements and fully understands the needs of the other party. It not infrequently happens that different companies operating in the same trade will programme and organise their work in very different ways, to the extent that a matter which is critical to one company will not be of any concern to another.

The facilities to be provided by the contractor are frequently expressed in very general terms, such as provision for 'the use of standing scaffolds' or 'shared use of tower crane'. The subcontractor needs to know that the facilities being provided will be sufficient for its needs and sufficiently available to enable it to carry out its work within the period set for it by the contractor. While shared facilities mean that there will be other users, the subcontractor needs an undertaking from the contractor that it has made the necessary calculations and had sufficient negotiations with its other trade users to be able to assure the subcontractor that the facility will be sufficient for its needs.

It is common practice for a contractor to hold a 'pre-contract' meeting. In an ideal situation the purpose of such a meeting will be to review the agreement reached and ensure that the understanding by both parties of the agreed terms is the same, and that there are no misunderstandings or matters to be finally resolved. It is good practice to have present at such a meeting the senior members of both the contractor's and subcontractor's construction teams, as well as the parties' negotiators.

Unfortunately, in many instances the 'pre-contract' meeting is used as a final opportunity for the contractor to impose its requirements on the potential subcontractor. While a pro forma agenda is useful to ensure all necessary matters are discussed, many such 'agendas' are written in the form of notes of the meeting and may in fact be headed as such. These forms are completed during the meeting, and at the end of the meeting the subcontractor may be expected to sign them as the record of agreement. Such forms are normally several pages long and a subcontractor should decline to sign until satisfied the matters stated reflect its intentions and not what the contractor wishes the subcontractor had agreed. In such a situation it may be advisable to require time to reflect on the contractor's notes before accepting them as a record of agreement.

The use of 'pre-contract' minutes in this way is to turn them into a new offer, which is accepted by the subcontractor at that time or upon commencement of its work, and is not a review of an existing agreement before it is finalised in formal articles. Also, such minutes are prone to fraud by matters not being accurately recorded but biased to the contractor. This is even more likely, if not probable, where the pro forma contains standard wording reflecting the contractor's requirements.

The subcontract

The various methods by which parties may reach a binding subcontract have been discussed in Chapter 2. Where there is doubt as to the existence of a subcontract, the attitude of the courts favours the existence of a subcontract where the work has been performed; where performance is resisted the courts will require a clearly enforceable agreement.

Frequently, therefore, it is performance of the subcontract that is the act of acceptance that forms the agreement and binds the parties. It is therefore good practice for there to be a senior manager or director within the subcontractor's organisation who is required to authorise the commencement of the work, thereby taking steps to ensure a subcontractor does not become bound to the contractor other than under circumstances and on conditions that it has understood and accepted.

As mentioned elsewhere, many subcontracts are commenced under letters of intent or other provisional subcontract arrangements, which may or may not be formalised at a later date. It must be fully understood that once there is a binding subcontract, whether provisional or not, there is no obligation for either party to accept alternative conditions proposed or requested by the other. If after commencing work on a provisional subcontract the subcontractor is subsequently sent a formal agreement, it will have no obligation to sign it. Such agreement will either be strictly limited to the terms that exist under the provisional arrangements, when the parties' situation will not be changed by the formalisation or the 'agreement', or will offer different terms and be an offer to amend the subcontract which will require the subcontractor's agreement prior to acceptance.

Subcontractors are frequently put under considerable pressure, if not duress, to sign such formal 'agreements', in many cases being advised that no further payments will be made until the signed copy is returned to the contractor. Should the subcontractor consider it is being placed under duress it must record the fact at the time, and in the event that there are adverse consequences the subcontractor must seek redress at the earliest opportunity (see the section 'Settlement' in Chapter 9).

It will be rare for there to be a formal signing of a Sub-Contract Agreement, where both parties sit down at a table and sign the two copies at the same time. In practice both copies, unsigned and undated, are sent to the subcontractor for signature, the stated intention being that when they are returned they will be signed and dated by the contractor and one copy returned to the subcontractor for its retention and records. Frequently this fails to happen and the subcontractor never receives its signed copy. It is therefore very important that the subcontractor keeps a photocopy of the returned documents together with a dated letter returning the Sub-Contract Documents to the contractor.

Although the 'agreement' as sent to the subcontractor is presented as being details of the agreement between the parties, unless there is already a concluded subcontract between them in those terms the subcontractor may at that late stage either withdraw its offer or seek changes. Should this be the subcontractor's intention, it should not just make alterations to the agreement and return the document without comment, but should advise the contractor what it has done.

Pre-site and off-site works

For almost all subcontracts there will be a number of matters to be formalised between the concluding of a Sub-Contract Agreement and the commencement of its work on site. Not least in importance will be the statutory health and safety requirements to prepare method statements for the work and for the control of substances hazardous to health. Some of these documents may require approval by the principal contractor and/or the planning supervisor. In addition, on certain sensitive sites there may be a need for all operatives and staff to undergo security checks, which may take some time. Finally, all operatives and staff will be required to attend an on-site induction talk, the extent of which will vary from project to project depending on the inherent risks.

In addition, differing trades may have trade- and/or project-specific activities which include:

1. the preparation and approval of detailed drawings and/or shop or fabrication drawings, for either or both temporary and permanent work;
2. the sourcing and approval of specific materials;
3. the preparation and acceptance of samples of both materials and workmanship and/or of work sections in advance;
4. manufacture of materials and/or fabrication of components off site;
5. site measurement and/or inspection and acceptance of the substrate;
6. arrangements for delivery to and/or distribution on site of special components.

As mentioned elsewhere, the planning of pre-site and off-site work, where required, is of major importance to the successful and timely completion of the subcontract work and probably of the project. While losses in time on site may be recoverable by employing additional resources and/or working longer hours, such options are less likely to be available for the production of specialised materials. In managing and programming pre-site and off-site work, it is therefore very important to understand the processes involved and to give special attention to those with a fixed time-scale or duration.

Many pre-site and off-site activities may, by reason of the specification, require acceptance or approval by either the contractor or, more frequently, a third party through the contractor. The inevitable possibility resulting from this is that acceptance or approval may be denied. Such denial may be on the basis that the matter offered is non-conformant with the subcontract requirements. Where this is so, the subcontractor may, if no or insufficient allowance has been made in its programme for such a possibility, be at risk of delaying the Works on site, both its own and that of others.

However, in many situations the reason for acceptance or approval not being given is not a clear failure by the subcontractor to comply with the specification, but the consequence of the accepting or approving person requiring something different to that offered. The effect of even minor changes can be considerably disproportionate to the apparent extent of the change required. It is therefore very important that all such changes are identified at the time and clearly instructed under the provisions of the subcontract. In the event that no formal instruction is issued, the subcontractor should identify at the earliest opportunity any adverse consequences of implementing the change and notify the contractor accordingly.

In this respect it is important to remember that a subcontract is between the contractor and the subcontractor; only in exceptional circumstances will a third party have a right to give instructions to a subcontractor and the subcontractor have an obligation to comply. A subcontractor normally will put itself at considerable risk by acting on instructions from a third party, such as a design consultant, without first obtaining the authorisation of the contractor. If the subcontractor does incorporate the suggested changes into its design without an express instruction to that effect, it may be held at a later date that this was in practice a design development by the subcontractor itself and that any consequences with respect to delay or additional cost are to its account.

In the event of instruction or change, the subcontractor must identify and advise at the earliest opportunity the consequences to both time and money. It is probably much easier to state this requirement than for it to be carried out. On the one hand, if the subcontractor advises that there could be extensive delays and costs it may be accused of attempting to cover up its own deficiencies under the smokescreen of a changed requirement. On the other hand, if the subcontractor understates the situation, then if there is a significant delay at a later date it will be more difficult for the subcontractor to attribute the whole delay to the early delay. The fact is that it will rarely be possible to predict accurately at the time of the instruction or change the actual effect, which may well in the event be further confused by other similar changes.

Whether or not the submission of samples and or work sections in advance is a requirement of the subcontract, it is good practice to seek acceptance of pre-site and off-site work at each stage and so avoid 'unpleasant surprises' at a late date. Where there is no obligation to receive approval or acceptance before completion, there will not normally be any obligation to accept or approve at an early stage either. But only in exceptional cases will the person responsible for accepting the project at completion not be prepared to examine and comment on samples, prepared and submitted along the way, and in many instances will welcome such involvement.

Whenever drawings, samples or work sections in advance are submitted for approval/acceptance, such offers should be made formally and fully recorded. On many projects there may be a specific procedure, which may require the use of a pro forma stating what is being offered and for what purpose, and recording the date of submission and the date of receipt. There may, in addition, be space on the pro forma for comments and for insertion of the date of return and receipt back. Whether or not a project system exists, the use of such a similar form will be good practice. Even though it may be obvious to the person offering the drawings, samples or work section in advance, why that is being done, unless the purpose is clearly stated it may not be so obvious to the receiving party. The offering party is advised to state the date by which it expects to receive comments, and to state what is dependent on the approval requested; for example, to state that approval is requested by DDMMYY date to enable manufacture to commence on DDMMYY date, will make clear the consequences of not responding quickly and/or responding requiring a change.

Just as important as recording the purpose of making submissions is recording the reason for rejection. This should set out both what is wrong and to be rectified and the justification for such rejection. Unless the approving person, by reference to a specific requirement within the subcontract, can justify such rejection it must accept that its requirements are in fact a variation to the subcontract, with all the consequences for delay and cost that are likely to flow from such a change.

Should the response to a request for approval require drawing amendments or the obtaining of further samples or the preparation of further work samples in advance, either on or off site, then the time of all personnel involved should be recorded as additional costs arising from any change and/or additional work that arises from the non-approval. It will be good practice to record such costs at the time of resubmission.

It is always possible that on the submission of drawings, samples or work sections in advance, these will be rejected by the approving person and the subcontractor will not accept the rejection and maintain that the design and/or quality required is outside the range specified and required by the subcontract. When this happens, the subcontractor must either accept the requirements of the approving person or challenge its decision, if necessary by adjudication or any other dispute resolution procedure within the subcontract. Where such a situation arises, it will be to the contractor's benefit to run such adjudication in conjunction with the subcontractor under the main contract, thereby ensuring that the decision is binding on the employer. While such combined action will be to the benefit of the contractor, should the contractor seek to rely on its right under a term of the subcontract, such as clause 3.24 of SBCSub/C, to charge the subcontractor with the costs, then the subcontractor may seek to adjudicate against the contractor in its own right. In such circumstances the contractor may need the support and assistance of the approving person to defend the subcontractor's claim.

A requirement to take site sizes prior to manufacture presupposes that the programme for the Works has sufficient time, between the completion of the first activity and the requirement of the installation of the components, for a survey to be made, drawings marked up accordingly and for manufacture and finishing of the components to be carried out in a timely and orderly manner. All building work requires a degree of tolerance within which it may be constructed and this should be reasonably practical having regard to the processes involved. It will not be a solution to insufficient programme time for the contractor to insist that the earlier trade subcontractor works to more limited tolerances, which are impractical and unachievable or only with extreme difficulty and expense. However, in some situations it may be possible to design into the second subcontractor's details sufficient flexibility to accommodate the first trade's tolerances.

For very good reasons there is a desire for more and larger components to be fabricated off site, reducing the amount of on-site work and enabling the work on those components to be carried out in the more controlled environment of a factory. However, the planning of the delivery and positioning of large components will need to be planned alongside the remaining construction work.

Where components comprise an abnormal load for road transport there will be the need to obtain police permission from the individual forces along its delivery route. Restrictions may be placed on the day and/or timing of delivery. In addition, such components may require special cranes for unloading and/or lifting if either very heavy or to be lifted to a significant height. On inner city sites this may require the obtaining of a road closure or other notification to the local authority. Where special cranage is to be employed there will be an additional need to coordinate all suppliers with similar requirements, where components may be not only from different manufacturers but to be supplied under different subcontracts as well.

Where large or heavy components have to be moved across the site, it may be necessary to leave down other work in order to provide a clear access route. In addition, there may be a need to provide additional propping below floors not designed to take heavy weights, and/or to lay down steel plates or similar to spread the load and/or protect the surface of the floors. Such planning not only involves the subcontractors whose materials are being delivered and the contractor, but may also place significant restrictions on the work of other trades.

Increasingly materials may be sourced in different countries, and in many situations different subcomponents or operations may be carried out in different countries. One risk, which the subcontractor must consider and make any necessary allowance for within its price, is the possibility of changes in exchange rates. Delay to the off-site work and procurement may place that work into a different currency market. There are both risks and benefits in paying for materials held off site, whether fully manufactured and ready for incorporation into the works or not. On the one hand, on payment title in the goods will normally pass to the contractor and at the same time the subcontractor's loss and/or expense, due to any delay between the subcontractor having the materials available and not being able to install them, will cease or be reduced. The risks to the paying party lie primarily in being uncertain as to whether the subcontractor has itself a good title to the goods. The more specialist the materials or components, including the fabrication for a specific purpose, the less the risk of misappropriation of such materials or components.

Before committing resources to site the subcontractor's contract manager should visit the site, preferably with his site foreman/chargehand. Such pre-start visits may be on a formal or informal basis as necessary, depending on the nature of the work and/or complexity or particular difficulties of the site. At such time the manager will need to satisfy himself that:

1. there is a sufficiently large clear area for an effective start to be made;
2. that all the attendances to be provided by the contractor are available;
3. that there is space to position its temporary accommodation and plant and to store materials.

At the same time he must establish and/or confirm the practical arrangements for communications, including the issue of instructions, distribution of mail, site progress meetings and the measurement, valuation and payment of the work.

On-site work

As described in Chapter 4, the degree to which differing trades can work independently of others varies considerably from trade to trade. The contractor's major management duty is therefore to integrate the work of all its subcontractors and directly employed operatives. Most subcontractors will be simultaneously engaged under several subcontracts for different contractors on different projects, and their duty will be to balance the deployment of their labour and other resources so as to reasonably meet the requirements of all their obligations.

An essential part of a management function is communication. While this may be considered obvious it is not reflected in the standard forms of subcontract or indeed in bespoke terms either. As mentioned before, SBCSub/C requires the contractor to give the subcontractor the period of notice for commencement of the work on site, as set out in the Sub-Contract Particulars item 5, and where agreed the period of notice to procure materials in preparation for a start on site, followed by possession of the works for the subcontract period; nothing is required of the subcontractor other than to complete the work in the period stated. What is not defined, either in SBCSub/C or most other subcontract terms, are the obligations of the contractor where the subcontractor's obligation for performance either becomes at large or becomes, as stated in SBCSub/C clause 2.3, a requirement to work 'reasonably in accordance with the progress of the Works'.

It is generally acknowledged that for efficient working there needs to be a 'clear space' between the work of one operative or gang and the trade in front. Production levels drop considerably where the next unit or place of work is not clear. There is a natural desire 'not to work oneself out of a job'.

Management of the work requires the continual review of the requirements of the project, both the short-term situation relating to the next section of work to be carried out and also the long-term requirements so that the supply or delivery of the necessary materials and other resources can be organised to suit both planned and changed requirements. Effective management is a two-way exchange whereby the contractor on the one hand is defining its requirements and/or predictions for the carrying out of the subcontract work, and on the other hand the subcontractor is defining its needs to perform its work so as to satisfy the requirements of the subcontract. If a subcontractor has undertaken to supply and install a stated number of units or complete an area of finished work in a given time, then it will generally be anticipated that that work will, in the absence of any express provisions to the contrary, be released at a steady rate over the period set for the subcontract work. In the event that the rate of release falls below the average or at least a significant way below the average, or a significant way into the subcontract period only a disproportionate amount of the work has been released to the subcontractor, then the parties must jointly review the situation and decide on the plan for the remaining work to completion. Whether under such circumstances the subcontractor will have a contractual obligation to notify the contractor formally that there is likely to be a delay to the works may be uncertain; however, under such circumstances the subcontractor needs to review with the contractor the contractor's requirements, which may in practice serve as a notice of delay or as a direction from the contractor.

Management is not only the planning of work and communicating the resultant requirements, it is also about doing it in such a way as to motivate the recipient into positive cooperation and compliance. The standard forms of subcontract, with their emphasis on the reporting by the subcontractor of breaches by and/or difficulties resulting from actions or inactions of the contractor, are divisive, with the result that the subcontractor is reluctant to issue notices as required since such action is likely to be seen as critical of the contractor and to destroy the relationship between the parties. However, it may be that the subcontractor by adopting a more positive or proactive approach may achieve the intended result of notification without creating an antagonistic response.

For example, a letter recording causes of delay and making a proposal as to the out-working of the remaining work, while at the same time seeking the contractor's instructions, comes across not as critical of the contractor but as seeking to assist. If the contractor fails to respond, it is possible that it would be considered to have accepted the proposal by allowing the subcontractor to proceed in the manner proposed. In reality the contractor is forced to respond, either by accepting the proposal or by issuing different instructions or directions.

Although the subcontract forms require the subcontractor to notify the contractor when the subcontractor's work on site is delayed or disrupted, in practice the contractor will or should be aware of such cause of delay or disruption before the subcontractor. Whether such delay or disruption is the result of a delay or disruption to an earlier trade or is the result of an instruction to vary the work, the contractor will know of it before the subcontractor and in all probability realise that there will be an adverse effect to the progress of the subcontractor's work. It is therefore considered that, irrespective of the lack of obligation under the subcontract, the contractor should at the earliest possible opportunity advise the subcontractor and issue directions and/or instructions as necessary, if for no other reason than to minimise the loss and/or expense that may be incurred.

The standard forms of subcontract only acknowledge completion of the subcontract works as a whole, or in the case of a subcontract for sectional completion, completion of a section. For most subcontractors their work will be taken over in the sense that others will work on or to it. Similarly, each subcontractor will take over the work of the preceding trade. On a well-organised project some formal record of inspection and acceptance by the subcontractor, and inspection by and handing back to the contractor of each work unit, should exist. Whether or not such a formal procedure exists, it will be beneficial to the subcontractor to keep records of when each unit became available to it, and on completion formally to offer it back to the contractor. Such a notice should state that the subcontractor's work is complete or identify work not complete with the reason. It may also be beneficial to record that the area has been left clear of materials and debris. Such a notice achieves a number of objectives. It records the dates each unit was completed and/or records why any work could not be complete, and it implicitly advises that there will be a cost of returning to the area to complete when the obstacle to completion is removed. Such notice allows the contractor to inspect and/or snag the work before the work of other trades. Such inspection will include ensuring that the area is in fact clear of the subcontractor's materials and debris.

It will be good practice for the contractor to issue weekly or regular short-term programmes identifying what work is anticipated to be made available to the subcontractor and when, and the date the contractor expects or requires the subcontractor to complete. Where such programmes are issued the subcontractor may have an implied obligation to advise the contractor if its requirements of the subcontractor are unachievable for any reason or are considered unreasonable.

Whether or not the contractor issues short-term programmes it will be good practice for the subcontractor's manager or foreman to review every two or three days its work ahead and, when appropriate, notify the contractor of its intentions and identify any work to be carried out by others to enable its work to proceed as defined. Such a notice might state that the subcontractor needs to commence in a certain area on a certain date and that it is noted that the following items have still to be completed. The notice should also record any apparent damage to the works of others and/or any material or debris obstructing the work.

Subcontractors, and indeed contractors, frequently seek to explain their failure to notify the other party by saying it results from practical difficulties in giving formal notices. It is suggested that there will rarely be a prescribed method of issuing such notices, and in practice the use of a simple triplicate book, whether prepared for the company's own use or just plain, will allow for handwritten notices and instructions or confirmation of instructions to be given at the time. If desired these can be issued more formally at a later date.

It is beneficial to all parties to record progress of the work on a regular basis. Various methods and systems exist but they seek to achieve two objectives. First, to measure and check that the rate of progress is as required, and if not, to identify the reasons for the shortfall so that the necessary investigations as to the cause can be made and/or the necessary action can be taken to increase the rate of production so as to recover the situation, or where this is not possible the required notifications can be made and instructions sought as necessary. Secondly, to record the work actually carried out during a particular period.

Traditionally this has been done by marking up programmes with the percentage of work completed against each programme activity and by recording detailed progress on site by marking up plans or using some form of tick book or unit schedule, where each activity in a unit of work is scheduled and the date it is completed is recorded. In addition, progress can be recorded photographically. For a subcontractor such records should include not only areas of their work in progress at any time but also other areas of the project, especially those not yet available to it and/or those completed but not occupied by the following trade.

Where either party believes that it may have a right to recover costs against the other and intends to rely on the actual hours spent as the basis of its charge, then there is likely to be an implied obligation, even if not an express one, that steps are taken to notify the party to be charged in due time so that it may monitor the actual time involved in the work. While most forms of subcontract or subcontract terms have express requirements for the subcontractor, where work is to be charged on a daywork basis the same provisions are rarely expressly set out for where the contractor intends to contra-charge the subcontractor.

Should the contractor carry out work which at a later date it seeks to contra-charge on a recorded hours basis, but has neither advised the subcontractor of its intention nor submitted the record of hours for agreement/acceptance, it may be considered that the contractor has thereby waived its right to recovery and/or is estopped from making such a claim. There are frequently informal agreements or understandings that the parties in certain minor matters bear their own costs rather than go through the formal procedures of recording and agreeing a series of minor claims and counter-claims. The subcontractor will naturally feel aggrieved if in such circumstances, having not itself kept records with which to raise a claim, it is at a later date contra-charged by the contractor.

Major difficulties arise on many projects where the finishing trades, for a range of possible reasons, become squeezed into a reduced period of time and are required or instructed to work prolonged hours on site and/or in more restricted areas and concurrently with other trades. Under such circumstances there can be extensive losses in productivity, to an extent rarely appreciated. It is submitted that general requirements to work longer hours will rarely if ever achieve the results sought and will inevitably lead to very significant additional costs, which will be extremely difficult to identify or justify. Any such proposals should therefore be rigorously resisted without some pre-agreement as to how such action is to be compensated.

However, where specific tasks can be identified and labour properly instructed, motivated and controlled, very significant output rates can be achieved. It follows that where there is a genuine desire to recover lost time the correct action is not a blanket set of instructions to all the subcontractors to work longer hours, but to plan in great detail the remaining work, identifying precisely what tasks are to be carried out by each trade and when, or setting short-term targets to be achieved. In carrying out such planning the contractor will need to consult each subcontractor and require each to 'buy into' the programme. This will be a situation where skilful management by the contractor, rather than the giving of authoritarian directions, will be most effective.

Payment

The primary objective of operating any business is to obtain payment for the work carried out and as a result make a profit. It is this that justifies the effort and risks undertaken by those investing capital in the business. However, it is very common on building projects for the valuation of work, and as a consequence payment in full for the work carried out, not to be done at the due time. This is especially true where the valuation includes variations and/or claims for delays, disruption and other losses.

The HGCRA requires the establishment of a payment mechanism such that a subcontractor's work is to be valued and paid within agreed periods of a payment due date. In practice the parties to a subcontract rarely have the payment due dates clearly defined. As mentioned in the section 'Payment mechanism' in Chapter 9, it is not considered that the default arrangement in SBCSub/C, that the first payment shall be due 'not later than one month after the date of commencement of the Sub-Contract Works on site', does in fact establish an effective mechanism. However, the issue of a first valuation by the contractor may establish the latest payment date for future payments.

Payment mechanisms under bespoke forms are generally even more complex, frequently trying to link payment to both an application by the subcontractor ahead of the date the contractor has in turn to make an application under the main contract, and to the date the contractor is to receive a certificate or valuation. The contractor will generally seek to make the payment period under the subcontract longer than the period for payment of the contractor under the main contract, with the intention of achieving in practice a 'pay when paid' clause.

Many of these arrangements lead to very long periods between the execution of the work and the receipt of payment. To ensure that there is no confusion or misunderstanding, before concluding the subcontract the parties should set out the payment mechanism in the form of a table.

Once the payment mechanism is established, it is for both parties to put procedures in place to ensure they fulfil their obligations in accordance with the agreed mechanism. For the subcontractor this will mean that all its subcontracts may be operating to differing procedures and time-scales and some form of daily prompt, by way of a wall chart or electronic programme covering all its subcontracts, will be a useful tool. For the contractor the desire to relate the valuation and payment of its subcontractors to the valuation and payment under the main contract means that it will have to value the work of all its subcontractors simultaneously. This may put a considerable workload on its surveying staff around the valuation date. Such a concentration of work will be exaggerated where payments are monthly, and the payment due date will be a different day of the week for different payment periods. Few contractors have payment procedures that make the compliance with their payment mechanism practical, and they rely on there being no sanction for failure to comply with the mechanism. This is a totally unsatisfactory situation.

The payment timetable described above should also be used to monitor actual payments. Having regard to the need, for the viability of a business, for it to receive payment on time for the goods and services provided, and with the introduction of the statutory sanctions for late payment, there is a need for both parties to monitor the payment of accounts and the receipt of money, and to keep accurate records.

In the event that payment is not received by the final date for payment, a subcontractor should have procedures for the prompt progressing of late payment, which can include a claim for statutory compensation, including the initial lump sum, introduced from 7 August 2002 by the Late Payment of Commercial Debts Regulations 2002.

An essential part of the payment process is the identification, valuation and application for payment of variations and other instructions, and loss and/or expense or damages. Despite the provisions of the HGCRA, making payment provisions conditional on payment by another ineffective, in practice few paying parties wish to commit themselves to the obligation to pay additional sums, and/or to the value of that additional sum or sums, until they have secured both a valuation and payment for themselves. The process will be greatly assisted and speeded up if all such items are identified at an early stage and payment applied for, even if only by way of an initial assessment of the amount and/or rate sought.

In order to succeed in an application for an additional sum it is necessary, in respect of each item, for the applicant to make its case justifying both the item itself and the sum sought, and the rate or prices used. It is suggested that for each item of additional work and or loss and/or expense there needs to be an introductory narrative stating:

1. the variation instruction or other matter giving rise to the costs claimed;
2. a detailed description of the work and/or other matter that flowed from the instruction or other matter;
3. the build up of the cost with, as necessary, measurements and calculations together with any supporting documents such as labour time sheets, invoices from suppliers and, where calculation is on a reasonable rate basis, justification for such rate or rates.

Only where very small organisations are involved will those seeking payment actually be those who were involved in the work. For this reason, if for no other, a detailed description of the work actually carried out is necessary. This is most easily recorded at the time. Where there is an instruction changing the detail of the work to be carried out, this is particularly true. The starting point must be to ascertain the state of the work at the time of the instruction. Work may have commenced either on site, off site or in the drawing office. Orders for materials may have had to be cancelled or varied and cancellation or small order charges may be involved. Work may have to be taken down and disposed of, which may or may not involve returning to the location and, depending on the timing of the change, the re-erection of access equipment.

Where a variation instruction arises from a problem identified during the course of construction, there may be significant delay and/or downtime and lost production while the problem is resolved. Such time may be limited to the relocation of labour to other work but may mean the postponement of a concrete pour or other significant event to a later date. The postponement of certain activities by even a few hours may cause days of delay to the work as a whole, while for others the delaying effect on the works may be regarded as insignificant.

The costs of managing variations is neither fully understood nor acknowledged. A foreman or manager, on receiving an instruction, has to analyse the full implication of the instruction and issue the necessary instructions down the line, such as ordering, cancelling or speeding up or slowing down the delivery of materials, diverting resources such as labour, plant and equipment, as well making such records as are necessary and issuing any notices required under the provisions of the subcontract. While such matters may be regarded as all part of a foreman or manager's duties, such work is additional to the foreman or manager's primary duty to manage its work, ensuring that it has sufficient resources deployed in the proper manner at the correct time and that the work is carried out correctly to the drawings and specification. Where the foreman or manager is diverted from such duties, whether by way of managing instructed changes and/or considering the resolution of site problems and/or difficulties not of its making or to its liability, then the control of the work on site will suffer and costs arise. Only very detailed records of site progress will identify the effect, if any, of such management issues or problems.

It frequently happens that the person preparing the valuation of either variation instructions or other claims believes the work involved to be so clear that there is no necessity to spell out all that is involved, or that because that person was remote from the actual work on site they did not in fact know what was entailed in implementing such instructions. If, for example, the removal of a light fitting involves isolating the power, collecting from a remote location the necessary tools, equipment and materials, setting up access scaffolds, removing and setting carefully aside diffuser, lamps or tubes, disconnecting electrical cables and removing fixing screws or bolts, then all this must be included in the description of the work, otherwise it may not be allowed for in the price, either as requested or valued and paid.

There is inevitably an expectation that a subcontractor valuing a variation will be seeking to claim much more than the work is in fact worth, with the resultant effect that the contractor will only allow a fraction of the sum claimed. It is only by understanding and considering everything involved in implementing the contractor's instruction that fair prices are likely to be agreed.

Completion of the works

A wide range of phrases have been used to describe or define the 'completion' of the subcontract works. These include acceptance, handover, occupation, beneficial use and practical completion, as well as completion itself. The situation in many subcontracts is further complicated by the contractor defining completion or practical completion as the point when completion or practical completion is obtained under the main contract. In such a situation a subcontractor will, or is required to, achieve completion before practical completion.

The reality is that for subcontracted work there are generally definable stages of completion which should be distinguished and catered for within the subcontract:

1. Completion of units of work will be the handover by the subcontractor and taking over by the contractor and/or following trade of such units. These may be small units, such as a room, and may be for one stage of the installation such as first or second fix or a large section, such as a house or flat, or on some projects a complete stage such as a steel frame.

2. The substantial completion of the works will be when all the work subcontracted for has been carried out, save only for any sections programmed or instructed to be omitted until a later date. Where a subcontract has a stated duration for the work on site, it is the achievement of this stage, which will normally equate to the subcontract duration.
3. The completion of all remaining works as described in 2 above. At this time the contractor will be satisfied that the subcontract work is complete and to a standard that can be offered to the employer and/or his representative for acceptance.
4. Completion of any snags and/or rectification of any defects identified when the contractor has offered the Works for acceptance.
5. Completion of any matters identified as necessary to be carried out prior to the issue of the certificate of making good defects.

However the subcontract is worded and whether or not these five stages are stated in the express terms of the subcontract, they will generally apply to all subcontracted work. For some trades one or more of the stages will be achieved simultaneously.

A further complication will be the right of the contractor to pass on instructions under the subcontract after the subcontract work has been completed. This was the thrust of the decision in *Costain Civil Engineering and Tarmac Construction* v. *Zanen Dredging & Contracting Company* (1996) 85 BLR 77, described in detail in the section 'Post-contract instructions' in Chapter 7. In order to ensure that the contractor is able to pass on late instructions issued under the main contract to the appropriate subcontractor without having to establish new and separate contractual arrangements with the subcontractor, the express terms of the subcontract must make provision for the issue of such instructions.

The physical work on site is dealt with above. Further matters which relate to completion of the subcontractor's obligations include:

1. the completion of the subcontractor's design and/or shop drawings;
2. the completion of off-site fabrication or manufacture;
3. the submission of operation and maintenance manuals and other documents for the health and safety file.

It is important that the parties have defined how the achievement of each of the stages affects their rights and obligations under the subcontract. Relevant rights and obligations include:

1. stage payments;
2. protection of the work and/or liability for damage;
3. liability for latent defects;
4. liability for delay;
5. requirements for insurance of the works;
6. release of retention monies.

Once completion of the work on site is reached and the subcontractor demobilises its construction team and/or removes any surplus materials and its equipment from site, communication between the parties becomes much more difficult.

Response times may also become considerably extended. The subcontractor may itself have sublet sections of the work, in some cases the whole of the installation on site, to others in specialist sub-trades. As a consequence, the subcontractor may not have direct access to the resources necessary to carry out final activities such as snagging or rectification of defects, and may require time to remobilise such resources. Even where there is no question of the subcontractor's work itself being sublet, it is frequently the case that the subcontractor is based a long way from the project if not in another European country.

While the contractor must understand the subcontractor's situation and organise itself accordingly, the subcontractor must also understand that it has ongoing responsibilities to the contractor, which may or may not have been precisely expressed within the subcontract terms and conditions. In the absence of express provisions there will be an implied term requiring a reasonable response to the contractor's requests. However, what is reasonable will depend on the circumstances, including the distance of the subcontractor's organisation from the project and whether or not the Works as a whole are close to practical completion.

In the simplest of subcontract arrangements completion of the subcontract work brings to an end the subcontractor's obligations and leads to an obligation for the contractor to pay the balance of the subcontract sum. However, sophisticated subcontracts may define a complex number of matters that flow from completion, some of which are to be carried out as a direct consequence of completion or a specific type of completion, and others are to be performed a stated period afterwards. It is necessary that the parties understand and have in place arrangements to carry out these obligations as required. Depending on the wording used in the subcontract, conforming strictly to these requirements may have fatal consequences to either party's rights.

It will be in the subcontractor's best interests to present its final account at the earliest opportunity. This should be fully supported by the necessary information, including calculations, measurements and documents. All too often such accounts are prepared and submitted with little or no supporting documentation. Whether or not the account includes a claim for payment against variation instructions or for loss and/or expense, the preparation of a detailed description of the work or loss involved will raise the chance of an early and successful agreement of the account. Full and accurate details of such matters will be most easy to ascertain shortly after the event rather than many months or even years after.

Where, as is generally the case, the subcontract includes for a defects liability period at the end of which it is intended that there will be a final inspection and the rectification of any defects identified, it may be in the subcontractor's best interests to seek, at the end of the defects period, clearance of further liability. Should the contractor not have sought clearance for the entire Works from the contract administrator, this may encourage it to do so. It will be in the interests of all parties, with the possible exception of the employer, to clear all such matters promptly and to recover any remaining retentions or other sums.

Chapter 6
Design Development

General considerations

A subcontractor can become responsible for design, either because of an express requirement of the subcontract or because the subcontractor, in the absence of specific requirements in the Sub-Contract Documents or subsequent instructions, makes a decision with regards to design and/or selection of materials and thereby assumes responsibility for that decision.

The general presumption, at law, is that the subcontractor warrants that its design or selection will be fit for purpose. Thus, if a joinery subcontractor selects the size and type of fixing and/or the number to be used to fix a component, then it will be liable if those fixings prove either unsuitable or insufficient, even if that failure is due to circumstances outside the subcontractor's knowledge and/or reasonable expectation.

However, where the subcontractor carries out design under the subcontract, it is normal for its liability to be restricted by the terms of the subcontract, to the express requirements of liability usual for that of a professional consultant designer, which is for 'reasonable skill and care'.

The subcontractor may be called upon to design an entire component, such as roofing or rainwater drainage, in which case it will be its responsibility to consider the likely wind forces, rainfall and other relevant maters, having regard to the location of the building and the use to which it is to be put, as well as any other relevant factors. Or it may be required to design to a performance specification, where others have made basic design considerations and/or calculations. In these circumstances, the subcontractor's liability becomes more limited but it will be obliged to satisfy the requirements of the specification, even if the fitness for purpose criteria would be satisfied by a lesser requirement. It will be no defence, if its design does not achieve the required performance requirements, for the subcontractor to show that the work as constructed is fit for purpose.

In pricing work on a design development basis, the subcontractor takes on the risk that the outline or indicative requirements may be inadequate, but it does not pre-contract carry out the design. There will rarely be a warranty that the design requirements will be achieved in a certain way. On the contrary there will be an implied term that the subcontractor will be allowed to develop the design in the manner most suitable to it, provided that the design satisfies the requirements of the subcontract, and if necessary other parts or components will be adjusted to conform. This was the effect of the decision in *Norwest Holst Construction Limited* v. *Renfrewshire Council* (20 November 1996, unreported), quoted and discussed in the section 'Instructions relating to subcontractor design' in Chapter 7. The subcontractor's exact obligation will be decided by the express and/or implied terms of the actual subcontract. Thus the duties and risks undertaken in one subcontract may differ considerably from those of another.

The right to develop the design

The subcontractor's scope to develop the design, economically and to its best advantage, is only limited by the express provisions of the subcontract and the associated drawings and specification. It not infrequently happens that a design, as developed by the subcontractor, is considered by the design team leader not to be entirely to his liking and/or wishes. Whether a change to the developed details for manufacture and construction will be a variation or not will depend on the subcontractor being able to demonstrate that its design, as issued for approval and acceptance, satisfied the requirements of the design brief without the amendments required, or the designer being able to demonstrate that the subcontractor's design development did not satisfy the design requirements of the subcontract.

NEC/Sub at clause 21.2 limits the reasons for not accepting a subcontractor's design to two: that it does not comply with the subcontract information, and that it does not comply with the applicable law:

> '21.2 The *Subcontractor* submits the particulars of his design as the Subcontract Works Information requires to the *Contractor* for acceptance. A reason for not accepting the *Subcontractor's* design is that
>
> - It does not comply with the Subcontract Works Information or
> - It does not comply with the applicable law.
>
> The *Subcontractor* does not proceed with the relevant work until the *Contractor* has accepted his design.'

It follows that under this condition the contractor must accept the subcontractor's design unless it can be shown to be non-conformant. If the contractor or design team leader requires changes that do not result from a non-conformant design, then it must still accept the design and concurrently issue its instructions to change the design requirements as necessary.

Contractors must be aware that under the SBC/XQ form of contract Schedule 1, Contractor's Design Submission Procedure, at paragraph 7, the contractor only has seven days from receipt of the architect's comments to disagree with such comments. No similar requirement is expressly incorporated into clause 2.6.2 of SBCSub/D/C and DBSub/C, and in any event it would require a very speedy response by the subcontractor to enable a contractor relying on such a response from the subcontractor to satisfy its obligation under section 1 of the JCT standard form. ICSub/D/C, unlike SBCSub/D/C, does not refer to the provisions of the main contract but to the design submission procedures in the Sub-Contract Documents.

Sub/MPF04 is different again and at section 12 sets out a detailed design procedure which requires the contractor to respond to design documents submitted by the subcontractor within 21 days of their receipt by the contractor or their programmed date for submission, whichever is later. Clause 12.3 makes the documents released for construction in the absence of comment from the contractor within the 21-day period. It will therefore be important for the contractor to monitor carefully any submissions to and responses from a third party, so as not to be caught by this default provision.

It is suggested that there will, in the absence of express provisions to the contrary, be an implied term giving the subcontractor the right to the following:

1. to develop the design in any way not incompatible with the Sub-Contract Documents and specification;
2. in the event that the design specification is unachievable, to amend the specification as necessary or for the adjustment of the design of the associated elements;
3. to design so as to comply with statutory requirements, including the Construction (Design and Management) Regulations (CDM);
4. that its design will be coordinated as necessary with the designs of others, by the contractor or design team leader;
5. to design within the requirements of any restrictions imposed by any product manufacturer or system provider.

It follows that the role of the specifier is a very important one. Specifiers must identify those matters they consider to be of significant importance, while at the same time not being overprescriptive. If it is necessary to have several specialist subcontractors producing products with the same appearance, it may be desirable to specify that the same sub-specialist carry out certain of the finishing work. On the other hand where minor differences in appearance are unimportant, then selection of such sub-trades should be left with the individual subcontractor.

It may be desirable, for cost-effective maintenance, to specify certain subcomponents that are compatible with other buildings owned by the same employer, thereby reducing the number of spare components need to be held by the employer's maintenance department. It is possible that such restrictions will limit the subcontractor's ability to design efficiently and/or most cost effectively. If the employer specifies a subcomponent, not normally used by the subcontractor, this may involve special orders rather than using components from stock, with the resultant prolongation and/or inability to react speedily to changed or additional requirements.

Where the outline design is, on detailed calculation, found to be insufficient there may be more than one way to resolve the design problem. For example, where an indicative design for, say, balustrading indicates a number of components of a stated size, then if they are insufficient a satisfactory design may be achieved either by increasing the size of those components or increasing their number. This, in the absence of specific requirements to the contrary, will be for the subcontractor's selection.

Methods of design vary from organisation to organisation. Consultant designers will generally have significant engineering skills and will carry out their designs based on those skills and the resultant calculations. Many specialist subcontractors, being from a trade or craft background, may base their designs on empirical methods such as set out in tables attached to British Standards and other reference sources. Two consequences may arise from this situation. First, the subcontractor may not be able to prove its design in the manner a consultant designer may anticipate, and secondly, the consultant may regard the subcontractor's solution as being overdesigned and, where this has led to an increased number or size of components, as described above, it may lead to a dispute over the need for change. In such a situation it will be for the consultant to decide whether to take back from the subcontractor its design obligation or to stay with the allegedly overdesigned solution. The real issue is that it is not cost-effective to pass on to subcontractors an obligation to design matters already designed by a competent consultant. Where the subcontractor does not have similar engineering design skills it must either buy in such skills, which will be costly, or rely on pre-designed methods.

Other design issues often arise out of the provisions under the CDM, such as the subcontractor's intention to reduce a component's size or weight so as to limit the risk of injury to its operatives while handling such components. Another design factor may be the need to provide stability to temporary works, such as scaffolding, and other means of access, which may require ties through the external finishes. The CDM allows for health and safety requirements to take account of, among other matters, the aesthetics of the design. However, it is likely that where a specifier has not identified an aesthetic requirement in the Sub-Contract Documents, then such aesthetic requirements may not be considered to be of such importance as to override the health and safety concerns of the subcontractor. In reaching this opinion, consideration has been taken of the design consultant's statutory duty to identify in the tender health and safety plan, as a residual risk, the matter which the subcontractor's design solution seeks to avoid.

The designer's responsibility to design out health and safety risks is limited by the approved code of practice to those 'which the designer could reasonably be expected to foresee'. It is suggested that this puts a more onerous requirement on a specialist designer than exists for general designers such as the architect or design team leader are likely to be. However, there is a statutory duty, under Regulation 13(2)(c), for any designer to cooperate with any other designer 'to enable each of them to comply with the requirements and prohibitions placed on him in relation to the project by or under the relevant statutory provision'. It follows that a design coordinator will need to have a significant reason or considerable justification for rejection of a design, developed by the subcontractor, that improves or reduces the risks to health and safety of its operatives and/or those subsequently to be employed in maintenance or demolition.

Limit of responsibility

As a general rule it can be said that a specialist subcontractor taking on a design or detailing function will only be responsible for that part of the design which it designs or details. The subcontract requirements may be either to detail to a performance level defined in the subcontract, or to develop a component or element of the entire building so as to be fit for purpose.

For example, where a subcontract requires a pipe system to be capable of withstanding a specified air pressure, should such a pressure prove to be too low it can be expected that the subcontractor will not be liable to the employer for the underdesign. Conversely, if the pipe system fails to meet the specified performance, it will not be a defence for the subcontractor to demonstrate that the system supplied was fit for purpose. An exception to this rule is likely to apply where the system, as installed, is dangerous to the extent that people or property are put at risk and the subcontractor knew or, with its specialist knowledge, should have known that the danger existed.

In certain circumstances there may be secondary factors which raise issues as to whether the subcontractor's design is fit for purpose. For example, the installation of a roofing and/or cladding system may satisfy the specification requirements in all respects and act satisfactorily so far as water and air resistance and thermal insulation are concerned, but in certain circumstances may vibrate causing an unwanted or annoying sound within the building. Whether such noise can be said to render the roofing and/or cladding unfit for purpose will no doubt depend on the level of noise, the nature of the building and all other relevant factors. Similar problems may arise with regards to appearance of the selected components.

Where, however, no performance level is specified, then it is for the subcontractor to ascertain all the relevant factors necessary to provide a satisfactory design. If, for example, the subcontractor is to provide a roof and drainage installation, then the subcontractor must size the gutters, downpipes, etc. so as to cater for the likely rainfall to be experienced at the location of the building.

This was one of the factors the subcontractor should have fully investigated and considered when designing a tall television transmission mast. In the House of Lords case of *Independent Broadcasting Authority* v. *EMI Electronics Ltd and BICC Construction* (1980) 14 BLR 1, where the primary consideration of the court was the liability of the main contractor to the employer, Lord Edmund-Davies said, at page 29:

> 'Again, although BICC were later to plead that the Emley Moor collapse was due to "freak" weather conditions (a plea not ultimately adhered to), they made no enquiries about likely meteorological conditions at any of the three sites. And they assumed that the fact that the Emley Moor and Belmont masts were to be 250 ft taller than the one at Winter Hill "would have made only marginal difference".'

Having found that the failure of the mast was due to a combination of relatively low wind speeds, causing the mast to vibrate, coupled with the ice on the guy cables not being shed as might have occurred at higher wind speeds, he said, at page 31:

> 'In the light of the available evidence, I find irresistible the submission of IBA's learned counsel that BICC paid no regard to the possibility of asymmetric ice loading occurring either by itself or in conjunction with vortex shedding, and thereby failed to act as a prudent designer should have acted.'

In this case BICC, who were specialists in mast design, were required to construct masts of a new, relatively untried, method of construction, namely a slender steel cylinder, rather than the common open lattice form of construction with which BICC were very familiar. As Lord Edmund-Davies said, at page 27:

> 'It was in that year that IBA initiated the project of erecting masts of a wholly different type, the lattice being almost wholly replaced by cylindrical structures which it was considered would have the advantage of facilitating all-weather maintenance. They were, as the Court of Appeal put it, "wholly new in concept". BICC themselves pleaded that no

such cylindrical steel masts had hitherto been designed or built by anyone in the United Kingdom, and none of the height projected had been built "anywhere in the world"; that accordingly there was "no available source of empirical knowledge"; and that the project involved "work which was both at and beyond the frontier of professional knowledge at that time".'

Lord Edmund-Davies then examined the authorities as to the level of skill and care required when designing at the fringes of knowledge, and concluded that although the employer must accept some risk that the design may fail, there is an increased duty on the designer to take all foreseeable steps to avoid such design failures, especially where the result of failure will have dangerous or fatal consequences. Lord Edmund-Davies said, at pages 27 and 28:

'Notwithstanding the absence of a contractual nexus between IBA and BICC, it is common ground that the latter owed the former a duty of reasonable skill and care in relation to the design of the masts. But they were not guarantors of success, for, as Mr Justice Erle directed the jury in *Turner* v. *Garland & Christopher* (1853) Volume 2 *Hudson's Building Contracts*, 4th edn, 1914, p. 1:

"... if you employ (an architect) about a novel thing, about which he has had little experience, if it has not had the test of experience, failure may be consistent with skill. The history of all great improvements shows failure of those who embark in them."

The project may be alluring. But the risks of injury to those engaged in it, or to others, or to both, may be so manifest and substantial, and their elimination may be so difficult to ensure with reasonable certainty that the only proper course is to abandon the project altogether. Learned counsel for BICC appeared to regard such a defeatist outcome as unthinkable. Yet circumstances can and have at times arisen in which it is plain commonsense, and any other decision foolhardy. The law requires even pioneers to be prudent.'

Further to the above, Lord Fraser of Tullybelton in his speech made the point that the collapse occurred because of two situations occurring at the same time, that one option had been considered in the subcontractor's design and the other had not, and that the subcontractor had been negligent in not considering the likelihood of both occurring at the same time. He said, at page 37:

'This accident was the result of two separate causes operating at the same time. By far the more important cause was aerodynamic stress against which BIC had made provision which, according to the best information available at the time, was adequate. The other cause was static stress from asymmetric ice loading, and BIC admit that they had made no provision at all against it. I have explained why I think the need for some provision was foreseeable. They were, in my opinion, therefore, negligent in failing to take precautions against one of the two causes which materially contributed to the collapse.'

In the *Independent Broadcasting Authority* case, even though an outline or conceptual design for the mast and its supports had been provided to the subcontractor, it was still the subcontractor's responsibility to have provided an effective design for a mast of the height etc. specified. The subcontractor's liability would not have extended to deciding the height to which the mast was to be constructed and the subcontractor would probably not have had any liability if the mast had been ineffective as a television mast.

In the more recent case of *Co-operative Insurance Society Limited* v. *(1) Henry Boot Scotland Limited (2) Henry Boot plc (formerly known as Henry Boot & Sons plc) (3) Crouch Hogg Waterman Limited* (2002) 84 Con LR 164, the judge had to decide where the design responsibility lay for a retaining wall which failed. The employer's engineers had produced a conceptual design but the contractor's designed portion included:

> '(i) Earthwork support to sub-basement excavations.
> (ii) Bored bearing piles to foundations and contiguous bored pile walls.
> (iii) Temporary propping to contiguous bored pile walls.
> (iv) Temporary supports and propping to walls of adjoining properties and to the retaining walls next to the footpaths at George Square and Hanover Street.'

As a consequence of the inadequacy of the design, as actually constructed, water and soil flooded into the sub-basement. The contractor claimed that its obligation for design was limited 'to prepare working drawings in respect of the concept devised by CHW' (the engineer). The employer claimed that the contractor, Boot, 'was responsible for the design of those elements under the contract'. Further, the employer claimed that Boot, under the conditions of the contract, had 'the obligation to co-ordinate elements of the design of the Works . . . and submitted that the responsibility for stability and structural integrity of the Works' was also an important part of Boot's obligation. Judge Richard Seymour QC found that Boot was responsible for the integrity of the complete design, at paragraph 68:

> 'In my judgment the obligation of Boot under clause 2.1.2 of the Conditions was to complete the design of the contiguous bored piled walls, that is to say, to develop the conceptual design of CHW into a completed design capable of being constructed. That process of completing the design must, it seems to me, involve examining the design at the point at which responsibility is taken over, assessing the assumptions upon which it is based and forming an opinion whether those assumptions are appropriate. Ultimately, in my view, someone who undertakes, on terms such as those of the Contract (that is to say, including clause 2.7) an obligation to complete a design begun by someone else agrees that the result, however much of the design work was done before the process of completion commenced, will have been prepared with reasonable skill and care. The concept of 'completion' of a design of necessity, in my judgment, involves a need to understand the principles underlying the work done thus far to form a view as to its sufficiency. Thus I reject the submission of Mr Baatz and Miss Doerries [Counsel for Henry Boot] that all Boot had to do in any circumstances was to prepare working drawings in respect of the bored pile walls. If and insofar as the design of the walls remained incomplete at the date of the Contract, Boot assumed a contractual obligation to complete it, quite apart from any question of producing working drawings. Thus I accept the submissions of Mr Coulson and Mr Bowdery [Counsel for Crouch Hogg Waterman].'

The above judgment shows that the extent of a subcontractor's design responsibility will be determined by the words of the subcontract but is likely to extend to ensuring that the element or elements for which the subcontractor assumes responsibility are complete and sufficient in themselves.

Design changes by the specifier

There may be doubt as to the requirement of the design brief. As always, the *contra proferentem* rule will apply against the author of the Sub-Contract Documents. Thus a line on a drawing, such as the indication of a glazing bar will only indicate an actual glazing bar in an actual location, as opposed to an indicative glazing bar, if it has been defined as in a certain position or at a defined spacing or aligning with other components. It is for the specifier of the design brief to define precisely his requirements. It is not for the subcontractor to be expected to second-guess the design leader's unspecified intentions.

Many design, manufacture and install subcontracts require the subcontractor's design to be to 'the complete satisfaction' of the contractor and/or architect. Some commentators suggest that such conditions put the subcontractor at risk and that 'the scope for abuse is endless'. It is suggested that in the absence of an express agreement that the subcontractor will incorporate any changes required by the contractor and/or architect, at no additional cost and no additional time, then the contractor's/architect's scope for satisfaction is limited by the subcontract specification, without there being a variation to the subcontract. This is the principle established in the case of *Dodd* v. *Churton* (1897) 1 QB 562. (See the section 'Instructions where there is no provision within the subcontract' in Chapter 7.)

Design changes are often made without a formal instruction, by drawings sent for approval being returned with comments. Although the comments on drawings issued 'for approval' will not be expressed as instructions, nor will they be listed as such on a formal instruction sheet, it is difficult to conceive that their effect is not that of a formal instruction. Whether or not they constitute a variation instruction will only be decided by comparing the required change against the subcontract specification. It will be wise for the subcontractor to confirm back all such instructions, identifying the change required and identifying the effect both as to time and/or price. The general use of computer aided design (CAD) systems has turned the drawing office into a process where output is limited by the resources available both in experienced personnel and/or the necessary computer hardware, software or workstations. Even a limited number of changes to the subcontractor's detailed design can create additional costs and overload the design facility, delaying work and/or requiring the working of additional hours at premium rates.

Disputes not infrequently arise as to the liability for the subcontractor to take on board design development changes. It is unlikely that the subcontractor will be liable for the cost and/or time effect of any changes not of its making and where it has not made the selection. The integration of services may be the responsibility of the general services subcontractor, and to his cost if there is a clash requiring the diversion or changed detail of one of the specialist services. However, it is likely that where each service is provided by a separate sub-subcontractor, the change will be a variation to the sub-subcontractor even if not to the general services subcontractor.

The individual subcontractor will probably be responsible for the design and provision of any necessary fixings, supports, hangers, etc. to connect its work to the structure. While generally the interface will be easily defined, there may be situations where intermediate 'structural' supports and or platform structures are necessary. It will be a matter of interpretation of the individual subcontract arrangements as to where the interface rests. Where confusion or ambiguity exists, again the requirements are likely to be interpreted against the contractor to the subcontractor's benefit.

Serious problems can arise with regards to design responsibility where the subcontractor is 'requested' or 'instructed' to make changes to its submitted design, since many specialised products or systems will have been the subject of extensive testing which in turn has led to their certification for quality and/or performance. It follows that the design submitted by the subcontractor should, unless qualified to the contrary, comply with all the design parameters to satisfy the test certificate. In the event that the contractor returns those drawings with changes made by the architect or design coordinator, such changes may justifiably be taken as instructions from the contractor. As previously mentioned, the subcontractor's obligation under most forms of subcontract, and SBCSub/C specifically at clause 3.5, is to comply with any instructions or directions 'forthwith'. The general situation in SBCSub/C must be distinguished from the requirements in SBCSub/D/C at clause 3.5.3, which require the subcontractor to notify the contractor where such an instruction would have an injurious effect on the efficacy of the design; a similar provision exists in ICSub/D/C at clause 3.5.2. Where the subcontractor realises or should realise that compliance with such instruction will invalidate any warranty or guarantee, then it is suggested that there is an implied duty, and under SBCSub/D/C clause 3.5.3 an express duty, for the subcontractor to warn the contractor of the consequences of complying with such an instruction. (See also the section 'Duty to warn that instructions may give rise to defective work' in Chapter 7.) Thus although Sub/MPF04 has no express term similar to that of SBCSub/D/C clause 3.5.3, it is likely that a similar obligation is implied into that form.

Interface of design responsibilities

The design of interfaces not only affects the final design but also any necessary requirements, whether temporary or not, to facilitate the safe construction by other adjacent trades. In the case of *Aurum Investments Limited and Avonforce Limited (in liquidation)* v. *Knapp Hicks & Partners and Advanced Underpinning Limited* (2000) 78 Con LR 114, the court had to consider the duty of a piling subcontractor undertaking the design and execution of underpinning work. The purpose of the underpinning was to lower the foundations of an existing building to facilitate the construction of a basement on the adjacent site. In the finalised design the basement would be formed with a vertical retaining wall, which would in turn support the underpinning work. The underpinning, as designed and installed, performed satisfactorily and supported the vertical loads from the walls above. The underpinning had not, however, been designed as a retaining wall capable of supporting lateral loads. There was no question of this being a requirement of the underpinning design. Furthermore, the underpinning was carried out in the traditional manner, as stated by the judge, Mr Justice Dyson, at paragraph 4: 'It was carried out in small sections followed by immediate backfilling'

Following completion of the underpinning work the subcontractor left site and basement excavation work commenced. No temporary support was provided to the underpinning, which failed by rotational failure:

'On 6 June 1997, the central section of the excavation, alongside the flank wall, collapsed as a result of rotational failure of the underpinning. The toe of the mass concrete underpinning was displaced horizontally into the excavation because of the lack of passive soil resistance. It is common ground that it would have been possible to have designed and installed some form of propping which would have prevented the failure of the underpinning, and which would have allowed the proposed basement to be constructed safely.'

It was accepted that there had been several safe methods of carrying out the basement excavation and construction work and the judge said, at paragraph 15:

'Mr Roberts [Engineer of Advanced Underpinning] also said that it would have been perfectly possible for temporary sheet piling to be installed between the bases and the retaining wall of the basement structure. Sheet piling would have retained the earth behind the bases, thereby reducing the lateral loads applied to them.'

This case was considered on the basis of whether an implied duty to warn existed between the subcontractor and the contractor; none was found. However, it can also be considered as a decision as to the extent of a subcontractor's design obligations. These were found to extend only to that which it had contracted to do. The judge summed up by saying, at paragraph 17:

'As I have said, in my view it is unreasonable to impose a duty to warn on Advanced in the circumstances of this case. The duty to 'warn' is no more than an aspect of the duty of a contractor to act with skill and care of a reasonably competent contractor. Reasonableness lies at the heart of the common law. As Lord Reid said in *Young & Marten Ltd* v. *McManus Childs Ltd* [1969] 1 AC 454, 465 "no warranty ought to be implied in a contract unless it is in all circumstances reasonable". Advanced was not asked to advise Avonforce what excavation techniques should be adopted, nor did Avonforce tell Advanced how it proposed to go about carrying out the excavation. It chose to employ a method that was negligent when suitable alternatives were available.'

The design of the interface between specialist trades may require some coordinated detailing. Not only will it be necessary for this detailing to be developed to the satisfaction of both trades, but also there will be a need to define which subcontractor is to be responsible both for the satisfactory design development and for installation of any specialist component at the interface. The development of such details may, in addition to decisions as to material selection, also affect the sequence of trades. Unless there is an express requirement to develop a detail that can conform to an identified installation sequence, the subcontractor will not be restricted in its choice of detail, which can have significant effects on the combined programme.

Construction Design and Management Regulations

The Construction Design and Management Regulations (CDM) place duties on designers to design so as to avoid risks to health and safety for operatives, not only when installing the work but also when maintaining and dismantling it as well. Typical health and safety issues include avoiding materials hazardous to health and large or heavy components, especially where they are to be transported and

installed in restricted locations. It is also necessary to provide a safe means of access for work to be carried out.

The regulations allow for there to be a balance between design considerations and the effect on the health and safety of construction workers. It has been said that the duty of the designer is limited to risks that the designer could reasonably be expected to foresee. It is suggested that this duty will be greater for the designer of a specialist subcontractor than for an architect or engineer with general design experience and responsible for the design of an entire building or project. Any decision as to the acceptable solution should be on the basis of a proper risk assessment balancing the level of risk, cost and the effect on design appearance. Denying the use of design solutions which reduce the hazards of the installation process may, as well as possibly being a breach of a statutory duty by the designer, by imposing such a restriction, almost certainly increase the installation cost as well.

The balance is to be made between designing so as to conform to the CDM and thereby reducing the risk to health and safety, and the need to achieve the design intention of the architect and or designer. It is suggested that where a specialist subcontractor believes that its design solution is necessary in the interests of health and safety, it will be for the party holding out that such a solution is unnecessary to demonstrate such, if necessary by reference to adjudication or some other independent tribunal such as the Health and Safety Executive. The statutory duty of the subcontractor designer will be persuasive in overruling all other considerations. It is suggested that it will not assist a consultant designer to seek to impose post-contract an aesthetic design limitation that he failed to identify pre-contract.

Aims and objectives of the subcontractor's designer

In passing the obligation for detailed design down the contractual chain to the specialist subcontractor, the design coordinator will seek to obtain one or more of the following benefits for the employer:

1. specialist knowledge;
2. additional design resources;
3. cost savings by designing to suit the subcontractor's methods of manufacture and/or installation.

The benefits to the subcontractor will generally be to satisfy one of two very different objectives:

1. as a service to obtain work for its core business of manufacturing components; or
2. to develop a design most suitable for the efficient use of its manufacturing/production facilities and installation equipment.

It follows that the greater the restrictions placed on the specialist subcontractor, the less will be the benefits to both the employer and the subcontractor. Further, the aims and objectives of one specialist may not be those sought by another subcontractor or in another situation. For example, in one situation it may be desirable to make subcomponents as large as possible thereby reducing the number of handling operations, while under other circumstances the object may be to limit size or weight so as to be transportable using a limited number of operatives or equipment of limited capacity.

Rarely, if ever, before the appointment of the subcontractor are the objectives of the subcontractor for the development of the design identified, considered or investigated. Thus when the subcontractor justifiably objects to changes required to the detail of its design and/or seeks significant increases to its subcontract price for such changes, there is no readily identifiable standard against which to consider such a claim. It must be realised that relatively minor changes may have a very significant effect on the subcontractor if such changes prevent the subcontractor using its intended method of manufacture or installation. A simple example is finishing treatments, such as galvanising of steelwork or toughening of glass, where the capacity of the equipment necessary for these treatments varies considerably from installation to installation. Where a variation requires an increase in component size, such as to exclude the use of the intended treatment facility, the work may become subject to very significant delays and/or increase in price; or if a particular finish is required, this may limit the number of competitive specialist suppliers to those with the appropriate facilities in-house. If, having accepted a specialist supplier with certain in-house facilities, a change is made in the specification that requires facilities which the selected specialist does not have, it will be likely to have a significant effect on both time and cost.

Subcontractor's design under JCT

The JCT forms of contract envisage two distinct arrangements under which the subcontractor may be required to carry out design or develop the detailed design. First, there is the situation where the contractor does not have a general design responsibility but is required under the terms of the main contract to obtain or provide the development of conceptual design for discrete elements of the work. This element of the work is described as the Sub-Contractor's Designed Portion. The alternative situation is that, where the full responsibility is taken on by the contractor under a design and build form of contract, the contractor may employ the subcontractor to perform part of that design function. Under these circumstances reference is made to there being Sub-Contractor's Designed Works.

Whether the subcontractor is carrying out its design as part of a standard building contract under the SBCSub/D/C or ICSub/D/C terms, or as part of a design and build contract under the DBSub/C terms, the actual requirements will be the same. The articles of agreement at the fifth recital refer to the subcontractor having supplied the contractor pre-subcontract with two types of documents: documents showing and describing the proposals of the subcontractor for design and construction of its portion or works, and an analysis of the portion of the subcontract sum that relates to the designed portion or works.

There must be considerable uncertainty as to what will be considered the necessary documents to be provided pre-subcontract that will adequately describe the Sub-Contractor's Proposals, and perhaps more importantly, as to how it can subsequently be said that the design development has departed from those proposals.

Indeed it is likely that the less detail given in those proposals, the more scope remains for the subcontractor to develop its design, while by giving specific details of its proposals the subcontractor will limit the development of its design within the constraints of its proposals.

Where the contractor has itself had to submit proposals pre-contract, it may need to ensure that its selected subcontractor has accepted those limitations before concluding a subcontract with it. Thus, if a contractor at tender stage considers several proprietary products and makes a selection which forms part of its offer to the employer, then at that time the contractor may be well advised to have a binding arrangement with the selected supplier or subcontractor for the proposed product. It is suggested that there will be few situations where a specialist trade contractor will need to make pre-subcontract proposals of the type envisaged in SBCSub/D/A and DBSub/A.

Further design information

There will be a general expectation that the subcontractor will develop its design so as to fulfil the requirements set out in the Sub-Contract Documents, and, as already mentioned, it will be for the subcontractor to make such design decisions as are necessary for it to complete its work. In practice, there will frequently be some need for a selection by others or to coordinate the subcontractor's design with the needs of and/or the design by others. These design interfaces need to be identified at an early stage and a detailed programme prepared showing both the design and installation interfaces. Where no such integrated programme exists, it is suggested that the subcontractor should identify its needs at the earliest possible opportunity, both as to the requirement for other details and the finalisation of its own workshop and/or installation drawings.

SBCSub/D/C at clause 2.7 and ICSub/D/C at clause 2.6 provide for both contractor and subcontractor to provide information to the other; however, there is no express requirement for the subcontractor to request such further information unless the subcontractor has reason to believe that the contractor is not aware of the time by which the subcontractor requires such information, when by reason of sub-clause 2.7.4 the party requiring the information shall, so far as reasonably practicable, advise the other party sufficiently in advance as to enable him to comply with this clause 2.7.

This sub-clause is modified by the requirement for the requesting party to provide drawings, details, information and directions when it is reasonably necessary for the recipient to receive them, having regard to the progress of the subcontract Works and main contract Works, as set out in sub-clause 2.7.3.

The requirement of the contractor to provide drawings, details, information and directions having regard to the progress of the Sub-Contract Works must include where appropriate the subcontractor's off-site works and not be limited to the subcontractor's work on site. The contractor will not be able to comply with this requirement unless it makes itself conversant with the subcontractor's intentions for its pre-site activities.

The provision under SBCSub/D/C is for such drawings, details, information and directions to be given at a time when, having regard to the progress of the works, it is reasonably necessary for either party to receive them. Where design is concerned, the timing of necessary further drawings and details must relate to the design programme, which may or may not have a direct relationship to the progress of the Works on site. The use of the word 'reasonably' in connection with the necessary time for the contractor to give directions must, it is submitted, mean reasonably in advance of the subcontractor's reasonable requirements. The subcontractor's obligations must relate not only to when the contractor may reasonably require such drawings, details, information and directions, but reasonably having regard to the date the subcontractor was itself instructed by the contractor.

The requirement is for either party to advise the other of its need for such further drawings, details, information or directions, when one party has reason to believe that the other party is not aware of the requirement. Where interface requirements are involved it is likely that the subcontractor will not know in advance the requirements for its design details to be incorporated within the details to be prepared by others, and consequently the contractor must be considered to have reason to believe that the subcontractor is unaware of the contractor's requirements. Further, the subcontractor will be unlikely to know the lead-time necessary for the contractor to obtain further drawings and/or details from others.

As well as the subcontractor needing design information in time for it to carry out and complete its design, so as to give time for the manufacture of its work to suit the installation programme, there may be a reciprocal need for the contractor to have the subcontractor's details in sufficient time to enable the carrying out of work by others. SBCSub/D/C at clause 2.7.2 requires that the subcontractor is to provide the contractor with working/setting-out drawings and other information necessary for the contractor to make appropriate preparations to enable the subcontractor to carry out and complete the subcontract works in accordance with the subcontract.

This clause attempts to place the onus on the subcontractor to provide its information to the contractor so as to enable the contractor to fulfil its obligations to the subcontractor, at least so far as is necessary for the contractor to make preparations for that subcontractor's work on site. It is therefore open for the contractor to say that it was not ready for the subcontractor to commence work on site because the subcontractor did not provide its working/setting-out drawings in time. This requirement for the subcontractor to anticipate the contractor's needs is subject to when the contractor is not aware, nor has reasonable grounds for believing, that the subcontractor is unaware of when the contractor requires such details. It might be held that, in the absence of reasonably detailed programming information issued by the contractor to the subcontractor for the entire works, the contractor would have reasonable grounds for believing that the subcontractor was not aware of its requirements. In fact, unless the contractor has specifically advised the subcontractor of his requirements, he must have reasonable grounds for believing that the subcontractor is unaware of those requirements.

SBCSub/D/C is a form of subcontract intended for situations where the subcontractor has a design obligation but the subcontract does not have specific provisions for dealing with the situation where there is a need to integrate the details of various trades under differing subcontracts. Such situations need positive coordination by the contractor and/or the lead designer. It is suggested that these requirements need to be incorporated by express terms of the subcontract and in excess of those of clause 2.7 quoted above. These may include for the subcontractor to provide preliminary interface details by a specified date, to attend design coordination meetings and to incorporate the requirements of others as instructed by the contractor.

Programming design development

SBCSub/D/C at clause 2.6 states in general terms the nature and extent of the design information which the subcontractor is to provide. This is described as only 'such as are reasonably necessary to explain and amplify the Sub-Contractor's Proposals' and 'all levels and setting out dimensions' which the subcontractor proposes for the carrying out of its designed portion. What is reasonably necessary to explain and amplify the proposals must depend on a number of circumstances and the needs of the specific project. However, by reference to the Contractor's Design Submission Procedure, referred to in clause 2.6.2, under item 1 of that procedure the necessary design information is to be as stated in the Employer's Requirements or the Contractor's Proposals.

In the absence of an express requirement for the subcontractor to prepare and submit its detailed design by a specified date, the subcontractor is under no obligation to provide its details to the contractor in advance of the date necessary for the subcontractor to obtain approval for its design, in time to satisfy its programme for off-site works, if any.

Where it is necessary for the detailed design of one subcontractor to be provided to another or to be integrated or coordinated with the design of other subcontractors or other parts of the design team, there needs to be an express requirement to that effect in the subcontract provisions. In this respect the provision of SBCSub/D/C clauses 2.6 and 2.7 and the Sub-Contract Particulars at item 5 of SBCSub/D/A are not considered to be sufficient.

Clause 2.6.2 of SBCSub/D/C sets out in general terms the requirements for the submission by the subcontractor of the Sub-Contractor's Design Portion. This is to be provided to the contractor as and when necessary from time to time to enable the contractor, in respect of the Sub-Contractor's Designed Portion, to observe and perform his obligations under clause 2.9.3 of the main contract and schedule 5. The subcontractor is not to commence any of the work to which such a document relates before that procedure has been complied with.

This clause, on the face of it, puts a heavy burden of responsibility on the subcontractor to provide the Sub-Contractor's Design Portion details to enable the contractor to carry out its obligations under the main contract. However, in retrospect this is limited by the reference to the Contractor's Design Submission Procedure. Should the subcontractor's design portion details be required at an earlier date for any purpose, whether by reason of the contractor's requirements or those under the main contract not covered by item 1 of the Contractor's Design Submission Procedures, the subcontractor will have no responsibility or liability under this clause.

It is a common situation that, following the detailed design of one subcontractor, the design coordinator will require the modification of that design either for reasons of coordination or otherwise. It is therefore advisable for the contractor to appoint its specialist subcontractors in sufficient time for such coordination to take place, and to make provision for the subcontractor's details to be provided by a certain date or within a specified time, a condition of the subcontract.

Contractor's Design Submission Procedures

SBCSub/D/C at clause 2.6.2 requires the subcontractor to provide to the contractor its design documents and other information to enable the contractor to observe and perform its obligations under the Contractor's Design Submission Procedure annexed to the main contract.

At item 1 the contractor is required to provide two copies of each design document to the architect/contract administrator as required by the employer's requirements or the contractor's proposals. Such documents are to be submitted in sufficient time to allow for the incorporation of any comments of the architect/contract administrator into those documents prior to the documents being used for procurement and/or the carrying out of the Sub-Contractor's Design Portion works.

Although not expressly stated, it is apparent that the contractor is to allow sufficient time for the Sub-Contractor's Design Portion details to be transmitted between the subcontractor and the contractor and between the contractor and the architect/contract administrator and back twice, and for twice the stated period for the architect/contract administrator's comments. To this must be added a period for the subcontractor to make the required changes. This may easily amount to a total period of eight to ten weeks from the completion of the subcontractor's drawings to the work being released for procurement and/or execution.

Item 2 allows the architect/contract administrator 14 days in which to return the design documents marked as necessary A, B or C, thereby indicating whether the details are cleared for procurement/execution or not. Item 5 sets out the meaning of the various statuses. Should the architect/contract administrator fail to respond in time, the design documents will be regarded as having A status.

Item 4 expressly requires the architect/contract administrator to identify by written comment why he considers the design document has failed to obtain A status and is not in accordance with the contract/subcontract. Should the subcontractor disagree with the comments of the architect/contract administrator, then the contractor is to notify the architect/contractor within seven days. This requirement places urgency on the contractor to ensure that the design details are returned to the subcontractor promptly and that both contractor and subcontractor are very proactive in responding to such comments of the architect/contract administrator. The architect/contract administrator then has a further seven days either to confirm or withdraw such comment.

The major consequence of item 7 is that the written comments of the architect/contract administrator are expressly not variation instructions and will not be considered as such unless challenged by the subcontractor/contractor and the change confirmed as such. A similar situation exists under Sub/MPF04 at clause 12.9.

SubMPF04 has similar arrangements for the submission for comments by the contractor, at section 12. No reference is made to either the procedures under the main contract or to any third party as in SBCSub/D/C. The period for the provision of comments by the contractor is 21 days from receipt of any design document, or 21 days from the expiry of the period for submission on any design programme, as contained within the requirements or proposals.

Payment for design

Design is defined, under section 104 (2)(a) of the HGCRA, as construction work. It follows therefore that for the design work in a design, manufacture and install subcontract, the subcontractor is entitled to be paid for its design under the provisions of section 109 of the HGCRA.

Where the subcontract involves design, the parties should agree the sums due to be paid for the design and the method of payment. Such method should include the method of valuation for the subcontractor to carry out design changes resulting from changed requirements, whether from design coordination or otherwise.

In this respect clause 4.9 of SBCSub/D/C is ambiguous. While sub-clause 4.9.1 makes the first interim payment due in relation to the commencement of the subcontract works, which suggests that payment for design work is to be paid ahead of work on site, it goes on to state that if there is no stated date for interim certificates under the main contract, then first payment shall in any event not be later than one month after the date of commencement of the 'Sub-Contract Works on site'. The reference to 'not later than' is considered to mean that in the absence of any other provision or requirement, the first payment due date will be one month after the commencement on site. It will always be possible, should the payment mechanism be inadequate, for the subcontractor to make an earlier application, so establishing the first payment due date.

Shop or fabrication drawings

In addition to the formalised requirements of subcontract forms such as SBCSub/D/C and DBSub/C, subcontractors may have to make decisions which impinge on the design or detail of the finalised installation, by way of the formal specification for their work or by reason of necessity. The specific requirements will vary from trade to trade and may or may not require prior review or approval before execution, whether for fabrication of components or insulation on site. For some trades the requirements may be to combine details obtained from different consultants' design drawings to produce a trade-specific drawing or drawings. For example, a detailed drawing for an entrance door or canopy may need to incorporate the requirements of the architect, the structural engineer and those of the services consultants.

It is now common for structural steel subcontractors to be required to design the connections between structural members from the detailed forces acting at each connection as provided by the structural engineer. Whether this is a requirement or not of the subcontract, the fabrication shop will require detailed engineering or setting-out drawings for each component, giving the exact lengths of the member, together with the positions and details of all cut-outs and holes required. Similar detailed drawings will be required for all engineered components, including balustrading and other architectural metalwork.

The consultant designers will rarely provide joinery manufacturers and shopfitters with more detail than the visual surfaces and the required provisions within the component for other items and/or services. The manufacturer will prepare its 'rods', which traditionally were full size and would give full detail of all the necessary and proposed carcassing structure and fixing/jointing details, both between subcomponents and between the components and the structure.

Stone or precast concrete components typically require the production of detailed unit drawings combining both the dimensional requirements obtained from the architect's general arrangement drawings of the elevations and detailed structural engineer's bracketry and fixing requirements. Such combined detail drawings will be necessary whether the design of the fixings is the responsibility of the subcontractor or of others. In the case of precast concrete the subcontractor may also need to incorporate components such as additional reinforcement and cast-in inserts to facilitate the handling and lifting of the components, both when being removed from the mould and for placing on transport and positioning on site.

Mechanical and electrical services contractors are often required to work from schematic drawings which require the subcontractor to position the pipe or ductwork in space and to provide for the necessary sets and bends to accommodate the structural features and or other services. Electrical services design intent drawings will frequently only indicate the circuits on which fittings are to be accommodated and it will be left for the subcontractor to decide the sequence of connection and the associated cable routes.

Decisions on the setting out of finishing materials, such as tiles to floors, walls and ceilings or panelling, may have been made and specified in the specification or on the drawings for the subcontract works. However, in many instances no detailed requirements are provided or instructions given to the subcontractor prior to the commencement of the works. In the absence of specific requirements, the subcontractor will anticipate being allowed to install the work in the most economic way, which may not provide a result compatible with the requirements of other trades or the design coordinator.

From these examples it can be appreciated that general rules are inappropriate and/or ineffective to cover the needs and requirements of all trades. It is therefore necessary to specify in the trade specification precisely what is required of the subcontractor and what is to be left to the subcontractor's discretion or choice. Where no limitations or restrictions are placed on the subcontractor, it will be for the subcontractor to carry out the work as best suits it, subject only to it being carried out in a workmanlike manner.

Supports and fixings

It is common practice for subcontractors to be responsible for the necessary brackets, hangers and fixings to connect their work to the main structural components. So long as such connections can be made generally at will, such as is generally the case when connecting to reinforced concrete structures or masonry walls, the subcontractor should have no problems in fulfilling such a requirement. However, difficulties arise where the adjoining surface or components are of an inappropriate nature for direct fixing and require intermediary members to span between suitable components.

Responsibility for such intermediary supports may or may not be adequately defined in the Sub-Contract Documents. Should there be doubt as to the liability for them, it is likely that the liability will not be with the subcontractor. A simple requirement to provide 'all necessary brackets, hangers and fixings' is unlikely to be held to include undefined secondary support works, unless the Sub-Contract Documents make very clear the nature and availability of suitable structural members to which a fixing may be made. The subcontractor's liability in respect of secondary steelwork, without a clear requirement within the Sub-Contract Documents to the contrary, is unlikely to include for any components installed for use by other trades, either jointly or otherwise, or requiring design input by others. Where supports are to be for the use of several trades or services it will be for the contractor to define which if any of the attaching trades or services should provide and install such common supports.

While the specification may limit or define the nature or type of brackets, hangers or fixings that may be used, in the absence of any such specification or requirement it will be for the subcontractor to select any suitable and satisfactory form of bracket, hanger or fixing that it chooses. Where fixings are to be made in areas where failure may give rise to serious safety risks, the subcontractor may be required to prove not only that the fixings chosen are satisfactory in themselves but that their installation is effective. This may be done by requiring that periodic 'pull-out' or other tests be carried out to monitor the standard of workmanship by the subcontractor.

Temporary works

The subcontractor may be required to provide the temporary works necessary for the carrying out of its work, or to be the subcontractor carrying out temporary works for the project as a whole on behalf of the contractor.

Temporary works will include:

1. temporary support works such as shoring, sheet piling, trench supports and false works for in situ reinforced concrete, brick or masonry and other major structural components that require positioning before connection to the structure, or where they do not work structurally alone, such as post-tensioned prestressed components;
2. temporary means of access and hoisting such as temporary roads, scaffolds, hoists, lifts and cranes;
3. temporary services such as water supply and drainage, electric power and lighting, evacuation alarms and emergency lighting arrangements, security and fire-fighting equipment;
4. site accommodation such as offices, changing and drying rooms, toilet and washing and messing facilities and in addition provision for the storage of materials.

Except for projects of a very simple nature and of limited duration, some aspect of temporary works will be required. At tender stage it is for the parties to identify which facilities are to be supplied by the contractor and which by the subcontractor. In a limited number of situations the temporary works may be designed by one party and carried out by another. Such temporary works might be, for example, shoring or sheet piling work where there is a need for early agreement by way of a party wall agreement. In other situations there may be conditions or restrictions placed on the subcontractor by reason of the design of the permanent works, for example the ability of the permanent floors to take the combined dead and live loads imposed when the slab above is being cast. Such limitations may relate to bearing points such as where there is a trough or waffle floor construction.

Where facilities are to be provided by the contractor there must be certainty as to what is being supplied. There may be a serious difference in the expectation of the parties if the contractor is only to allow the subcontractor 'free use of standing scaffolds', a phrase which on its own guarantees nothing to the subcontractor. However, should the contractor undertake to provide scaffolds 'as necessary' for the subcontractor's use, then this can be seen as an open cheque for the subcontractor's requirements.

Access to the work is an essential requirement for the carrying out of the work. At tender stage the subcontractor needs to identify its likely needs and possibly make a contingency allowance for the unidentified and unforeseen. However, delays to the subcontract works and/or changes by way of variations or changed sequence may lead to associated changes in scaffold and access needs. Such additional costs should be recoverable by the subcontractor either by way of the pricing of the variation or as loss and/or expense under the subcontract conditions.

The design of temporary works must take into consideration the method of both installation and dismantling and removal. Whilst at the erection stage there may be an open site or floor with easy access by crane, by the time of removal the temporary works may be enclosed within the confines of the building. Such restrictions will affect the choice of equipment to be used, limiting the size and/or weight of the components, or may mean that the temporary works cannot be recovered intact but must be cut up into handleable pieces, with the consequential loss of the capital value.

Not infrequently the temporary works may require the adjustment of the design of the permanent works, or the insertion of fixings into such work. Any such interface with the permanent work will require coordination with, and agreement from, the consultant designer if not the employer as well. Such matters will range from the creation of crane bases within basement slabs, sometimes including the incorporation of lost mast sections or the provision for tie-backs for crane or hoists, to the provision of drilled inserts for scaffold ties and/or through bolts as shutter ties in reinforced concrete walls.

Temporary works such as scaffolds and temporary services will normally be installed on a progressive basis and once installed will be subject to frequent adjustment and alteration as permanent works proceed. Unless the original provider of the temporary installation carries out all subsequent extensions and alterations there will be a division of responsibilities, which may cause difficulties if any failure in the installation occurs. Subcontract arrangements for such installations should include for regular inspections by the providing subcontractor, as well as detailed arrangements with regards to response times and payment for the carrying out of alterations and adjustments.

Some temporary works require the provision of purpose designed and manufactured equipment, which is likely to become the property of the contractor on the supply or supply and installation of that equipment. In many instances standard equipment will be used, such as scaffold or shoring and bridging equipment.

The payment for such equipment will vary with the subcontract arrangements. The most common arrangement will be for the subcontractor to price for its supply, erection, dismantling and removal from site, and its price will include for the hire up to a stated number of weeks, with a weekly hire rate in the event the equipment is required for a longer period. Other arrangements more suitable for long duration projects may include the sale of the equipment to the contractor with or without an agreement to buy it back at the time of removal.

Consultant designers as subcontractors

The contractor may employ consultant designers, as subcontractors, to carry out design activities where the contractor has undertaken the design and/or the design development. Alternatively, consultants may be employed as sub-subcontractors providing a design service to subcontractors under the main contract who have a design development responsibility. In either case it will be for the commissioning contractor or subcontractor to decide the extent of the design service required.

Design services will include any or all of the following:

1. outline details and specification;
2. details sufficient for statutory planning consent;
3. details sufficient to satisfy building control requirements;
4. indicative design requirements for development by others;
5. full design details capable of construction without further detailing;
6. provision of details only as and when required;
7. supervision and inspection of the as constructed work.

A designer as a subcontractor will generally have a restricted brief to provide expertise not available within the contractor's organisation and/or sublet to other specialists. There will therefore normally be a considerable difference in responsibility between, for example, an architect providing full architectural services to a building owner and design services as a subcontractor to a contractor under a design and build contract.

In the Court of Appeal decision in *Bellefield Computer Services and Others* v. *E. Turner & Sons Limited and Others* [2002] EWCA Civ 1823, the court had to consider the liability of an architectural practice in tort to the ultimate building owner, Unigate, following a serious fire where the fire spread within the building due to a defectively constructed fire partition wall. Although this case is based on a liability in tort, most of the relevant principles apply in contract as well. Lord Justice May made the point, at paragraph 73, that the architect's responsibilities were those set out in the agreement between it and the contractor:

> 'The significance of the distinction between the duty in tort and the duty in contract usually relates to questions of limitation. The scope of the duty depends on the express and implied terms of the architect's contract of engagement. Those terms may be affected by the extent to which others – e.g. engineers, contractors, specialist subcontractors – are also engaged to do things in relation to the building, the scope of their responsibilities and their interrelationship with the architect.'

He went on to describe how design responsibility may become divided between differing parties as a consequence of the provisions of individual contracts and/or the assumption of liability. He said, at paragraph 76:

> 'The extent of an architect's responsibility for the detailed working out of construction details for which he has provided an underlying design depends again on the express and implied terms of his engagement and its interrelation with the responsibility of others. The scope of any such responsibility depends on the facts of each case. There is a blurred borderline between architectural design and the construction details needed to put it into effect. Borderlines of responsibility cannot be defined in the abstract. A carpenter's choice of a particular nail or screw is in a sense a design choice, yet very often the choice is left to the carpenter and the responsibility for making it merges with the carpenter's workmanship obligations. In many circumstance, the scope of an architect's responsibility extends to providing drawings or specifications which give full construction details. But responsibility for some such details may rest with other consultants, e.g. structural engineers, or with specialist contractors or subcontractors, depending on the terms of their respective contracts and their interrelationship. As with the carpenter choosing an appropriate nail, specialist details may be left to specialist subcontractors who sometimes make detailed "design" decisions without expecting or needing drawings or specifications telling them what to do. In appropriate circumstances, this would not amount to delegation by the architect of part of his own responsibility. Rather that element of composite design responsibility did not rest with him in the first place.'

Lord Justice May defended the architect's position in that case, at paragraph 77:

> 'In the present case Watkins were engaged, not by the building owner, but by the contractor. They were not the supervising architect. The extent of the duty for which they were engaged was circumscribed as found by the judge. There were specialist subcontractors for the fire lining engaged by Turner.'

Just as it is for the contractor to ensure that all elements of construction work are allocated to defined subcontractors or retained for execution by its directly employed workforce, so with design. Where the contractor takes over a partly developed design and sublets the completion of that design to a subcontractor, the nature of the liability passed to the specialist subcontractor may or may not be that which the contractor has taken on.

If, for example, as part of an enquiry for a design and build project, the contractor receives drawings of the steel frame structure which indicate the number and size of components required by the design, this may or may not be a warranty by the employer that such details will satisfy the overall design brief. Should such drawings be indicative only, the outworking of a detailed design may or may not increase the amount of steel required to satisfy the design brief.

In the case of *Blyth & Blyth Ltd* v. *Carillion Construction Ltd* (2001) 79 Con LR 142, where a consultant design engineer was novated to the contractor, it was held that the contractor could place no reliance on the design prepared pre-novation, against which the contractor had priced the work. The contractor had undertaken to provide the building on a design and build lump sum basis and it was for the contractor to satisfy itself at tender stage that the proposed design and/or assessment of material requirements was sufficient to satisfy the employer's requirements. Once subcontracted to the contractor, the engineer was required to provide design details which satisfied the employer's requirements. The extent to which these required materials and/or labour were above that indicated in the details provided at tender stage was irrelevant to the contractor's obligations to the employer.

A contractor in such circumstances, not wishing to rework at tender stage the information provided, could seek a warranty pre-contract from the engineer to be novated that it could design a satisfactory structure within the parameters detailed in the tender documents.

One difficulty for consultant designers required to be novated from the employer to the contractor will be the fundamental change in objective. A consultant employed by the building owner has a duty to the building owner. On novation to the contractor the consultant's duty is to the contractor. So far as design is concerned that becomes a duty to provide the most cost-effective design which is compliant with the employer's requirements as defined in the contract. Should it be found that the consultant's design is significantly more than the minimum required, then he may become liable to the contractor for the excess cost incurred by the contractor.

Finalisation of the Sub-Contract Agreement

A practical issue that will need to be addressed in all subcontracts involving the provision of design and/or design development services, is whether the subcontract has fully defined what is to be provided by the subcontractor. It is important to know at any time during the design process whether design by the subcontractor is being provided as an inducement to obtaining the work. The extent to which a concluded agreement can be said to have been made, or that the arrangements are only provisional at that moment because of the number of issues still to be resolved between the parties, will determine the contractor's right to require performance and/or adherence to a time restraint by the subcontractor. While most subcontractors are prepared to provide some detailing work pre-subcontract, there will be an increasing reluctance to expend resources providing further or alternative details, without some assurance that the cost of such expenditure will be recovered.

In most situations, changes during the detailing of the installation will be resolved by negotiation post-subcontract, leading either to acceptance by the contractor of the subcontractor's details, whether by way of Sub-Contractor's Proposals under either SBCSub/D/C or DBSub/C, or otherwise. In other situations, especially where the requirements are at the leading edge of technology, it may not be possible to achieve the full requirements of the specification or not reasonably within the limits of the subcontract period or price. Whether the subcontractor or the contractor will have to stand the loss arising from such a failure will depend on the nature and provisions of the subcontract.

In circumstances requiring advanced design, the contractor or consultants may have reservations in recommending or advising on acceptance on the basis of price alone and may set out a number of requirements to be satisfied before tender acceptance. In the event that such a subcontract is concluded by way of acceptance of the tender or agreement prior to the full satisfaction of the tender requirements, then such a subcontract may well be provisional in nature only. In such a situation the subcontract may have an implied term to the effect that, should the subcontractor be unable to satisfy the outstanding technical matters or to satisfy them in such a way that the other conditions of the subcontract can be met, then under such circumstances the subcontract may be terminated or the terms amended to allow performance as reasonably necessary or possible.

Where the design brief is vague and uncertain and the contractor is heavily reliant on the specialist subcontractor to provide proposals for consideration, before the conclusion of a subcontract for the construction work it may be appropriate for the work to be let in two stages: design and construction. In such circumstances, once an acceptable design has been achieved, the subcontractor will be paid the agreed sum for the design development and may or may not be selected to carry out the construction work itself.

Chapter 7
Instructions and Variations

The right to change

The right of a contractor to vary the requirements and/or conditions of a subcontract once it is concluded is a right under the contract itself. Where there is no such right, then there is no obligation for the subcontractor to accept any such instruction. The parties may agree at any time to vary the contract, and in many instances this is effectively what happens, even though this may not be by way of an express agreement. Thus where a contractor requests or instructs a change and the subcontractor responds to that request, there is in effect a change to the subcontract to the effect instructed. The major problem will be to define the result of such changes and this is discussed below.

The standard forms of construction subcontracts, and indeed most bespoke forms, will make express provisions for both the subcontract works and the conditions under which they are to be executed to be varied. For such provisions to be effective, the subcontract must state who is empowered to make changes, the allowable nature of such changes and how the effect in both time and money is to be considered and the subcontractor compensated. There are many ways in which this may be achieved and the parties must be clear what those procedures etc. are, otherwise difficulties will arise. Bespoke forms of subcontract frequently grant the contractor rights and impose on the subcontractor obligations for which no remedies are set out or defined. In such situations it is not infrequently suggested and/or claimed by the contractor that the subcontractor has no remedy. Without express agreement to that effect it is suggested that the subcontractor will have a remedy for complying with any changed requirements of the contractor, and where these are undefined they will be on a reasonable or fair price and/or time basis.

Both the SBCSub/C and ICSub/C forms of subcontract make provision for the issue of directions of the contractor, at clause 3.4. ShortSub has a similar provision within clause 9.1, but Sub/MPF04 has no such express provision. This clause defines the nature of instructions as any reasonable direction in writing in regard to the Sub-Contract Works and states that such instructions will not vitiate the subcontract. Clause 3.5 requires the subcontractor generally to comply with such instructions or directions, as does Sub/MPF04 at clause 3.1. Clause 3.7 of SBCSub/C deals with instructions, other than in writing. Strangely no other forms allow for the confirmation of oral instructions by the subcontractor and only ShortSub allows for the confirmation of oral instructions by the contractor.

It is acknowledged that the casual approach by many contractors to the issue of instructions is to be deplored. This may reflect similar sloppy procedures by many consultants under main contracts. However, the omission of the provision for requiring subcontractors to confirm instructions other than in writing is to be commended. However, it is likely that many subcontractors will be intimidated into accepting oral instructions for some time to come, regardless of the express provisions of the subcontract to the contrary.

Failure to provide provisions within the subcontract terms for the right to vary the subcontract works and/or the method and/or sequence in which they are to be carried out, may put the contractor in breach when it fails to comply with the Sub-Contract Agreement. The subcontractor will in such circumstances either be entitled to the resultant loss by way of damages as a consequence of the contractor's breach; or, it is submitted, by accepting the contractor's instructions and/or directions the subcontractor will be entitled to compensation for both time and money on a fair and reasonable basis.

Nature of change

The subcontract should state clearly what changes may be made under its provisions. It is often asked whether the subcontractor must take on extra work, as instructed, even if it will put a severe strain on the subcontractor's resources or when such work is outside the scope of the subcontract and/or the subcontractor's experience or expertise. The answer will depend on what the subcontract provides for. There may be provision for a subcontractor to reject an instruction on specific grounds, but there may not.

Clause 5.1 of both SBCSub/C and ICSub/C makes a distinction between, at clause 5.1.1, variations generally and, at clause 5.1.2, instructions of the architect under the main contract.

Clause 5.1.1 of SBCSub/C defines a variation as 'the alteration or modification of the design, the quality or (except where the Remeasurement Basis applies) the quantity of the Sub-Contract Works' and specifically includes the following:

1. the addition, omission or substitution of any work;
2. the alteration of the kind or standard of any of the materials or goods to be used in the Sub-Contract Works;
3. the removal from site of any work executed or materials or goods brought on site by the subcontractor for the purposes of the Sub-Contract Works other than work, materials or goods which are not in accordance with this subcontract.

Clause 5.1.2 refers to an imposition in an instruction from the architect/contract administrator issued under the main contract. Such impositions are any obligations or restrictions in regard to the matters set out in this clause 5.1.2, or the addition to or alteration or omission of any such obligations or restrictions so imposed or required by the numbered documents and the Schedule of Information and its annexures. Four express impositions are listed as follows:

1. access to the site or use of any specific parts of the site;
2. limitations of working space;
3. limitations of working hours;
4. the execution or completion of the work in any specific order.

The subcontractor by reason of clause 3.5 is to comply with all directions 'forthwith' except where the subcontractor makes a reasonable objection to a variation issued under clause 5.1.2.

Clause 3.5 refers to all directions issued to the subcontractor under clause 3.4 and makes two exceptions:

1. Where such a direction requires a variation by reason of an imposition by the architect, the subcontractor need not comply, to the extent that he makes reasonable objection to it in writing to the contractor.
2. Where a direction for a variation is given under the pre-pricing provisions of clause 5.3.1 which requires the subcontractor to provide a schedule 2 quotation. Then the variation shall not be carried out until the contractor has issued either written acceptance or a further direction under clause 5.3.2.

The word 'forthwith' cannot be taken to refer to the physical work on site but to the incorporation of the direction into the subcontractor's planning, management and procurement activities. It will be for the contractor, in issuing its directions, to advise the subcontractor when it requires the work to be carried out. Failure by the contractor to make clear its requirements as to the timing of the work on site may make time for this activity at large.

SBCSub/C at clause 5.1 by definition distinguishes variations from other directions of the contractor. There are no express provisions within SBCSub/C, similar to those for variations, as to how other directions are to be managed, valued or paid.

While minor instructions may be of little effect, instructions considerably increasing the amount of the subcontract work and/or of specific types of work, and/or directions to delay or accelerate the works, may have a major effect on the subcontractor's resources, which must be considered when assessing the effect on the subcontractor of the instruction and/or direction.

There is no express provision for a subcontractor under SBCSub/C to object to the directions of the contractor other than where the direction is a variation under clause 5.1.2. However, it is considered that there will be an implied term that a subcontractor may refuse a direction on the basis that it is unreasonable or in conflict with any duty of the subcontractor either under statute or at common law. Such duties would include health and safety issues and/or directions which could adversely affect the quality of the subcontractor's design or work.

Sub/MPF04 has much simpler provisions at clauses 3.1 and 3.2:

1. The subcontractor is generally to comply with all written instructions issued by the contractor in connection with the design, execution and completion of the Sub-Contract Works.
2. Instructions can be given that are not change instructions, when the subcontractor will not be entitled to any further payment or to an adjustment to the period for completion; nor is the subcontractor to be relieved of any of its obligations under the subcontract.

Under these provisions the subcontractor is to comply with all written instructions issued by the contractor. It is implied by clause 3.1 that instructions other than in writing are of no effect and it is entirely for the contractor to make his instructions effective by putting them in writing. The implication of clause 3.2 is that except where specifically provided within the subcontract, all instructions are to be regarded as a change and to be paid for and a time adjustment made as appropriate to the period for completion.

Instructions outside the subcontractor's competence

If the instruction is outside the competence of the subcontractor, then it must either expressly or impliedly be a part of that instruction that the subcontractor shall take all necessary steps to implement such an instruction to the standards required by the subcontract. It is likely that one test as to what standard would be required is that it would be the standard which would normally be found in someone holding themselves out as being in the business that they practise. It is unlikely that, if objection is made to the instruction, there could be any implied warranty beyond those expressed or implied in the subcontract as executed. It cannot be the case that, by imposing an instruction on a subcontractor to carry out work for which it states it has no expertise and/or suitable resources, a duty can be imposed on the subcontractor that it did not have prior to that instruction. If the subcontractor is employed on a reasonable skill and care basis, it is likely that this skill and care will extend only to that of a person practising the trade or profession of the original subcontract and necessary for executing the work subcontracted for, and only work of such nature.

In any event, the subcontractor will have a duty to engage such expertise and/or skilled labour as it identifies as necessary to carry out any work, including any instructions it undertakes, but will be limited to that which it identifies as necessary or should have identified as necessary. Again, what it should have identified will be limited to the expertise of that trade or profession and not extend to the wider knowledge and expertise of a general contractor. There is also a duty on the subcontractor to consider the effect of carrying out work as instructed. There are apparent conflicts in reported cases, which may be distinguished on grounds of safety rather than product performance. (See also the section 'Duty to warn that instructions may give rise to defective work' in Chapter 7.)

Pre-priced variations

A contractor may at any time request a subcontractor to provide a price for either additional or varied work. Without express provisions within the subcontract there will be no obligation for the pre-pricing of varied work; however, it will generally be to the subcontractor's benefit for it to price and obtain agreement to such price in advance of a firm instruction to carry out such additional or varied work. Where additional work is required there is likely to be no obligation for the contractor to instruct the subcontractor and the contractor will be at liberty to instruct others if the contractor so wishes. Once the contractor has sought and obtained a price from the subcontractor ahead of giving a firm instruction, it will be unlikely that the contractor will subsequently be able to insist on the subcontractor accepting such an instruction on a less advantageous basis to the subcontractor.

At some stage the employer's agent may seek to select a colour. In doing so, he must be aware that his choice may be limited if there is not to be a cost penalty. There will rarely be any express requirement for a subcontractor to indicate a change in price before carrying out an instruction. Indeed the contrary is more usual in that there will be an express requirement to carry out instructions 'forthwith'; see SBCSub/C condition 3.5. There may however be an implied obligation, if asked to provide samples for selection, to identify if any are outside the range contracted for and/or for which an additional price will be required or which are on long delivery. Should the subcontractor, without a specific request from the contractor, seek a selection from a range offered by the subcontractor then such an implied term will be even more likely. It is suggested that in the absence of an indication to the contrary, there will be a presumption that all samples offered are of a cost and/or availability allowed for in the subcontract.

Who may instruct

The subcontract may limit the right to vary the subcontract to a specific person named or indicated by position in the Sub-Contract Documents. In practice, especially on a large project, even when the subcontract limits such right to one individual, other members of the contractor's management team may give instructions. Even if no express change to the subcontract conditions is made it is unlikely that, where generally such other staff members are directing the work and/or issuing instructions, these will not be effective. In such circumstances, the contractor will be deemed to have waved compliance with the strict interpretation of the subcontract arrangements. However, where in general all instructions are in writing and are issued or at least signed by a specified person, then such a literal interpretation could be upheld. Of course, such restrictive clauses apply both ways and a subcontractor will be within its rights to insist that all instructions are properly issued before compliance.

SBCSub/C at clause 3.4 requires the contractor to issue instructions in writing and does not seek to limit this to a specific member or class of the contractor's team. A requirement that all instructions be issued in writing by a specific person creates particular difficulties in connection with the issue of directions concerning the work actually in progress, which require both an immediate instruction from the contractor and action by the subcontractor. To refuse to respond to a reasonable request because the instruction is not in writing and/or signed by the named authority, will almost certainly be seen as obstruction on the subcontractor's part. To avoid the possibility of such circumstances arising, the contractor must be careful in the use of restrictive clauses and/or the unreasonable enforcement of the strict requirements of the subcontract.

Post-contract instructions

The right to issue instructions and vary the work will normally die on completion of the subcontract works. This will in many cases be before the completion of the main contract works. This could create the situation where the contractor has an obligation under the main contract to carry out additional work, relating to or involving that of the subcontractor, without the subcontractor having a reciprocal obligation to the contractor. In addition, there will normally be a period for making good defects after completion of the works, during which additional and/or alteration works may be requested. It will be quite common for the employer, on occupation of a building, to require further changes and/or additional works to be carried out and it will naturally seek for this to be done under the main contract, and the contractor in turn under the subcontract.

Unless there is express provision for such instructions in the subcontract, such work, if carried out, will not necessarily be under the terms of the subcontract and unless pre-priced will probably be subject to payment in restitution on a fair price basis. A subcontractor undertaking such work should make it quite clear, in advance, how it expects to be paid, and any other relevant matters such as access to the work etc. Such a letter would be an offer and unless countered will be subject to the normal rules of contract formation.

The case of *Costain and Tarmac* v. *Zanen*, mentioned in the previous chapter, concerned a dredging and earth moving subcontract in connection with the construction of a tunnel under the Conway river, as part of the North Wales coast road improvements. As part of the project the tunnel sections were precast in a temporary dry dock excavated by the subcontractor. The head contract allowed for various options for the use or reinstatement of the dry dock. In the event, the contractor negotiated a separate contract with a different employer to convert the dry dock into a marina. The contractor then instructed the subcontractor to carry out further works. At that time the subcontractor queried whether those works came within the ambit of the existing subcontract and the contractor gave assurances to the effect that they did.

Following an award in arbitration, matters of law were appealed to the High Court where Judge David Wilcox decided that the additional works related to a different head contract and that therefore the additional work was not carried out under the existing subcontract. He said, at page 92:

> 'The reality is that both Mr Ramsey [Counsel for Zanen] and Mr Fernyhough [Counsel for Costain and Tarmac] were able to present to the arbitrator a detailed analysis of the contractual position and he concluded, in my judgement correctly, that the works undertaken by Zanen, the respondent to this appeal, were outside the contract. The finding that it was more likely than not that the employers for the marina works and its development were the Crown Estate Commissioners and Pearce Developments is consistent with the evidence cited in the award and earlier referred to. It is consistent with the analysis of the contractual position as I have found it to be.
>
> The instructions from the joint venture described in paragraph 8.23 of the amended points of claim did not constitute authorised variations of the subcontract works within the meaning of clause 8(1) of the subcontract made between the parties on 1 June 1987, because such instructions required work to be done outside the scope of the respondent's obligations under the subcontract.'

As a consequence of there being no subcontract the subcontractor was to be paid on a *quantum meruit* basis.

Instructions relating to subcontractor design

Specialist subcontractors are frequently appointed to provide design development services as well as manufacturing and installation work. Such services may be less clearly defined than those for manufacturing and installation. Further, the extent to which the outcome of that design development will be a variation to the subcontracted work will depend on the express provisions of the subcontract, including the documents included within it such as pricing schedules, drawings and specification.

At its simplest level, the obligation of the subcontractor will be to take the design information supplied to it and develop a compliant working solution. If the subsequent design can be shown to be non-compliant, then manifestly the subcontractor must adjust his design as necessary and rectify it so as to make it compliant. Any other change required by the contractor will be a variation instruction and subject to all the consequences of a variation instruction with regards to time, money, etc.

Since the employer's outline design will be just that, argument frequently arises as to what is or is not a change requirement. A drawing showing glazing bars, for example, does not necessarily, just because a certain number have been drawn, mean that this is a specified number. If the employer requires glazing bars in specific positions, it is submitted that he must expressly say so, otherwise the number will be subject to the subcontractor's design development considerations. As with the written provisions of the subcontract, so with the drawn information; the *contra proferentem* rule will apply and where there is uncertainty and/or ambiguity, then the interpretation of the subcontractor is most likely to prevail.

In contracting to do design development, it is unlikely that the subcontractor is warranting that the design can be developed or can be developed in a certain way. The subcontractor, however, probably is undertaking to carry out the necessary work to resolve any unforeseen design difficulties, but only to the extent its work is to be changed. Any changes necessary to the adjoining work will not normally be for the subcontractor's action or to its cost. Conversely, the subcontractor is unlikely to be liable for any additional work resulting from changes arising from an adjoining subcontractor's design development. There is inevitably a residual risk, which unless expressly allocated within the subcontract arrangements will rest with the contractor or employer.

The case of *Norwest Holst Construction Limited* v. *Renfrewshire Council* (20 November 1996, unreported) concerned the design and construction of piles to support the abutments to a railway bridge. These were to be constructed within a cofferdam, which restricted the working space. The contractor was able to design the piles to specification but these could not be constructed because of the limited working space. The problem was resolved by changes to the design of all these elements: piles, cofferdam and abutments. The court considered the issues in relation to the specific conditions of the contract; these included a specific condition 76, which provided:

> 'Save in so far as it is legally or physically impossible the Contractor shall construct complete and maintain the Works in strict accordance with the Contract.'

This condition was found not to apply to the design obligations of the contractor. In this regard the opinion of Lord Prosser is most clear, at page 1:

> 'His general obligation under clause 8(1), to "construct, complete and maintain" the works will in such circumstances be an obligation to construct, complete and maintain the works designed and specified by others, and unless and until there is a design, the contractor will not know what it is that he has to construct, complete and maintain. When presented with a purported design, and called upon to construct, complete and maintain the works in terms thereof, his obligations will crystallise: he has been told what to do, and is obliged to do it, unless the opening words of clause 13(1) provide him with an escape, on the basis that it is legally or physically impossible for him to construct, complete and maintain the works as specified by others.
>
> In those cases, such as the present, where by express provision the contractor is responsible for some element of the design or specification of the permanent Works, it appears to me that there is potentially a genuine problem as to what is or may be "impossible". I am satisfied that the opening words of clause 13(1) relate to impossibilities in constructing, completing or maintaining physical works. Such works cannot be works which have not yet been designed. Any impossibility, or indeed difficulty, in design will therefore already have been overcome, by the stage of construction obligations which is contemplated in clause 13(1). But where the obligations of both design and construction are laid upon one person, the contractor, in relation to any one part (or indeed all) of the works, I think there may well be situations in which it is difficult (or indeed impossible) to draw a simple line between his design responsibilities and his construction responsibilities. If he finds it impossible, as designer, to design any form of works which (without changes in the contract) it is physically possible to construct, I suspect that in many situations it would be quite meaningless to argue whether the impossibility was one of design, or one of construction.'

Later in his opinion, Lord Prosser, at page 3, distinguishes between a situation where it is impossible to achieve a design solution which conforms to the specification and a design solution which cannot physically be constructed:

> 'The "impossibility" is not therefore properly to be seen as an impossibility of design but an impossibility of construction.'

Even where there are no express provisions such as existed in the above contract, it is likely that there will be an implied term, to similar effect, that it will be possible to reach a design which satisfies all the requirements of the specification. Where this is not possible, such changes will be made and/or allowed to facilitate an achievable design solution.

Further, problems may arise where the employer's agent instructs change to the subcontractor's design proposals. Under what circumstances and to what extent will the employer's agent relieve the subcontractor of its design responsibilities and take on those himself? It will be the general rule that design liability will be limited to those design decisions taken by each party to the design. Thus if the design consultant has decided on a performance requirement such as a fire rating, rate of flow, etc., then the specialist contractor's responsibility is limited to providing that performance and not for whether that performance is adequate or appropriate. Conversely, it will not be a defence to an inadequate design to maintain that it is fit for purpose if it has not achieved the performance standards contracted for, however overdesigned those standards may be.

If a specialist subcontractor develops the design in a certain way and the design consultant, on consideration of the design solution, instructs that an element is to be changed, it is likely he will take on the design responsibility for that element unless he makes it quite clear that this is not so. It is perfectly possible for the design consultant to present the change either as an instructed change or as a proposal for consideration and/or acceptance by the specialist subcontractor. Therefore, whether the design responsibility moves or not from the specialist subcontractor to the design consultant will depend on the wording of the instruction. In addition, the subcontractor's liability may be subject to any implied or express duty to warn that the design is rendered non-compliant with the specification and/or not fit for purpose by the instructed change.

Both the SBCSub/D/C and ICSub/D/C forms of subcontract, which are specifically intended for use where the subcontractor has a design obligation, limit at clause 3.5.3 for SBCSub/D/C and clause 3.5.2 for ICSub/D/C the effect of a contractor's direction where the subcontractor considers it injuriously affects the efficacy of the design. If the subcontractor is of the opinion that compliance with any direction of the contractor would injuriously affect the efficacy of the design of the Sub-Contractor's Designed Portion, then the subcontractor is, within five days of receipt of the direction, to give a notice in writing to the contractor specifying the injurious effect. In those circumstances the direction shall not take effect unless confirmed by the contractor.

This is an express requirement for the subcontractor to consider the effects of a contractor's direction on the efficacy of the subcontractor's design and to notify the contractor within five days should the subcontractor consider the direction will have such an effect. The subcontractor's notice renders the contractor's direction of no effect unless and until such direction is confirmed by the contractor. Thus the subcontractor, on the receipt of a direction affecting its design, must consider it and must either implement it forthwith or issue a notice under clause 3.5.3 and continue as though no such direction had been given.

The use of the phrase 'injuriously affects the efficacy of the design' may lead to much debate as to its true meaning. A subcontractor, acting as a specialist designer, seeks in the development of its design to achieve both the requirements of the specification, whether by way of fitness for purpose or to achieve a specified performance, and a design that is efficient as regards manufacture, installation, maintenance and demolition. While the satisfaction of the requirements of the specification is an express requirement of the subcontract, there are also implied obligations through the CDM Regulations, which are in turn expressly referred to in the subcontract. In addition, there is an implied term which allows the subcontractor to develop a design which is both the cheapest and most efficient for it to provide and install, subject to satisfying the express and implied requirements of the specification. A direction of the contractor may affect any of the above, and indeed any direction appertaining to the design is likely to impinge on some aspect of the design solution so as to be considered to injuriously affect the efficacy of that design.

It is not considered that clause 3.5.3 as drafted can be considered to limit the subcontractor's right and requirement to issue a notice under this clause, only to directions that affect the design's compliance with the requirements of the specification.

If this conclusion is correct the subcontractor has an express requirement under this clause to issue a notice to the contractor setting out all the injurious effects of the direction and proceed as before unless and until the subcontractor receives a further direction in response to the subcontractor's notice. Indeed failure by the subcontractor to issue a notice under clause 3.5.3 may be taken by the contractor as acceptance by the subcontractor that the instruction has no effect on the subcontractor or its design, with the inference that there are no direct cost or time effects.

Instructions to vary programme

Under most subcontract arrangements, the subcontractor will be given a specified and agreed period of time within which to carry out and complete the subcontract works. For various reasons, it may not be possible for the subcontracted work to be completed within the agreed period and the programme for the subcontract Works will need to be changed. Even if the period for the Works is not altered, there may be a need to change the sequence and/or rate at which it is carried out. Subject to there being the necessary provisions within the subcontract, the contractor will be able to issue instructions or directions to that effect.

The form of such an instruction or direction from the contractor may be by way of an amended programme, either for the subcontractor's own work or for all trades. Such a programme may be on a short-term basis or to contract completion. Alternatively, such instructions or directions may be through minutes of site meetings. The extent to which all these methods of communication can be considered as 'instructions in writing' and/or satisfy any other express requirements of the subcontract will be a matter of construction of that subcontract. If in doubt the subcontractor should confirm back to the contractor, as a formal confirmation, any matter considered by the subcontractor to be an instruction. Where there is no formal instruction, it is suggested that all such written communications are likely to be regarded as instructions in writing.

Where the contractor prevents the subcontractor from carrying out its work in accordance with the period or periods defined in the subcontract, and the contractor fails to issue appropriate instructions or directions as to its changed requirements and/or grant an extension of time, then it is likely that time will become at large and the subcontractor's obligation will become one to complete in a reasonable time. It will be in the subcontractor's interests to obtain such an instruction from the contractor and where necessary to confirm such instructions or directions, if not issued in writing by the contractor. This is so that the contractor cannot, at a later date, claim that the resultant delay or acceleration of the subcontractor's work is not attributable to the contractor's changed requirements and directions. (See also the section 'Acceleration agreements' later in this chapter.)

The effect of instructions

The issue of an instruction will not normally be an acknowledgement by the contractor that the instruction is a change or other matter for which the subcontractor is entitled to have its contractual obligations with regards to time adjusted, and/or an admission that the subcontractor is entitled to an adjustment to the subcontract sum. It will therefore be for the subcontractor to identify the nature and detail of any change arising from an instruction, and notify the contractor to that effect. While some subcontracts will require the subcontractor to give a formal notification to the contractor that its direction is regarded as a variation, such notification will be good practice in any event. Where no notification is given that a contractor's direction will have a cost effect, a claim for payment by the subcontractor may, especially if made promptly, act as a confirmation of instruction as well as an application for payment.

It will normally be for the subcontractor to demonstrate the extent to which, by implementing the instruction or direction, it has a right to a change to the contract completion date and/or for additional payment, and for it to demonstrate the amount of additional time and/or payment that flows from that instruction. It will be of great assistance to the subcontractor to keep accurate records of all aspects of the work, both when operating as anticipated and/or efficiently, and when implementing the contractor's post-subcontract directions and instructions. Being able to demonstrate the output achieved before an instruction or direction, and comparing it with the output for similar work after the implementation of the direction, is one way in which the subcontractor will be able to demonstrate the effect of the instruction.

Implementation of instructions

The subcontract and/or the instruction may indicate when to carry out an instruction. Any requirement must be subject to a test of reasonableness. Many standard forms refer to instructions being carried out 'forthwith'. It is considered that forthwith cannot refer to the physical carrying out of the work, but to the commencement of the necessary management action, which commences with the assessment of the effect of the instruction and such associated pre-site tasks including procurement, drawing production, component manufacture and labour resourcing. In any event, an instruction may in many cases relate to work not to be carried out for some time. Indeed, there is an obligation on the contractor to issue necessary instructions in time so as not to hinder the subcontractor in the carrying out of its work.

The introduction of changed and/or additional work is likely to have a detrimental effect on the other work under the subcontract, by diverting management and other resources from it. Such detrimental effects may include a reduced level of labour control and supervision, with the resultant loss of productivity and/or quality. Again, it is advisable for subcontractors to record in detail their management and administration activities and involvement, in addition to those of their operatives.

Instructions requiring a change to work already carried out

Where a drawing is revised and issued without express comment, and on examination it affects work already commenced and/or completed, should it be considered an instruction to undo or discard the work completed and construct again to the amended detail? In theory the instructing person should be fully aware of the state of work at the time of instruction. In practice this will not be so, especially with regards to work being carried out off-site. It is suggested that an express instruction is required to discard or demolish work already carried out correctly under the subcontract. This is especially true with regard to materials purchased and/or manufactured but not yet installed, and for which payment has been received. Even where payment has not been received but will be sought, express disposal instructions should be obtained. Pending such instructions the subcontractor will be responsible for the protection and security of such materials and/or components. Again, where such protection, storage and/or security will involve expense, then the contractor must be so advised under the loss and/or expense provisions of the subcontract. It is likely that in the absence of an express instruction to undo or discard work, such an instruction should be treated as any other conflicting information and will require a further instruction under the subcontract.

It is suggested that the subcontractor's obligations will normally be to act on the instructions 'forthwith'. Where such action would conflict with other directions of the contractor, then the subcontractor's action should be to seek clarification. If and when such an instruction is confirmed following the seeking of clarification, then the subcontractor's obligation is, in addition to complying with that instruction or direction, to advise the contractor that loss and/or expense will be incurred by it and, if necessary, that the progress and/or completion of the subcontract work is likely to be delayed.

Where the contractor, following the subcontractor's request for clarification, instructs that the work is to continue to the previous directions and for the work not to be undone or discarded, there is likely to have been a disruption and consequential delay to the subcontract works. Again, this delay should be notified to the contractor and further instructions sought as to the contractor's requirements for the recovery of the delay.

Necessary instructions

While the right to issue instructions for varied work and/or additional work is derived from an express term of the subcontract (SBCSub/C clause 3.4), there may also be an express term to require the issue of instructions necessary for the carrying out of the work, over and above the details contained within the Sub-Contract Documents. Such necessary instructions will include such matters as the selection of colours and/or instructions as to the expenditure of provisional sums.

Where there is no express provision for the issue of necessary instructions, there is likely to be an implied term requiring that such instructions will be issued and at a time reasonably necessary for the subcontractor to fulfil its obligations under the subcontract. This matter was considered in the case of *J. & J. Fee Ltd* v. *The Express Lift Company Limited* (1992) 34 Con LR 147. Judge Bowsher QC said, at page 14:

> 'The only issue remaining between the parties as to implied terms is as to the approved drafting of the implied term set out in the defence and counterclaim.
>
> Which, with my markings, is as follows:

"provide the defendant with correct information concerning the works in such a manner and at such times as was reasonably necessary for the defendant to have in order for it to fulfil its obligations under the subcontract."

Mr Stewart [Counsel] for the plaintiffs submitted that the contested words are inconsistent with clause 11.10.6 of the DOM/2 terms and therefore ought not to be implied into the contract.

Clause 11.3 provides that in certain circumstances the contractor (in this case the plaintiffs) shall give an extension of time to the subcontractor (in this case the defendants). One of the matters for the contractor to consider under that clause is whether there has been an "occurrence of a Relevant Event". Relevant events are defined by clause 11.10. Clause 11.10.6 relied on by the plaintiffs defines only one of those relevant events as follows:

"11.10 The following are the Relevant Events referred to in clause 11.3.1:

. . .

11.10.6 the Contractor or Subcontractor through the Contractor, not having received in due time necessary instructions decisions information or consents from the Employer for which the Contractor or Subcontractor through the Contractor specifically applied in writing . . ."

For the plaintiffs it is submitted that that clause requires that the defendants should specifically apply in writing for information and no term should be implied which requires that they should be entitled to have information without applying in writing. But clause 11.10.6 only refers to an event happening between the employer and the companies employed (the contractor and the subcontractor) and it does not impinge on the way in which the contractor and the subcontractor necessarily have to cooperate if the work is to be done. There is nothing in this contract, DOM/2, which puts on the contractor an express duty to provide necessary information. For that reason it is agreed by the parties that a term should be implied into the contract that information should be supplied. The contract cannot be made to work unless the information is supplied: it is essential. Moreover, there is no point in supplying that information unless it is supplied in such a manner and at such times as is reasonable for the defendants to have it in order for the defendants to fulfil their obligations under the subcontract. It is necessary and it is reasonable that the whole of the implied term for which the defendants contend should be implied into the contract. It is not necessary or reasonable that it should be implied that the defendants should only be given essential information if they ask for it in writing. I see no inconsistency between such an implied term and the express terms of DOM/2.

The dispute under the fourth issue must be resolved in favour of the plaintiffs.'

The provision at clause 2.7 of SBCSub/C, 'Further drawings, details, directions', makes express provision for the issue of necessary drawings and details and generally reflects the provision of clause 2.12 of the JCT 2005 form, with the exception of there being no reference to an information release schedule. It will be for the contractor to ensure that the needs of the subcontractor do not conflict with the express provisions of any agreed information release schedule under the main contract.

SBCSub/C makes provision for the issue of further drawings, details and directions which are reasonably necessary, at clause 2.7, by way of four sub-clauses which make provision as follows:

1. The contractor is to issue without charge such further drawings or details that are necessary to explain and amplify the numbered documents in the subcontract and such directions as are necessary for the subcontractor to be able to carry out and complete the subcontract works.

2. The subcontractor is to provide the contractor with the information necessary for the contractor to make the necessary preparations for the subcontractor to be able to carry out and complete its work in accordance with the subcontract.
3. The details/information defined above are to be provided when it is 'reasonably necessary for the recipient party to receive them'. This is to have due regard to the progress of the main contract works.
4. If the recipient party has reason to believe that the other party does not know when the information is required, then it is to notify that party providing the further drawings, details or information, when those items are required.

In ICSub/C the provisions at clause 2.5 are limited to the first two sub-clauses, and clause 2.6 incorporates the further requirements of SBCSub/C clauses 2.7.3 to 2.7.4. Section 10 of Sub/MPF04 provides for the provision of both information to the subcontractor and design details to the contractor and consequently is more far-reaching than clause 2.7 of SBCSub/C. Section 10 of Sub/MPF04 allows for a pre-agreed programme for the provision of details by the contractor to the subcontractor.

The above clauses are vague and woolly. In practice, the need for exchanges of information between the subcontractor and contractor, and in turn the contractor and the consultant designers and/or other subcontractors, will vary considerably from trade to trade. The preparation of detailed drawings and exchange of design solutions are not matters to be left to general contractual provisions such as 'reasonably necessary' or 'has reason to believe'. These are matters which require full detailed and integrated programming, as discussed under the section on 'design development' and 'programming of the subcontract works'. Where the contractor has not set out clearly in the Sub-Contract Documents both its intentions under clauses 2.7.1 and 2.7.2 and its requirements under 2.7.3, then where, for the execution of the subcontractor's work, the provision of further drawings and details is necessary by specific dates, in advance of the execution of the work on site, it is important that the subcontractor makes such requirements known at the earliest possible time.

It is advised that no reliance be placed on there being no reasonable grounds for believing that the contractor is unaware of what is required and when. This will be even more the case where the subcontract includes for the procurement of special materials, off-site manufacture and/or design. In this respect the phrase 'in accordance with the Sub-Contract' has little if any meaning, since there is no provision within the Sub-Contract Particulars to define the timing of such activities. In addition, the contractor may require details from the subcontractor for incorporation into the design of others, about which the subcontractor will have no knowledge and will be unable to ascertain from the progress of the main contract Works.

Instructions resulting from discrepancies within the documents

Where a subcontract incorporates a number of documents, whether physically incorporated or incorporated by reference, there are likely to be discrepancies relating to the work that the subcontractor is required to do or the method of carrying it out. Most standard forms of subcontract have a clause to the effect that, as and when the subcontractor identifies a discrepancy between the documents, it is to notify the contractor of the discrepancy and seek instructions.

Normally such obligations will be qualified by a phrase such as, 'as soon as it shall become apparent'. Clauses making it an obligation for the subcontractor to identify any discrepancies that may exist, rather than to notify the contractor when they are discovered, are extremely onerous on the subcontractor and should be resisted. Even where such an obligation exists, there will rarely be a requirement for such discrepancies to be identified by a specific date, nor will there be consideration of the effects and/or sanctions of failure.

The effect of the discovery of a discrepancy will depend on the decision of the contractor as to which option is to be adopted. It can be assumed that the subcontractor has allowed, within its price and programme, for completing the subcontract work fully in accordance with one of the options that is the subject of the discrepancy. Thus, should that option be the one the contractor adopts, there will be no effect to the subcontractor other than by reason of any delay by the contractor in issuing its instruction. The consequences of instructing an alternative option will depend on the nature of the discrepancy and the timing of its discovery. In an extreme case, this may be when the subcontract work is well advanced and will require removal and replacement of the subcontractor's work, with the inevitable delays to the Works and additional costs.

Timing of instructions

While in theory and for contractual purposes all instructions to the subcontractor are issued by the contractor, in practice instructions as to the nature of the Works will have been given in the first place by the employer's agent. In the absence of any agreed information release schedule or other agreement as to the timing of instructions, it is likely that there will be an implied term that necessary instructions will be required to be issued so as not to delay the works, or in the words proposed in *J. & J. Fee*, 'at such time as was reasonably necessary for the defendant to have in order for it to fulfil its obligations under the subcontract'. This is a more onerous provision than that contained within SBCSub/C at clauses 2.7.1 and 2.7.3, in that there is to be no relaxation to the timing of the issue of the instructions as a result of delays to the works, whereas SBCSub/C relates such provision to the 'progress of the Sub-Contract Works and main contract Works'.

It will be for the contractor when entering into a subcontract to ensure that the planned dates for issue of further instructions which are reasonably necessary for its proposed subcontractor are no more onerous than that which was or could be considered necessary under its contract with the employer. Where instructions only affect work on site, requirements will be unlikely to differ. However, where there is off-site design, development and manufacture, this is less likely to be so certain. A specialist subcontractor's ability to fulfil its obligations may depend on being able to carry out off-site manufacture during a certain period or at a specific time which may or may not be directly related to the on-site programme. The loss of the planned period for the production at the subcontractor's works may make a significant difference either to the cost of performance and/or the subcontractor's ability to perform by the due date.

The decision in *Hadley* v. *Baxendale* (1854) 9 Exch 341 sets out two considerations leading to a claim in damages:

1. those matters that flow naturally from a contract;
2. those matters that are in the contemplation of the parties at the time of entering into the contract.

Specialist subcontractors should make sure that the attention of the contractor is drawn pre-contract, to all relevant matters that may affect the subcontractor's ability to perform. These matters should be recorded on a document forming part of the Sub-Contract Agreement. It is not easy for a specialist subcontractor to identify such matters, which it often takes for granted but which may not be apparent so far as the contractor is concerned, as flowing naturally from failures and/or delay by the contractor. Further, other specialists operating in the same field may have considerably differing constraints on their operations. As a simple example, a steel fabricator who buys from stock will generally have less time constraint than one who buys from the mill and is constrained by the rolling sequence of several weeks.

It is generally held that there will be an implied term to the effect that the contractor will not hinder or prevent the subcontractor from carrying out its obligations in accordance with the terms of the subcontract, or from executing the work in a regular and orderly manner. Such an implied term derives from the judgment in *London Borough of Merton* v. *Stanley Hugh Leach Ltd* (1985) 32 BLR 51, in which the following implied terms were held to be incorporated:

> '(i) [Merton] would not hinder or prevent Leach from carrying out their obligations in accordance with the terms of the contract [and from executing the works in a regular and orderly manner].
>
> 1. [Merton] would take all steps reasonably necessary to enable [Leach] so to discharge their obligations [and to execute the works in a regular and orderly manner].'

Mr Justice Vinelott had to consider a number of terms claimed by Leach to have been implied into the contract. First he considered the nature of implied terms and concluded, at page 76, that:

> 'the process of implication is one of spelling out or deducing what is implicit in the contract in the sense of being part of the legal context appropriate to contracts generally or to contracts of the particular type under consideration and so inherent in the legal relationship between the parties created by the contract.'

In other words, there are certain matters which affect all contracts or at least all contracts of a certain type, such as for specific industries. In construction contracts and similarly in construction subcontracts, the contractor is not to hinder the subcontractor in performing its obligations in accordance with the terms of the subcontract. In addition, the contractor is to take all steps reasonably necessary to enable the subcontractor to discharge its obligations and execute the works in a regular and orderly manner.

Mr Justice Vinelott said, at page 80:

> 'As regards the second of these two terms it is well settled that the courts will imply a duty to do whatever is necessary in order to enable a contract to be carried out. The principle was expressed in a well-known passage in the speech of Lord Blackburn in *MacKay* v. *Dick* [(1881) 6 App Cas 251] where he said:
>
> > "Where in a written contract it appears that both parties have agreed that something should be done which cannot effectively be done unless both concur in doing it, the construction of the contract is that each agrees to do all that is necessary to be done on his part for the carrying out of that thing though there may be no express words to that effect"
>
> However, the courts have not gone beyond the implication of a duty to cooperate.'

This implied obligation is strengthened by the quotations from previous judgments, quoted at page 80. While relating to main contracts under the JCT terms, there can be little doubt that similar obligations exist between a contractor and its subcontractor:

> 'Again it is recognised in all the leading text books that a contract incorporating the JCT standard terms falls within the category of contracts in which a requirement that a building owner will "do all that is necessary to bring about completion of the contract" will be implied'

In *Holland Hannen and Cubitts* v. *WHTSO* (1983) 18 BLR at 117, it was conceded that under the contract at issue (which incorporated similar terms):

> '"the building owner would do all things necessary to enable the contractor to carry out the work" and Judge Newey clearly thought that that concession was rightly made.'

Instructions other than in writing

It is a frequent requirement of a subcontract that instructions from the contractor are in writing, such as SBCSub/C clause 3.4. Manifestly, where there is such an express restriction, if instructions are then given other than in writing there will be no obligation for the subcontractor to implement them and no sanction if he does not. However, problems may arise if, contrary to the requirement of the subcontract, the contractor instructs the subcontractor other than in writing and the subcontractor carries out the work in the manner so instructed. In many subcontracts there will be provision, such as in SBCSub/C clause 3.7, for the subcontractor to confirm to the contractor the instructions or purported instructions given other than in writing. There is often provision that such confirmations are not effective for a specified period, usually seven days. The subcontractor is expected to make such a confirmation before implementing the instruction; however, there is rarely any specified bar to confirmation being after the carrying out of the instruction. Thus it is possible that an application for payment might be considered both an application for payment and confirmation of an instruction other than in writing, and certainly could be phrased by the subcontractor in such terms under such circumstances. Such an application would become an effective instruction at the end of the specified period for dissent by the contractor.

SBCSub/C, at clause 3.7, makes provision as above with seven days' lapse between confirmation and the instruction being effective. At clause 3.7.2 it makes provision for confirmation of instructions by the contractor up to the final payment. This clause requires a subcontractor receiving a direction that is not in writing to confirm the direction to the contractor within seven days, and provided that the contractor does not dissent within seven days of that confirmation, then the instruction becomes effective. However there are two further provisions:

1. If the contractor confirms the instruction, then the subcontractor does not have to confirm the instruction and the instruction has immediate effect.
2. If neither party confirms the instruction but the subcontractor nonetheless carries it out, then the contractor may confirm that instruction at any time up to the time of the final payment.

Where there is no express provision for the confirmation of instructions issued other than in writing, as for example ICSub/C and SubMPF04, it is likely that such a term would be implied either on the basis of trade custom and practice or to give business efficacy to the situation. However, where the subcontractor either ignores an instruction other than in writing or refuses to implement it until properly issued, it is unlikely that the contractor would have any sanction against the subcontractor for failing to take any action on such an instruction.

Where the subcontractor carries out work having been instructed other than in writing, and where the subcontract has an express provision that all instructions are to be in writing, the situation may become complex. One approach may be to regard the instructions as being outside the subcontract and that the provisions of the subcontract do not apply and the subcontractor will have to seek payment in restitution, in which case it will be entitled to a fair valuation of the work. An alternative possibility is that since the contractor is himself in breach of the subcontract in not issuing the instruction in writing, and the law will not allow the contractor to benefit from its own breach, the contractor will be likely to be required to make payment as if an instruction in writing had been made.

This situation was considered in the case of *Total M & E Services Limited* v. *ABB Building Technologies Ltd (formerly ABB Steward Limited)* (2002) 87 Con LR 154. In this case there was a written contract with no provision for variations. The valuation of the Works was referred to adjudication and the adjudicator made his decision including the valuation variations. The paying party challenged the decision on the basis that the variations, which had been issued orally, were not part of the subcontract in writing but had been a separate contract not in writing and therefore were not adjudicable. The court decided that the variations arose out of the single written contract and that the HGCRA applied, notwithstanding that the subcontract was partly in writing and partly oral. The situation between the parties can be summarised as follows:

> '26. The subcontract was for a lump sum of £250,000. It is common ground that there were considerable additional works performed by the claimant and the adjudicator's award reflected the value of such additional work. The defendants contend that the adjudicator had no jurisdiction to base an award upon the additional works. They submit that there is no evidence as to the true contractual basis upon which the work was performed and in respect of which payment was due. Mr Coulson [Counsel for ABB] submits that it could not be under a variation of the original subcontract in the absence of a written mechanism for variation within the subcontract . . . He contends that the adjudicator had no jurisdiction to reach a series of decisions on a series of oral agreements.

> . . .
>
> 28. The additional work performed by the claimant was the same type of work which was the subject of the original subcontract and the only inference to be drawn is that the scope of the subcontract work was enlarged.
>
> 30. Mr Coulson goes on to submit that since a construction contract must be in writing, it could not be varied orally.'

The judge then said:

> '34. What has to be considered here is not the enforceability of the contract but whether the statutory adjudication scheme can be invoked in relation to a particular construction contract. That is governed by section 107 of the Act. There is reference in the sub-section 3 to an agreement made otherwise than in writing; such an agreement, provided it refers to terms which are in writing is an agreement in writing. In my judgement, the adjudicator made his decision on the basis of dispute arising out of the single written construction contract as varied orally by the parties. The contract as varied is clearly within the provisions of section 107. Notwithstanding that it is a contract evidenced partly in writing and partly oral. The adjudicator therefore had jurisdiction to make determinations as to the additional works.'

In any event, where the contractor issues instructions to the subcontractor regardless of the express terms of the subcontract, there is likely to be an implied promise to pay. Both *Keating* and *Hudson Building Contracts* state:

> 'When there is a condition in the contract that extras shall not be paid for unless ordered in writing by the architect . . . and the employer orders work which he knows, or is told, will cause extra cost, a judge or arbitrator may find that there was an implied promise by the employer that the work should be paid for as an extra and especially so in cases where any other inference from the facts would be to attribute dishonesty to the employer'

This principle was considered by Mr Justice Humphreys in the case of *S. C. Taverner & Co Ltd* v. *Glamorgan County Council* (1941) 57 TLR 243, but was rejected by reason of a matter of law. However, the judge did not rule out the possibility that on the facts there might have been an entitlement in equity, where to decide otherwise the action of the employer would be fraudulent. In that case counsel for the plaintiffs had quoted from *Hudson Building Contracts* as follows:

> '. . . "When there is a condition in the contract that extras shall not be paid for unless ordered in writing by the architect . . . and the employer orders work which he knows, or is told, will cause extra cost, a jury or an arbitrator may find that there was an implied promise by the employer that the work should be paid for as an extra, and especially so in cases where any other inference from the facts would be to attribute dishonesty to the employer." I entirely agree with that statement of the law, which is based on two decisions. In a particular case a jury may of course find on evidence that there was an implied promise by the employer that the work should be paid for as an extra.'

In the subcontract situation, this principle probably applies equally to instructions given by the employer or architect directly to the subcontractor, as to instructions by the contractor given contrary to its stated procedures, although the obligation to pay may be directly from the employer and not the contractor.

Instructions by third party

Because of the very fragmented nature of the construction process and especially the involvement of a large number of separate organisations in the design, supply and installation chain, it frequently happens that instructions and/or agreements are made between parties involved in the project who have no direct contractual relationship. The basic premise must be that any such instruction and/or agreement is of no immediate effect and will not create any right or benefit to the subcontractor, or any obligation or duty on the contractor under the subcontract.

In a contract between contractor A and subcontractor B to provide goods or services to employer C, a request or agreement from, or agreement with, consultant D for B to provide varied and/or additional services to C, will not change or add to A's obligation to B. The same will be true if the request comes from C in person.

If the specialist subcontractor meets with the specialist designer, with the full knowledge and agreement of both the contractor and employer's agent, it may mean that both act as agents of the contractor on the one hand and the employer's agent on the other. Even if this is the true situation it will not change the contractual requirements for instructions to be issued in writing or to be confirmed in writing. Thus where the issue of instructions to the subcontractor is limited to the contractor's site manager and to be in writing, such requirement will remain and the subcontractor must ensure that this requirement is fulfilled.

Subcontracts frequently have a provision for the confirmation of instructions other that in writing, such as SBCSub/C clause 3.7, which might be held to be effective for confirmation of an instruction by a third party if the agency principle could be substantiated. In any case such a confirmation would not become of immediate effect, either under the terms of subcontract such as in SBCSub/C or by reason of custom and practice, unless the subcontract expressly made such provision. In the absence of any express provision for the confirmation of instructions, the seven-day rule would probably be implied by custom and practice. However, if the third party is not regarded as an agent for the contractor, either directly or indirectly, such instructions can be of no effect and the subcontractor must either await instructions through the contractual chain or seek such instructions from the contractor before taking any action.

The risk for the subcontractor acting on the oral, or even written, instruction of a third party is two-fold. First, the subcontractor may have difficulty in being paid for such work and/or any claim for loss and/or expense associated with such an instruction. Secondly, such work would at least technically be in breach of its subcontract and the contractor could require such work to be undone at the subcontractor's expense.

It may be that work arising from such instructions can be distinguished into classes. One class will be where instructions are necessary, particularly when in response to a formal request by the subcontractor, and may be regarded as a response to such a request or as the fulfilment of an obligation of the contractor to provide all necessary information for the carrying out of the Works. Action in accordance with and in reliance on such instructions is less likely to lead to the difficulties identified above.

Another class would involve a request by the consultant designer to vary the specification and/or to carry out additional work. This is a very different situation.

Even when the consultant designer has the authority of the employer and/or his agent to issue such instructions, such authority can only be to instruct the contractor. The subcontractor would require specific instructions from the contractor, either as a provision of the subcontract or issued to it post-subcontract, to accept oral instructions from a design consultant. Even where such an effective provision exists, for the subcontractor to accept oral instructions from the design consultant, there will probably be an implied requirement if not an express one to notify the contractor prior to proceeding.

A further class may arise where the specialist subcontractor for some reason seeks permission to diverge from the strict requirements of the specification. The acceptance post-contract of a change to the nature and/or quality of the work will normally require the authority of the employer. For this reason written authority is required, under the appropriate provisions of the subcontract, before any such proposal and/or acceptance in principle can be effective. It will be important for a subcontractor seeking a relaxation or change, away from the specification, to do so in good time or at least make clear its reason for requiring a quick decision. Since such circumstances frequently arise out of a problem of availability of materials from a specified supplier and/or source of the specified materials, the request is with the intention of avoiding or at least reducing the consequential delay to the work, when only a prompt reply will assist and a delayed decision will be of no benefit.

If there is no effective instruction there can be no obligation under the subcontract for the subcontractor to be paid by the contractor. This is not a question of 'pay when or if paid', but a question of no entitlement. There can be no implied promise to pay, as described above, nor can failure to pay be found to be dishonesty on the part of the contractor or an act of fraud on his part, since he has not given the instruction relied on by the subcontractor. It is, of course, open for the contractor to seek a retrospective instruction under the main contract, which if obtained will require the work to be valued and paid for.

It is possible that payment might be made to the contractor without a formal instruction, for the change carried out by the subcontractor. In those circumstances, it is suggested that an obligation lies with the contractor to pay for the varied work in accordance with the provisions of the subcontract, which may or may not be the same or similar to the payment provisions under the main contract, and in any event no more than a sum proportional to that which it has itself received.

It is, of course, also possible that the specialist subcontractor may have a claim directly with the employer under some quasi-contract, but consideration of this possibility is outside the scope of this book.

Directions to cease work

Where the contractor suspends performance under the main contract, it is necessary that the contractor has the contractual right to instruct the subcontractor to conform. SBCSub/C, at clause 3.21, and ICSub/C, at clause 3.19, make provision for this eventuality by requiring the contractor to notify the subcontractor should the contractor give the employer a written notice of his intention to suspend the performance of his obligations under the main contract.

Condition 3.21 requires the contractor to copy the subcontractor with any seven-day notice of intention to suspend performance that it may serve on the employer. This will have the effect of putting the subcontractor on notice that it may be required to suspend performance. However, the contractor is not required to direct the subcontractor to take any action itself and without a suitable express provision there is no right for the contractor to instruct or require the subcontractor to suspend performance. In fact, even clause 3.22 of SBCSub/C and clause 3.20 of ICSub/C only use the phrase 'may direct' the subcontractor to cease the carrying out of the subcontract works. However, the contractor shall issue such further directions as may be necessary, following such cessation.

If, having suspended its performance, the contractor resumes its performance under the main contract, then the contractor shall, if he has directed the subcontractor to cease the carrying out of the subcontract works, direct the subcontractor to recommence and may issue further directions in regard to such recommencement.

However, where such an instruction to 'cease the carrying out of the Sub-Contract Works' is issued, the contractor may issue such further instructions to the subcontractor as may be necessary. The use of the word 'may' appears strange since it must be the case that a subcontractor having had its performance stopped must need instructions as to how to proceed. This requirement is more relevant than may at first be supposed. The ceasing and restarting of the Sub-Contract Works will involve the subcontractor in expense arising from that instruction, which the contractor will wish to keep to a minimum.

The contractor may need to issue instructions to stop and/or suspend the Sub-Contract Works for various reasons, which may include:

1. the site becoming inaccessible to the contractor for any reason;
2. work on site not being available due to the lack of progress of preceding trades;
3. the contractor having suspended performance under the provision of section 112 of the HGCRA.

It will be necessary for the contractor to issue to the subcontractor instructions as necessary. In the event that the contractor does not give instructions or directions, the subcontractor must both advise the contractor of the action being taken by it and request either agreement to such action or the contractor's instructions or directions for such other action required.

It may be considered desirable to maintain the skilled workforce so as to be available for an early resumption of the work, by paying a retention wage, or the contractor may instruct that the workforce be redeployed with the resultant probability that they will need to be replaced by other labour on recommencement; such replacement labour may be less skilled and/or more expensive and in addition will not be conversant with the specific project and so will need a learning period before maximum performance will be achieved. The supply of materials may also need to be interrupted or stored off site, involving costs of both storage and double handling. In any event, it is for the contractor to issue directions as necessary. In the absence of express directions the subcontractor will have to take action on its own initiative, since some action will have to be taken immediately, on suspension of the Sub-Contracted Works.

Duty to warn that instructions may give rise to defective work

The duty of a subcontractor to warn with regard to a design believed to be defective or insufficient, was discussed in great detail in the Court of Appeal case of *Plant Construction Plc* v. *Clive Adams Associatiates, JMH Construction Services Limited* [2000] BLR 137, and in the judgment of Lord Justice May. The case concerned the collapse of a single-storey building while underpinning work was being carried out to one of the supporting stanchions. Although the subcontractor was responsible for the temporary works design, the temporary support works had in fact been designed and instructed by the employer's engineer. The subcontractor had objected to the instruction since it considered the instructed design to be inadequate, as it in fact proved to be.

This case goes further than just a simple matter of a duty to warn, but raises the question as to what extent may, or indeed must, a subcontractor resist the implementation of an instruction which it believes will lead to defective or inadequate work.

The *Plant Construction* case, in reviewing previous decisions, can be said to have identified three distinct categories of contract relevant to a duty to warn by the subcontractor; these give rise to an ever higher obligation or duty. The first category is where the employer or contractor has the services of a separate consultant and where the effect of the defective or inadequate design will not lead to an immediately dangerous situation. In such a situation there is probably no duty to warn and in any case no sanction on the subcontractor for not warning.

The second category concerns subcontracts where there is no professional consultant or adviser employed by the employer. In this category the builder probably has a duty of care to the employer to advise if any of its instructed work is likely to be inadequate and/or defective. In such cases a subcontractor may have, in turn, a similar duty to the contractor as its specialist.

The third category is where the works as instructed are considered to be dangerous or unsafe. In such situations the duty may extend not only to where the subcontractor believes there is a danger but also to situations where it should have been aware that there was a danger.

The Court of Appeal in *Plant* found, at page 148, that the third of these categories applied in that case. The conclusion as to what the subcontractor should have done, in the circumstances of an obligation under the subcontract to comply with an instruction believing it to be dangerous, was stated as follows:

> 'to Mr Stow's [Counsel for JMH's] question, what more could JMH have done? Generally speaking, the answer is that they could have protested more vigorously.'

This seems very weak. It is suggested that, for any subcontractor to carry out or allow to be carried out work it believed to be dangerous, is in any event an unlawful act. To the extent that the provision of a subcontract makes the carrying out of instructed work, which is known to be dangerous and certainly when that danger is believed to be immediate, a condition of that subcontract, it is suggested the subcontract must itself be unlawful and such an instruction will be unenforceable against the subcontractor.

The decision in *Plant* was considered in the case of *Aurum and Avonforce Limited* v. *Knapp Hicks and Advanced Underpinning*, mentioned in the previous chapter. This was a case where underpinning installed by one specialist subcontractor failed by rotational failure when the adjacent area was excavated by the contractor or another subcontractor without the provision of any temporary works or alternatively carrying out the excavation in sections. Mr Justice Dyson recorded, at paragraph 15:

> 'it would have been perfectly possible for temporary sheet piling to be installed between the bases and the retaining wall of the basement structure. Sheet piling would have retained the earth behind the bases, thereby reducing the lateral loads applied to them.'

The judge, at paragraphs 11 and 12, summarised the judgment in *Plant* in the following words:

> 'It has now been held by the Court of Appeal that if the duty to warn arises, it is part of that duty to act with skill and care of an ordinarily competent contractor. What is to be expected of such a contractor will depend on the particular facts of the case. The facts of the *Plant* case show that, where a contractor is asked to do work, he is likely to be under a duty to warn his client if he knows that the work is dangerous, and that duty will not be negatived by the fact that the client is being advised by a professional person who knows, or ought to know, that the work is dangerous.
>
> Thus, if Advanced had been instructed to carry out underpinning work which it knew to be unsuitable and dangerous, it would seem to follow from *Plant* that it would have been under a contractual duty to warn Avonforce, notwithstanding that Avonforce was being advised by KHP. No reasonably competent contractor would have failed to warn in such circumstances. It is interesting to note that at page 148, Mr Justice May left over for future consideration circumstances where (a) the contractor did not know, but arguably ought to have known, that the design was dangerous, and (b) where there was a design defect, of which the contractor knew or ought to have known, which was not dangerous. This shows the cautious and incremental approach that has been adopted in this area of law.'

The judge, at paragraph 16, found that the subcontractor did not have a duty to warn about the effect of carrying out the excavation without proper temporary support works or by another method:

> 'It is true that on the facts of *Plant*, the fact that the instructions had been given by the clients' engineer did not negative the duty to warn. But in that case, JMH was aware that what it was instructed to do was dangerous. Where, as here, the contractor is not aware of what is proposed, and at its highest, the case is that it ought to have known that what occurred might have been proposed, it seems to me that the position is quite different. In such a case, I consider that it is relevant to the question of whether there is a duty to warn that the client is being advised by an independent professional person. Why should the contractor assume that the client will act negligently, particularly when he is being independently advised by an engineer?'

Duty to install to manufacturer's instructions

A duty to warn must not be confused with a duty to install or carry out work correctly. Where such a duty exists, a letter advising that there is a problem and seeking instructions will not relieve the subcontractor of its liability if it incorrectly installs equipment or material products.

In the Court of Appeal case of *Six Continents Retail Ltd* v. *Carford Catering Ltd and R. Bristoll Ltd* [2003] EWCA Civ 1790, Carford Catering had been employed for the design and installation of kitchen equipment in a building designed and built by others. One wall of the kitchen was a timber wall lined with plywood and tiled. Carford arranged for a spit roast rotisserie to be positioned on brackets on this wall.

After the completion of the kitchen the suppliers of the rotisserie, Bristoll, were called to make adjustments to the equipment and on removing it from its brackets discovered that some tiling had come away from the wall. On completion of their work, Bristoll notified Carford of the missing tiles. Carford in turn advised the employer's construction department manager. Neither party took any urgent action and some weeks later there was a serious fire as a consequence.

The court found that the direct cause of the fire was that the rotisserie had not been installed in accordance with the manufacturer's instructions, when fitting to a combustible surface, which required:

> 'that the back of the appliance should be separated from any combustible surface by (a) a shield of non-combustible material at least 25 mm thick or (b) an air space of at least 75 mm.'

Since Carford had provided neither of these, they were liable for the resultant fire. The court reversed the decision of the judge at first instance. Lord Justice Laws considered the effect of Carford's letter:

> '22. The true question here is whether the appellants' failure to respond to the letter of 21 January 1997 ought to absolve the respondents of what would plainly otherwise be their responsibility for the fire. For my part, I think it plain that the risk of a fire of this kind was, on the facts, well within the scope of outcomes which the respondents' contractual duties were intended to avoid. So much appears, I think, from the catalogue of obligations cited by the judge at paragraph 38 which I have set out. Accordingly, even if the letter of 31 January 1997 and the enclosed fax did constitute a warning of a risk of fire of the kind which occurred on 9 January 1997, a question to which I will come in a moment, it was a warning of an outcome which the respondents themselves should have prevented happening. I find it very difficult to see how the giving of such a warning ought to transpose the burden of avoiding that very outcome from the respondents, who owed a duty in effect to prevent it, to the appellants who were the beneficiaries of that duty.
>
> 23. But in any event I do not consider that the letter was a warning – certainly not a sufficient warning – that there was a risk of fire happening as this fire happened. The wording of the letter did not in terms amount to a warning at all. Indeed the expression "please advise us what action, if any, you wish us the builders to take", suggesting that action was optional rather than necessary, is all but inconsistent with the notion of a warning; and the enclosed fax is, to say the least, indefinite as to fire risk. To constitute a proper warning the respondents must have drawn attention – in terms, or at least very plainly – to the fact that the unit was fixed directly on to a combustible surface; and that, as I have foreshadowed, would have been a warning of the respondents' own breach of contract.'

Lord Justice Buxton stated the true cause of the fire and therefore the liability:

'29. Carford were in breach of their contractual obligations by not installing the rotisserie in accordance with the manufacturer's instructions – the issue before her was therefore in what circumstances, if any, would it be possible for a notification or warning by Carford of its own breach discharge its continuing liability for damage caused by that breach.

30. For that to be achieved, any warning would as a matter of law have to be overwhelming and plainly effective before it could excuse Carford.'

It is likely that had the rotisserie been installed correctly and the tile failure been due to poor tiling work, then a letter from Carford to the employer might have been all that was required of it, especially if it had stressed that the lack of tiles breached the manufacturer's installation requirements and that consequently the equipment should not be used until repairs had been carried out.

The liability of the supplier, Bristoll, in this instance was for the supply only of the equipment. It is however submitted that had they been the installers, then they would also have had an obligation to ensure that there was sufficient air space behind the equipment, should they have had any reason to believe the wall was combustible in nature.

While the consequences of failure to install to the manufacturer's instructions in the Carford case were very serious, there is no reason to believe that the courts would apply a less stringent duty to install to a manufacturer's instructions where the inherent risks are less. It is not the level of danger resulting from the failure which makes the failure a breach of the subcontract. Even if there were very clear directions from the contractor that installation was to be contrary to the manufacturer's instructions, it is likely that the subcontractor would still be in breach of its obligations if compliance with such express directions was likely to create a danger to the health and safety of either the subcontractor's operatives and/or others.

Instructions where there is no provision within the subcontract

As mentioned above, the right to issue instructions and to change the requirements of the subcontract arises out of the subcontract itself and not as a dissociated right. Since many subcontracts are concluded without any agreement as to terms, frequently by way of a preliminary subcontract such as a letter of intent or a simple acceptance of an offer for goods and services, there may be no express right for the contractor to vary the work and/or timing of the subcontracted work.

In such circumstances, should the contractor require variations to the subcontracted work, it is likely to be considered as an act of prevention and to limit or nullify the effect of any express provisions of the subcontract, such as to complete by a certain date. In the case of *Balfour Beatty* v. *Chestermount Properties* (1993) 62 BLR 1, Mr Justice Colman stated, at page 21:

'A variation instruction would, in the absence of contractual provision to the contrary be an act of prevention: see *Dodd* v. *Churton* [1897].'

The reference to *Dodd* v. *Churton* [1897] 1 QB 562 was probably to the words of Lord Esher MR, who in that Court of Appeal case said, at page 566:

> 'The principle is laid down in Comyn's Digest, Condition L (6.), that, where one party to a contract is prevented from performing it by the act of the other, he is not liable in law for that default; and, accordingly, a well recognised rule has been established in cases of this kind, beginning with *Holme* v. *Guppy* [[1838] 3 M&W 387] to the effect that, if the building owner has ordered extra work beyond that specified by the original contract which has necessarily increased the time requisite for finishing the work, he is thereby disentitled to claim the penalties for non-completion provided for by the contract.'

In *Dodd* v. *Churton* [1897] 1 QB 562, it was stated that the builder had in fact undertaken to perform 'any extra work that may be ordered'. The court had to decide whether that was to be construed as being within the original contract time and the court decided it was not to be so construed. Lord Esher MR said, at page 567:

> 'The whole question here is whether on the construction of this contract, by which undoubtedly the builder has undertaken to perform any extra work that may be ordered, he has agreed to take upon himself the burden which the builder had taken upon himself in *Jones* v. *St John's College* [(1870) LR 6 QB 115]; in which case, however foolish and unreasonable such an agreement may be, he must stand by it. One rule of construction with regard to contracts is that, where the terms of a contract are ambiguous, and one construction would lead to an unreasonable result, the court will be unwilling to adopt that construction. In *Jones* v. *St John's College* the court had no opportunity of construing the contract really made. The demurrer admitted the statement on the pleadings that the builder had entered into the unreasonable agreement alleged. I cannot construe the contract in this case as containing such an agreement by the builder.'

Thus, where a subcontract makes no provision for adjusting the obligations of the subcontractor in the event of variation instructions, unless there is a very clear agreement that the subcontract work will in such an event still be completed to the original time, the subcontractor will be relieved of its obligations under the subcontract.

Acceleration agreements

The need or desire to accelerate the works may arise either where the rate of progress has for some reason fallen behind that which was planned, with consequential anticipation that completion will be delayed or late, or for some reason the contractor and/or employer desires that the subcontracted work is to be completed ahead of schedule. Where the work is behind programme there may be uncertainty or a dispute as to the cause of the delay, and where there is more than one cause, the actual or dominant cause.

Where the cause of delay is matters for which the subcontractor is at risk, then it is for the subcontractor to take the necessary steps to recover the time lost or bear the consequences, by way of damages for breach of the subcontract. However, where matters at the contractor's or employer's risk have caused the delay, then the subcontractor has no authority to spend money on the recovery of such time without a direction or instruction from the contractor.

Instructions and/or agreements to accelerate may come in one of two basic forms: first, for the subcontractor to carry out certain specified actions such as to obtain more plant and/or labour resource or to work longer hours; secondly, a simple requirement for the lost time to be recovered or for the work to be completed as stated or agreed on an earlier date. In the second situation it is anticipated that the subcontractor will decide what steps are necessary to comply with the instructions and may expressly or impliedly accept that it will be in breach of the subcontract if it fails.

Where the agreement or instruction is to carry out express accelerative actions, such instructions will imply that the subcontractor will be paid either the pre-agreed rates or sum, or a sum to be ascertained by the valuation procedures for instructions under the subcontract. Normally the contractor and employer will not have a claim against the subcontractor if the outcome does not fully achieve the goal required or anticipated.

An agreement to complete on time will generally put the onus on the subcontractor to decide the actions to be taken, and the subcontractor will remain liable for the consequences of failure. This is certainly true where it has been agreed in advance that the subcontractor will receive a certain sum by way of consideration for achieving the revised completion date. Such an obligation may itself be removed by further acts of hindrance by the contractor/employer.

It is considered unlikely that a simple instruction to complete the subcontract works early and/or recover lost time for which the subcontractor has no liability, will be effective in imposing an obligation on the subcontractor to achieve the objective set by the contractor in such an instruction, since it would be a unilateral attempt to amend the Sub-Contract Agreement between the parties. A change to the agreement will require express agreement to be binding, and in addition will require consideration.

Sub/MPF04 provides at section 17 for agreement to accelerate the subcontract works. Clause 17.1 allows the contractor, should it wish to investigate the possibility of achieving practical completion before the expiry of the period for completion, to invite proposals from the subcontractor. The subcontractor shall either:

1. make such proposals accordingly, identifying the time that will be saved and any additional costs that would be incurred; or
2. explain why it is impracticable to achieve practical completion at an earlier date.

Clause 17.2 allows the contractor to accept any proposals made by the subcontractor or seek revised proposals. If the contractor accepts any proposals it shall issue an instruction identifying the agreed adjustment to the period for completion and the additional costs, and that instruction shall be treated as giving rise to a change.

Clause 17.3 restricts the contractor's right to reduce the period for completion of the subcontractor's work to two express provisions under clauses 17 (Acceleration) and 22 (Cost savings and value improvements).

It is to be noted that under these arrangements it is for the subcontractor to make proposals for acceptance by the contractor, to achieve practical completion ahead of the expiry of the period for completion. The difficulties of defining completion and/or practical completion under subcontract arrangements have been discussed in the section 'Completion of the works' in Chapter 5; however, it is surprising that these provisions do not envisage the contractor defining its requirements or objectives and seeking a response and/or proposal from the subcontractor related to those requirements. It is to be noted that under Sub/MPF04, the contractor cannot instruct, without an acceleration agreement, the subcontractor to achieve practical completion before the expiry of the period for completion. At any time the period for completion must be the period for completion as adjusted for all existing relevant events. Given the provision for the contractor to take 56 days from notification by the subcontractor to making a decision as to the additional period to be added to the period for completion, there may at the relevant time be uncertainty as to the true date for completion of the subcontract works.

Several practical difficulties exist in reaching any agreement over acceleration measures. Not least of these is that the ability to recover lost time rapidly diminishes as time progresses. This is for the simple reason that the effective employment of additional resources becomes less and less possible as the amount of remaining work for them to be deployed on diminishes. A simple example might be the decoration of a building or series of units. If a painter can be effectively employed on a group of rooms or in a single unit, then the number of painters that can be so employed will be proportional to the number of groups of rooms or units. As such groups or units are completed, the employment of more painters will mean that the unit of work becomes smaller, with the consequence that the additional painters must work alongside others in the same group or unit of work, with the resultant loss of efficiency. If the subcontractor is to pre-price the additional resources and any associated loss of efficiency involved to achieve an early date, it must be able to relate this to the last date for it to receive such an instruction.

Another difficulty is that there will rarely be an unreserved acceptance by the contractor and/or employer that the subcontractor has no liability for some of the costs of recovery. In extreme situations neither side may admit any liability for the delay that has arisen. If the goal of completion to a date, ahead of the anticipated date for completion, is to be achieved, then it is suggested that the parties will best achieve their objectives by agreeing the actions to be taken and a method of recording those actions, together with their effect and the costs incurred. Then on completion of the work, such records can be analysed to ascertain a fair compensation to the subcontractor, if not by agreement then by recourse to a third party. The final assessment must, as well as considering the costs of the acceleration, also review the causes of and liability for the delay pre-acceleration.

In considering the pre-existing delay it will be necessary to take into account the extent to which the subcontractor's original programme had some float or slack time to allow for it to recover its own delays. Alternatively, the extent that recovery of the delay, for which the subcontractor was at risk, would have been achieved by a smaller increase in resources must be considered. Thus it may be considered that a proportion of the additional resources should be to the subcontractor's cost and by expending this cost it has recovered its own delays and eliminated any liability for damages.

It will be important for the parties to agree the nature of the agreed additional payment. At least three possibilities exist and it will be for the parties to decide which to adopt. First, there can be an agreement for the subcontractor to adopt a specified course of action for which it is to be paid, and when the employer will obtain the benefit of any and all of the resultant acceleration.

Secondly, the subcontractor may agree to recover or improve on the completion date for a stated sum. Such agreement will effectively revise both the subcontract sum and subcontract completion date, with all the other terms, obligations and rights remaining in place. This will mean, in the event that the amended completion date is not achieved, the subcontractor will be liable for damages in the normal way.

Thirdly, the subcontractor may agree to achieve an earlier completion date for a stated sum by way of a bonus. Such an arrangement is likely to be regarded as a collateral subcontract and will not change the subcontract completion date, and damages will not flow from the subcontractor's failure to achieve the earlier date. Instead, the subcontractor will lose the entire benefit of the bonus, however close it may come to success, unless it can demonstrate an act of hindrance by the contractor which prevented completion by the subcontractor by the revised date. In such circumstances it might be expected that the date would be adjusted using the normal extension of time methods of assessment.

In the case of *John Barker Construction Ltd* v. *London Portman Hotel Ltd* (1996) 83 BLR 31, Mr Recorder Toulson QC, in a main contract case, adopted a very different approach by disallowing the bonus altogether and granting the contractor damages by way of 'loss of chance'. He said, at page 64:

> 'It is impossible to tell whether, as a matter of probability, the plaintiffs would or would not have finished by 26 August 1994, but for those changes. They would have had a reasonable opportunity of doing so, but they could easily have failed for all manner of reasons. In those circumstances I would hold that the plaintiffs are entitled to damages for loss of that chance equal to 50 per cent of the agreed performance bonus, or £10,000.'

Instruction to omit the remaining work

Most building subcontracts will have a provision for the contractor to vary the work, including the instruction of additional work or the omission of work. Such wide-ranging rights may include being able to omit all remaining work beyond a certain stage or at any time. The express terms of the subcontract will normally deny the subcontractor any right to recover any loss, or more reasonably any loss of overheads and profit, from such a reduction in the subcontract work. An exception may be the right to recover such loss against the variation. The consequential increased costs to the work as carried out may include the costs of special plant or mobilisation costs where these have been spread over the unit rates, and/or labour and plant hire costs resulting from the early demobilisation.

Such rights are unlikely, except under express provisions of the subcontract, to allow the contractor to omit work and engage others to carry out work to completion. In the case of *Abbey Developments Limited* v. *PP Brickwork Limited* [2003] EWHC 1987 (Technology), Judge Humphrey Lloyd QC was required to consider whether the terms of a labour-only brickwork subcontract gave such rights. Having reviewed a number of authorities and the commentary of Mr I. N. Duncan Wallace QC, editor of *Hudson on Building Contracts*, the judge made the following general comments:

> '45. The justification for these decisions is in my judgement to be found in fundamental principles. A contract for the execution of work confers on the contractor not only the duty to carry out the work but the corresponding right to be able to complete the work which it contracted to carry out. To take away or to vary the work is an intrusion into and an

infringement of that right and is a breach of contract. (The work has to be defined sufficiently for there to be a right to execute it.) Hence contracts contain provisions to enable the employer to vary the work in order to achieve lawfully what could (not) be achieved without breaking the contract or (only) by a separate further agreement with the contractor. By entering into a contract with a variations clause such further agreement is obviated as the contractor's consent to changes in the works is in the primary contract. So such clauses enable an owner to remove work from a contractor, just as they oblige the contractor to carry out additional work or to make alterations in the work, none of which could be achieved without the consent of the contractor.

46. Provisions entitling an owner to vary the works have therefore to be construed carefully so as not to deprive the contractor of its contractual right to the opportunity to complete the works and realise such profit as may then be made. They are not in the same category as exemption clauses. They have been common for centuries and do not need to be construed narrowly. In developed forms they now offer contractors opportunities to participate actively in the success of the project and to enhance their returns (e.g. by way of "value engineering" or the application of concepts such as "partnering").

47. However the cases do show that reasonably clear words are needed in order to remove work from the contractor simply to have it done by somebody else; whether because the prospect of having it completed by the contractor will be more expensive for the employer than having it done by somebody else, although there can well be other reasons such as timing and confidence in the original contractor. The basic bargain struck between the employer and the contractor has to be honoured, and an employer who finds that it has entered into what he might regard as a bad bargain is not allowed to escape from it by use of the omissions clause so as to enable it then to try and get a better bargain by having the work done by someone else at a lower cost once the contractor is out of the way (or at the same time, if the contract permits others to work alongside the contractor).'

Instruction to use supplementary labour

In practice the problem often faced by the contractor is that work on site is not proceeding at the rate required. It may not be perceived as relevant by the contractor's management whether the rate required by it is or is not that agreed under the terms of the subcontract. The rate of production the contractor now requires may be a faster rate than the subcontractor undertook to work at, when the Sub-Contract Agreement was concluded. Alternatively, owing to the contractor's delays, the subcontractor may now be required to work at a different time and as a consequence no longer has the resources available that it had or anticipated having at the time programmed for it to carry out the work, in the Sub-Contract Agreement. Also, because of the delays by the contractor not releasing work to the subcontractor as planned, the work is now concurrent with other obligations of the subcontractor. However, the contract delay will almost certainly be a failure by the contractor that will relieve the subcontractor from its original obligation to carry out the work to a specific programme or period. If, as a consequence, there is a significant rescheduling of the subcontractor's work, time may become at large or such that the subcontractor's obligation will change to one to complete in a reasonable time having regard to all the circumstances.

Although the contractor's inclination is frequently to dismiss the first subcontractor and obtain a second one, usually at a much higher rate or cost, with the intention of contra charging the increased cost against the first subcontractor, such action will rarely be lawful and permissible under the subcontract. Even if such action were permissible under the subcontract, the delay is likely to be further aggravated by the time lost disengaging the first subcontractor and re-sourcing and setting to work the replacement.

An alternative, rarely adopted, is to instruct the subcontractor to supplement its labour with labour provided to it by the contractor from another source. While it is considered that such action would be allowable under most variation and/or instruction clauses, there will be considerable scope for dispute as to the valuation of such an instruction. The contractor will not, it is suggested, be able to recover a higher labour rate than that which the subcontractor is itself paying to its directly employed operatives. Conversely, the subcontractor may be able to recover from the contractor for any proven lower rate of productivity and performance by the supplementary labour.

If, however, the contractor is able to demonstrate that the delay was the subcontractor's, then while the effect of giving the instruction will probably not be as described above, the contractor might be able to recover such additional costs by way of set-off for the subcontractor's default. It is considered that the contractor's primary obligation, however, is to value and pay for the effect of its instruction and subsequently demonstrate its right to set-off.

Chapter 8
Valuation of the Works

Introduction

The intention in all subcontracts is for the subcontractor to provide certain defined goods and/or services to the contractor for a stated sum. On completion of the subcontracted work that sum will become due to the subcontractor. The subcontractor will be entitled to the full subcontract sum and, if relevant, additional payment, subject to the provisions of the subcontract, and will be subject to adjustment for the failure by the subcontractor in its obligations and/or breach by the contractor.

Thus the valuation of the subcontract works will comprise a number of parts and any or all of these parts may be present whether the valuation is of an interim application for payment or of the final account. These component parts will comprise the sum of:

1. works contracted for;
2. varied and/or additional or omitted work;
3. works to be paid against a schedule of rates and/or hourly rate;
4. work to be paid on a reasonable price basis;
5. loss and/or expense for events as identified under the subcontract;
6. loss arising from defaults by the contractor;
7. any other sums due to the subcontractor.

Less:

8. any retention and/or discounts allowable to the contractor;
9. any abatement of the subcontract sum as a consequence of defective work etc.;
10. any set-off or loss arising from defaults by the subcontractor;
11. any other sums due to the contractor from the subcontractor.

The HGCRA requires, under section 110, that 'every construction contract shall provide an adequate mechanism for determining what payment becomes due under the contract and when'. The nature of the mechanism is for the parties to decide and may be the usual periodic valuation of work actually carried out or some other, such as stage payments or payment based on progress against a payment graph.

Further, the HGCRA requires the issue of a notice not later than five days after the payment due date, 'specifying the amount (if any) of the payment made or proposed to be made, and the basis on which that amount was calculated'.

Such notice is to be what:

'payment becomes due from him (the payee) under the contract, or would have become due if –

(a) the other party had carried out his obligations under the contract; and
(b) no set-off or abatement was permitted by reference to any sum claimed to be due under one or more other contracts.'

Several issues arise from the requirements of section 110. The first is the nature and extent of the detail necessary to satisfy the requirement to provide 'the basis on which that amount was calculated'. It is suggested that such a notice should at least distinguish between the sums allowed and those to be deducted, and these should conform to the general headings set out above. Where the subcontractor has made a detailed application, such notice should, it is suggested, as a minimum list all areas of difference or disagreement, since the purpose of this notice is to enable the payee to challenge the valuation if necessary.

The second difficulty with section 110 relates to the meaning of 'if the other party had carried out his obligations under the contract'. Such obligations include both procedural obligations, for example with regards to the method and/or timing of the seeking of payment, confirming instructions, and submitting day-work sheets, and obligations regarding the execution of the work, both to specification and to time.

It is submitted that the contractor's obligation, under the requirements of the HGCRA, is to value the works to the best of its ability having regard to any details, notices or applications made by the subcontractor, and that such obligations are not removed where the subcontractor, contrary to the requirements of the subcontract, fails for example to make a timely application for payment.

Having calculated the value of the work in accordance with section 110, the contractor may wish to issue a notice of intent to withhold payment under section 111. This also requires valuation to be effective and at subsection (2) must specify:

(a) the amount proposed to be withheld and the ground for withholding payment; or
(b) if there is more than one ground, each ground and the amount attributable to it.

From this it is apparent that the contractor must identify the various grounds against which it seeks to withhold payment. Again, the purpose of this notice is to inform the subcontractor so that, in the event of a dispute, it can make an effective challenge. Thus where the contractor gives vague and/or global grounds and/or details of the amount to be withheld, it is likely to invite a challenge unnecessarily, whereas a well-detailed schedule of grounds and details of the amounts is less likely to be rejected out of hand.

The provision within section 111 requires the contractor to specify the sum or sums it intends to withhold. It must be considered whether the use of the word 'specify' as opposed to 'state' or similar implies an obligation to provide calculations and/or supporting evidence to justify the sum to be withheld. Inevitably, the more justification given by the contractor for its intention to withhold sums otherwise due, the less likely there is to be conflict and dissatisfaction by the subcontractor. However, the issue of notices of intent to withhold payment gives to the subcontract a two-stage valuation procedure whereby, dependent on the periods within the payment mechanism, a subcontractor anticipating a payment on account only learns at a very late date that the payment due to it is to be denied.

The valuation of the subcontractor's work and other matters is dealt with by SBCSub/C, at clauses 5.2 to 5.12. Provision is made at clause 5.11 for the adjustment of the rate or rates where work is not carried out under similar conditions to those set out in the bills of quantities and/or other documents comprising the Sub-Contract Documents. This may include:

1. work to isolated areas;
2. work carried out either earlier or later than the main body of the work;
3. procurement of materials in small quantities;
4. working prolonged hours and/or in conjunction with other trades.

The conditions for the valuation of the subcontractor's work are set out at clauses 5.2 to 5.12 of SBCSub/C. These comprise the subcontract works and, specifically at clause 5.2, set out the general rules for the valuation of the work. SBCSub/D/C also includes provision for the valuation of Sub-Contractor's Designed Portion Works. Clause 5.2 requires the contractor to value the work described at items 1 and 2 below and in accordance with the valuation rules set out in clauses 5.6 to 5.12 ('the Valuation Rules').

Sub-clause 1 lists the items to be included where the work is to be valued on an 'Adjustment Basis' as:

1. all variations including any sanctioned by the contractor in writing but excluding any to which clause 5.3.3 applies;
2. all work which under these conditions is to be treated as a variation;
3. all work executed by the subcontractor in accordance with the directions of the contractor as to the expenditure of provisional sums which are included in the Sub-Contract Documents;
4. all work executed by the subcontractor for which an approximate quantity has been included in any bills of quantities.

Sub-clause 2 requires that where work is to be valued on a remeasurement basis, then the valuation is to include all work executed by the subcontractor in accordance with the Sub-Contract Documents and the directions of the contractor, including any direction requiring a variation or in regard to the expenditure of a provisional sum included in the Sub-Contract Documents.

These clauses are strictly limited to the valuation of the work and do not provide for any entitlement of the subcontractor under any other provision of the subcontract, including preliminaries, if priced separately, and loss and/or expense due to the regular progress of the subcontract works being affected by any of the relevant subcontract matters.

The provisions in ICSub/C are in all relevant aspects the same as those for SBCSub/C. The equivalent provisions with Sub/MPF04 are set out at clause 25.5, which requires that each interim payment shall include:

1. the proportion of the subcontract sum to which the subcontractor is entitled, calculated in the manner set out in the pricing document;
2. the value of any charges executed by the subcontractor;

3. the amount of any reductions made as a consequence of the provisions of clause 12.6; and
4. any amounts that either party is liable to pay the other in accordance with the provisions of the subcontract.

From this sum is to be deducted the value of payments previously made, giving the sum to be paid from one party to the other.

These are more general and inclusive terms, especially by reference to clause 25.5.4, when compared to those of SBCSub/C.

ShortSub defines the amount due to the subcontractor at clause 12.2 as being 'the value of the work properly carried out', together with any loss and/or expense due under clause 10.3.

Valuation of variations

Most standard forms of construction subcontract make express provision for the issuing of instructions by the contractor to the subcontractor to vary the work to be carried out. In addition, there will also be express arrangements for the valuation of such variations. If no express provision for valuation is incorporated within the subcontract, then payment is likely to be by way of an implied term to value varied or additional work on a fair price basis or in restitution on a *quantum meruit* basis.

SBCSub/C defines a variation for the purposes of the valuation of variations at clause 5.1 and sets out the rules for the valuation of variations and other matters at clauses 5.6 to 5.9. Similar provisions are made in ICSub/C at clauses 5.2 to 5.6.

Clause 5.6 sets out the rules for the valuation of variations for measurable work on an adjustment basis. Principally, additional or substituted work is to be valued by measurement using the five rules set out in five sub-clauses as follows:

1. Where the additional or substituted work is of similar character to, is executed under similar conditions as, and does not significantly change the quantity of, work set out in any bills of quantities and/or other Sub-Contract Documents, the rates and prices for the work so set out shall determine the valuation.
2. Where the additional or substituted work is of similar character to work set out in any bills of quantities and/or other Sub-Contract Documents but is not executed under similar conditions thereto and/or significantly changes its quantity, the rates and prices for the work so set out shall be the basis for determining the valuation and the valuation shall include a fair allowance for such difference in conditions and/or quantity.
3. Where the additional or substituted work is not of similar character to work set out in any bills of quantities and/or other Sub-Contract Documents, the work shall be valued at fair rates and prices.
4. Where the approximate quantity is a reasonably accurate forecast of the quantity of work required, the rate or price for that approximate quantity shall determine the valuation.
5. Where the approximate quantity is not a reasonably accurate forecast of the quantity of work required, the rate or price for that approximate quantity shall be the basis for determining the valuation and the valuation shall include a fair allowance for such difference in quantity.

Clause 5.6.2 provides for the pricing of omitted work and requires the use of rates and prices for that work as set out in the bills of quantities or other Sub-Contract Documents.

Clause 5.7 sets out the rules for the valuation of measurable work where the work is to be valued on a measurement basis. The measurement is to be carried out and valued using the following three rules:

1. Where the work is of similar character to, is executed under similar conditions as, and does not significantly change the quantity of, work set out in any bills of quantities and/or other Sub-Contract Documents, the rates and prices for the work so set out shall determine the valuation.
2. Where the work is of similar character to work set out in any bills of quantities and/or other Sub-Contract Documents, but is not executed under similar conditions thereto and/or significantly changes its quantity, the rates and prices for the work so set out shall be the basis for determining the valuation and the valuation shall include a fair allowance for such difference in conditions and/or quantity.
3. Where the work is not of similar character to work set out in any bills of quantities and/or other Sub-Contract Documents, the work shall be valued at fair rates and prices.

Clauses 5.6 and 5.7 of SBCSub/C, above, reflect the two basic methods of adjusting the subcontract sum for variations: 'adjustment basis' and 'remeasurement basis'. It will be for the subcontractor to know and establish which method applies. Where the Articles of Agreement, SBCSub/A, are completed and accepted, either by both parties signing the form or otherwise such as by performance, then a selection of either article 3A or 3B will have been made. Where there is no express provision, either because no Articles of Agreement, SBCSub/A, have been completed or no reference as to whether clause 5.6 or 5.7 is to apply is made in the Sub-Contract Documents, then it is likely that the provision of the main contract will be deemed to apply. In the event that no pre-selection can be established, then it is suggested that the adjustment basis is likely to apply.

The effect of the difference between the adjustment basis and the remeasurement basis may be insignificant by reason of clause 2.9, which allows for any error in description or in quantity or any omission of items within the bills of quantities to be corrected and be treated as a variation. On either basis the subcontractor is to be paid for the work it actually carries out, the difference being that in the adjustment basis it will be for the party seeking to have the bills of quantities adjusted for error in description, quantity or omission, to demonstrate such error exists.

Adjustments are to be made to the rates for any item where:

1. there is a significant change to the quantities;
2. the work is not executed under similar conditions; or
3. where approximate quantities are included, the approximate quantity is not a reasonably accurate forecast

as set out in the bills of quantities and/or the other Sub-Contract Documents.

A significant change in the quantity must, it is suggested, refer to any change in quantity which has a significant effect on the cost to the subcontractor, either by reason of the cost of materials, the utilisation of plant, including the cost of bringing it to and from the site, and the ability for the labour to operate efficiently, including the effect of any learning curve etc.

'Significant' is a matter of degree. What is significant to one subcontractor in one circumstance may not be significant to another subcontractor or to the same subcontractor in differing circumstances. It is therefore likely that a change will be significant if the subcontractor considers it justifies the preparation of a case for an allowance for such change in the valuation.

A change in conditions may relate to changes including the time of year and associated changes in both weather and labour availability, the market place, access to the work, handling and installation method and/or equipment, and site accessibility and/or congestion. It must be noted that this provision relates to changed conditions that flow directly from the variation and not any direct loss and/or expense flowing from any effect on the regular progress of the subcontract works (see clause 5.12 below). In this respect a distinction will be made between an instruction which, because it is issued late has an effect on the regular progress of the works, and an instruction which, because of the nature and detail of the variation, causes a delay and/or prolongation to the works.

Where an approximate quantity is not reasonably accurate, the valuation is to include a fair allowance for the difference. What will be considered not reasonably accurate will depend on the actual effect the difference has on the subcontractor. The effect of clauses 5.6.4 and 5.6.5 may lead to either an increase or decrease in the quantity of work instructed as compared to the approximate quantity. Manifestly, an allowance may in fairness need to be made to the sum as adjusted by reference to the rate or price, which may be either an addition or a reduction to that sum. It will be for the party seeking to make such adjustment to demonstrate that it is fair to do so, and the sum that should be fairly added to or deducted from the valuation calculated from the rate or price. SBCSub/C at clause 5.8 sets out the general rules for the making of such adjustments:

1. Where there are bills of quantities, measurement shall be in accordance with the same principles as those governing the preparation of those bills, as referred to in clause 2.8.
2. Allowance shall be made for any percentage or lump sum adjustments in any bills of quantities and/or other Sub-Contract Documents.
3. Where the adjustment basis applies, an allowance, where appropriate, shall be made for any addition to or reduction of preliminary items of the type referred to in the Standard Method of Measurement, provided that no such allowance shall be made in respect of compliance with a contractor's direction for expenditure of a provisional sum for defined work.
4. Where the remeasurement basis applies, any amounts priced in the preliminaries section of any bills of quantities shall be adjusted, where appropriate, to take into account any variations or any contractor's directions for the expenditure of provisional sums for undefined work included in the Sub-Contract Documents.

The general rules at clause 5.8 of SBCSub/C relate to the valuation of variations both on an adjustment basis and on a remeasurement basis. Clause 5.8.2 requires that allowance be made for any percentage or lump sum adjustments provided for in the subcontract. Such a situation commonly arises as a result of post-tender negotiations when the subcontractor agrees to the adjustment of its offer. Where this is done, it is essential for the subcontractor to define whether such adjustment is to be regarded as a general adjustment to the subcontract price and consequently subject to the provisions of clause 5.8.2, or as a once and for all adjustment, and that variations are to be valued on the tendered rates, contrary to the principle of clause 5.8.2.

Clauses 5.8.3 and 5.8.4 of SBCSub/C make provision for the adjustment of the preliminaries element of the bills of quantities affected by the variation, which may be adjusted up or down as necessary. Such allowance should include for all preliminary costs directly affected by the variation, whether the subcontractor has made allowance for its preliminary costs within the schedule of rates and/or prices or as a separate schedule of preliminary items.

SBCSub/C at clause 5.9 provides for the valuation of work that cannot be valued by measurement to be valued on a daywork basis using the schedule of daywork rates within the Sub-Contract Documents. In the event that there are no rates stated within the documents, then those published by the Royal Institution of Chartered Surveyors are to be used; alternatively rates published by an appropriate specialist trade employers' association may be used.

The use of daywork is subject to the subcontractor providing vouchers for verification by the contractor, not later than the Wednesday of the week following the week in which the work was carried out. Such vouchers are to give the daily time spent, the workers' names and details of plant and materials used.

Clause 5.9 of SBCSub/C allows for the valuation of variations where work cannot be properly valued by measurement, to be valued on a prime cost basis. The clause does not define who is to decide when this clause is to apply, or provide or require either the contractor or the subcontractor to notify the other in advance of the subcontractor carrying out the work that it considers daywork to be the proper method of valuation for such work, and not for valuation and payment under clause 5.6.1.3 on a fair rate or price basis. In the absence of such an express requirement, it is suggested that such action represents good practice and should in any event be carried out by the subcontractor.

There are, however, express requirements for the submission of daywork vouchers giving the daily time spent on the work and giving details of the workmen's names, and of plant and materials used. Such vouchers are to be submitted to the contractor not later than the Wednesday following the week in which the work was carried out. There is no express provision for the contractor to accept or sign the vouchers; however, unless the contractor challenges such vouchers soon after their submission, there will likely be a presumption that the contractor considered, at the time, that they were a true record of the time and materials involved.

Although clause 5.9 expressly requires the submission of daywork vouchers by the Wednesday following the week in which the work is carried out, it is submitted that the subcontractor will not, as a consequence of this express provision, lose all right to payment for such work should the subcontractor fail to issue the vouchers on time. Should the subcontractor's failure to issue vouchers on time result in the contractor being unable to recover on a similar basis or being involved in additional expense, then the subcontractor's entitlement may be reduced.

It must be realised that this clause of SBCSub/C only refers to work relating to variations. Separate requirements may exist for work related to other directions of the contractor and or requests from the subcontractor to the contractor. In the absence of express provisions, the pre-notification and early submission of vouchers recording the resources involved should be regarded as necessary to support any subsequent claim for payment.

SBCSub/C sets out the valuation rules for where there is a change of conditions for other work and additional provisions, at clauses 5.11 and 5.12. ICSub/C has a similar provision at clauses 5.7 and 5.8.

SBCSub/C at clause 5.11 provides for the adjustment of the rates within the Sub-Contract Documents where there is a substantial change to the conditions under which the work is to be carried out as a result of any of the following four circumstances:

1. compliance with any direction requiring a variation (except a variation for which a schedule 2 quotation has been accepted or where the variation is one to which clause 5.3.3 applies) or (where the remeasurement basis applies) any direction as a result of which work included in the Sub-Contract Documents is not executed;
2. compliance with any direction as to the expenditure of a provisional sum for undefined work;
3. compliance with any direction as to the expenditure of a provisional sum for defined work, to the extent that the direction for that work differs from the description given for such work in any bills of quantities; or
4. where the adjustment basis applies, the execution of work for which an approximate quantity is included in any bills of quantities, to the extent that the quantity is more or less than the quantity ascribed to that work in those bills.

Clause 5.11 of SBCSub/C allows for the situation where the compliance with a variation in one part of the subcontractor's work has an effect on the conditions under which other parts of the subcontractor's work are to be carried out. The comments above in respect to changed conditions under clauses 5.6 and 5.7 are also applicable to this clause. However, the change under clause 5.11 must be substantial and not just any change, and again it is suggested that the test for 'substantial' is whether the party seeking an allowance for changed conditions regards the making of a claim for such an allowance as justified and necessary.

Clause 5.12 of SBCSub/C makes two additional provisions. The first provides for the situation where the other valuation rules cannot reasonably be applied when a fair valuation is to be made. The second prohibits the valuation of loss and/or expense owing to an effect on the regular progress of the subcontract works, being made under the valuation rules, where the subcontractor is to be reimbursed under other provisions of the subcontract.

Clause 5.12 of SBCSub/C is a general term which at clause 5.12.1 acknowledges that there may be a need to make a valuation for matters arising from a variation which do not relate to the execution or omission of work or cannot reasonably be valued under any other provision. It will be for the subcontractor to identify any such matters which may be considered under such a description but might include the need to manage and/or respond to variation instructions, such as the issue of revised and/or amended drawings.

The second part of this clause, at sub-clause 5.12.2, expressly excludes, from the valuation under a variation, the effect on the regular progress of the subcontract works. There will inevitably be the possibility of a fine distinction between a change in the conditions for the execution of work as allowed for under clauses 5.6.2, 5.7.2 and 5.11, and a claim resulting from the regular progress of the works being disrupted. The less direct the effect, the more likely such costs will be regarded as excluded from valuation as a variation and will be regarded as loss and/or expense due to the regular progress of the work being affected. The effect of variations on the off-site and factory-based work should be regarded as subject to this provision, although delay or disruption at the factory may not normally be considered an effect on the regular progress of the Sub-Contract Works.

The provisions of SBCSub/C, as discussed above, may be regarded as the norm for all building subcontracts and to represent trade custom and practice. Where subcontracts incorporate different express provisions those will apply in the normal way but may be interpreted generally as discussed above.

SBCSub/D/C contains an additional clause, 5.10, relating to the valuation of the Sub-Contractor's Designed Portion, ICSub/D/C, which has the same provisions as clause 5.9. DBSub/C has a similar clause 5.10 for the valuation of the Sub-Contractor's Designed Works. Under these provisions the valuation of the work is to take account of the following four provisions:

1. Allowance shall be made in such valuations for work involved in the preparation of the relevant design work.
2. The valuation of additional or substituted work (adjustment basis) or executed work (remeasurement basis) shall be consistent with the values of work of a similar character set out in the subcontractor's design portion analysis, making due allowance for any change in the conditions under which work is carried out and/or any significant change in the quantity of work so set out. Where there is no work of a similar character set out in the subcontractor's design portion analysis a fair valuation shall be made.
3. The valuation of the omission of work set out in the subcontractor's design portion analysis shall be in accordance with the values therein for such work.
4. Clauses 5.8.2, 5.8.3, 5.8.4, 5.9 and 5.11 shall apply so far as is relevant.

Clause 5.10 of SBCSub/D/C deals with the valuation of variations affecting the Sub-Contractor's Designed Portion, and specifically at clause 5.10.1 the valuation of work in the preparation of the relevant design work. This clause provides no express procedures for the recording, notification and valuation of additional design work such as is provided for the work itself, which is likely only to be capable of valuation on a daywork basis, and clause 5.9 may be considered to apply. It will therefore be for the subcontractor to keep such accurate records of the work carried out in connection with each variation, as well as, it is suggested, the name of and time spent by each member of staff involved in the design process and associated production, printing and distribution of drawings and other information. The design process may require the designer or other members of the subcontractor's team to attend meetings and/or visit site or the offices of specialist suppliers as part of this process, in addition to the direct activity of redesign.

The underlying principle of the valuation rules in SBCSub/C, and generally where those terms do not apply, is that valuation will give the subcontractor an equal opportunity to profit as it would have had if the subcontract had not been varied. To this end any rates included in a pricing document, whether a bill of quantities or schedule of items, should form the basis of the valuation of variations. Where no such rates are incorporated into the subcontract, then reliance may be put on the subcontractor's tender build-up.

ShortSub provides for the valuation of variations at clause 10.2 on a fair and reasonable basis and with reference to any rates and prices in the pricing documents.

Sub/MPF04 at clause 23 under the section on valuation and payment sets out the procedures for the valuation of changes, and at clause 23.2 whereby changes are either to be pre-valued in accordance with clause 23.5 or by a fair valuation under clause 23.6. Clause 23.2 also makes reference to the possible need to adjust the period for completion.

Clauses 23.3 to 23.5 provide for the pre-pricing of instructed changes, which may be considered to be the preferred procedure, under this form of subcontract. However, clause 23.6 deals with the situation where no agreement as to valuation is reached prior to the implementation of the change instruction. Such a situation will either arise where, following a request under clause 23.3, there is no agreement, or where the change is instructed by the contractor without a request under clause 23.3. Clause 23.6, as ShortSub, requires a fair valuation and lists four matters to be considered in making a fair valuation.

Clause 23.6 of Sub/MPF04 sets out the procedures for the valuation of a change for which a pre-agreement under clause 23.5 has not been reached. In those circumstances a fair valuation is to be made by the contractor based on the following four rules:

1. the nature and timing of the change;
2. the effect of the change on other parts of the Sub-Contract Works;
3. the prices and principles set out in the pricing document, so far as applicable;
4. any loss and/or expense that will be incurred as a consequence of the change, provided always that the fair valuation shall not include any element of loss and/or expense if that element was contributed to by any cause other than a change or a matter set out in clause 24.2.

There is no express provision for changes to be priced on a prime cost or daywork basis, but such general terms must include such a pricing method where appropriate.

Sub/MPF04, unlike SBCSub/C or ICSub/C, at clauses 23.7 to 23.9 provides a timetable for the valuation of changes as follows:

1. The subcontractor is to provide to the contractor the subcontractor's proposed valuation of any change identified by either party, within ten days of that identification. In addition the subcontractor is to provide such information as is necessary for the fair valuation of the change by the contractor.
2. Within 21 days of the contractor receiving the details of the subcontractor's valuation and the associated supporting information, the contractor is to notify the subcontractor of the contractor's valuation of the change. That notice is to be in

sufficient detail for the subcontractor to identify any differences between the valuations.
3. Within 28 days of practical completion the subcontractor is to provide details of any further valuation that the subcontractor considers it should be given in relation to any change. The contractor is to review those particulars and within 56 days notify the subcontractor of any further valuation that the contractor considers is due to the subcontractor.

Clause 23.7 requires the subcontractor, within ten days of the identification of a change, to provide to the contractor 'details of its proposed valuation of the change'. The phrase 'details of its proposed valuation' will normally require the subcontractor to provide a valuation with associated calculations within the ten day period. However, where such details are not available, such as where the subcontractor is dependent on obtaining quotations from its suppliers and/or sub-subcontractors, then such proposals could define how such components of the valuation are to be valued. Indeed, it will be open to the subcontractor to propose any method of valuation it believes to be fair including, it is submitted, a prime cost or daywork basis if appropriate.

Manifestly, the contractor's obligation under clause 23.8 to respond to the subcontractor's valuation within 21 days must, where appropriate, include agreeing to a method of valuation which is dependent on a future event, such as daywork.

In the case of *Weldon Plant Limited* v. *The Commission for New Towns* [2000] BLR 496, Judge Humphrey Lloyd QC found that the effect of an instruction, issued under clause 51 of the ICE form of contract, gave rise to the need for work to be valued on the basis of a fair valuation because the contractor was being denied its rights under the contract by means of the variation, and he set out the objective of a fair valuation as being to ensure that the contractor was no worse off than if the variation had not been made. Judge Lloyd said, at paragraph 3:

> 'That option was removed and consequently a fair valuation is I consider appropriate and I accept the claimant's argument that it should not be worse off as a result of having complied with the instruction. It would be improper for Weldon to suffer financially because of the situation imposed upon them.'

Judge Lloyd made the important point, at paragraph 15, that under the rules for the pricing of variations contained in the ICE conditions at clause 52 and which, it is suggested, is not different in effect to the pricing provisions of clauses 5.6–5.12 of SBCSub/C:

> 'The variation must be valued in accordance with clause 52, the provisions of which are clearly directed to seeing that the contractor will not have to bear the costs of the variation, except to the extent that, where Rules 1 and 2 apply, the contract rates or prices were inherently insufficient, or to the extent that the costs incurred are not reasonably or properly to be treated as forming part of the valuation. The contractor therefore takes the risk that its rates and prices for the work may not cover its costs of carrying out a variation which is the same as or comparable to contract work, just in the same way that the employer takes the risk that by having second thoughts more might have to be paid for the work than might otherwise have been the case had the need been known prior to the date of contract and rates or prices for the work in question then obtained.'

Thus, where the valuation is just to make an adjustment for a change in the quantity of work to be constructed, as a result of minor adjustments to the design, the subcontractor is at risk to the extent he has underpriced the work or rates in his accepted offer.

However, where because the bill rates cannot be used and the work is to be valued on a fair rate basis, then, also in paragraph 15, Judge Lloyd said:

> '. . . a fair valuation which would ordinarily be based upon the reasonable costs of carrying out the work, if reasonably and properly incurred . . .'.

He then added the important rider:

> 'Clearly if, in the execution of the work, cost or expenditure is incurred which would not have been incurred by a reasonably competent contractor in the same or similar circumstances, then such costs would not form part of a fair valuation.'

It will be a matter of fact which, it is suggested, it will be for the contractor to demonstrate, whether or not expenditure was such as would not have been incurred by a reasonably competent subcontractor, acting reasonably in the same or similar circumstances. This does not mean, it is suggested, that any avoidable expenditure is to be ignored and discounted in making a fair valuation. Any subcontractor is likely, at some stage in the construction process, to have a certain amount of waste and/or inefficiency and to have to rectify work incorrectly and/or defectively constructed. It is suggested, therefore, that in making such a counterclaim, it will be necessary for the contractor to submit evidence as to what is a reasonable level of competence and to show that this was not achieved, or that the amount of waste significantly exceeded that which could reasonably have been expected and, probably in addition, for the works as a whole and not limited to the work in connection with the variation being valued.

On that basis it might be shown that the rate of waste far exceeded the average throughout the subcontract works as a whole and therefore the prime cost should be reduced to no more than the average. Even such a premise will not be universally applicable. An error in interpretation of an instruction may lead to a significant loss or waste relative to the value of that instruction, but will flow directly from the issue of that instruction since the loss would not have arisen but for the change or variation required by the contractor. It may be that the contractor will have to show that the subcontractor was grossly negligent in its management of the variation or that no reasonably competent tradesman would have so misunderstood the relevant instruction, before the subcontractor would become disentitled to such costs.

Judge Lloyd then analysed the nature of a fair valuation and referred to the various commentaries. First, he suggested that normally a fair valuation will be on a cost plus basis but did not rule out the use of comparable market rates. Quoting from a book by Mr Max Abrahamson, he said, at paragraph 16:

> '"Fair Valuation" will normally mean cost plus a reasonable percentage for profit, (but not contingencies if the work is being valued after it has been carried out on actual not estimated costs) with a deduction for any proven inefficiency by the contractor, but if there is proof of a general market rate for comparable work it may be taken into consideration or applied completely.'

He stated that there should not be an allowance for contingencies, which is not surprising since they would be an allowance in an estimate for matters over and above the expected direct cost of the item. He made reference to 'proven inefficiency', which has been discussed above. It is again suggested that a fair valuation must make allowance for such problems and difficulties that a subcontractor will normally encounter on a reasonably efficient site and which would be allowed for by way of a contingency if the work were being pre-priced. These must either be allowed for on an actual cost basis or by a suitable addition. The practical difficulties in showing the amount by which the actual costs should be reduced for inefficient working, and then adding back a suitable percentage for site difficulties, could defeat the purpose and intent of a payment on a cost plus basis.

There is no assistance given as to when a market rate might be used in preference to a cost plus valuation. Prices obtained from published price books at best are only accurate at the time of going to press and therefore represent an average rate at that time and make no adjustment for either project-specific matters and/or changes in the market place and/or seasonal adjustments. Similarly, prices quoted and/or obtained for similar work on other projects will not be project specific. It is suggested that where market rates are to be used, a project adjustment must be considered and agreed. Then such rates should be universally applied; it will not be a fair valuation if either party feels justified to pick and choose the method of valuation between items depending on which method is most advantageous in each case.

Judge Humphrey Lloyd found that a fair valuation must include all elements ordinarily included in a contract rate or price, also at paragraph 16:

> '. . . the proposition that a fair valuation had to include each of the elements which are ordinarily to be found in a contract rate or price; elements for the cost of labour, the cost of plant, cost of materials, the costs of overheads, and profit. In my judgment a fair valuation has not only ordinarily to include something on account of each of those elements, but also it would not be a fair valuation within the meaning of the contract if it did not do so.'

The judge finally considered whether a fair valuation should include an allowance for overheads and profit. He had no difficulty in finding that a valuation without an allowance for overheads and profit would not be a fair valuation, in the following words, at paragraph 18:

> '. . . so there is, in my judgment, no reason why on the facts found by the arbitrator a fair valuation should exclude profit. Indeed in my judgment a fair valuation must, in the absence of special circumstances, include an element on account of profit. First, a contractor is in business to make a profit on the costs of deploying its resources, and accordingly an employer must under clause 52(1) pay profit in a valuation made under any rule, (via the rates or otherwise on a fair valuation) or costs for a valuation under clause 52 would not otherwise be a fair valuation within the meaning of those words. Secondly, a valuation which did not include profit would not contain an element which is an integral part of a valuation under Rules 1 and 2. A fair valuation under Rule 3 would not be in accordance with the principles of clause 52 if it did not include all relevant elements to the value or represented in some significant manner in a valuation under that clause.'

He concluded, at paragraph 19, by deciding that overheads and profit, in relation to a fair valuation, do not need to be proven as would be the situation where claiming loss and/or expense but may only be an approximation by reference to records of previous work or years as appropriate:

> 'In my view, a valuation which in effect required the contractor to bear that contribution itself would not be a fair valuation, in accordance with the principles of clause 52(1) which are intended to secure that the contractor should not lose as a result of having to execute a variation (except, as I have stated, to the extent its costs etc. are of its making). Unlike overheads such as time-related overheads, it is not necessary to prove that they were actually incurred for purposes of fair valuation (although their approximate amount must of course be established, e.g. by deriving a percentage from the accounts of the contractor including where appropriate associated companies that provide services or the like qualify as overheads).'

In the case of *Henry Boot Construction Ltd* v. *Alstom Combined Cycles Ltd* [1999] BLR 123, Judge Humphrey Lloyd QC again had to consider the valuation of variations and restated the law as to the definition of a fair valuation, and said, at paragraph 36:

> 'A fair valuation when used as an alternative to a valuation by, or by reference to, contract rates and prices generally means a valuation which will not give the contractor more than his actual costs reasonably and necessarily incurred plus similar allowances for overheads and profit – for anything more would confer on him an additional margin for profit and would not be fair to the employer. Fairness is an objective test which takes into account the position of both parties.'

Difficulties have arisen where the subcontract rates are either exceptionally good or poor, whether by design or in error. The Court of Appeal, in the case of *Henry Boot Construction* v. *Alston Combined Cycles* [2000] BLR 247, considered this problem. Although this concerned a civil engineering contract under the ICE conditions, the valuation principles are similar. In this case some piling work had been priced high in error. The Court of Appeal upheld the decision, at first instance, that the contract rates were required to be used whether high or low. The arbitrator had recognised that the piling price contained an error and put the sum to one side when valuing the additional work. In doing this he had had to consider the requirements under the ICE conditions, not present in the SBCSub/C conditions, that the contract rates were to be used 'so far as may be reasonable'. This was held to mean that the use of the rates had to be reasonable, not the rates themselves. The Court of Appeal upheld the judgment of Judge Humphrey Lloyd QC, which is quoted in the speech of Lord Justice Ward in these words:

> '34. . . . the same approach (as applies to Rule 1) must apply to Rule 2 for that is no more than a continuation of Rule 1 to deal with the position where the factors mentioned in Rule 1 are not present – similarity of work or conditions. If the varied work is of a dissimilar character or to be executed under dissimilar conditions, then the contract clearly maintains the principle that a valuation ought to be made if there is a contract rate or price applicable or which could be used as a basis for valuing the variations. The fact that the result of the use of the contract rate or price might not be reasonable is as irrelevant as it is under the first principle.'

SBCSub/C, at clauses 5.6.1.2 and 5.7.2, requires the use of the rates in the bills of quantities and other contract documents for the pricing of variations, but where the work is of a similar character but is not executed under similar conditions or has significantly changed the quantity of work, then the valuation shall include a fair allowance for the difference.

Thus, while the work is to be priced using the subcontract rates, an additional sum by way of an allowance for the different conditions or quantity is to be valued in addition. The effect of this clause is not to change the basis for payment for the work included within the original subcontract requirements, but only to adjust the valuation of the additional work or where the work is carried out in changed conditions, all such work.

The significant difference between the provision under the SBCSub/C and the ICE conditions is that, under the SBCSub/C terms, in the stated circumstances the subcontract rates are to be used and an allowance made for the change, and that allowance must be fair, rather than the use of a fair valuation with no specific reference to the contract rates. In other words, it is a fair valuation of the changed circumstances that the contractor is required to add to the existing subcontract rate. There is no provision for total re-rating.

Again, it is likely that 'fair' means where it is fair to make an allowance because of the different circumstances. Such a circumstance may result from a reduction in quantity an or increase in unit production costs, as for example when the usage of moulds required for concrete work changes. This was the case in *Tinghamgrange Ltd* v. *Dew Group Ltd and North West Water Ltd* (1995) 47 Con LR 105. In this case, an order for special pre-cast blocks was cancelled when only part of the order had been completed. The supplier claimed for 75% of the mould cost, which would have been recovered spread over the cost of the cancelled blocks. Eventually the mould cost was accepted; however, the supplier sued in addition for loss of its profit and the contractor was required to pay the supplier for this. However, in a third party action the contractor failed to recover these costs from the employer, and appealed.

Lord Justice Stuart-Smith considered the nature of the product and the cancellation, at page 116:

> 'It seems to me that it may well be a question of fact and degree. To take one example, if the engineer issues a variation to change a substantial quantity of blockwork to brickwork, both of which I will assume are readily available, one would not expect the contractor to have committed himself to purchasing 10m blocks, so that in the event of cancellation he would have to pay his supplier a loss of profit claim. I would expect the contractor to ensure that a supply was available, but place orders as and when required. On the other hand, if the contractor had subcontracted for the construction of a highly sophisticated piece of equipment, which may involve great capital cost on the part of the subcontractor to fabricate, I can readily see that such a subcontractor might not be at all willing to agree to a cancellation clause such as Mr Naughton [Counsel for North West Water] envisages, or at least without adequate protection in the form of increased price. If he is to bear the risk of cancellation, he will require in those circumstances a higher price, and this might be commercially unrealistic. I appreciate that this example is not apt for a contract such as this where the contract price depends upon measured quantities applied to the contractor's rates stated in the bills of quantities. I merely use it as an example.'

Sir John Megaw, who allowed the appeal, considered what is a 'fair valuation' and related this to the unilateral nature of variations, in the following words at page 117:

> 'Clause 51 of the contract ('the head contract') between North West Water Ltd ('the employer') and Dew Group Ltd ('the contractor') gives the employer the right to make any variation comprised within the wide scope of clause 51(1). It is unnecessary to decide in this appeal whether a contract containing such a clause would have been legally binding – one party having the right to vary any term of the contract – if it had not been for its inclusion of a further provision, expressed in clause 52(1), whereby a 'fair valuation' has to be made. The 'fair valuation' of the variation is intended to provide fair compensation to the contractor for any adverse financial effect upon it, resulting from a unilateral variation. We may assume, therefore – the contrary was not argued – that the validity of the head contract is not affected by clause 51, because of the provision for fair compensation in the event of loss resulting to the contractor from a unilateral variation.'

The concept of the contractor having the right unilaterally to make variations and requiring the valuation process to provide fair compensation to the subcontractor supports the contention that where contract rates are manifestly underpriced and the issue of a variation would cause the subcontractor a major further loss, the contractor may not be able to hold the subcontractor to its contracted prices for the additional work. In such circumstances the subcontractor may be able to hold that it would be unreasonable to use such rates, to use the ICE terminology or to require an additional fair allowance for a change in conditions or significant change in the quantity, to use the language of SBCSub/C.

However, in the case of *Galliford (UK) Ltd* v. *Aldi Stores* (8 March 2000, unreported), in an appeal from an arbitrator's award, it was held that a negotiated reduction in the contract sum, which had been effected by changing the rates for disposal of soil off site to nil, was not to be adjusted when the requirement to remove soil from site was reintroduced post-contract, and despite the quantity of contaminated soil being considerably increased.

In the contractor's original offer, it had priced for removal of excavated material from site against bills of quantities which indicated that there was 50% contaminated and 50% clean material to be disposed of. Post-tender, it was agreed to seek to redesign the project so as to eliminate the need to remove soil from site by adjusting the levels of the landscaping. Initially, the contractor offered a lump sum adjustment by way of the saving for not removing any spoil and instead spreading it over the site. Subsequently, the contractor amended the bills of quantities by putting a rate of nil against the quantities for removal from site, rather than changing the quantities to nil or deleting the items altogether. As a result, Judge Browsher QC found that under the valuation rules of the contract, the contractor was to bear the costs of carting spoil to tip:

> '12. At its simplest, what was required was that work should be redesignated from one heading in the bills of quantity to another. Work was moved from work on uncontaminated soil to work on contaminated soil. The rate for both was £0.00 per cubic metre so the change in category should, on the appellant's case result in no change in the rate. Clause 3.7.3(a) provided that "for work of a similar character to that set out in the priced document the valuation should be consistent with the values therein". Here, the

appellants say, the work was not just of a similar character, it was of the same character as work set out in the bills, therefore the same charge should be made, namely, £0.00.

13. What is the justification for departing from the rates set out in the bills of quantity? One can look only to the words in clause 3.7.3 "making due allowance for any change in the conditions under which the work is carried out and/or any significant change in the quantity of the work so set out".

14. The arbitrator made no finding about the conditions under which the work was carried out. There is no question of the work moving from summer conditions to winter conditions, or from normal hours working to weekend and overtime working or anything of that sort. The work was simply moved from one category in the bills of quantity to another.

15. Equally, the arbitrator made no finding regarding the bills of quantity rate of £0.00 related to any change in the quantity of work.

16. What the arbitrator has done is to make what in his view was a fair valuation of the work. He did that by going back to his tender and considering the negotiations between the parties. I think that his approach very probably did produce a valuation that was fair between the parties, and it would have been extremely attractive if the parties had given to the arbitrator the jurisdiction to decide what was reasonable. But that is not what the parties agreed. Clause 3.7.4 provides that:

> "a fair valuation shall be made – where there is no work of a similar character set out in the priced document . . ."

Here there was work set out in the priced document which was not just of a similar character but was of the same character and the arbitrator had no jurisdiction to do what he did.

17. While I have much sympathy for Galliford on the facts of this case, I regret that I find that the arbitrator was wrong in his decision.'

The judge in this case found that the valuation was not to be made on a fair valuation basis but under the valuation rules in the contract. In adjusting the pricing of the bills of quantities the contractor had not made the correct amendments by adjusting the rates to nil rather than adjusting the quantities to nil, and as a consequence was not entitled to any payment for removing spoil to tip when the design reverted to the original intention, which was made worse by the fact that the quantities were wrong in the first place.

Contractor's directions

Throughout SBCSub/C there are references to contractor's directions. The right for the contractor to issue directions is set out in clause 3.4. This allows for the contractor to issue directions in writing which are reasonable, including those which are variations to the subcontract. Also, the contractor may instruct the subcontractor in respect of any instruction issued to the contractor under the main contract that affects the subcontract works.

Clause 3.4 defines directions as including but not limited to, the issue of variations, which are defined for the purpose of SBCSub/C at clause 5.1. This definition of a variation does not include directions by the contractor which have a similar effect to, but are not the impositions of, the architect under the main contract, as set out in the second part of that definition and which include the addition to or alteration or omission of any obligations or restrictions set out in the Numbered Documents in regard to:

1. access to the site or use of any specific parts of the site;
2. limitations of working space;
3. limitation of working hours;
4. the execution or completion of the work in any specific order.

Clauses 5.6 and 5.7 concern the valuation of variations and by definition exclude the valuation of contractor's directions that are not variations. In the absence of express provisions for the valuation of contractor's directions not variations, it is suggested that there will be an implied term that the consequences of such directions are to be valued on a fair valuation basis. A fair valuation will not be less than, but may be held to be greater than, that allowable under the valuation of variations clauses where the effect of the direction is similar to that which would have been the case if the direction had been an imposition by the architect under the main contract.

While some directions may involve clearly definable additional activities by the subcontractor, others will affect production rates in general, when clause 2.17, delay to the progress of the subcontract works, may apply. Deciding a fair method of valuation for the effect of such directions will rarely be easy where they lead to an effect on the subcontractor's achievable outputs. In such circumstances the keeping of accurate records of labour output achievements throughout the duration of the work will be of great assistance to the assessment of the effect of the direction.

SBCSub/C at clause 3.5.1 states that where such a direction requires a variation of the type referred to in clause 5.1.2, the subcontractor need not comply to the extent that he makes reasonable objection to it in writing to the contractor.

The scope of this clause is strictly limited to the imposition of an instruction of the architect under the main contract, as defined in clause 5.1.2, and does not state what may be regarded as grounds for reasonable objection. It is suggested that reliance on this clause will only be lawful where the carrying out of the direction would be unlawful or unsafe or so difficult to comply with as to lead the subcontractor to believe that it could not be achieved or carried out. It is not considered that a direction leading to a significant cost variation would give the subcontractor grounds for not complying, if it had received from the contractor a variation instruction fully detailing and defining what is required.

Pre-priced variations

Variations may be pre-priced before execution, either under the provisions of the subcontract or unilaterally by the subcontractor. Where such a pre-price is made under the subcontract, then that valuation will have the effect allowed for in the subcontract. Where there is no provision or no request for the subcontractor to pre-price changed and/or additional work, the effect of such pre-pricing will depend on whether or not the subcontract contains provision for the change to be pre-priced. Where the procedure for the valuation of variations is uncertain, such as where valuation is to be on a fair price basis and there is no pre-pricing provision for the change required, then the pre-priced offer is likely to be an offer for a collateral agreement, which may or may not be accepted by the contractor.

SBCSub/C, at clause 5.3, sets out provision for a subcontractor's quotation in compliance with an instruction to provide a price, and for its acceptance prior to the implementation of the instruction. The reference to schedule 2 is to the schedule at page 46 of SBCSub/C, which is a fairly complex and lengthy process which allows 14 days for the pricing of the variation by the subcontractor and a further 14 days for acceptance. In practice, where the subcontractor has itself to obtain prices from others and the contractor is to obtain prior acceptance from the employer, 28 days is likely to prove insufficient time. While SBCSub/C makes this provision, it is not believed that it is used often in practice, if at all.

In a subcontract such as SBCSub/C where there is a provision under the subcontract for both the giving of instructions for variations and other directions to the subcontractor by the contractor, and provisions for the subcontract duration and subcontract sum to be adjusted as a consequence of such variations and directions, the providing of an unsolicited pre-priced sum by the subcontractor can be no more than an indication of what is likely to be due to the subcontractor under the subcontract.

In any event, the subcontractor may not use undue pressure to seek agreement as to the price before complying with the instruction, and in the event that the contractor is able to demonstrate that such agreement was given under undue duress, it is likely that such agreement will not be upheld.

The situation may be different if the requirements of the subcontract are for the subcontractor to advise the contractor of the likely cost arising from an instruction and/or direction, before commencing implementation of the instruction or within specified time of such an instruction, especially where this is an express condition of payment of any additional sum. Even under such conditions, where the contractor can be regarded as having allowed the varied or additional work to proceed in the knowledge of the subcontractor's requirements for implementing such work, the pre-pricing of the variation may not, it is submitted, alter the agreed method for payment under the conditions of the subcontract. Conversely, if the subcontract requires acceptance of the price for the variation before proceeding, and after a reasonable period the subcontractor proceeds with the full knowledge of the contractor, then the contractor may be deemed to have accepted the price by performance.

Such rules, either requiring the pre-pricing of all variations before implementation or a refusal to agree at an early stage the cost of significantly varied work, are considered to be inflexible and to lead to confusion and dispute. If instructions or directions require prompt or urgent action, a requirement for such instructions or directions to be pre-priced will prevent such prompt action. On the other hand, if instructions are to be carried out regardless of cost and without any pre-notice that considerable costs are involved, disputes will frequently arise when the subcontractor seeks payment of what it considers is a proper valuation. It is therefore advisable to ensure that flexible terms are incorporated into a subcontract requiring the notification by the subcontractor of cost implications, when urgent action is not necessary, prior to implementation but in any event when considered to be significant for notice to be given within a stated period of the issue of the instruction.

There is no express requirement under SBCSub/C for the subcontractor to give advice and/or notice as to the cost effect of directions other than for loss and/or expense under clause 4.20. However, even where there is no such provision, the giving of a price by the subcontractor may indicate what is a reasonable price where such is the appropriate criterion. In the case of *Bentley Construction Ltd* v. *Somerfield Property Company Ltd and Somerfield Stores Ltd* (2002) 82 Con LR 163, Judge Richard Seymour QC said, at paragraph 59:

> 'Each request to undertake work was, on the evidence in my judgment, merely an offer which, if acted upon, gave rise to an entitlement on the part of Bentley to be paid a reasonable sum for the work in question, up to any limit specified in the relevant form, unless a price was specified in the form, in which case the entitlement was to be paid that price. . . . What was a reasonable sum would not depend upon what rates had been notified for particular work executed in particular circumstances, although any rates notified might be evidence of what was a reasonable sum to charge.'

The use of the verb 'to charge' rather than 'to pay' suggests that by submitting a price, prior to executing the work, the subcontractor may be limiting its entitlement to payment unless there is a subsequent change in the requirements or circumstances in which the work is to be carried out.

Enhanced rates

The standard forms of subcontract contain express rules for the valuation of variations, which allow for the adjustment of a rate where work is carried out under changed circumstances. While this is a principle that is easy to write into an agreement, it is extremely difficult to apply in practice. The sort of circumstances that may justify an enhanced rate include:

1. working longer hours;
2. working in more restricted circumstances and/or in conjunction with other trades;
3. intermittent working involving repeated visits to the same work area;
4. working in a different period of the year, giving rise to changed weather conditions and/or labour availability.

One approach is to use one of the relatively few published loss of productivity studies, such as *Effects of Accelerated Working, Delays and Disruption on Labour Productivity* by R. M. W. Horner and B. T. Talhouni, published by the Chartered Institute of Building, which shows dramatic reductions in productivity for relatively little disruption. Indeed, a situation can soon be reached where it is likely that the productivity rate per hour will drop to such an extent that the total output will be less by working long hours than would have been achieved in a shorter working week.

A common approach is to seek to be paid on an actual cost basis or cost plus. It is suggested this is not a correct method for such a valuation, unless there is an express agreement to this effect between the parties. Indeed, it is likely that such an arrangement, by way of a supplementary agreement if necessary, is likely to be the most straightforward and controllable method of valuation. Such an agreement might be to pay a fixed sum for each operative who worked more than a defined number of hours per day or per week. Alternatively, it might be agreed to pay a defined percentage premium for all work carried out when the labour force was working a stated average number of hours per week or in congested or restricted areas.

Conversely, over short periods, very high outputs can be achieved either as a result of a high level of pre-planning and organisation or by the use of suitable incentives. An agreement to pay a bonus for completion of a defined amount of work or a specific task by a stated time can be a very effective way to encourage the operatives to participate and cooperate to achieve the required result.

A further method is for the subcontractor to demonstrate that he was able to obtain a certain rate of productivity when working under the conditions as contracted for, and demonstrate that a lower output was obtained under the changed circumstances. Strictly this will not be an enhanced rate but a payment of actual loss. While such an exercise may demonstrate that the changed conditions should require a review of the rates, the actual result is to demonstrate a loss rather than to establish an enhanced rate.

Loss and/or expense

Loss and/or expense, unlike a claim for damages, is an entitlement under an express provision of the subcontract and as a consequence is limited to the circumstances for which that provision is stated to apply. A clear definition of the phrase 'loss and/or expense' was given in the case of *Robertson Group (Construction) Limited* v. *(First) Amey-Miller (Edinburgh) Joint Venture; (Second) Amey Programme Management Limited and (Third) Miller Construction (UK) Limited* [2005] BLR 491, where the meaning of a similar phrase 'all direct costs and directly incurred losses' was discussed by Lord Drummond Young:

> '7. Both in the present contract and in the JCT forms, the two basic elements of cost and loss are qualified by the words "direct" or "directly". The significance of the word "direct" as used in the JCT forms of contract has been the subject of a number of judicial decisions, both in Scotland and in England. The result of these can be summarised in two propositions. First, the word "direct" in the expression "direct loss and/or expense" is concerned with remoteness of loss, not causation. Secondly, the word denotes that the loss or expense in question must flow naturally from the contractual event relied on by the claimant, in the sense of the first rule in *Hadley* v. *Baxendale* (1854) 9 Ex 341.'

What flows naturally from an event will depend on the facts and pre-knowledge of the parties. The immediate loss to a contractor caused by delay may be that plant and/or labour either stands without any work to do or is underutilised. In other cases a significant delay may mean that the resource required for the Works is no longer available or that the work has to be carried out under more adverse conditions, such as winter working or with a greater overlap with other trades.

Such a situation was considered in the case of *Ellis-Don Limited* v. *The Parking Authority of Toronto* (1978) 28 BLR 98. In that case an employer delay led first to the delay in the mobilisation of specialist subcontractors, then to a further delay due to the non-availability of a large crane. The further result of all these delays was to push concreting work into the winter period. Mr Justice O'Leary summarised the delays as follows, at page 112:

'Because of the original delay in the obtaining of the excavation permit and the consequent delays related to the excavation subcontractor and the crane problem, concrete had to be poured through much more of the winter than originally contemplated and the accompanying loss of efficiency delayed the completion of the project a further 3 weeks. In fact $14^1/_2$ days were lost because the job had to be completely shut down on various days because of bad weather.

In the result Ellis-Don was delayed in completion of the project as follows:

Delay in obtaining excavation permit	7 weeks
Consequent delay in commencing excavation	$1^1/_2$ weeks
Consequent delay in obtaining crane	6 weeks
Time lost by extended winter work	3 weeks
	$17^1/_2$ weeks

As a consequence the judge calculated Ellis-Don's losses on the following basis:

'The on-site cost to Ellis-Don of the $17^1/_2$ weeks' delay in completing the work'

Ellis-Don was decided on the analysis of the effect of a delay and associated loss incurred by the contractor on a specific project, when the contractor was awarded its project-related losses. However it is unlikely that such a situation will be considered to have resulted in losses to the contractor's business, such as head office overheads or profit.

For specialist subcontractors, whose work is naturally affected by changes in weather conditions, such as piling, earth works, structural frames, asphalt, etc., while their obligation to complete in a reasonable time, following a delaying event for which the contractor is liable, may need to make allowance for the changed weather conditions resulting from such delay, the subcontractor may not be entitled to payment for those losses since such trades will anticipate a certain amount of lost time due to weather in the normal course of their business.

For example, a road surfacing specialist contractor would normally, during the course of a year, anticipate sometimes being prevented by adverse weather from performing at all, or performing only at a reduced performance level. Such losses will not be attributable to a specific project but to the business as a whole, unless the subcontractor is able to demonstrate that the facts are otherwise.

However, a delay to the progress of work may involve lost production during a period of good weather at the time of the delay itself, when should the subcontractor be able to demonstrate that it had no alternative employment for its resources during the whole or part of the delay period, then the costs of retaining plant and operatives during the period of delay may be shown as a loss to the subcontractor. As always it is necessary to look at the actual losses incurred by a subcontractor and not rely on rule of thumb claims.

Where, for whatever reason, a subcontractor's work is severely reduced or cancelled, its direct loss may include the anticipated loss of contribution to the business overheads and profit. This was considered by Lord Drummond Young in *Robertson Group (Construction) Limited*:

'9. Both the contract under consideration in the present case and the JCT forms use the word "loss" or "losses". The general significance of the word "loss" is that a person does not have something that he had or would otherwise have had but for an event with legal significance. In cases involving breach of contract the general principle is that the

> innocent party is entitled to be put in the same position as he would have been in if the contract had been performed; the difference between that position and the actual position in which the innocent party finds himself is accounted a loss. It is obvious that a loss of that nature may include both a loss of profit and a contribution to general overheads. The present case, of course, does not involve a breach of contract; it rather involves a specific contractual entitlement to remuneration. Nevertheless, it is of the nature of a contract that the party who supplies goods and services expects, or at least hopes, to make a profit. The intention to make a profit lies at the heart of all, or nearly all, commercial activity, and the law must recognise that elementary economic fact. I am accordingly of opinion that the failure of a contractor to make a profit should be accounted a "loss" not only in calculating damages for breach of contract but also in construing contractual terms relating to payment for goods and services. The same is true of earning a contribution to general corporate overheads; indeed, until such a contribution has been earned it probably cannot be said that any profit has been generated.'

It was a significant feature of the dispute in the *Robertson* case that it related to payment for work carried out under a letter of intent, when the works contractor did not complete the Works because the parties failed to reach an agreement as to the contract price.

Where the subcontractor seeks payment for loss and/or expense, like a claim for damages, it will be for the subcontractor to prove its loss and its entitlement will be limited to its actual loss. Unlike a variation involving more or more expensive work, where there will normally be an entitlement to a percentage addition for company overheads and for profit, with a loss and/or expense claim there will normally only be an entitlement to the net loss. There is an anomaly in the provisions of SBCSub/C clauses 5.6.2 and 5.7.2, the carrying out of the subcontracted work under changed conditions, which require an assessment or an adjustment to the valuation on a fair allowance basis for what is frequently regarded as a loss and/or expense matter. Under the SBCSub/C agreement the subcontractor will receive an additional overhead and profit allowance for variations only increasing the cost of the work, when there is in fact no justification on traditional grounds for additional overheads since there is no more work nor has the work itself increased in value.

For a loss and/or expense claim, the burden of proof on the subcontractor is to demonstrate the relevant event allowed for under the subcontract conditions which has caused the subcontractor loss and the direct link to the costs claimed. Further, it will have to show that the losses were in fact losses incurred by it. The difficulty in satisfying this burden of proof cannot be exaggerated, nor can the importance of this principle be overemphasised.

Occasionally there may be clearly defined costs that relate solely to a relevant act by the contractor. If, for example, the subcontractor is instructed to stop work on site or work stops because of restrictions forced on it by the contractor, then it should be possible to demonstrate the costs incurred by the subcontractor such as, for example, the cost of hired-in plant during the period of the stoppage. It will be much more difficult to demonstrate the loss for the same plant if work does not stop but extends over a longer period, since it will need to be shown that the extended period is solely due to the contractor's default, or where there is more than one cause of the extended period, to show that period which is due to the contractor's default.

However, the level of proof to be given by the subcontractor is 'on the balance of probabilities'. Provided the subcontractor can satisfy the requirement to demonstrate the nature of and existence of its loss incurred and can show causation by relating actual loss to an identifiable relevant event or breach by the contractor, then the requirement for the contractor to ascertain and/or agree that loss may require positive action by the contractor. In *Alfred McAlpine Homes North Ltd* v. *Property & Land Contractors Ltd* (1995) 76 BLR 59, Judge Humphrey Lloyd QC defined 'to ascertain' by the much-quoted phrase 'to find out for certain'. Many commentators in analysing this phrase have latched on to the word 'certain' and sought to infer there was a requirement for the claimant to demonstrate its entitlement absolutely. However, the phrase, which applies to the duty of the contractor as well as an independent tribunal, contains the words 'find out', which requires active participation on behalf of the contractor. In the judgment in *How Engineering Services Ltd* v. *Lindner Ceiling Floors Partitions plc* (1999) 64 Con LR 67, it was held that the arbitrator was entitled to exercise judgment where the facts were not sufficiently clear. Mr Justice Dyson said, at page 8:

> 'Reference was made to *Alfred McAlpine Homes North Ltd* v. *Property and Land Contractors Ltd* (1995) 76 BLR 59, a decision of Judge Humphrey Lloyd QC. At page 88, the judge said:
>
> "Furthermore 'to ascertain' means 'to find out for certain' and it does not therefore connote as much use of judgment or formation of an opinion as had 'assess' or 'evaluate' been used. It thus appears to preclude making general assessments as have at times to be done in quantifying damages recoverable for breach of contract."
>
> . . . I do not understand Judge Lloyd to be saying that there is no room for the exercise of judgment in the process of ascertainment. I respectfully suggest that the phrase "find out for certain" might be misunderstood as implying that what is required is absolute certainty. The arbitrator is required to apply the civil standard of proof.
>
> . . . I would hold, therefore, that, in ascertaining loss or expense, an arbitrator may, and indeed should, exercise judgment where the facts are not sufficiently clear, and that there is no warrant for saying that his approach should differ from that which may properly be followed when assessing damages for breach of contract.'

SBCSub/C, at clause 4.19.3, requires the subcontractor to provide such information in support of its claim that the contractor reasonably requests to ascertain and agree the amount of direct loss and/or expense.

Judge Thornton QC, in the case of *Norwest Holst Construction Ltd* v. *Co-operative Wholesale Society Ltd* [1998] All ER (D) 61, had to consider an appeal from an arbitrator's award, where the arbitrator had used a formula to ascertain the value of the subcontractor's loss. In a lengthy judgment the judge repeatedly found that the arbitrator had found, as a matter of fact, that the subcontractor had suffered a loss. The judge then held that it was for the arbitrator to quantify that loss, to the best of his ability, using the evidence before him. The judge states the difference between proving loss and evaluating it, in these words:

> '362. However, NHC's submissions confuse proof of loss and the quantification of that loss once its existence has been established. Judge Lloyd's judgment is not concerned with the latter. It is concerned with showing that it is not axiomatic that because delay occurred, loss of additional overheads must also have occurred. In this award, the arbitrator has made unimpeachable findings of fact that the heads of loss were incurred. It was then open for him to quantify those losses in general terms without CWS having to prove the precise number of hours worked by management or the precise number of units of electricity used up.'

It must also be noted that SBCSub/C, at clause 4.19, requires the subcontractor to make an application to the contractor, once it becomes apparent that it is suffering or is likely to suffer loss and/or expense. At the time the subcontractor becomes aware of the situation, i.e. that it is incurring or is likely to incur costs or expenditure of money over and above that for which it is contracted to spend, it can be argued that it should avoid incurring further additional costs and/or not continue to expend further additional monies where possible, without an instruction, direction or authorisation from the contractor.

Where the subcontractor incurs costs or expense for any of the reasons set out in the subcontract, it may in addition to those actual costs or expense also be entitled to the associated financing costs. These will include any expenditure other than that directly related to a variation, such as in circumstances where materials and/or plant have been procured and, for reasons such as a delay to the works, cannot be incorporated and/or employed and the capital expenditure recovered, as anticipated, by the subcontract. For subcontractors manufacturing components off site, they will include the entire cost of the manufactured components where there is no provision for paying for materials prior to delivery to site.

The principle of paying financing costs as part of a valuation of direct loss and/or expense was decided by the Court of Appeal decision in *F. G. Minter* v. *Welsh Health Technical Services Organisation* (1980) 13 BLR 1. In this case, the action was by the nominated subcontractor under the name-borrowing arrangements of the JCT form of subcontract for nominated subcontractors, which explains any references to the subcontractor Drake and Scull Engineering. Mr Justice Stephenson summarised the matter to be decided at page 7:

> 'The question is easier to put in a nutshell than to keep there. It can be compressed into this: does "direct loss and/or expense" in the context of the printed conditions include interest which the contractor has paid on capital which he has borrowed as a result of the events specified in those clauses and interest which he has been prevented from earning on his capital as a result of those events?'

Lord Justice Stephenson found that the relevant words of the contract related to the contractor having been involved in direct loss and/or expense, for which he would not be reimbursed, by payment in respect of a valuation of a variation made in accordance with the rules contained in the contract. He then considered, at page 16, the proposition that the purpose of this provision was to ensure that the subcontractor was not:

> '. . . worse off as a result of actions for which the respondents are responsible and that (an) unjust result is reached by limiting the natural meaning of direct loss and/or expense to exclude losses which were just as clearly within that meaning as the sums which the architect has allowed.'

Lord Justice Ackner put the situation and the question of interest at page 23:

> 'By the terms of this sub-clause the employer clearly agrees to pay to the contractor his "direct loss and/or expense" incurred by him as a result of his complying with variation instructions given by the architect. For the purpose of this consultative case, we are to assume that as a result of the contractor complying with a given variation he has incurred

> a primary loss or expense e.g. the payment of hire on machinery which has had to remain idle for a period while the variation instructions have been carried out, and that he has incurred a secondary expense or loss by raising the necessary capital to meet the primary expense. He may have had to increase his overdraft at his bank or for instance drawn money out of his deposit account thereby losing the interest which it was earning. The first question which arises is – is such secondary loss or expense a direct loss or expense within the meaning of the sub-clause?'

In passing, Lord Justice Ackner observed that 'building contractors in the ordinary course of things, when they require capital to finance an operation, either have to pay charges for borrowing that capital, or if they use their own capital, lose the interest which it otherwise would have earned'; and secondly, that 'what the appellants here are seeking to claim, is not interest on a debt, but a debt which has as one part, of its constituent parts, interest charges which have been incurred'. Lord Justice Ackner then found that financing costs are to be included within direct loss and/or expense, at page 25:

> 'Upon the true construction of the building contract the amounts which have been certified and by which the contract sum has been adjusted by virtue of the conditions of contract shall also include any sums in which the claimants may have been involved by way of finance charges upon the amounts otherwise certified and paid or payable thereunder and/or by being stood out of their money (if established) for the following period only, viz. between the loss and/or expense being incurred and the making of the written application for reimbursement of the same as required by the said contract.'

Any further loss, as a consequence of either the delay and/or failure of the contractor to value the loss and/or expense and or/pay the sum as valued, will not be loss and/or expense directly related to the delay or other relevant event. This judgment therefore draws a clear distinction between the cost of borrowing on capital expended prior to an application for payment and a delayed payment following such an application. It follows that it is the application for payment that creates the debt, not the valuation by the paying party which will, if not paid on time, attract interest where there is such an entitlement.

For many companies and especially smaller companies, which many subcontractors are, there will be a proportional low limit on their ability to obtain credit and/or an overdraft from the bank. In extreme cases a subcontractor's ability to continue with the work may be prevented either by a failure by the contractor properly to value the work, including variations and any loss and/or expense caused by the contractor's actions, or by an excessive increase in the subcontractor's work or required rate of production.

As mentioned above, loss and/or expense provisions are express terms of the subcontract and where no such provision is made the subcontractor will normally have a right to damages for those events or matters for which there will generally be a right to loss and/or expense under the subcontract. However, where a subcontract makes specific reference to such events and requires the contractor to grant an extension of time, it will be necessary to consider whether this is to be regarded as the sole remedy to which the subcontractor is entitled or whether in the absence of a loss and/or expense provision the subcontractor may seek recovery of any direct loss and/or expense by way of general damages.

It is considered, in the absence of very express words to the effect that an extension of time is the subcontractor's sole remedy in consequence of any relevant event listed in the subcontract, that the subcontractor will have a right to general damages as a consequence of any such event.

This situation was considered in the case of *George Trollope & Sons and Colls & Sons Limited* v. *Washington Merritt Grant Singer* (1913) Hudson's Building Contracts, 4th edn, Vol. 2, p. 849. Although this case concerned a contractor and a house owner, it is considered that the same principles would apply in a subcontract situation. In the case, the contractor had been considerably delayed by reason of variations, late receipt of details from the architect and the employment of other contractors on the site by the employer. In addition, the employer had occupied part of the incomplete works and restricted access to the contractor to a significant extent. The contract contained provisions for the granting of an extension of time and the architect in fact issued a certificate extending the contract completion date to the actual date for completion. The contractor sought considerable damages and the dispute was referred to arbitration. The arbitrator found for the contractor and his award was referred to the courts by way of case stated.

Mr Justice Channell could find no reason for holding that the provision within the terms of the contract for extending the contract period was an agreement that this was the sole remedy the contractor was entitled to. He said, at page 857:

> '"It was contended on behalf of the employer that upon the true construction of the first contract the contractors were not entitled to damages in respect of the delay in the completion of the same attributable to any of the causes enumerated in clause 25 of the same." That is the principal point which Mr Sankey [Counsel for Mr Singer] has argued; that the power to extend time which is given in certain events by clause 25 has the consequence of precluding a claim for damage if there is an extension of time, that is to say, that the extension of time is to be deemed to be taken as satisfying all damage. Now, I cannot think that that is what it means. There appears to me to be no reason for putting that interpretation upon it. The clause specifies a considerable number of events, in the happening of which the architect is empowered to extend the time, and he is not only empowered, but it says that he shall make in those events a fair and reasonable extension of time for completion in respect thereof.'

The judge then considered the list of relevant events and divided these into three: *force majeure* and exceptionally inclement weather; authorised extras or additions; and a breach of contract by the employer. The judge examined the third category, at page 858:

> 'Therefore all three classes of matters which are in clause 25 are all matters as to which it is desirable to give the architect the power of extending the time, and why is that a reason for depriving the contractor of his remedy in damages for one of these three, the only one which gives him the remedy in damages? I can see no reason whatsoever for doing it. There is ample reason for the clause, but there no words in it which appear to me to have the effect contended for. If delay affects the contractor not merely in the time he has to take in order to complete the whole work, but also affects him pecuniarily in the way of damages, why, because the time has been extended, the employer should not pay damages I cannot see. It is a simple case, that by reason of the delays of giving orders to go on with the work in some particular, this part of the work was idle, or the clerk of the works or somebody on the spot, whom the contractor has to pay, was idle. The extension of his time

> will prevent the contractor from having any difficulty about time, it will prevent his being liable for not doing the work by the contract date, and give him time to do it, but it will not put back into his pocket the damage which he has sustained by reason of having the men there idle and paying them. It seems to me that he has a right to have those damages, and therefore I must decide against that contention of the employer.'

A similar situation arose in a dispute between a contractor and a subcontractor in the case of *The Jardine Engineering Corporation Ltd and Others* v. *The Shimizu Corporation* (1992) 63 BLR 96. In this case there was a conventional loss and/or expense clause within the main contract but no such express provision within the subcontract. The contractor claimed that the subcontractor's sole remedy was for an extension of time. The subcontractor's case for reimbursement of its direct loss and expense was put in a number of ways.

First, it was suggested that because the subcontract was intended to be back to back, the loss and expense provisions of the main contract were incorporated by reference into the subcontract. The judge did not accept this argument.

Secondly, the subcontractor claimed that there should be implied into the subcontract a number of terms, specifically to the effect that the subcontractor should have equivalent rights to the contractor, that the contractor was to indemnify the subcontractor against loss and that the contractor should not hinder or prevent the subcontractor from carrying out its work. The judge reviewed the law on implied terms and could not find that the first two of the proposed terms were to be implied, but found that the final one was consistent with general law.

Thirdly, the subcontractor claimed damages for breach of contract, which the judge accepted. Mr Justice Kaplan said, at page 118:

> 'If a party is given 100 days to complete a piece of work and the other party tells him to suspend work for 10 days then he has only been given 90 days to complete the work. True it is that he will get an extension of time for 10 days and this is designed to assist the employer because without the extension of time the liquidated damages clause would go on the doctrine of employer's prevention. The extension of time provision is very much to the benefit of the employer. However, on this factual scenario, the extension of time does not cure all damage suffered because although it gives 10 days' extension it does not compensate the contractor for the extra 10 days' cost involved. A 100-day contract becomes a 110-day contract and this, in all probability, would be a more expensive proposition. If this is correct and subject to causation and remoteness, I do not see why the plaintiffs cannot recover it as a matter of law.'

Prolongation

Where the subcontractor's time involvement with the project as a whole, as opposed to a specific task or activity, is extended, this is generally referred to as prolongation. Where this happens not only will the subcontractor incur additional on-site costs but may also incur additional costs with, for example, visiting and other staff. Since this is a special type of loss and/or expense the subcontractor will be required to show proven actual loss.

A further consideration is that prolongation may be due to more than one cause. For example, there may be additional work or a late variation, to which may accrue additional sums by way of site preliminary adjustment either within the rates or by way of true valuation. Further, there may be delays for which the subcontractor must bear the consequences. Sometimes these occur together and may be considered concurrent. Generally, the subcontractor will not recover its additional costs if it can be shown that in any event it would have incurred them through its own breach and/or difficulties.

The concept of concurrency is complex. The concept of a concurrent event must not be confused with concurrent effect. Thus two delays, on site at the same time, will be of concurrent effect but will be unlikely to be as a result of concurrent events. A variation instruction may affect work executed several weeks after the instruction. If at the time of the implementation of that instruction there is a shortage of labour and/or inclement weather which affects other activities on site, they will have concurrent effect but not be concurrent events and the effective or dominant cause of the delay will be the issue of the instruction. Put another way, if by the time of the labour shortage or inclement weather the work would have been delayed in any event by reason of an earlier relevant event, not at the subcontractor's risk, and is not further delayed by the events at the subcontractor's risk, then the subcontractor will have a right to a revised completion date and the associated loss and/or expense.

To demonstrate a claim of loss and/or expense by reason of prolongation, it will be necessary to be able to produce contemporaneous records showing, for each item of expense, the reason for that expense at that time. While the requirement to provide evidence of costs, and their links to the delays referred to, applies to all expense, it most certainly applies to costs which are less directly associated with the delay. Thus, where the work is prolonged it will be reasonably certain there will be additional costs for a site supervisor, but if the site supervisor is a working foreman then a record of his non-working activities will be necessary. Similarly, visiting staff, such as contracts managers and surveyors, will need to record any additional involvement considered as justifying payment by the contractor and such records should not be just that they visited the site but detail of what they visited for and/or did while there.

Under certain circumstances it may also be possible to claim a contribution for the operational costs of the subcontractor's company. Such claims, for subcontractor's head office and overheads, have frequently been allowed, based on the use of one of several formulae available for that purpose. In recent decisions the courts have required that the subcontractor demonstrate that it actually incurred a loss due to the delay to, or extended period for, the work. To do this, it is necessary to provide evidence such as showing that the subcontractor actually had turned away work as a consequence of the ongoing commitment to that project.

Underutilised resources

For a variety of reasons a subcontractor is frequently required by the contractor to commence work when only a limited amount of the subcontractor's work is available to it. In some instances this will be an express requirement of the subcontract, but in many instances this will be a changed requirement. It may be to the contractor's benefit for a trade to commence work on the programmed date, even if only on a restricted basis. Alternatively, the contractor may be keen to demonstrate progress by starting trades with less than the programmed lag time.

In other circumstances a subcontractor, having commenced its work, does not have further work released to it by the contractor at the anticipated time or rate. Where this happens and the previous trade accelerates, the subcontractor may subsequently itself be required by the contractor to achieve a higher output than anticipated or allowed for in the subcontract.

Many resources, such as items of plant and equipment and supervisory staff, will involve the subcontractor in the same cost regardless of the work output, up to a finite maximum, after which the subcontractor will require a further item of that plant or need to engage a further supervisor. Thus, if the work period becomes extended the subcontractor is likely to suffer loss by way of underutilisation of its resources. Should the work accelerate, the subcontractor will have the benefit of higher than anticipated utilisation, provided such acceleration does not exceed the maximum capacity of such resource or require working in two or more locations on the site at the same time or leads to unproductive working for some other reason.

Such losses can be assessed by comparing planned output and/or expenditure against actual, and can and should be calculated and claimed for at the time, with every periodic payment application. If it can be clearly shown that production output is less than anticipated by reason of the contractor's breach, then for the period of low productivity it may be possible to establish a pro-rata loss for the associated site costs. Further such losses are less likely to be challenged on the basis that there are concurrent delays by the subcontractor.

In the unreported case of *Whittal Builders Company Limited* v. *Chester-le-Street District Council* [1985] No. 1980 W 3060, Mr Recorder Percival QC had to consider the loss and/or expense claim of the builder due to the slow release of local authority flats for refurbishment and repair. The judge found as a fact that for a significant part of the contract period the amount of work available to the contractor was considerably less than that contracted for, and that during that time the contractor 'was grossly hindered in the progress of the work and that ordinary and economic planning and arrangement of the work was rendered impossible'.

The judge had to consider differing approaches to the valuation of such loss. In considering a comparative approach he said, at page 8:

> 'Several different approaches were presented and argued. Most of them are highly complicated, but there was one simple one – that was to compare the value to the contractor of the work done per man in the period up to November 1974 with that from November 1974 to the completion of the contract.
>
> It seemed to me that the most practical way of estimating the loss of productivity, and the one most in accord with common sense and having the best chance of producing a real answer, to take the total cost of labour and reduce it in the proportions which those actual production figures bear to one another – i.e. by taking one-third of the total as the value lost to the contractor.'

This process of calculating from recorded production during a period of undisturbed work the actual cost of production on that site at that time, and comparing the cost so calculated with the cost in the period of disrupted work, is frequently referred to as 'the measured mile' method. Since for many trades the period of disrupted work may occur late in the project, it is advisable for subcontractors to keep records of hours spent and production achieved throughout the duration of all projects.

Delay

Delay can be distinguished from prolongation, as being where the work contracted for is carried out in the same way over a similar period but at a later date. For many subcontractors the effect of a simple delay will have little or no effect. The most common effect may be to displace the work to a time of year where the working environment is less good. Such matters include:

1. less good weather;
2. shorter daylight hours;
3. holiday periods.

A second consequence may be that the works cost more as a result of changed labour, plant or material costs. These may arise either because of a periodic price or rate review or because the resource that was intended for the project is not now available and/or that those resourses now have to be made available to other projects or existing obligations of the subcontractor, thus necessitating the subcontractor obtaining the necessary resources from other sources, which are more expensive. Yet a further reason for a change in cost due to delay may be changed market conditions. Labour resource costs in particular can vary considerably over a short period due to fluctuating demand.

A third consequence may be where materials are obtained and/or manufactured to the anticipated programme, and when the work on site becomes delayed, the subcontractor will incur financing costs arising from the delay. In addition, there may be storage and/or double handling costs. Further, certain factory activities may become postponed leading to factory losses, both due to downtime during the original manufacturing period and overtime costs during the new period of manufacture.

Again, it will be for the subcontractor to prove its loss. The use of formula indices, such as commonly used and incorporated in many subcontracts where there is a contractual provision for the payment of inflation costs, will not in any way prove actual loss. The subcontractor will need to demonstrate that the resource was available at the time subcontracted for, at the rate claimed, and that the additional costs incurred were a direct consequence of the delay and no other cause.

For subcontractors involved in off-site or pre-site work, there may be other expense and/or loss as a result of project delay. Two basic situations may arise. Either the subcontractor is able to proceed with the pre-site/off-site work such as design and/or procurement/manufacture to programme, but is unable to carry out the work on site as anticipated; or the delaying event causes delays or prevents all or some of the pre-site/off-site work proceeding as programmed.

In the event that work off site proceeds as intended but delay on site prevents delivery/payment for the work, then the subcontractor may incur loss and/or expense in storing the manufactured goods and/or financing costs due to the delayed payment.

In the event that the work is delayed and only proceeds to conform to progress on site, then the subcontractor's losses may initially include loss due to undercapacity and return on factory overheads, staff and labour costs and/or followed by the additional costs for acceleration. It is possible that acceleration will in part or in whole cancel losses for undercapacity. Factory production is generally much more restricted in relation to a change in the rate of output, than is work on site, and will be restricted by reason of space and/or the number of machine, etc. It will rarely be possible for the factory management to so balance production that finished products will only become available on a 'just in time' basis for all projects. However, in a well-organised company there will be a total integration of the planning of off-site and on-site work, which may be severely affected by significant changes in on-site availability.

Acceleration

Acceleration can be defined as the taking of measures with the intention of carrying out a certain activity and/or activities in a shorter period of time than previously intended. Even where there are express terms in the subcontract for the subcontractor to 'use its best endeavours' or 'to cooperate with the contractor and other subcontractors to achieve the contract completion date', it is not considered that this will extend to the subcontractor unilaterally expending money to recover a situation not of its making or at its risk.

Acceleration, it is submitted, relates to the situation where the subcontractor is required to complete more work per unit of time than originally anticipated, often because of delays earlier on in its subcontract period. While this may be the natural meaning of the word acceleration, it is also applied to the situation where one trade is required to work in an area of the project before completion of the work by the preceding trade, and/or to allow the succeeding trade to commence prior to the first subcontractor's work being complete in that area and handed back to the contractor. Both situations may lead to the subcontractor incurring additional costs which, depending on the degree of acceleration, may or may not be significant. The analysis by R. M. W. Horner and B. T. Talhouni, referred to earlier, shows clearly how the rate at which loss is incurred increases progressively with the amount of overtime working and/or congestion on site.

Whether or not such requirements to work alongside and in close proximity to other trades or to make several repeat visits to the same location will strictly be regarded as acceleration or disruption, they will be changed conditions unilaterally imposed on the subcontractor by the contractor. In either event the contractor will need to issue its directions or instructions to the subcontractor defining its precise requirements, which may need to include the contractor's requirements with regard to the timing of return visits.

The right to recover additional payment for acceleration will require an express or implied requirement to do so, coupled with the associated right to be paid the resultant costs. While SBCSub/C, at clause 4.19, requires the subcontractor to advise the contractor as soon as it has become apparent that it is suffering loss due to the regular progress of the work being affected, it does not expressly require the contractor to issue directions or the subcontractor to take action so as to limit such loss. However, it is submitted that such terms are to be implied into the subcontract and it will be for the contractor to issue directions, as necessary, to the subcontractor, that the contractor either requires the subcontractor to take or continue with accelerative measures, or to proceed in the manner envisaged in the subcontract and for the completion date to be delayed as necessary. Since clause 2.18.1 of

SBCSub/C requires the contractor to review the period or periods for completion, acceleration is not envisaged under this form of subcontract. Further, it may be implied that where the subcontractor fails to notify the contractor, under clause 4.19, that it is taking or being required to take actions involving it in loss and/or expense, then the contractor will be justified in assuming that the subcontractor is not incurring such loss and/or expense and the subcontractor may be estopped or prevented from making any such claim at a later date.

In the judgment in *Ascon Contracting Limited* v. *Alfred McAlpine Construction Isle of Man Limited* (1999) 66 Con LR 119, Judge Hicks QC was very scathing as to the meaning of the term acceleration and the resultant entitlement to additional payment:

> '50. "Acceleration" tends to be bandied about as if it were a term of art with precise technical meaning, but I have found nothing to persuade me that that is the case. The root concept behind the metaphor is no doubt that of increasing speed and therefore, in the context of a construction contract, of finishing earlier. On that basis "accelerative measures" are steps taken, it is assumed at increased expense, with a view to achieving that end. If the other party is to be charged with that expense, however, that description gives no reason, so far, for such a charge. At least two further questions are relevant to any such issue. The first, implicit in the description itself, is "earlier than what?" The second asks by whose decision the relevant steps were taken.'

From this point of view it is clear that there needs, in the first place, to be a very clear agreement as to what will be achieved without the acceleration measures and the likely outcome by taking them. Secondly, if the measures are agreed and/or instructed, there may be a further dispute as to whether the party taking the measures warrants success or has merely agreed to take a specific action, at the other party's cost, which is hoped or at best expected to recover the lost time.

In *Ascon*, the subcontractor pleaded its costs by way of mitigation for its loss. The judge found, at paragraph 56, that since there was an existing remedy for the subcontractor under the subcontract by way of an extension of time and payment of loss and/or expense, such a plea was not valid:

> 'It is difficult to see how there can be any room for the doctrine of mitigation in relation to damage suffered by reason of the employer's culpable delay in the face of express contractual machinery for dealing with the situation by extension of time and reimbursement of loss and expense. However that may be as a matter of principle, what is plain is that there cannot be both an extension to the full extent of the employer's culpable delay, with damages on that basis, and also damages in the form of expense incurred by way of mitigation, unless it is alleged and established that the attempt at mitigation, although reasonable, was wholly ineffective.'

There is no room for a subcontractor unilaterally to take acceleration measures to recover delays due to matters for which it is not at risk, and where, as in SBCSub/C at clause 2.18.1, there is an express provision for a remedy by way of an extension of time. While it may be totally contrary to the contractor's requirements that the period for the subcontract works be extended and the subcontractor be paid the associated loss and/or expense, when the contractor's obligation to its employer is to recover its lost time, the law is clear; without express agreement between the parties to amend the subcontract provisions, there can be no acceleration with certainty of payment to the subcontractor.

It is considered that requirements such as those in SBCSub/C at clause 2.18.6.1, to the effect that the subcontractor shall use his best endeavours to prevent delay however caused or to do all that may reasonably be required, to the satisfaction of the contractor, to proceed with the subcontract works, do not mean that the subcontractor shall be required to take actions beyond those directly required by the subcontract, without agreement or directions from the contractor.

The same might apply to similar general requirements, such as to do everything necessary to ensure that the subcontract works allowed the completion of the main contract works to programme; or as in SBCSub/C clause 2.3, which requires that the subcontractor shall carry out and complete the subcontract works in accordance with the programme and reasonably in accordance with the progress of the works.

Although, no doubt, various meanings can be given to this clause, it probably means no more than that the subcontractor is to proceed with the work in accordance with the programme, but that that is to be adjusted to conform to the actual progress of other works on the site. This clause makes no reference to the subcontractor having to speed up or slow down its production rate, but merely allows for and requires the subcontractor to change the programme for its work, i.e. the timing of parts of the work, reasonably in accordance with on-site progress. Were the speed of the work to change significantly from that envisaged under the subcontract, then the subcontractor would be bound to give the contractor a notice to that effect under clause 2.17 of SBCSub/C.

Where there are delays, some of which are the liability of the subcontractor and others of the contractor, any action by the subcontractor to recover the lost time must first be set against its own failures. The only exception will be where the subcontractor takes express action on the instructions of the contractor and in the anticipation that it will be paid for such action by the contractor. Even where the subcontractor has experienced difficulties of its own during the execution of the work, it may claim that its programme had sufficient slack or float to accommodate those problems and that the overrun is as a result solely of the contractor's acts or omissions. However, should directions to accelerate be given by the contractor, it may be difficult to distinguish actions taken by the subcontractor for its own purposes and those taken in response to the contractor's directions.

SBCSub/C, at clause 2.18.2, requires the contractor, following receipt of a notice of delay from the subcontractor, to fix such revised date for completion as soon as reasonably practicable but in any event within 16 weeks from the receipt by the contractor of the notice and particulars required from the subcontractor. In any event, the contractor is to endeavour to reach a decision before the expiry date for the relevant period for completion. This means that, in practice, the subcontractor will rarely know during the duration of the subcontract work, and certainly not at the time action should be taken, whether the contractor is accepting liability for any delay and if the lost time is to be recovered. Contractors hiding behind the 16-week rule may be found not to be managing the works and/or making clear their requirements and issuing such necessary directions arising from such delays, and as a consequence to be failing in their obligations to the subcontractor, whether express or implied.

Global claims

As mentioned in the section 'Presenting a claim' in Chapter 17, a subcontractor making a claim for loss and/or expense under the subcontract or for damages, or a contractor making a similar counter-claim, will generally need to identify each and every event which it claims has caused it loss and to demonstrate the loss that flows from each event. However, there are frequently situations where loss is suffered by reason of a significant number of factors to which specific loss cannot be directly related. In such cases the claiming party will inevitably have to resort to what is termed a global claim.

Typical situations where claims on a global basis may be appropriate include those of prolongation costs and/or the need for additional resources to manage a significant number of variations or other matters. In such circumstances the claimant will have two hurdles to overcome. First, it will have to demonstrate that it actually incurred the costs claimed as a loss, and secondly, that those costs were actually incurred as a consequence of the cause or causes relied on.

Most staff will be permanently employed and many items of plant and equipment will be owned by the company. In such circumstances, unless the claimant can demonstrate that it hired in additional staff, plant or equipment, either for that project or another, to meet a shortfall in resources directly related to the event or events relied on, there will have been no loss. While such a counter argument may be sustainable for a short period, it would be reasonable to assume that after an undefined period such prolonged use of resources would be a loss, since it would be considered that an employer would after a while cease to employ surplus staff and would dispose of unwanted plant and equipment. Manifestly the period of expense to be considered as no loss to the claimant is unlikely to be a matter of fact and will be a matter of assessment on the basis of the balance of probability.

The law relating to the causation of loss and global claims was defined in the opinion of the Scottish Court of Session delivered by Lord Drummond Young in the case of *Laing Management (Scotland) Limited* v. *John Doyle Construction Limited* [2004] BLR 295. He set out the three matters to be proved to succeed in a loss and expense claim:

> '10. For a loss and expense claim under a construction contract to succeed, the contractor must aver and prove three matters: first, the existence of one or more events for which the employer is responsible; secondly, the existence of loss and expense suffered by the contractor; and, thirdly, a causal link between the event or events and the loss and expense. (The present case involves a works contract concluded between a management contractor and a works contractor; in such a case the management contractor is obviously in the position of the employer and the works contractor in the position of the contractor.) Normally individual causal links must be demonstrated between each of the events for which the employer is responsible and particular items of loss and expense. Frequently, however, the loss and expense results from delay and disruption caused by a number of different events, in such a way that it is impossible to separate out the consequences of each of those events. In that event, the events for which the employer is responsible may interact with one another in such a way as to produce a cumulative effect. If, however, the contractor is able to demonstrate that all of the events on which he relies are in law the responsibility of the employer, it is not necessary for him to demonstrate causal links between individual events and particular heads of loss. In such a case, because all of the causative

events are matters for which the employer is responsible, any loss and expense that is caused by those events and no others must necessarily be the responsibility of the employer. That is in essence the nature of a global claim. A common example occurs when a contractor contends that delay and disruption have resulted from a combination of late provision of drawings and information and design changes instructed on the employer's behalf; in such a case all of the matters relied on are the legal responsibility of the employer. Where, however, it appears that a significant cause of the delay and disruption has been a matter for which the employer is not responsible, a claim presented in this manner must necessarily fail. If, for example, the loss and expense has been caused in part by bad weather, for which neither party is responsible, or by inefficient working on the part of the contractor, which is his responsibility, such a claim must fail. In each case, of course, if the claim is to fail, the matter for which the employer is not responsible in law must play a significant part in the causation of the loss and expense. In some cases it may be possible to separate out the effects of matters for which the employer is not responsible'

Lord Drummond Young then goes on to state that a subcontractor/contractor's claim may be divided up and the subcontractor/contractor seek to recover its loss on a global basis for certain aspects of its loss, such as delay, while claiming other elements of its loss on the conventional causation method. He says:

'11. . . . A modified total cost claim is more restrictive, and involves the contractor's dividing up his additional costs and only claiming that certain parts of those costs are the result of events that are the employer's responsibility. This terminology has the advantage of emphasising that the technique involved in calculating a global claim need not be applied to the whole of the contractor's claim. Instead, the contractor can divide his loss and expense into discrete parts and use the global claim technique for only one, or a limited number, of such parts. In relation to the remaining parts of the loss and expense, the contractor may seek to prove causation in a conventional manner. This may be particularly useful in relation to the consequences of delay, as against disruption. The delay, by itself, will invariably have the consequence that the contractor's site establishment must be maintained for a longer period than would otherwise be the case, and frequently it has the consequence that foremen and other supervisory staff have to be engaged on the contract for longer periods. Costs of that nature can be attributed to delay alone, without regard to disruption. Moreover because delay is calculated in terms of time alone, it is relatively straightforward to separate the effects of delay caused by matters for which the employer is responsible and the effects of delay caused by other matters. For example, delay caused by late instructions and delay caused by bad weather can be measured in a straightforward fashion, subject only to the possibility that the two causes operate concurrently.'

At paragraphs 16 and 17 Lord Drummond Young set out the reasons for apportioning loss where there are concurrent delays. While in this case he is referring to causes at the employer's risk and those not at the employer's risk, it is suggested that the same principles might equally be applied where a contractor can demonstrate that its loss is attributable to more than one subcontractor causing concurrent delay.

'17. Apportionment in this way, on a time basis, is relatively straightforward in cases that involve only one delay. Where disruption to the contractor's work is involved, matters become more complex. Nevertheless, we are of opinion that apportionment will frequently

> be possible in such cases, according to the relative importance of the various causative events in producing the loss. Whether it is possible will clearly depend on the assessment made by the judge or arbiter, who must of course approach it on a wholly objective basis. It may be said that such an approach produces a somewhat rough and ready result. This procedure does not, however, seem to us to be fundamentally different in nature from that used in relation to contributory negligence or contribution among joint wrongdoers. Moreover, the alternative to such an approach is the strict view that, if a contractor sustains a loss caused partly by events for which the employer is responsible and partly by other events, he cannot recover anything because he cannot demonstrate that the whole of the loss is the responsibility of the employer. That would deny him a remedy even if the conduct of the employer or the architect is plainly culpable, as where an architect fails to produce instructions despite repeated requests and indications that work is being delayed. It seems to us that in such cases the contractor should be able to recover for part of his loss and expense, and we are not persuaded that the practical difficulties of carrying out the exercise should prevent him from doing so.'

This approach in valuing the loss can be contrasted with the attitude of Mr Justice Forbes in *Tate & Lyle Food Distribution Limited and Silvertown Services Lighterage Limited* v. *Greater London Council and Port of London Authority* (22 May 1981 unreported), where he was deciding a claim for management time supervising the berthing of shipping, interrupted by dredging works which the port authority had failed to carry out. In that case he found that since the dredging work was necessary and would have been carried out in any event, with the associated necessity to schedule carefully that work with the arrival, discharging and departure of shipping, there would have been a significant management task in any event, which might have been no different to that actually incurred. He said, at page 4:

> 'I think there is evidence that managerial time was in fact expended on dealing with remedial measures. There was a whole series of meetings, in which the first plaintiffs' top managerial people took a leading role, the object of which was to find out what could be done to remedy the situation and to persuade the defendants to do something about it. In addition, while there is no evidence about the extent of disruption caused, it is clear that there must have been a great deal of managerial time involved in dealing with the dredging required and in rearranging berthing schedules and so on at the Raw Sugar Jetty to enable the delivery of material to the refinery to proceed without interruption. I have no doubt that the expenditure of managerial time in remedying an actionable wrong done to a trading concern can properly form a subject matter of a head of special damage. In a case such as this it would be wholly unrealistic to assume that no such additional managerial time was in fact expended. I would also accept that it must be extremely difficult to quantify. But modern office arrangements permit of the recording of the time spent by managerial staff on particular projects. I do not believe that it would have been impossible for the plaintiffs in this case to have kept some record to show the extent to which their trading routine was disturbed by the necessity for continual dredging sessions . . . While I am satisfied that this head of damage can properly be claimed, I am not prepared to advance into an area of pure speculation when it comes to quantum. I feel bound to hold that the plaintiffs have failed to prove that any sum is due under this head.'

It would appear that in the *Tate & Lyle* case, the judge made no award in their favour because, while the facts presented to the court showed that there had been a loss for which Tate and Lyle were entitled to compensation, they had been unable to provide any proven basis for the assessment of the value of such loss. If, for example, Tate and Lyle had been able to produce records of staff time involved in the additional tasks relating to the loss they were claiming, then the judge might have been prepared to consider such costs and make a judgment as to whether all or only a part of such costs were attributable as costs Tate and Lyle would have incurred in any event, even if there had been other cause of expense allowable as loss which they were entitled to recover as damages.

In the recent case of *Skanska Construction UK Limited (formerly Kvaerner Construction Limited)* v. *Egger (Barony) Limited* [2004] EWHC 1748 (TCC), Judge David Wilcox was faced with a global claim of a different nature. In this case the judge was satisfied that the increase in staff had been employed to deal with difficulties of the employer's making. However, the contractor, in a similar manner as was described in the *John Doyle* case, could identify a large number of causes of action which had in total led to a need for the contractor to engage additional management staff. The judge found that it was not necessary to allocate specific time or individual staff to separate acts by the employer or his agent, since the judge was satisfied that those additional staff had all been brought to site solely for the purpose of managing those acts.

The judge made his finding of fact and said:

> '346. The changes to the site layout summarised in the A.J.S.P [Architect's] letter and enclosures of 15 April of 1997 and the drawings that accompanied it in my judgment led to a significant change of emphasis, compared with the tender proposals and rendered much of the crucial early preparatory work of lesser value. This is not the kind of project where long lead-in times were practical. In the early stages of such a contract, it was vital to establish the basics such as the layout, affecting as it does, site preparation measures, and the incorporation of complex plant and design and other priorities. I accept the evidence of Mr Bradley and Mr Frodzicki [the quantity surveyor for Skanska and Skanska's structural engineer] as to the extent of the impact of these significant changes which arose out of the revised layout. It is evident that in order to keep the momentum and in order to achieve "first board" when planned, that the advised additional staffing was made necessary. I am satisfied that because of the piecemeal release of information by Egger, analysed in the liability judgment, and their failure to properly coordinate the activity of their process plant contractors, it was necessary for SCL to appoint an additional planner in order to achieve that coordination.
>
> 347. I am satisfied on the evidence before me that the personnel particularised in the schedule at T1 page 38 were employed for the period shown and that the cost of their employment was as shown in Schedule T1 38. . . . I am satisfied that there was no material contribution to the necessity of employing additional staff other than that which was the responsibility of Egger. The only causally significant factors responsible for the additional staffing costs were those for which Egger was liable.'

Where a subcontractor or contractor as a consequence of an event or events is entitled to recover loss and/or expense under the subcontract or as damages, then the early notification, such as required under SBCSub/C clause 4.19.1 for the subcontractor or clause 4.21.1 for the contractor, should be given, and where such loss has in fact led to the provision and employment of additional resources of a specific type, such notice should give details of those additional resources. Unless such a notice is given it may be difficult for the claiming party at a later date to make a convincing case that at the time of the event it took a positive management decision to bring on to the project such resources as an addition, for the purpose of managing the event or events relied on. If no such notice was given at the time, there will be grounds for considering that such additional resources were imported on to the project for some purpose other than subsequently claimed by the subcontractor.

However, claimants must not regard this apparent relaxation of the burden of proof as making claims for loss and/or expense easy. Quite the contrary, as at the same time the courts have tightened up the need to demonstrate that there was actual loss and not just a possibility that there were additional costs. In the case of *Weldmarc Site Services Limited* v. *Cubitt Building & Interiors Limited* [2002] HT-01–408, Cubitt was found not to have demonstrated any loss in site staff costs due to the project overrun, and in the case of *Alfred McAlpine* v. *Property & Land*, there was no loss where the contractor owned the equipment that was retained for an additional period on site. (See the section 'Recovery of preliminary costs' in Chapter 13.)

From the above cases it would appear that, provided the subcontractor/contractor can demonstrate that it has suffered loss and that such loss has resulted from a cause for which the other party is liable, then the court or tribunal will assess that loss as best it can on the facts available and will not be deterred from making such an assessment because the basis of valuation may be inaccurate. However, claims on a global basis may lead to a win all, lose all situation and any claimant is advised so far as possible to present its claim on a cause by cause basis.

Quantum meruit

Under certain circumstances, the subcontractor may be entitled to have its work paid for on a *quantum meruit* (how much he has deserved) basis. Such circumstances will arise where the subcontractor carries out work on a contractor's instruction but without any pre-agreement as to the method of valuation and/or rate at which the subcontractor is to be paid.

Four basic possible methods may be used for calculating or arriving at the sum due to the subcontractor:

1. a sum based on an offer made by the subcontractor but not accepted by the contractor;
2. a sum based on rates generally available or obtainable in the market place at the relevant time;
3. a sum based on the actual cost incurred with a suitable addition for administration/overhead costs and for profit;
4. a sum related to the benefit gained by the contractor or employer.

It is considered that, subject to the subcontractor being able to demonstrate the costs claimed, the subcontractor is entitled to be paid by the method most beneficial to the subcontractor. A contractor will not be able to rely on a cheaper sum, on the basis that, had it accepted the subcontractor's offer or agreed to pay on the basis of the subcontractor's actual costs or on prices generally available at the time or in reliance on the actual benefit obtained, then the sum due would have been less.

In the case of *Monk Construction Ltd* v. *Norwich Union Life Assurance Society* (1992) 62 BLR 107, the Court of Appeal considered what was the correct sum to be paid to the contractor Monk who commenced work under a letter of intent which limited the amount to be paid to it in the event that no contract was concluded, to its 'proven costs incurred'. In the event no contract was concluded and Monk proceeded to complete the works. Monk sought payment of a reasonable sum, including profit. In considering the remuneration to be paid to Monk, Lord Justice Neill, at page 123, quoted from the decision of Mr Justice Goff in the case of *British Steel Corporation* v. *Cleveland Bridge* [1981] 24 BLR 94:

> 'In that case a dispute later broke out as to whether the plaintiffs' standard terms or the defendants' standard terms were to apply to the supply of goods. It is clear that Mr Justice Goff regarded the nature of this dispute as a significant reason for holding that no "if" contract had been concluded because "it would be an extraordinary result if, by acting on [the buyer's] request in such circumstances, [the seller] were to assume an unlimited liability for his contractual performance, when he would never assume such liability under any contract which he entered into".'

Although the *Cleveland Bridge* case primarily concerns the time for delivery of materials and the resultant claim for damages, it gives a strong indication that where there is no agreement as to price, but one party is requested to perform prior to a concluded subcontract, it will be held that the subcontractor will have no greater obligation to the contractor and will be entitled to no lesser sum than that it would have been prepared to accept had a concluded subcontract been entered into. Consequently, the sum due to the subcontractor for the performance of that service will not be less than that offered at the time of the unconcluded negotiations.

The benefit to the contractor or employer situation was considered in the case of *Costain Civil Engineering Ltd and Tarmac Construction Ltd* v. *Zanen Dredging and Contracting Company Ltd*, discussed in the section 'Post-contract instructions' in Chapter 7. In this case, the subcontractor was on site when asked to do work outside the scope of its subcontract. The subcontractor claimed against the contractor for a valuation, over and above its cost of carrying out the work required, on one of two grounds. First, that it was entitled to a share in the very significant profit being made by the contractor, on which Judge David Wilcox said, at page 93:

> '£2.55 m was paid to the joint venture by the Crown Estate Commissioners in respect of the marina development. The respondent claimed £725,000 in addition to its primary claim as being its entitlement to a share of that sum. It estimated the profit element to be £1,459,843. Its claim represented a half share of that profit. If there was an entitlement to such profit and it were divided in proportion to the cost of works carried out by each party as found by the arbitrator, then the respondent's share would be £0.583 m.'

Secondly, Zanen's claim was on the basis of the saving of the mobilisation costs which resulted from its employment for the work and which would have been incurred had another contractor been employed, and which the existing subcontractor could have added to a competitive bid had the contractor sought competitive tenders, together with the time saving by it being immediately available to commence work:

> 'The respondent in the alternative claimed a further £386,000. That sum was awarded "as being such sum as Zanen might well have assessed as being commensurate with the

mobilisation/demobilisation charges its competitors would have had to allow for had competitive tenders been sought for the marina works".'

Judge David Wilcox reviewed a number of authorities and commentaries and concluded, at page 96:

'I turn to the specific questions.

May the determination of a *quantum meruit* for such work:

(a) be valued by reference to any profit allegedly made by the Costain Tarmac Joint Venture within the sums paid to them for carrying out that work; and/or
(b) be valued by reference to charges (such as the costs of mobilisation and demobilisation) which a competitor for the said work would have incurred but which Zanen in the circumstances would not have incurred; or is such determination confined to the value of work done and the materials supplied with a fair and reasonable percentage uplift in respect of overheads and profit?

In my judgment the approaches above are not mutually exclusive, they are all elements that may be taken into consideration by the arbitrator in arriving at his judgment as to what is fair and reasonable.'

This judgment, it is suggested, is of major significance, first because it makes the saving of costs to the contractor by the continuing involvement of a subcontractor resident on site, a matter to be considered in deciding a reasonable price. Secondly, because whichever reasoning is adopted, the subcontractor received a significantly higher sum than it would have done under any of the other three methods of valuation listed earlier. The only reasonable conclusion is that the correct method of valuation in any specific case will be that to the best advantage of the subcontractor as payee.

Pre-valued prolongation costs

One way by which a subcontractor can avoid some of the difficulties of proving its actual losses is to agree its prolongation costs as part of its accepted offer. Known as pre-ascertained costs, they act in the same way as LADs in that they are what they are: whether an over- or underestimate, that is what the subcontractor will be entitled to be paid for those matters to which the pre-estimate refers, irrespective of its actual costs.

It will be necessary to define clearly what costs are included and the matters covered in the defined cost or loss. Such a weekly sum might be limited to establishment costs, including visiting staff, being those costs most likely to be difficult to demonstrate as a loss. Such costs need not be expressed as costs for prolongation but as a weekly-related cost in the offer, instead of either being lost in the rates or expressed as a percentage addition. Even where pre-priced prolongation costs have been accepted and incorporated into the subcontract arrangements, there will still be a requirement for the subcontractor to demonstrate the period of prolongation, due to causes entirely at the contractor's risk and cost.

One common use of pre-priced prolongation costs is where the supplier and installer of temporary works allows within its lump sum tender for the stated hire period only, with a further weekly rate to be applied should the temporary works be required for a longer period. The provision of scaffolding is typically priced on this basis.

Disputes will frequently arise as to the period for which additional hire is due. Scaffolds may be provided and installed complete in one operation, such as for use in repair or refurbishment work, or on a progressive basis as construction work proceeds. Where a scaffold is erected complete in one operation, there is unlikely to be grounds for dispute as to the hire period. However, where the scaffold is erected progressively, the hire period may be considered to commence from the end of the period of notice to commence work or from the date the final lift is complete. It is of course advisable that the parties make clear in their concluded agreement the basis on which such prolongation costs are to be calculated. In the absence of a detailed agreement, the additional hire rate must be regarded as payment to the subcontractor for its materials not being available for it to use on other projects during the extended period.

On the above basis there may be different considerations for each of three stages of the subcontract duration. In the event that erection is prolonged because of delays in the construction of the works, the scaffolder will have had to allocate and keep available the necessary material to meet the original construction programme, and it is therefore not unreasonable to anticipate that the subcontractor should be paid for this delay. There can be little doubt that the period during which the scaffold is fully erected will count as full hire. It may be considered that the subcontractor is entitled to additional hire by comparing the tendered period with the total of the actual periods from commencement of erection to the commencement of dismantling, plus the planned period for dismantling.

In the event that the operation of striking the scaffold is itself prolonged by the contractor's actions or requests, and as a consequence takes longer than planned, then the rate for the further additional hire should, it is suggested, be reduced proportionally to the percentage of scaffolding material remaining erected, since the dismantled material is now released to the subcontractor for other work.

Works carried out by others

As mentioned in Chapter 11, once the parties have contracted for the subcontractor to carry out work for the contractor, the subcontractor is to be allowed by the contractor to carry out that work and the subcontractor is to carry it out. The arrangements, timing and speed with which this is to be done are as set out in the Sub-Contract Agreement and may define the procedures for the giving of directions or instructions for the further requirements of the contractor. The general remedy in the event of a breach of that subcontract will be damages to be paid by the party in breach to the other.

A contractor may, not infrequently, find itself in the very difficult position where its subcontractor is unable to carry out work as required by the contractor, either as required by the terms of the subcontract or due to changed requirements imposed on the subcontractor by the contractor. In these circumstances it is still necessary for the contractor to satisfy its obligations under the main contract. In either of the situations referred to, the contractor may have little option but to arrange for another subcontractor to carry out some or all of the remaining subcontracted work, so that the contractor does not breach the requirements of the main contract. However, by such action, in the absence of express terms in the subcontract to the contrary or by a supplementary agreement, the contractor will thereby be in breach of the subcontract.

The general rule is that where, other than by agreement between the parties or under the express provisions of the subcontract, work is removed from the subcontractor and given to another, then the contractor will be liable to the subcontractor for any loss the subcontractor suffers, by way of damages. This may include payment to its suppliers and sub-subcontractors and/or loss of profit and head office overheads resulting from the lost work. Conversely, where the subcontractor is in default to the extent that that default gives rise to an express provision in the subcontract for work to be removed from the subcontractor, or the default can be regarded as a repudiatory breach, then it is likely that the contractor will become entitled to recover the additional cost of carrying out the work with a replacement subcontractor and/or labour.

A special case will arise where work carried out by the subcontractor and paid for by the contractor is found to be defective. Defects may be found either during the carrying out of the works, i.e. before completion, or at the time of completion, inspection and acceptance of the Works, generally described as snagging or after completion of the works as latent defects. The law acknowledges that in construction work there is generally no breach of contract if a subcontractor carries out defective work, known as a 'temporary disconformity', once such a defect is identified. Provided the subcontractor rectifies it within the period for the subcontract works or the period programmed for such work, or where appropriate the period for the making good of defects, then it should not be in breach of the Sub-Contract Agreement.

For a number of reasons the contractor and/or the employer, after the completion of the subcontractor's work or indeed during the period for the making good of defects, may wish or need to take action quicker than can be done through the contractual chain. In taking such action, it is generally acknowledged that the sum chargeable against the subcontractor shall not be more than it would have been had the subcontractor attended to the temporary nonconformity or latent defect itself.

Costs arising out of the preparation of an account

It will be a natural part of the subcontract process, by way of an express requirement such as in SBCSub/C clause 4.6.1, for the subcontractor to prepare an account and/or provide the necessary details for the contractor to assess the final subcontract sum. In SBCSub/C clause 4.6.1 the subcontractor is to provide the contractor with all documents necessary for the contractor to calculate the final subcontract sum, within 4 months of practical completion.

It therefore follows that the costs associated with this work are part of the costs of the subcontract works, and it will be considered that such costs are allowed for and included within the subcontractor's price.

While, generally, the subcontractor's final account will be settled by agreement, reasonably within the time-scale anticipated by the parties, there will be instances where the contractor either feels unable to agree to sums claimed by the subcontractor or that the presentation of the subcontractor's account does not justify payment as claimed. In such cases, the subcontractor may employ a specialist consultant to assist in the preparation of that account and its presentation to the contractor or to a third party tribunal. While the subcontractor may consider that such further presentations have been unnecessary and may seek to recover the cost of its specialist consultant, such attempted recovery is unlikely to succeed.

The case of *James Longley & Co Ltd* v. *South West Regional Health Authority* (1983) 25 BLR 56, was an appeal against the taxation of the parties' costs in an arbitration by the taxing master. In dismissing the appeal and upholding the decision of the taxing master, the judge stated, at page 62, that the master had found that the work of the contractor's consultant, Mr Short, could be divided into three categories:

> 'work done in preparing the claimants' final account for submission to the architect, work done in preparing the case for arbitration, and work done in the course of the arbitration as general adviser.'

The master had disallowed the first and third categories and thus rejected the costs of preparing the claimants' final account. Although the matter of the appeal was that none of Mr Short's fees should have been allowed, there is no suggestion in the decision of the court that the fees under the first category should have been allowed.

The contention that the costs of preparation of a claim are generally not allowable is further supported by the judgment of Judge David Wilcox in *Skanska Construction* v. *Egger*, where he had to decide whether the extended period of involvement of a Mr Tranter on the project was to be charged to the employer Egger. He concluded, at paragraph 351, that since most if not all of Mr Tranter's time had been involved in preparing Skanska's claim, his time could not be charged:

> 'In my judgment in the post-February 1998 period, Mr Tranter was primarily engaged in the logistic exercise of preparing a claim. I am not in a position whereby I can apportion what, if any, time Egger might be responsible for, save that it would be very minimal.'

Costs in pricing variations

Where a subcontractor is instructed to price a proposed variation to its work, ahead of an instruction being given to carry out the work, such as may be requested under SBCSub/C clause 5.3 and schedule 2 of that form, the subcontractor will incur costs in preparing and submitting that price. Then, where the contractor does not proceed to instruct the execution of the work, priced under such an instruction, it is likely that the subcontractor will be entitled to a fair valuation of the work involved in providing the instructed quotation, including the costs of obtaining quotations from its sub-subcontractors and suppliers. Such costs are distinguishable from the valuation of variations or preparing the subcontractor's applications for payment, whether interim or as a final sum, and must be regarded as the cost of the instruction to provide a quotation. Thus, requests under SBCSub/C clause 5.3 and the related schedule 2 are two-part instructions and contemplate expenditure, first for the cost of providing the quotation and secondly, if the quotation is accepted, for the cost of carrying out the resultant instruction.

Chapter 9
Payment

Requirements of the HGCRA

Sections 109–111 of the HGCRA require all construction contracts to have provisions for payment compliant with these clauses and that the provisions of the Scheme are to be incorporated where the subcontract arrangements are non-compliant. These sections state:

'Payment

109 (1) A party to a construction contract is entitled to payment by instalments, stage payments or other periodic payments for any work under the contract unless –
- (a) it is specified in the contract that the duration of the work is to be less than 45 days, or
- (b) it is agreed between the parties that the duration of the work is estimated to be less that 45 days.

(2) The parties are free to agree the amounts of the payments and the intervals at which, or circumstances in which, they become due.

(3) In the absence of such agreement, the relevant provisions of the Scheme for Construction Contracts apply.

(4) References in the following sections to a payment under the contract include a payment by virtue of this section.

110 (1) Every construction contract shall –
- (a) provide an adequate mechanism for determining what payments become due under the contract, and when, and
- (b) provide for a final date for payment in relation to any sum which becomes due.

The parties are free to agree how long the period is to be between the date on which a sum becomes due and the final date for payment.

(2) Every construction contract shall provide for the giving of notice by a party not later than five days after the date on which a payment becomes due from him under the contract, or would have become due if –
- (a) the other party had carried out his obligations under the contract, and
- (b) no set-off or abatement was permitted by reference to any sum claimed to be due under one or more other contracts,

specifying the amount (if any) of the payment made or proposed to be made, and the basis on which that amount was calculated.

(3) If or to the extent that a contract does not contain such provision as is mentioned in subsection (1) or (2), the relevant provisions of the Scheme for Construction Contracts apply.

111 (1) A party to a construction contract may not withhold payment after the final date for payment of a sum due under the contract unless he has given an effective notice of intention to withhold payment.

The notice mentioned in section 110(2) may suffice as a notice of intention to withhold payment if it complies with the requirements of this section.

(2) To be effective such a notice must specify –

(a) the amount proposed to be withheld and the ground for withholding payment, or

(b) if there is more than one ground, each ground and the amount attributable to it,

and must be given not later than the prescribed period before the final date for payment.

(3) The parties are free to agree what that prescribed period is to be.

In the absence of such agreement, the period shall be that provided by the Scheme for Construction Contracts.

(4) Where an effective notice of intention to withhold payment is given, but on the matter being referred to adjudication it is decided that the whole or part of the amount should be paid, the decision shall be construed as requiring payment not later than:

(a) seven days from the date of the decision, or

(b) the date which apart from the notice would have been the final date for payment,

whichever is the later.'

Application of the HGCRA in practice

While the provisions of the HGCRA appear straightforward enough, a number of problems arise both under standard as well as contractor's own forms of subcontract. Much commentary has been written on the adequacy of the payment provisions of the standard forms of subcontract to determine when payments become due and/or the use of pay-when-certified clauses.

Although many subcontracts follow what is perhaps the norm in the construction industry, of paying on the basis of the value of work completed by the valuation date, the HGCRA makes provision for other payment mechanisms by use of the words 'instalments, stage payments or other periodic payments'.

Payment by instalment may be as simple as an agreement to pay a certain sum at the end of each period. Such a mechanism needs some control to allow for non-payment if work does not proceed at the rate or in the manner subcontracted for.

Stage payments have been the traditional method of payment for design services and have been applied to construction work in some instances. Under this system the paying party may resist payment by demonstrating that the work to that stage is not complete. Failure to complete may be because a section of works is to be left out on a temporary basis, such as windows, where the opening is left for access from a site hoist or other purposes, or glass is left out pending the removal of scaffold ties. Alternatively, a unit may be incomplete because of minor defects and/or omissions which will be corrected or completed as part of the snagging and/or final commissioning and cleaning of the Works.

The meaning of 'complete' is discussed in detail in the section 'General principles' in Chapter 10, Completion. One definition may be that a section is complete when the next trade is able to carry out its work. It is suggested that the decision in *Hoenig* v. *Isaacs* [1952] 2 All ER 176, where it was decided that in a lump sum contract there was an entitlement to be paid, albeit a sum less than the full sum, when there were still 'some defects or omissions', supports this contention. This principle, it is suggested, will apply with even more force for stage payments than it does for entire performance of a lump sum contract. In that case Lord Justice Denning said, at page 180:

> 'This case raises the familiar question: Was entire performance a condition precedent to payment? That depends on the true construction of the contract. In this case . . . the essential terms were set down in the letter of 25 April 1950. It describes the work which was to be done and concludes with these words:
>
> "The foregoing, complete, for the sum of £750 net. Terms of payment are net cash, as the work proceeds; and the balance on completion".
>
> In determining this issue the first question is whether, on the true construction of the contract, entire performance was a condition precedent to payment. It was a lump sum contract, but that does not mean that entire performance was a condition precedent to payment. When a contract provides for a specific sum to be paid on completion of specified work, the courts lean against a construction of the contract which would deprive the contractor of any payment at all simply because there are some defects or omissions.'

In the subcontract situation where the contractor, either with its own labour or by that of other subcontractors, carries out or commences the work of following trades, which is directly connected to or makes use of the work of the first subcontractor, then to deny any payment to the first subcontractor for that section of its work on the basis that it is incomplete could be considered as being conversion or trespass on the work of the first subcontractor.

The payment graph system used to be popular with Government sponsored projects and involves a pre-agreed graph linked to the programme for the work. Progress is measured against the programme, in programme weeks, and periodic payments are decided by reading off on the payment graph the sum due, against the achieved progress. Adjustments to the calculated sum then have to be made for additional or varied work.

The HGCRA does not impose periodic payments where the duration of the work is specified in the subcontract to be less than 45 days, or the parties have agreed that it is estimated that the duration will be less than 45 days. It follows that to resist making an interim payment on this basis, there must either be a specific agreement or statement in the subcontract that the duration of the work is less than 45 days. Again, this section specifically relates to the 'duration of the work' and cannot by that phrase be considered to be limited to the on-site time, even if this were to be stated in the subcontract to be 45 days or less. It must be noted that no such restriction is made in SBCSub/C, so where this form of subcontract is used payment becomes due on a monthly basis as detailed above, regardless of the estimated duration of the subcontract works.

Where the subcontractor has prepared and issued a programme for its works, it is submitted that in the absence of express comment from the contractor, this would constitute agreement as to the estimated duration of the subcontract work. There is no provision, either in the HGCRA or in the Scheme, to cover the situation where the progress of the subcontract works becomes interrupted and/or extended by variation. It is submitted that where the subcontract works become extended beyond 45 days under these or similar circumstances, then by reason of the instruction or direction, the subcontract changes from one of less than 45 days' duration to a longer one and periodic payments become due. This would not be the case where the delay is due entirely to the failure of the subcontractor, when payment will only become due on completion of its works.

The Scheme at paragraph 5 of Part II refers to 'The final payment under a relevant construction contract'. In theory, the subcontractor's work is being fully valued on a periodic basis and such valuations are expected to include both the valuation of the work and other entitlements such as variations and loss and/or expense directly incurred as a consequence of that variation and/or as a result of matters affecting the regular progress of the subcontract works or by way of damages for the contractor's default. In practice, these other matters are both applied for and valued at an extended interval beyond their occurrence. An effective mechanism must allow for ongoing valuations and assessment and the subsequent payment of the final sum. Paragraph 5 sets the mechanism to be adopted, by default, for this payment as:

(a) 30 days following the completion of the work; or
(b) the making of a claim by the payee;

whichever is later.

Payment mechanism

The statutory payment provisions of section 110 of the HGCRA require an adequate mechanism for determining what payments become due under the subcontract and the final date for payment of such sums. In addition to these two dates, i.e. the payment due date and the final date for payment, the HGCRA requires a notice of payment to be given by the contractor to the subcontractor not later than five days after the payment due date, and at section 111 the HGCRA allows for the incorporation of the provision for the giving, by the contractor to the subcontractor, of a notice of intent to withhold payment.

The Scheme has the concept of making an application by the subcontractor as the trigger for payment. This is not to be confused with the requirement, in some subcontracts, for the subcontractor to make an application by way of its valuation of the works completed on a day a set period before the payment due date. It is important for the terms to be used correctly otherwise confusion will arise as to the intentions of the parties. Typically the term 'payment due date' is confused with the 'final date for payment'. The events in a typical payment mechanism are considered here.

Events in typical payment mechanism

The HGCRA does not require an application for valuation by the subcontractor for there to be a requirement for the contractor to value the work. There must be an express requirement in the subcontract for the subcontractor to submit its valuation by a certain date on a periodic basis or by a stated number of days prior to the relevant payment due date to relieve the contractor of its obligation to value the work on a periodic basis. Some subcontracts make such an application a condition precedent to being paid in that period. Such an arrangement presupposes that the payment due dates have been effectively defined so that the subcontractor is fully aware of its obligations with regard to the last dates for its applications. Further, it is doubtful whether such a condition, however expressly written, can override the contractor's obligation under the requirements of section 110(2) of the HGCRA, to issue a notice of payment to the subcontractor within five days of the payment due date, in any event.

The payment due date is not the date on which the subcontractor is to be paid as sometimes thought, but the date on which the contractor becomes obliged to pay a sum to the subcontractor and the payment period commences. For a payment mechanism to be adequate, the subcontract must clearly either define the dates on which payment is to become due or provide a mechanism by which such dates can be calculated with precision. The default provision in SBCSub/C, at clause 4.9.1, that 'the first payment shall in any event be due not later than one month after the commencement of the Sub-Contract Works on site', does not on its own define the first payment due date and is therefore not an adequate mechanism by which such a date can be calculated with precision.

In the absence of an adequate mechanism for determining the payment due dates, the Scheme defines the making of a 'claim by the payee' as being the payment due date. It may be debated whether the claim is made on the date the claim is sent to the contractor by the subcontractor or on its receipt, but in either circumstance the contractor is not in control of the payment timings.

It is common for payments to become due either on a monthly basis or at intervals of a specified number of weeks. Such periods are defined without any consideration of future circumstances, such as weekends, holidays and especially the long breaks at Christmas and Easter. SBCSub/C, in part overcomes this difficulty by adjusting the payment due date to the nearest Business Day, at clause 4.9.2, but this will still frequently result in the date for the issue of the notice of payment falling on a non-working day. There is no requirement for payment due dates to be at regular intervals and it may well be in the parties' best interests to plan these through for the duration of the subcontract, to avoid the problems created by the use of a monthly payment cycle.

Contractors have traditionally sought to time the making of an application by their subcontractors to coincide with their need to make an application under the main contract. This is to an extent understandable, but the statutory requirement to issue to all their subcontractors a notice of payment five days after the payment due date makes having the same payment due date for all subcontractors on one project of questionable sense. The potential benefit of not paying subcontractors before receiving payment under the main contract is not seriously affected by having different payment due dates, but by the relevant length of the payment periods.

The payment period is usually defined in the Sub-Contract Documents. Where the intention is to use the provisions of a standard form of subcontract but to change the payment period, then this must be clearly stated to avoid subsequent dispute. In the event that no payment period is defined and/or agreed, then Part II of the Scheme, paragraph 8, will make the payment period 17 days. Further problems can arise over interpretation where a discount is offered for a reduced payment period. Depending on the wording of the agreement these may or may not be considered to be linked. In the event of late payment, the contractor is likely to lose its right to the discount as allowed in the agreement, and in addition may become liable for interest and/or compensation on the late payment. SBCSub/A, at page 14, defines the payment period as 21 days for interim payments and 28 days for the final payment.

Where the HGCRA applies to a construction contract, section 110(2) requires the giving of a notice not later than five days after the date on which a payment becomes due from the contractor, under the subcontract, or would have become due if:

1. The subcontractor had carried out its obligations under the subcontract.
2. No set-off or abatement was permitted by reference to any sum claimed to be due under one or more other contracts.

The notice of payment is manifestly a statutory requirement, yet in practice it is rarely provided either in time or at all. Further, it is generally accepted that the subcontractor has no sanction if such a notice is not provided. Only where there is an express term in the subcontract to such effect does the application by the subcontractor become the sum due in the absence of a notice of payment from the contractor.

The effect of the issue of a notice of payment is to define the sum to be paid by the contractor to the subcontractor. Until such a notice is issued, the sum due is undefined. Where the notice issued by the contractor is for a significantly lesser sum than that sought and/or expected by the subcontractor, then this difference will be a dispute between the parties for which the remedy will be adjudication where negotiation fails. This may be of major significance for recovery of interest on unpaid sums.

Section 110(2) of the HGCRA also requires the notice to state the basis on which that amount was calculated, as well as the amount of the proposed payment. It is unclear what extent of detail is necessary to satisfy this subsection. It is, however, suggested that where an application for valuation has been made by the subcontractor, a notice by the contractor in less detail would not satisfy this requirement. Sub/MPF04, at clause 25.2, makes it an express requirement that the contractor's interim payment advices are prepared in similar detail to the application for payment by the subcontractor. Further, the crossing through in the valuation of individual items and the insertion of different sums of substantially reduced value, without any explanation, calculations or substantiation in any other way, is unlikely to be a compliant notice.

The obligation to issue a notice of payment expressly stands, where the subcontractor has not carried out its obligations under the subcontract, by the words 'or would have become due if the other party had carried out his obligations under the contract'. It is difficult to know exactly what is intended by this section. At least two possibilities exist:

1. where there is defective and/or late work the contractor is to value the work completed at its true value and separately value any deductions;
2. where the subcontractor fails to comply with any express procedures under the subcontract, such as the making of an application for valuation, the contractor must nevertheless value and pay for works as carried out.

A notice of intent to withhold payment, as set out in section 111 of the HGCRA, expressly makes allowable the withholding of money from a valuation, subject to notice. Lawfully, to reduce payment against a valuation, the contractor must issue a notice the required number of days before the final date for payment expressly stated in the subcontract or, where no such period is defined, seven days before, as required by paragraph 10 of Part II of the Scheme.

The HGCRA requires the notice to state the amounts to be withheld, with the grounds for that withholding. Failure to issue a notice in time or in the appropriate form will render such withholding unlawful.

It is considered that this obligation to issue a notice by the contractor to the subcontractor does not give the contractor any right to withhold sums due to the subcontractor where no such right exists either under the terms of the subcontract or at law. This requirement provides a shield to the subcontractor, not a sword to the contractor. Sub/MPF04, at clause 26.1, makes this point clear by listing the grounds on which a notice of withholding may be issued as follows:

1. any amount that the subcontractor is liable to pay the contractor in accordance with the terms of the subcontract;
2. any sums owed to the contractor by the subcontractor as a consequence of any breach of the subcontract; and/or
3. where a final payment advice has been issued, a proper estimate of the cost to the contractor of rectifying any defects referred to in clause 21.3.1.

For example, SBCSub/C at clause 4.21 makes provision for the contractor to recover loss and/or expense due to the regular progress of the main contract works being materially affected by any act, omission or default of the subcontractor. However, clause 4.21.2 makes the amount that may be deducted 'any amount agreed by the parties'. It is submitted that where it is considered by the contractor that the subcontractor is liable under such circumstances, the contractor must first obtain agreement before being entitled to issue a notice of withholding under the contract, or in the absence of agreement must obtain an adjudicator's decision to that effect.

Where there is an express payment mechanism within the terms of the subcontract this must be complete and where necessary provide a default provision so that the requirements can be achieved with certainty. For example, the subcontract may require the pre-agreement of a valuation prior to the subcontractor making its application for payment; unless there is a default provision dealing with the circumstances where agreement is not reached, then such a mechanism may not be effective.

This was one of the matters considered in the Scottish case of *Maxi Construction Management Limited* v. *Mortons Rolls Limited* [2001] CA39/01. In that case the judge found that the contract envisaged a two-stage payment mechanism, the first stage being the agreement of a valuation between the contractor and the employer's agent, and the second stage being an application to the employer for payment. Lord Macfadyen held that this mechanism was ineffective because there was no default provision in the event that no agreement was reached at the first stage. At paragraph 28 he said:

> 'The contract thus made a clear distinction between two procedural stages, namely (i) the agreement of the valuation and (ii) the subsequent application for payment of the sum

> agreed to be due. . . . There is, in my opinion, force in Mr Cowie's [Counsel for Maxi Construction's] submission that the requirements of paragraph 2.5.20 are inconsistent with the requirements of section 110(1)(a). A requirement that a valuation be agreed by the employer's agent before a claim for payment can be made is not necessarily, in my view, incompatible with section 110(1)(a), provided a timetable for the process of agreement, and a means of resolving a failure to reach agreement are provided. But paragraph 2.5.20 makes no such provision. Failure on the part of Bucknall Austin [the employer's agent] to agree a valuation could hold up the making of a claim for payment indefinitely. That, in my view, means that the contract does not provide an adequate mechanism for determining when payments become due under the contract.'

Whatever the payment mechanism, the subcontractor's applications must be made in the correct form if a dispute over entitlement to payment is to be avoided. In an era where much paperwork is generated using standard forms held within computer software, the risk of falling foul of specific requirements must be significant. It may well be prudent for a subcontractor to obtain express agreement, at an early stage in the subcontract, to the precise documentation to be used by way of application for payment and other processes. Such agreement is likely to avoid the problem that arose in the *Maxi Construction Management* case. There Maxi had submitted its 'Interim Valuation No. 10', which was suggested by Maxi to be a valid application for payment. Lord Macfadyen did not agree and found that no effective application for payment had been made. Again at paragraph 28, he said:

> '(Maxi) made no attempt, at the time when Interim Valuation No. 10 was presented, to argue that paragraph 2.5.20 was invalid and that they were entitled to claim payment on the basis of their own valuation without the agreement of Bucknall Austin. Instead they submitted Interim Valuation No. 10 under cover of a request (albeit erroneously expressed so far as it referred to "certification") for agreement of their valuation in terms of paragraph 2.5.20. In these circumstances, I am of opinion that Bucknall Austin were entitled to take Interim Valuation No. 10 and its covering letter at face value, and treat them as constituting no more than a request for agreement of the valuation, as a preliminary to the subsequent making of a claim for payment.'

It is true that in that contract there were distinct stages for (1) agreement of the valuation with the employer's agent and (2) application to the employer by the subcontractor for payment.

However, subcontractors should be careful to use the phraseology of the specific subcontract terms. Where a standard pro forma is used it is suggested that any difference in phraseology may be corrected by a covering letter. It will be arguable that where a contractor has on previous occasions accepted documents, misnamed, that it will be estopped from claiming on a future occasion that such a document is invalid for the purpose for which it had previously been accepted.

Many subcontracts require that an application for payment sets out in detail the basis of the application including, where necessary, calculations. This is expressly relevant where an application includes for additional matters such as variations and loss and/or expense. Paragraph 12 of Part II of the Scheme, for example, defines the requirement for a claim by the payee as follows:

'claim by the payee' means a written notice given by the party carrying out work under a construction contract to the other party specifying the amount of any payment or payments which he considers to be due and the basis on which it is, or they are calculated.

Regardless of the express provisions of the subcontract, it is good practice for all applications to be as full and detailed as possible. In the *Maxi Construction Management* case Lord Macfadyen found that the requirements of that contract had not been satisfied. He said, at paragraph 29:

'It is not, in my view, appropriate to demand of an application for "interim" payment that it set out in full detail the basis of calculation of items already paid for under earlier applications. But paragraph 12 does, in my view, require specification of the basis of calculation of the new matter included in the application in question. In my opinion many of the items in the "Variations/Contract Instructions" part of Interim Valuation No. 10 cannot be regarded as specifying the basis on which they are calculated.'

Sum due to the subcontractor

The definition of the sum due to the subcontractor has been questioned on the basis that the sum certified by the contractor by way of its notice under section 110(2) of the HGCRA is the amount (if any) of the payment to be made. However, section 110(1)(a) requires every subcontract to have 'an adequate mechanism' for determining what payments become due under the subcontract. For most subcontracts the only mechanism comprises the application by the subcontractor and the certification by the contractor.

Such a mechanism contrasts with the arrangements under most main contracts where the valuation and certification of the work is carried out by a third party or parties and the obligation of the employer is to pay the sum certified.

It is to be noted that SBCSub/C, at clause 4.10, refers to the contractor's gross valuation, which in turn refers to clause 4.13, which is described as 'Ascertainment' and does not expressly use the phrase 'sum due' but talks of 'the Gross valuation by the Contractor'. By reference to clauses 4.10.1.1 and 4.10.1.2 this valuation is made less retention and previous payments. Sub/MPF04, at clause 25.3, uses the phrase 'amount of payment proposed to be made' and at clause 25.5 lists the component parts of the payment advice. There is therefore an apparent intention that the 'sum due' be the contractor's net valuation.

Judge Richard Harvey QC, in his judgment in the case of *Emcor Drake & Scull Ltd* v. *Sir Robert McAlpine Ltd* [2004] EWHC 1017 (TCC), at paragraph 71 rejected a suggestion that 'the sum stated in the certificate was the amount due':

'71. I reject Mr Thomas' [Counsel for EDS's] submission that the amount of the proposed payment stated in a paying party's notice under section 110 is ipso facto the amount due under the contract. Mr Thomas further submitted that the certification by SRM made the amount due. I was referred to the recent case of *Rupert Morgan Building Services (LCC) Limited* v. *David Jervis and Harriet Jervis* [2004] BLR 18, CA. In that case Lord Justice Jacob (with whom Lord Justices Sedley and Schiemann agreed), approving the reasoning of Sheriff Taylor in *Clark Contracts* v. *The Burrell Co.* [2002] SLT 103, drew the distinction between the case where it was the certificate that made the sum due, and the case where it was the doing of the work that made the sum due. In the case in question, Lord Justice

Jacob referred (paragraph 11, p.21) to clause 6.33 of the relevant contract, which provided that the employer should pay to the contractor the amount certified within 14 days of the date of the certificate, subject to any deductions and set-offs due under the contract. Notwithstanding that that contract provided that no certificate should be taken as conclusive evidence that the work, materials or goods to which it related were in accordance with the contract, the sum was due and the clients had to pay the builder "for the present" (paragraph 16, p.22). The point is brought out most clearly in Lord Justice Jacob's quotation from Sheriff Taylor's analysis (paragraph 12, p.21). Sheriff Taylor distinguished a case where "there had been no calculation of the sum sued for by reference to a contractual mechanism *which gave rise to an obligation under the contract to make payment*" (my emphasis). Lord Justice Jacob, referring to the case distinguished by Sheriff Taylor, said:

"The contract there had no architect or system of certificates. The builder simply presented his bill for payment. *The bill in itself did not make any sums due*. What, under that contract, would make the sums due is just the fact of the work having been done." (Emphasis added)

72. I conclude that the question I have to decide is whether there was a contractual obligation on SRM to pay the amount of the certificate, by virtue of its having been certified. Apart from documentary evidence of the procedure adopted, there was scarcely any evidence before me of the contractual status of the system of certification that had been adopted. Mr Mellor [The financial manager for McAlpine] was cross-examined about it (Day 7, p.94). He said that the relevant person in SRM certified whatever he thought right as being due to EDS. In my judgment, that is insufficient to show any legal obligation on SRM to pay an amount certified by virtue of its certification. It was (the) fact of the work having been done (assuming that it was) that made the sum due. Notwithstanding the certificate . . .'

In this case the subcontractor was seeking to be paid the sum certified by the contractor, on the basis that the certificate established the sum due. Since this was judged to be wrong at law, it must be even more certain that a contractor's certificate will not establish the sum due where the subcontractor claims that such certificate has been undervalued. In part, the judge based his judgment on there being no evidence that the certificate had been prepared in accordance with the contract. The statement that the person preparing the certificate 'certified whatever he thought right' clearly did not convince the judge that there was a correct valuation. On this basis, were a contractor's valuation to be supported by an accurate measurement and/or detailed calculations, then such a certificate might be held to be the sum due.

Specific provisions in the SBCSub and ICSub forms

The standard form of subcontract, SBCSub/C, sets out the provisions for the payment of the subcontractor, in conditions 4.7 to 4.12.

Under clause 4.9.1, SBCSub/C sets out the procedure for ascertaining the date for the first payment under the subcontract. This requires the first interim payment to be due on the date that the first interim payment under the main contract becomes due, after the commencement of the subcontract work. In the event that the main contract particulars do not identify the payment due dates, then the first payment due date for the subcontract works is to be no later than one month after the commencement of the subcontract works on site.

It will be seen that this clause relates the first payment to the subcontractor to the date for issue of the next interim certificate under the main contract, following commencement of the subcontract works. It must be noted that this is not the date when the next certificate *is issued* but the date *for issue* under the main contract. This is important since the date on which certificates are issued may not be strictly in accordance with those provisions. For this reason a condition relating payment to a certificate, as or when issued by a third party, would not be an effective payment mechanism. There is no provision within the Articles of Agreement SBCSub/A, or elsewhere, for listing of the dates for certification under the main contract to be stated. It is therefore left to the subcontractor to obtain details of the dates for issue of such certificates in order to know the details of the payment mechanism. Only if these are expressly defined in the main contract will there be an effective mechanism. In that circumstance the first payment due date is to be no less than one month after the commencement of the works on site.

The defining of the first payment due date in these circumstances becomes open to post-subcontract action or agreement, since there is no defined default arrangement or mechanism. Where the subcontractor takes the initiative by making an application for payment, this is likely to establish the first payment due date by reason of paragraph 4(b) of Part II of the Scheme, which makes the payment due date the making of a claim by the payee. It is open for the subcontractor to make an early application, which will be especially important if the subcontract work has been delayed and the subcontractor has a claim for loss and/or expense, as a consequence, for such matters as financing charges for work carried out off site to comply with an earlier programme, or for the procurement and payment of materials and/or storage costs, etc.

In the absence of an application for payment by the subcontractor within one month of the commencement of the subcontract works on site, it is a requirement of this clause that the contractor makes a valuation no later than five days after the end of that month. Subsequently, payment will become due at monthly intervals.

SBCSub/A, at page 14, sets out a summary of the payment provisions. Those for interim payments relate the payment due date under the subcontract to the issue of the interim certificates under the main contract, but in any event it is to be within a month of the commencement of the subcontractor's work on site. Further payments are to be made monthly. Those payment due dates are to be adjusted so that the payment due date is a business day, and/or for the situation where the first payment is due on a date that does not occur in short months. The contractor is to issue the statutory notice of payment within five days of the payment due date and the payment period is defined as 21 days.

In addition, SBCSub/C at clause 4.10 sets out the details of the payment mechanism in five sub-clauses:

1. The amount of each stage payment is defined as being the gross amount as calculated under the valuation procedures described in clause 4.13, less:
   1.1 any retention that the subcontract allows the contractor to deduct;
   1.2 the total amount due under previous interim payments.
2. The contractor is to give the statutory five-day notice of payment, which is to state the sum to be paid and the basis of calculation, specifying to what the payment relates.

3. The contractor is allowed to issue a withholding notice in writing up to five days before the final date for payment. This is to state the grounds and against each ground the amount to be withheld.
4. The contractor is to pay the net sum due on the final date for payment.
5. In the event that the contractor fails to pay the sum due by the final date for payment, then the contractor is in addition to pay simple interest at the interest rate. The clause further states that paying interest is not to be construed as a waiver to the subcontractor's rights to be paid properly, suspend work or to terminate the subcontract under the appropriate clauses.

The reference to clause 4.15 at sub-clause 4.10.1.1 is a reference to retention, which is discussed in detail under the section on retention.

Sub-clause 4.10.2 requires the contractor to issue the statutory notice of payment five days after the payment due date. Neither SBCSub/C nor the HGCRA identifies any effect or sanction of a contractor ignoring this clause and failing to issue any notice of payment. The courts have made it clear that sums do not become due to a subcontractor just because it makes an application for those amounts of money. In practice, where a contractor fails to give a notice of payment it will rarely, at some time in the future, be possible for an adjudicator or other tribunal to ascertain with any certainty what sum was actually due at the time of the application, and the only tangible evidence will be that of the application made by the subcontractor. Further, it may well be considered by a tribunal that the fact that the contractor did not challenge or correct the subcontractor's application at the time, by the issue of a notice of payment, is itself evidence that the contractor did not disagree with the subcontractor's own valuation.

Sub-clause 4.10.3 sets the period for the issue of any notice of withholding as being five days before the final date for payment, which is a reduction from the seven days provided for at paragraph 10 of the Scheme. No justification can be identified for such a change from that considered fair by the Minister of State. Such a change is even less justifiable since the payment period defined in SBCSub/A is greater than that provided for within the Scheme. Further, it is not at all unusual for contractors to seek both to extend the payment period and reduce the period for the issue of a notice of withholding. It is suggested that such actions can only be regarded as evidence that a contractor intends to act unreasonably in the payment of its subcontractors.

Sub-clause 4.10.4 requires the contractor to make payment no later than the final date for payment. No reference is given as to the payment period, or where such period is defined or any other method of determining the final date for payment. SBCSub/A makes such payment period for interim periods 21 days, which must be compared with the 17 days provided by paragraph 8(2) of the Scheme. Since SBCSub/C does not define the payment period, unless such a period is expressly fixed either by reason of a signed agreement SBCSub/A, or otherwise, the payment period of 17 days defined in the Scheme will be implied into the subcontract by default.

Sub-clause 4.10.5 requires that in the event of the contractor failing to pay the amount due to the subcontractor by the final date for payment, the contractor is in addition to pay interest. The rate of interest is described as 'the Interest Rate for the period'. Again, this clause does not identify a rate of interest or state where such is to be found. SBCSub/C at section 1, Definitions and Interpretation, states:

'Interest rate a rate 5% per annum above the official dealing rate of the Bank of England current at the date that a payment due under this Subcontract becomes overdue.'

It will be noticed that not only is this rate less than that provided for in the Late Payment of Commercial Debts (Interest) Act 1998 (Late Payment Act) but it makes no provision for a lump payment in addition. Main contractors frequently seek to reduce still further the rate of interest to be paid if its payment is late. The effectiveness of such reduced rates is discussed more fully in the section 'Interest' later in this chapter. In the event that the contractor's rate is ineffective, the provisions of the Late Payment Act will apply.

Although SBCSub/C, clause 4.9.1, does not expressly define the commencement of the subcontract works as being the works on site, unless there is an express provision for payment of pre-site works by way of condition 4.13.3, which requires the listing of such items in the Sub-Contract Particulars (item 7) in agreement SBCSub/A, this clause will not give an entitlement to payment for such works. However, where there is such a provision, this will determine the first payment due date for those works. Condition 4.13.3 requires the contractor's valuation to be the total of the amounts referred to in clauses 4.13.1 to 4.13.6 less the total of the amounts referred to in clause 4.13.7, applied up to and including a date not more than seven days before the date when the interim payment becomes due. This includes, by way of sub-clause 3, the total value of any listed items. Payment of the value of the listed items is subject to the provisions in clause 4.14 having been fulfilled.

Clause 4.14 sets out the four conditions with which the subcontractor must comply before the payment of listed items will be made. These conditions are:

1. The listed items are in accordance with this subcontract.
2. The subcontractor has provided the contractor with reasonable proof that the subcontractor has a vested interest in the property such that when the contractor has paid the subcontractor those items will become the property of the contractor and that the subcontractor has insured the listed items from the time they become the property of the contractor until they are delivered to site.
3. At the premises where the listed items have been manufactured or assembled or are stored, there is in relation to such items clear identification of:
   (a) the contractor and the employer to whose order they are held;
   (b) their destination as the main contract Works.
   There is a further requirement that these materials are set apart and clearly marked as required.
4. If the Sub-Contract Particulars (item 7) require a bond then such bond has been provided by the subcontractor, as specified in the Sub-Contract Particulars and in terms set out in Part 1 of Schedule 3.

Provisions, such as set out in clause 4.14 of SBCSub/C above, will only be of any practical application in very exceptional circumstances. In practice, a subcontractor will require payment on account for materials held off site and work done to them, either because its total financial commitment to the project would otherwise be too large for the subcontractor to bear, or because delays on site have led its financial commitment to the project to be considerably increased. In the first case there will be a need for the contractor to take on the necessary financial risks which the provisions of clause 4.14 are intended to eliminate, since the situation at works will be a moving target and it is unlikely that the subcontractor would be in a position to satisfy the requirements of a clause such as 4.14. In the second situation, since such circumstances will not have been foreseen and as a consequence there is unlikely to be a Listed Item within the Sub-Contract Particulars (item 7), it will be necessary to make special arrangements for such payment.

In either of the above situations, there may be a possibility that the risk of the subcontractor's insolvency may override the associated risks of early payment. Where the subcontractor's ability to bring materials to site and be paid is a consequence of delays on site, then early payment will defeat any potential claim from the subcontractor for interest or financing charges by way of loss and/or expense.

Strangely there is no provision for the early payment of design and/or survey work in the SBCSub or SBCSub/D forms of subcontract. Clause 4.13.1, under both SBCSub/D/C and DBSub/C, expressly limits the scope of interim valuations to the Sub-Contract Work on site properly executed. It is at least arguable that these forms of subcontract fail to provide adequate payment mechanisms for the design and/or site survey elements and therefore fail to satisfy the requirements of section 110 of the HGCRA, which makes both design and surveying work construction activities. Thus it is suggested that under the SBCSub forms of subcontract, an application for design fees may be effective in establishing a payment due date for that work, under the provisions of paragraph 4(b) of Part II of the Scheme by default. By expressly limiting the payment provisions to work on site, none of the provisions in clause 4.10 of SBCSub/C will apply and consequently all the provisions of the Scheme, including the lesser payment period, will be implied into the subcontract for the payment of design and/or survey work.

Where design is part of the Sub-Contract Works, it is likely that the design coordinator will seek for such design work to be carried out at the earliest possible time and not at a time directly related to the fabrication and/or installation of the work on site. The risks relating to transfer of title, involved in early payment for goods or materials, do not arise with regard to design development work, where there is likely to be an express right to use the design in relation to the relevant project. Where a subcontractor is required to expend resources preparing design details and drawings significantly ahead of the subcontractor's own needs, and without a promise of early payment, then without very express terms within the subcontract to that effect it is unlikely that where the subcontractor, at the contractor's request, prepares such design details and drawings, the contractor will be able to deny payment to the subcontractor.

It will not normally be difficult to establish when a subcontractor commences the Sub-Contract Works on site, since this phase will normally be taken to mean the commencement of the permanent works or in some cases the temporary works. Where there are extensive temporary or site establishment works, the commencement of these works would normally fall into the description of subcontract works on site. Less clear would be the commencement of surveying or setting-out work and the taking of measurements for off-site fabrication, which would normally be considered part of the off-site works.

Condition 4.9.2 makes subsequent payments due at monthly intervals related to the first payment due date. That date is to be adjusted as necessary to be a business day and for short months where no similar date exists. These arrangements are up to practical completion of the Sub-Contract Works. After practical completion, further amounts are to be paid as they are ascertained and shall become due on the same date in the month as for interim payments before practical completion.

This condition requires the contractor to issue monthly notices of payment 'up to and including the month following the date of practical completion of the Sub-Contract Works'. Where contractors vary the terms of this form to make completion of the Sub-Contract Works the date of completion of the Works, they are theoretically duty bound to issue monthly payment certificates to each subcontractor until the completion of the Works.

Clause 4.9.2 of SBCSub/C acknowledges that the final sum due to the subcontractor may not be established within one month of the completion of the Sub-Contract Works and has defined that such further sums 'as and when further amounts are ascertained as due' will become due on the 'same date in every month'. This makes it quite clear that where further sums are agreed as being due, they become due on the next monthly date related to the date on which the first payment became due. However, there is no express obligation for the contractor to consider and ascertain such further sums within any express period of an application by the subcontractor, before the issue of the final account. In the absence of such express provision it is suggested that payment will become due on the next date following the application by the subcontractor, and the contractor will have a further five days in which to issue a notice of payment. Such notice may or may not be stated to be an interim or 'on account' payment.

The importance of being able to establish from the Sub-Contract Documents the 'payment due date' cannot be overemphasised. Much turns on this since these are the key dates for establishing:

1. dates for notice of payment;
2. final dates for payment;
3. whether notices of intent to withhold payment are effective;
4. whether cash or prompt payment discounts are allowable;
5. whether interest on debts or other compensation for late payment is due;
6. the right to suspend performance.

Under SBCSub/C, if the first payment due date can be established, then all the following dates may be said to be at monthly intervals, unless there is clear agreement that the payment interval is something different, such as two or four weeks, when it will be necessary to establish exactly what are the payment due dates.

Clause 4.10.1 is subject to 'any agreement . . . as to stage payments'. While traditional building trades with no design or off-site work will normally be content to be paid on a measured work basis, other subcontractors, especially those operating in trades such as supplying and installing of furniture or those providing professional services, may require payment on a stage basis, including but not necessarily a percentage payment with order. The difficulty for the contractor is that it is unlikely to be able to obtain back to back payment in such circumstances and may as a consequence have to finance the work for a longer period than when paying on a monthly measured account basis. The term 'stage payment' however may also be used to describe the payment of an agreed sum as and when defined stages of the work on site are completed. Such agreements can lead to disputes as to what constitutes completion of a stage, especially where small elements are left to be completed at a later date because of restraints in the construction process.

SBCSub/A, in the Summary of Interim and Final Payment provisions at page 14, makes the final payment due seven days after the issue of the final certificate under the main contract, with the final date for payment 28 days after that. The contractor is to issue the statutory payment notice five days after the payment becomes due.

It must be questioned as to whether this is an effective payment mechanism since, unlike the provisions for interim payments, there is no reference to 'the date for issue of the certificate', which relates the payment to a predetermined date or dates. Clause 4.6 of SBCSub/C provides a timetable for the issue by the contractor of a statement of the calculation of the final subcontract sum, which is in two parts:

1. Within four months of practical completion of the subcontract work, the subcontractor is to provide to the contractor all the documents necessary for the purpose of calculating the final subcontract sum.
2. Within eight months of receipt of the documents from the subcontractor, and before the issue of the final certificate under the main contract conditions, the contractor is to prepare and send to the subcontractor the contractor's statement of the calculation of the final subcontract sum.

The effect of the combination of the summary given in SBCSub/A and clause 4.6 of SBCSub/C, mentioned above, is that while any additional sum less retention will become due to the subcontractor, by reason of clause 4.9.2, at the next payment due date after such valuation, the final release of retention may not.

Clause 4.12 of SBCSub/C sets out the procedures for the payment of the final sum due to the subcontractor in four sub-clauses:

1. The amount to be paid by the contractor to the subcontractor is the amount calculated in accordance with the calculation procedures set out at clause 4.3 or 4.4 as appropriate, less the amount previously due as interim payment.
2. The final payment becomes due seven days after the issue of the final certificate under the main contract and the contractor is to issue a payment notice five days after the final payment to the subcontractor becomes due. The notice of the final payment is to be in writing and to be sent by either special or recorded delivery. The final date for payment is 28 days after the final payment becomes due to the subcontractor.
3. The contractor may, five days before the final date for payment, issue a notice of withholding, which is to state the amount to be deducted and the grounds, and against each ground the amount to be withheld. Subject only to such a notice, the contractor is to pay the subcontractor the balance of the final account sum by the final date for payment.
4. Should the contractor fail to pay the sum properly due to the subcontractor by the final date for payment, the contractor is to pay in addition interest at the interest rate. Such interest is to treated as a debt and is not to be construed as a waiver to the subcontractor's right to be paid properly under the subcontract.

It is questionable as to whether either the provision within SBCSub/A, Final Payment, or clause 4.12.2 of SBCSub/C provides an effective payment mechanism by the use of the phrase 'not later than 7 days after the issue of the Final

Certificate under the Main Contract'. First the phrase 'not later than' is itself uncertain, and secondly the issue of the Final Certificate under the Main Contract is itself dependent on actions by third parties and specifically the completion of the rectification of defects following the end of the period of liability for defects.

However, when read in conjunction with the provisions of clauses 4.6 (Calculation of Final Subcontract Sum) and 4.15 (Retention) it is difficult to find any relevance in clause 4.12 since it must be assumed that most subcontractors will have cleared any outstanding defects ahead of the issue of the Final Certificate under the Main Contract.

Some contractors seek to relate the release of each subcontractor's retention to the issue of a certificate of rectification of defects under the main contract. Because this relates to an action by a third party, such a provision must be considered an ineffective payment mechanism.

While it is easy to state that in the absence of an effective mechanism for payment of the final subcontract sum, the provisions of the Scheme will apply, it is more difficult to see how this can easily be applied in practice, since the Scheme has no provision for rectification of defects and/or retention. Paragraph 5 of the Scheme states that the final payment shall become due on the expiry of:

(a) 30 days following completion of the work; or
(b) the making of a claim by the payee

whichever is the later.

It is suggested that in the case of SBCSub/C, the completion of the work would relate to the end of the rectification of defects period or the rectification of defects by the subcontractor, whichever is later.

Clause 4.6 of SBCSub/C sets out a strict timetable for the submission of documents by the subcontractor and the calculation by the contractor of the final subcontract sum: 'not later than 4 months after practical completion of the Sub-Contract Works the subcontractor shall send to the contractor all the documents necessary', for the purpose of the calculation of the final subcontract sum. No doubt the subcontractor will, in practice, have been setting out its requirements for further payment during the course of its work and will have submitted documents within the four-month period such that any further documents and calculations will only be to support the evidence of the subcontractor's valuations and claims. Within eight months thereafter the contractor is to respond and specifically before 'the Architect/Contract Administrator issues the final certificate under clause 4.15 of the main contract conditions'. In the event that the contractor requires further details from the subcontractor, then the contractor may be regarded as required to identify such requirements in sufficient time for the subcontractor to provide such details and for the contractor to complete its calculations within the eight-month period.

Two issues arise from this condition. First, how fatal is a failure to comply with the express periods set out in clauses 4.6.1 and 4.6.2? Secondly, the trigger for this process is the practical completion of the subcontract works.

Some subcontract conditions make the date for practical completion of the subcontract works, the completion of the Works. In cases where the Scheme applies, this should not affect the date for making the final payment since the Scheme relates to 'completion of the work' and not 'practical completion' as in

SBCSub/C. However, if practical completion of the Sub-Contract Work is defined as practical completion of the Works, then there may be a significant period between the completion of the actual work and practical completion. The consequence is that the subcontractor may seek monthly valuations during the intervening period which must be considered and evaluated by the contractor. However, the eight-month finalisation period will not commence until four months after the completion of the Works, by which time agreement may have been reached or a dispute crystallised.

While failure to issue documents within the four-month period stated in clauses 4.6.1 may not be fatal, the use of the phrase 'not later than' may have such an effect. In any event, should the subcontractor not issue details in accordance with these conditions, at least two possible situations may arise. First, it may have become too late for the contractor to make a claim under the main contract for the sums claimed by the subcontractor. In such circumstances it is likely that the consequence of the subcontractor's breach will be that, while the sum may be due to it, the damages suffered by the contractor as a result of that breach will be that any sum due cannot now be recovered from the employer and thereby lead to a loss to the contractor equal to the sum due from it to the subcontractor.

The second consequence of a late submission by the subcontractor may be to relieve the contractor of the obligation to respond within the eight months expressly set by the subcontract conditions. As a consequence, the time for finalisation of the adjustments to the subcontract sum will become at large and the contractor's obligation changed to a response in a reasonable time.

ICSub/C contains similar provisions to SBCSub/C at clauses 4.9 to 4.14, although in a slightly condensed form. It must be noted that there is no express retention clause in ICSub/C. However, by reason of clause 4.10.1 and by reference to the Sub-Contract Particulars (item 7), the contractor is only required to pay the stated percentage of the value of the work. The main consequence is that there is no provision for paying the whole of the value of the work until the final certificate, which is stated to be due no later than seven days after the issue of the final certificate under the main contract.

Specific provisions under other standard forms

ShortSub, in common with SBCSub and ICSub, makes the first interim payment due no later than one month after commencement of the subcontract works on site. ShortSub provides for a 28-day payment period, and a period for the giving of a notice of withholding of five days before the final date for payment.

ShortSub allows for the contractor to pay 95% (or such different percentage as set out in the Sub-Contract Documents) of the value of the work properly carried out, and, 17 days after practical completion of the Sub-Contract Works, 97.5%. This is effectively a retention provision, although not expressly referred to as such. Clause 12.5 makes the balance of the total subcontract sum and the amount previously paid due seven days after the expiry of the defects liability period under the main contract, or seven days after the completion of any defects in the subcontract works. The reference to defects liability period needs to be changed to rectification period to conform to the current main contract forms.

ShortSub makes interest for late payment 5% over base rate.

Sub/MPF04, at section 25, sets out the procedures for payment to the subcontractor. Under clause 25.1 the subcontractor is to make an application seven days before the subcontractor considers an interim payment advice should be issued by the contractor. This application is to set out the amounts it considers should be included within a payment advice and the amount that it considers due.

It should be noted that clause 25.1 makes reference to the interim payment advice and not to the date payment becomes due. However, by reference to clause 25.4, the date of issue of the payment advice is stated to be the payment due date and is subject to clause 35.4.4 where the subcontract has been terminated by the contractor. The final date for payment shall be 21 days after the amount to be paid becomes due.

Clause 25.4 therefore only becomes an effective payment mechanism provided the contractor issues the payment advice and issues it on time. In the event that the contractor either fails to issue the payment advice or issues it late, then either the payment due date is seven days from the subcontractor's payment application, or the Scheme will apply and the subcontractor's payment application will be an application for payment and the payment due date will be the date of such application. However, appendix A makes the date for the issue of the payment advice by the contractor either the date in each month as stated or the 25th by default.

Clause 25.5 sets out in four sub-clauses the items to be included within the payment advice:

1. the proportion of the subcontract sum to which the subcontractor is entitled, calculated in the manner set out in the pricing document;
2. the value of any changes executed by the subcontractor;
3. the amount of any reductions made as a consequence of the provisions of clause 12.6;
4. any amount that either party is liable to pay the other in accordance with the provisions of the subcontract.

From these items is to be deducted the sum of the previous payments, which will determine the sum due from one party to the other.

The subcontractor under clause 25.1 is to submit a detailed application and the contractor under 25.2 is to issue the payment advice in similar detail.

Sub/MPF04 does not expressly refer to retention in section 25. Clause 25.6, however, relates the final advice to the expiry of the rectification period and the remedy of all work instructed to the subcontractor for rectification. It appears that there is a need for express provision elsewhere within the Sub-Contract Documents if the contractor is to be allowed to withhold a percentage of the value of the works as retention against latent defects.

Section 27 of Sub/MPF04 provides for interest at 5% per annum in excess of base rate. In addition, clause 27.2 is a statement of agreement that the 5% provision is a substantial remedy for the purposes of section 9(1) of the Late Payment Act. While there may be agreement that 5% in excess of base rate constitutes a substantial remedy for the purposes of section 9(1), it is submitted that it will only be the actions of the contractor in not making late payments that will demonstrate whether or not it is substantial for the purposes of this section, since section 9 sets the test as to whether the remedy deters late payment or not.

Retention

It is common practice in construction contracts for a percentage of the value of work to be withheld as retention against either default by the subcontractor or the identification of latent defects. The normal arrangement is that half of the retention is to be released either on completion of the subcontract works or on the completion of the Works. For some subcontractors the earlier date of release will be of significant importance, while for most finishing trades the earlier date for release will make little or no difference. The second half of retention will be repaid when the outstanding works or defects, apparent at the end of the defects liability period, are complete. Again, this may relate to either the works of the subcontractor or the entire Works.

SBCSub/C makes provision for retention at clause 4.15, and for a subcontractor's bond in lieu of retention at clause 4.16. Clause 4.15 allows the contractor to deduct and retain a retention calculated on the following basis:

1. Where the Sub-Contract Works or such works in any section have not reached practical completion, as defined in clause 2.20, the retention which the contractor may deduct and retain as referred to in clause 4.10.1.1 shall be the percentage stated in the Sub-Contract Particulars (item 8) of the total of the amounts referred to in clauses 4.13.1 to 4.13.3.
2. Where the Sub-Contract Works or such works in any section have reached practical completion as defined in clause 2.20, the retention which the contractor may deduct and retain as referred to in clause 4.10.1.1 shall be one half of the amount that would have been deductible under clause 4.15.1.1.

Clause 4.15.2 requires the contractor to repay the balance of retention to the subcontractor at the end of the rectification period for defects under the main contract, provided there are no defects in the subcontractor's work and such repayment shall be at the next interim payment date.

Clause 4.15.3 provides for where there are defects in the subcontractor's work at the end of the rectification period for defects under the main contract. In these circumstances the balance of retention is to be repaid on the next interim payment date following the making good of such defects by the subcontractor. This clause has two default provisions for repayment of the balance of retention to the subcontractor: a decision of an adjudicator, or the issue of a certificate of making good defects under the main contract.

SBCSub/C, at clause 4.15.1.1, provides for retention prior to practical completion of a section, and clause 4.15.1.2 allows for release of half the retention on completion of 'any section', where the subcontract makes provision for completion by section. Similarly, clauses 4.15.2 and 4.15.3 refer to the release of retention at the end of the 'Rectification Period for a Section'. Thus if there are defect-free sections at the end of the rectification period, then retention on those sections is to be released with the next interim payment and progressively on a section by section basis. However, unless the main contract provides for the contrary, the defects liability period will run from the completion of the last section, thereby giving only one date for the release of the second half of the retention. Both ICSub/C and ShortSub make similar provisions to those of SBCSub/C.

Clause 4.15.3 of SBCSub/C is a much more just clause than the equivalent clause in DOM/1, 21.5.3, which made the release of the second half of retention the issue of the Certificate of Completion of Making Good Defects, under the main contract:

> '21.5.3 where the Sub-Contract Works have been the subject of a Certificate of Completion of Making Good Defects under clause 17.4 of the Main Contract Conditions or a certificate under clause 18.1.3 of the Main Contract Conditions, any Retention retained by the Contractor shall be included in an interim payment to be paid at the end of the monthly period in which the relevant Certificate is issued.'

It is, however, suggested that clauses such as DOM/1 clause 21.5.3 do not provide an effective payment mechanism for the release of the second half of retention since they rely on actions by a third party.

The right to retention will only arise where there are express provisions in the subcontract for such action. In the absence of such an express provision there will be no such right even, it is suggested, by incorporation where such a right exists under the main contract. The HGCRA requires the subcontract to contain provisions for payment and where there are no such provisions or they are defective, then by reason of section 110 (3) of the HGCRA, the Scheme will apply. Since there is no provision within the Scheme for retention, there can be no incorporation by general reference. However, any express reference to retention being as in the main contract would, if sufficiently clear, be effective. Coupled with the right to withhold retention goes the obligation to repay. It is suggested that in the absence of an effective mechanism for repayment, a retention clause is ineffective and the provisions of the Scheme would apply.

It is necessary to know the method for the calculation of the value of the retention. SBCSub/C, at condition 4.15.1.1, relates this to the amounts referred to in clauses 4.10.1 to 4.13.3. This comprises the gross valuation of the works but excludes additional payments for items allowable under clauses 4.13.4 to 4.13.3, including loss and/or expense and fluctuations, where applicable. SBCSub/C sets the retention percentage as that stated in item 8 of the Sub-Contract Particulars in agreement SBCSub/A, which allows for the insertion of any rate of retention but states that, 'The percentage is 3 per cent unless a different rate is stated'. Thus it is possible that where no formal Sub-Contract Agreement has been concluded, it is likely that there will be no agreed percentage for retention; however, it is at least arguable that the default position will stand in any event, whether there is a formal agreement or not.

Difficulties of construction of a subcontract arise where there is reference to a standard form of subcontract or to incorporation of the main contract terms, which may have provisions for retention and it is claimed that they are thereby incorporated within the subcontract. It is suggested that where no mention of retention is made in the express terms of the subcontract, such a right will not be effectively incorporated by reference. However, where a retention percentage is expressly agreed within the subcontract but no effective procedures are stated for its release, then the procedures incorporated by reference will probably be incorporated, provided they are themselves an effective mechanism that can be applied directly to the subcontract arrangements, which for reasons stated elsewhere, under incorporation of terms, are likely to be highly problematical, since those arrangements will generally be both reliant on the performance of the contractor and/or other subcontractors and the certification by a third party. It is considered that had SBCSub/C, within clause 4.15.1.1, incorporated the default provision rather than stating it in item 8 of the Sub-Contract Particulars within SBCSub/A, then it would have been more certain to be incorporated as an effective retention provision than has been achieved in the circumstances where no formal agreement SBCSub/A is concluded.

Since retention is held against each valuation progressively and repayment is made at a date related to the completion of the subcontractor's work or the Works as a whole, and for the final percentage the completion of the rectification of defects after the end of the rectification period, it follows that if the duration of the work on site becomes extended, then the date for the repayment of retention becomes delayed and the subcontractor will be required to finance the cost of the retention for the additional period. Such cost is a loss arising out of the regular progress of the works being delayed, and such loss should be recoverable under the provisions of clause 4.19 of SBCSub/C, or similar provisions in other forms of subcontract.

The delay referred to may fall into one of two categories or a mixture of the two. The completion of the works as a whole may be a longer than anticipated period after the completion of the subcontractor's work. Such a situation will exist when an early trade, such as piling, proceeds generally to the anticipated programme, but there are delays to the finishing works, giving rise to a prolonged period and a delay to the completion of the Works. Alternatively, there may be delays to the subcontractor's work, owing to matters directly or indirectly concerning those works which lead to the subcontract works taking longer than anticipated in the subcontract. In this case the period for which the subcontractor is required to finance the retention over and above that anticipated when the subcontract was concluded will vary through the actual period of the subcontract works. Where there is no agreed programme for the subcontract works within the subcontract, it may be very difficult to demonstrate that there has in fact been a loss, or at least to define the actual loss incurred. One method of calculation may be to take the anticipated period between the programmed start of the subcontract works and the completion of the works, and project this period back from the actual completion of the Works to define the theoretical start date. That work carried out by the subcontractor prior to the theoretical start date will have been financed for an additional period.

The release of retention is generally subject to specific performance in contrast to the payment of construction work generally. This point was made by Lord Denning in the case of *Hoenig* v. *Isaacs* when he said, at page 181:

> 'A familiar instance is when the contract provides for progress payments to be made as the work proceeds but for retention money to be held until completion. Then entire performance is usually a condition precedent to payment of the retention money, but not, of course, to progress payments. The contractor is entitled to payment pro rata as the work proceeds, less a deduction for retention money. But he is not entitled to the retention money until the work is entirely finished, without defects or omissions.'

A further difficulty arises where the release of retention is subject to the actions of a party or parties other than the contractor. Such arrangements may be held not to be an effective payment mechanism for the purposes of section 110, or to be a conditional payment under section 113 of the HGCRA. It is suggested that a condition making release of retention under the main contract, which is dependent on the issue of a certificate by an architect or employer's agent, cannot effectively be incorporated by reference. A condition making release of retention on the subcontract sum dependent on completion of defects other than those related to the subcontract works will also be ineffective.

It is a very common arrangement for the release of the second half of the retention under the subcontract to be made dependent on the certificate of making good defects under the main contract. This means that all defective work, including where the quality of other work may be the subject of a dispute between the employer's representative, the contractor and another subcontractor, has to have been completed to the satisfaction of the architect before the contractor would be required to release any of the second half of the retention to any subcontractor.

Where a subcontractor has accepted a term to the effect that the second part of its retention will not be released until after the issue of the certificate of rectification of defects under the main contract, there is probably an implied term requiring the contractor to do all that is reasonably necessary to obtain such a certificate within a reasonable time after the expiry of the defects liability period. It will be anticipated that, following the expiry of the rectification period, there will be an inspection by the architect/contract administrator, followed by a time during which those having work to do rectify such work. There will then be a further inspection and only if all is satisfactory will the necessary certificate be issued. What period should be considered reasonable for such action will depend on the circumstances of each case, but a period in the order of four months is likely to be considered reasonable or at least not excessive.

In the event that the contractor fails to obtain a certificate of making good defects within a reasonable period following the end of the defects liability period, then the subcontractor may have a cause of action in damages against the contractor, to recover the remaining retention or at least for financing costs related to the period between the time such retention could reasonably have been expected to be released and the actual release of the balance. Any such entitlement will be on a monthly basis from the date it is considered such entitlement commences.

The obligation of the contractor to the subcontractor to obtain a certificate of making good defects under the main contract, as required as a prerequisite to the release of the second half of retention under the DOM/2 conditions of subcontract, was considered in the case of *Pitchmastic Plc* v. *Birse Construction Limited* (19 May 2000, unreported). The court held that there could be a right for a subcontractor to recover retention held against its account even where the express requirements of the subcontract had not been achieved by the contractor, such as obtaining a certificate under the main contract, but only if it could be shown that the contractor had prevented the issue of such certificate or achievement of such other express condition.

Mr Justice Dyson at paragraph 6, having dismissed Pitchmastic's contention that the court was empowered under the subcontract to determine any issue relating to any certificate issued under the main contract, stated:

> 'On the other hand, if it could be shown that Birse prevented the issue of the certificate, then on well-established principles, the absence of the certificate would not operate as a bar to recovery of the money: see *Roberts* v. *Bury Commissioners* [1870] LR 5 CP 310, 326, and *Panamena Europea Navigation* v. *Frederick Leyland* [1947] AC 428.'

In that case the judge was not prepared to find that the contractor had prevented the issue of the certificate of making good defects:

> 'But the problem with this is that the question whether Birse prevented the issue of the certificate of making good defects is essentially one of fact. Paragraphs 11 and 12 [of] amended statement of claim merely assert an entitlement to the retention money. There is no allegation that Birse prevented the issue of the certificate of making good defects. The reason why the certificate was not issued was not explored in evidence, beyond Mr Harding's [Counsel for Birse's] eliciting of the fact that there were no defects in Pitchmastic's work. On this exiguous material, I am not prepared to find that Birse "prevented" the issue of the certificate within the meaning of the *Roberts* v. *Bury Commissioners* principle. In my view, it is not sufficient for Pitchmastic merely to show that the architect withheld the certificate because there were outstanding defects in the work that had been carried out by Birse or other subcontractors. Provided that Birse (and its subcontractors) were proceeding with reasonable diligence to make good the defects, Birse was not preventing the issue of the certificate within the meaning of the *Roberts* v. *Bury Commissioners* principle. Some defects take longer to put right than others. There may be genuine difficulties in deciding what remedial work is required. For these reasons, the claim for £33,651.66 must be rejected.'

It is apparent from this judgment that it will be for a subcontractor seeking release of its retention ahead of a precondition stated in the subcontract, such as the obtaining by the contractor of a certificate of rectification of defects, to demonstrate the contractor's acts of prevention. This is unlikely to be easy since the subcontractor will not be privy to the actions being taken by the contractor to achieve such a certificate. However, the longer the contractor fails to obtain such a certificate, the greater the presumption will be that the contractor or others for whom it is responsible have prevented the issue of the relevant certificate.

In any event, there may be a time limit for which retention may be withheld after the date for the rectification of defects, beyond which the outstanding matters cease to be a delay in the rectification of defects and become a breach of the contract by reason of the failure of the contractor to complete the rectification and/or of the employer to make a claim on the contractor under the main contract for damages for the defects that have not been rectified.

The incorporation of retention clauses into subcontracts for work such as demolition, where on completion nothing remains, is of questionable value other than to give the contractor use of the subcontractor's money for an extended period, without any justification. Similarly, there is little or no justification for the deduction of retention on temporary works such as support works and/or scaffolding, which are finally removed from site. If retention is to be deducted it must be clear whether it is installation of the works or their final removal from site that constitutes completion of the subcontract works for the purposes of retention.

Special requirements, either by way of retention or otherwise, may apply in some demolition subcontracts where the subcontractor will recover at an early stage valuable fixtures and fittings or materials. Within the subcontract price will be an allowance for the value of such items, which may mean that at an early stage the subcontractor will in fact make a net gain without any payment from the contractor. Conversely, the work to completion will be costly and unattractive to the subcontractor. A practical solution may be to agree that monies are held until the work is complete, thereby ensuring that the subcontractor completes the work in its entirety.

Where there is a valid retention clause, repayment will be subject to the terms of the subcontract, which will generally relate to the completion of the rectification of defects at the end of the rectification period. Where work is accepted prior to full compliance with the detailed requirements of the subcontract specification, and monies are withheld either by way of retention or abatement of the value of the work, then the payment of the balance of the subcontract price will normally be subject to entire performance of the outstanding and remedial work required to complete performance of the work.

Discounts

Discounts may be offered for various reasons. One common form is a general contractor's discount. This custom of allowing a discount probably derives from the use of nominated subcontractors, or other arrangements where the subcontractor tenders directly to the architect or employer and was required to make provision for allowing a discount to the main contractor to cover the contractor's costs and expenses.

In the case of *Team Services plc* v. *Kier Management & Design Ltd* (1993) 63 BLR 76, which concerned a subcontract let under the provisions of a management contract, the Court of Appeal had to consider the meaning and effect of the phrase 'cash discount' in the following term of the subcontract:

> 'Within seven days of the receipt of payment due under the principal agreement pursuant to a certificate which showed a sum in respect of the subcontract works the management contractor taking into account any deductions ordered by the employer's engineer in respect of works not in accordance with this subcontract and any bona fide estimates made in accordance with clause 16.7 hereof of amounts to be paid or allowed by the subcontractor to the management contractor, the management contractor shall discharge such sum less:
>
> (i) retention money at the rate specified in Appendix A,
> (ii) a cash discount of 2½% (two and one-half per cent) on the difference between the said total value and the said Retention Money; and
> (iii) the amounts previously paid.'

The decision of the court appears to have turned on two facts established in that case. First, that:

> 'The documents show that the plaintiffs arrived at their tender figure by adding 1/39th, or 2.5 per cent, to their price, so as to cover the 2.5 per cent cash discount.'

Secondly, and perhaps more significantly, that the management contractor relied on the discount for most if not all of its remuneration. Lord Justice Lloyd said, at page 89:

> 'But so far as the discount is concerned, the defendants were nothing more than a conduit pipe, since the prime cost payable by the employer took account of all discounts,

including cash discounts, obtainable by the defendants. It would be an odd result of the plaintiffs' construction that, if the defendants were to be a few days late in paying each of the instalments, they should find themselves out of pocket to the extent of 2.5 per cent of the contract price, an amount which would comfortably exceed the whole of their remuneration under the contract.'

The subcontractor had argued that the phrase 'cash discount' could not have its literal meaning:

'Cash, in the phrase "cash discount" must, he says, mean something. It cannot bear its primary meaning of a discount for payment in cash, as opposed to payment by some other means such as a cheque, credit card or bank transfer. Therefore, it should be given its secondary meaning of discount for prompt payment.'

The court, however, did not accept this argument, by the following words of Lord Justice Lloyd, again at page 89:

'As for the argument that 2.5 per cent cash discount is meaningless, unless it imports a time element, the explanation may well be the reluctance of commercial men to abandon old forms and usages, even when they are inappropriate in a new context. It must be many decades since freight or hire under a charter party was actually paid in cash. Yet cash is what is required by many well known forms of charter parties still in use today. Whatever may be the explanation for the continuing adherence by the parties to the traditional 2.5 per cent discount, it cannot carry great weight against the arguments advanced by the defendants.'

The decision in this case was not unanimous and, it is suggested, was decided very much on the facts of the case, including the use of the phrase 'cash discount' rather than a more accurate description such as either 'management contractor's discount' or alternatively 'discount for prompt payment'.

In the Scottish case of *Wescol Structures Limited* v. *Miller Construction Limited* (8 May 1998, unreported), Lord Penrose held that where, as in *Team Services*, the phrase 'cash discount' was used and there was also clear evidence that a 1/39th had been added to the original tender sum to allow for the $2^1/_2$% discount, in this case the subcontract incorporated the subcontractor's letter, which contained the following words:

'Our price is based on monthly valuations, payment to be received 17 days thereafter. The 2.5% cash discount is for payment to these terms.'

Because these sentences expressly linked the discount to the payment provisions, Lord Penrose decided that the discount was for prompt payment and held that:

'I consider that the terms of Wescol's letter of 14 September do form part of the contract, and, in the absence of any relevant counter-stipulation by Miller, must be taken to have been agreed by the incorporation of the letter into the letter of intent dated 14 November 1994.

The incorporation of the terms for payment is determinative of the right of Miller to deduct the discount. In *Team Services plc* v. *Kier Management & Design Limited* (1993) 63 BLR 76, the majority of the Court of Appeal construed the particular contract under consider-

> ation as providing for a price reduction irrespective of prompt payment. There are aspects of the correspondence in this case which might have been thought to point in the same direction. Wescol's original pricing structure was 'net' and the arithmetical adjustments made to inflate the quotations to provide for the anticipated 'cash discount' might have suggested that Wescol's interests would have been fully served by payment net of discount without any particular stipulation as to time of payment, so long as the DOM 1 or 2 payment provisions were applied. But that would have ignored the specific references back to monthly valuations and to payment 17 days thereafter in the letter of September for example.'

With the general introduction of interest on late payments by reason of the provisions of the Late Payment Act, the concept of a discount for payment on time is, it is considered, no longer needed but the custom persists. An alternative would be a discount for early payments. For example, payment may be at 21 days after which interest will accrue, but a discount might be offered for payment at say seven days.

Where there is a discount for prompt or early payment, it is often suggested that in the event of a failure to pay in the agreed manner, the contractor will still be entitled to the discount since the subcontractor will only have suffered very minor loss as a consequence. Since there is now a universal right to interest at the statutory rate, which will generally be above the actual rate of loss suffered either by way of interest on the subcontractor's overdraft with its bank or interest on capital held on deposit, it is suggested that this is a wrong analysis. Unless it can be demonstrated to the contrary, the purpose of the discount is to make 100% certain that payments will not be made late or will be made in a more advantageous manner to the subcontractor. Where the construction of the subcontract shows that giving a discount is not solely for improved payment arrangements, but is in addition to those arrangements, then it is likely to have the same effect as the giving of a bonus to a subcontractor for early completion.

In *Leslie & Co Limited* v. *The Managers of the Metropolitan Asylums District* [1904] J of P Vol LXVIII 86, CA, it was held that where the contractor was to receive a bonus if the work was completed early but was delayed by nominated subcontractors, it was not entitled to that bonus. The words of Lord Justice Romer are significant:

> 'It also follows, as the terms on which the defendants offered a bonus to the plaintiffs for early completion of the hospital have not been fulfilled, through no default of the defendants, or for which they are responsible, that the plaintiffs cannot claim the bonus.'

The salient points in this judgment appear to be, first, that the bonus was for doing something over and above that required by the contract, i.e. to complete early, and secondly, in the absence of a default by the party giving the bonus, no bonus could be due if the terms of the bonus were not fulfilled. It would therefore appear in the analogous situation of a discount, even where the discount is offered and accepted in consideration for more favourable payment terms, that such discount will only be allowable where those more favourable terms are fulfilled.

It is uncertain whether by the failure to meet just one payment date on time, the contractor will forfeit the discount on all payments or just on that one. Each case will be decided on its facts, but if the discount is for reliability of payment, then it must be assumed that just one failure will deny the contractor the right to any discount on any payment, subject only to some failure by the subcontractor. It is suggested that in this situation the bonus agreement may, in effect, be a collateral contract, with the discount being consideration for the assurance of prompt payment and/or early payment, throughout the duration of the subcontract.

Discounts are generally taken, as work proceeds, against each interim payment. The provision for discounts for prompt and/or early payment may be made allowable for each payment due under the subcontract; however, it is suggested that in the absence of express agreement to that effect, such discount will be for prompt and/or early payment on all payments even if deducted progressively during the entire subcontract period.

There is no reason why a discount should not be made conditional on final account settlement and for the agreement only to allow the deduction of such discount with the payment of the final subcontract sum. Subcontractors offering discounts should consider this approach seriously, since many subcontractors will experience most difficulty in obtaining payment of the final sums due to them from the contractor. Such sums will include, but not be limited to, repayment of retention.

There is now a general recognition that prompt payment of subcontractors is of major importance to the industry. The introduction of the provisions of both the HGCRA and the Late Payment Act demonstrate the changing attitude to the need, in commercial dealings, for the prompt payment of sums as they become due. As was famously stated by Lord Denning MR in the Court of Appeal case of *Gilbert-Ash (Northern) Ltd* v. *Modern Engineering (Bristol) Ltd*, which was quoted with approval in the subsequent House of Lords judgment, (1973) 1 BLR 73, in the speech of Viscount Dilhorne:

> 'There must be "cash flow" in the building trade. It is the very lifeblood of the enterprise.'

It is suggested that this changing approach to, and regard for, prompt payment will not favour a slow paying contractor in breach of a payment mechanism which has linked to it a discount to secure frequent and/or early payments.

A further complication relates not only to the timing of each payment but also to the valuation of interim payments. There is a natural reluctance to overvalue work in progress, which leads to a culture of intentional undervaluation. This is despite the fact that, by the time payment is made, the value of work actually completed will generally have increased, thereby safeguarding any possible overvaluation of the subcontractor's work. However, it is the sum due which is to be paid and the sum due will be that accurately decided in accordance with the valuation procedures of each individual subcontract. Undervaluation is just as much a breach of the payment provisions as is late payment of a certified amount. This will be true whether it is the value of the actual work which is wrong, or it is a wrongful deduction by way of a notice to withhold payment.

Although such undervaluations are less easy for the subcontractor to demonstrate, it is suggested that failure to pay because work has not been properly valued, even if paid promptly, is just as much a failure to pay on time as not paying money certified after the final date for payment. The contractor should not be able to have the benefit of its undervaluation and the associated use of the subcontractor's money while at the same time claiming a discount for prompt payment, without incurring the sanctions either expressly or impliedly provided for in the subcontract conditions.

Whatever the discount being offered, the purpose of the discount and the arrangements for it should be clearly stated. Sometimes a general discount will be given during the pre-contract negotiations as a revised offer. It must be made quite clear whether such a discount is a once-only lump sum against the tendered sum or is to be considered a general reduction to the contract sum, which will result in a percentage reduction on the rates, since this will affect the pricing of variations. If the intention is that the rates are to be adjusted, this should be done in the contractual documents so that there will be no misunderstanding or dispute when the subcontractor's account comes to be adjusted and finalised.

It is unlikely that a discount can be implied into a subcontract by reference to another contract, since in the absence of an express payment mechanism the Scheme will apply and this makes no provision for discounts of any nature.

It is often questioned as to whether a discount should be deducted from the value of the work gross or net of retention. It is suggested that this will depend on the nature of the discount. A general discount is in effect a reduction in the subcontract sum and should therefore be taken before retention. However, discount for prompt payment, being a discount on the sum due, should be deducted from the net value of the payment, i.e. after deduction of retention.

Capped price or expenditure

Where the subcontract terms and/or price are not agreed, such as when a contractor requests a subcontractor to commence work under a letter of intent, then the contractor may state that the work to be carried out is not to exceed a stated sum. In many instances, the actual work carried out by the subcontractor will eventually exceed the value stated as not to be exceeded, or as revised by the contractor. In such circumstances, the contractor may deny that it has an obligation to pay more than the stated sum. Whether or not the contractor will be able to refuse payment for the value of work properly carried out, in excess of the stated sum, will depend on the facts of each case.

It is likely that, where there are express terms that the subcontractor is not to carry out work to a value more than that stated in the subcontract, then without further instructions from the contractor to proceed further, the contractor may be able to avoid payment. This, it is suggested, will be more likely where there is a stated reason for such a cap. Such a reason might be because funds for the entire project were not available at the time of the request to commence work. Certain projects, such as those for charities or for restoration works, may proceed progressively only as funds become available. In other words, if it was clearly in the contemplation of the parties at the time the provisional contract was accepted, that this was the situation and/or the reason for imposing the cap, then the cap should be enforceable. Even in these circumstances it is possible that the subcontractor remains entitled to be paid the value of the work it has carried out, but only when funds become available.

A further effect of a cap is to define the work to be carried out under the provisional arrangements, while an agreement is finalised. Before the expenditure of the stated sum or the inclusion of the finalised agreement, both parties remain free not to proceed further with the work. The letter of intent may, by express words, define the procedures to be fulfilled for further work to be instructed and executed. However, if there is no such express requirement or procedure, should the project proceed without further agreement, it is likely that the contract price will revert to the subcontractor's offer as amended by it at the time of the request by the contractor for the subcontractor to commence work. In the Court of Appeal case of *Monk* v. *Norwich Union*, it was held that where there was a restriction for the contractor to be paid 'proven costs' only, if no agreement was signed and there was a stated 'maximum expenditure of £100,000.00', then since that sum had been exceeded, the limitation for the contractor to be paid only 'proven costs' by the employer died. In that case the words of the letter of intent were:

> 'Our client has instructed us to confirm that this letter is to be taken as authority for you to proceed with mobilisation and ordering of materials to a maximum expenditure of £100,000.
>
> In the event that our client should not conclude a contract with you, your entitlement will be limited to proven costs incurred by you in accordance with the authority given by this letter.'

Lord Justice Neill found that the provisional payment provisions only had effect up to the time work on site commenced. He said, at page 124:

> 'It seems to me perfectly clear that these two paragraphs were directed to a problem which might arise if for some reason no contract for the execution of the sub-structure phase II work was ever concluded between the parties. It was important that Monk should establish a presence on site before the contract works began and that they should be able to order materials. As far as Monk were concerned, however, this expenditure would be wasted if in the end they did not get the contract. Accordingly the letter of 11 July authorised Monk to proceed with "mobilisation" and to order materials "to a maximum expenditure of £100,000". The letter further provided that if Norwich did not conclude a contract with Monk, Monk could claim, but could only claim, for the "proven costs" which they had incurred in accordance with the authority given to them to mobilise and to order a certain quantity of materials.
>
> For my part I do not find it necessary to come to a conclusion on all the various arguments addressed to us by Monk. In my view the provision for recovery of "proven costs" was intended to deal with the situation if no contract was signed and if no work on the sub-structure phase II project (apart from mobilisation and the ordering of some materials) was carried out by Monk. In that event Monk were not to be out of pocket because they had brought huts and stores etc. onto the site and had ordered materials for the contract in advance. I agree with the judge that the "proven costs" formula was never intended to apply to the execution of the main contract, nor was it intended to apply if work on the main contract began. It is true that Mr Archer [the project manager for Monk] referred to the letter of intent when he wrote to Schol [the construction managers] on Friday 2 September. I am quite unable to regard this letter, however, as a statement on behalf of Monk that they were intending to carry out the contract work, or indeed any of it, on a "proven costs" basis.'

Two conclusions can be drawn from *Monk* as to the effect of a cap on payment, under a provisional subcontract arrangement. First, such a cap may be limited to the activities described in the provisional subcontract. In the *Monk* case, this was the mobilisation and the ordering of materials. Secondly, the cap may define the last moment at which the subcontract can be cancelled and the provisional nature of the subcontract ceases. At that time the terms and conditions as agreed become the basis of a binding agreement.

In *Monk* the main issue was not whether the sum to be paid to the contractor remained capped, but whether the contractor's work was to be valued on a 'proven costs' basis. Where a subcontractor, under a letter of intent or similar arrangement, proceeds with work beyond the cap, it is likely that the effect of the cap will be held to have been removed. In the event that the contractor issues a further letter seeking to cap the sum to be paid to the subcontractor at a higher value, then again its effect will depend on both the wording and timing of such a letter. It is likely that its effect will depend on a number of issues.

First, this may be an offer by the contractor to amend the subcontract arrangements, again on a provisional basis. It is probable that the subcontractor's proceeding further with the work will be evidence of the subcontractor's acceptance of that offer. Alternatively, a letter from the contractor raising the cap may merely indicate that the contractor acknowledges that the agreed cap has been or is about to be exceeded, and that the contractor is doing no more than acknowledging an obligation to pay more to the subcontractor.

Secondly, the contractor may seek to hold that such a letter was an agreement either that the subcontractor would not carry out work exceeding the value stated or that the contractor need not pay for work carried out to a value exceeding the value of the revised cap. In such a case, who is to be the judge that the stated value has been reached? Given that valuations are carried out retrospectively by both the subcontractor's applications and subsequently by the contractor's valuations, there will be no certainty when the work executed actually reaches the cap as imposed.

Thirdly, the contractor will inevitably have a presence on site and will, or should, be aware of the likely value of the work being carried out by any one subcontractor, and will therefore be in a position to issue instructions to stop work should it so require. Further, the contractor will almost certainly be issuing directions on a daily basis to the subcontractor with regards to the sequence of the works and other coordination matters, and consequently may be held to be revising its requirements. There will be an implied obligation that where a contractor requires a subcontractor to carry out work, it will pay for such work either in accordance with the subcontract or on the basis of a fair price. Where the contractor denies that work was carried out under the subcontract because the cap was exceeded, it may be held, as in *Monk*, that payment is to be made on a fair price basis.

It is suggested therefore that a capped value provision is unlikely to be effective in denying a subcontractor payment of the value of work properly carried out on the instruction of, or to the requirements of, and in the knowledge of the contractor. (See also the comments in the section 'Capped price or expenditure' in Chapter 2.

Withholding payment

Section 111 of the HGCRA makes express provision for the contractor to withhold payment against the sum due under the subcontract after giving an effective notice of that intent. For a notice of intent to withhold to be effective, it will be necessary for the contractor to issue a formal notice to the subcontractor the stated number of days prior to the final date for payment. Such notice must set out the grounds on which the contractor seeks to withhold payment and the sum sought against each ground. SBCSub/C, at clause 4.10.3, makes such provision and states that the period for the giving of a notice of withholding shall be five days before the final date for payment. The Scheme, however, in paragraph 10 makes the prescribed period seven days before the final date for payment.

A notice of withholding must be a response to each application for payment and/or each certificate of payment. A pre-existing notice will not suffice, since the grounds on which withholding is made may, in the interim, have been removed and/or found to have been overvalued. It is probably sufficient in a valid notice of withholding to make reference to previous notices or grounds detailed in a previous notice, without having to detail such grounds fully again. This principle was stated in the Scottish case of *Strathmore Building Services Limited* v. *Colin Scott Greig t/a Hestia Fireside Design* [2000] CA18/00, where the judge Lord Hamilton said, at paragraph 14:

> '14. The second matter raised was whether a notice effective for the purposes of section 111 could be a communication in writing sent earlier than the making of the relevant application. Mr d'Inverno [Counsel for CSG] pointed out that, while section 111(2) provided that any notice must be given not later than a particular time, it did not provide that it required to be given after any particular time, i.e. there was no *terminus a quo*. The letter of 17 August, albeit sent prior to the invoice of 27 November, was (or was arguably) a notice of intention to withhold payment within the meaning of section 111. I am unable to accept that argument. The purpose of section 111 is to provide a statutory mechanism on compliance with which, but only on compliance with which, a party otherwise due to make a payment may withhold such payment. It clearly, in my view, envisages a notice given under it being a considered response to the application for payment, in which response it is specified how much of the sum applied for it is proposed to withhold and the ground or grounds for withholding any amount. Such a response cannot, in my view, effectively be made prior to the application itself being made. It may, of course, be that the matter of withholding payment of any sum which might in the future be applied for has previously been raised. In such circumstances a notice in writing given after receipt of the application but which referred to or incorporated some earlier written communication might suffice for the purpose – though I reserve my opinion on that matter. But such an earlier communication, whether alone or referred to subsequently in an oral communication, cannot, in my view, suffice. This is, as a matter of statutory interpretation, in my view, unmistakeably the case.'

Failure to issue such a notice, whether late or at all or with insufficient detail, will not deny the contractor the right to the sums sought for the reasons stated, but will make it unlawful for the contractor to withhold such sums at that time. Where a subcontractor successfully obtains payment for the sums, which the contractor has sought to withhold but on examination the notice is found to be defective, it is likely that the contractor will seek to recover such sums by way of a separate action or at the next stage payment.

Where a contractor, in breach of contract, fails to issue a notice of payment as required by section 110(2) of the HGCRA, and the sum due to the subcontractor is referred to the decision of an adjudicator or other tribunal, then the sum due to the subcontractor must take into account any related sums due from the subcontractor to the contractor under the subcontract, as well as those due from the contractor to the subcontractor. However, such sums that the contractor may deduct from the subcontractor's valuation must be those directly related to the value of the work as carried out, and may not include any counter-claim for losses incurred by the contractor.

Any doubt as to the effect of a failure to issue a notice under section 110 was discussed and decided by Lord Macfadyen in the case of *SL Timber Systems Limited* v. *Carillion Construction Limited* [2001] BLR 516. In this judgment Lord Macfadyen made it clear that the valuation of the work carried out requires the claimant to demonstrate its entitlement to that sum, and that a mere statement that the work is of a certain value will not be sufficient. However, where the contractor seeks to set off money for some unrelated matter, then it will not be able to do so unless it has issued a satisfactory notice either as a 110(2) notice or as a separate 111 one. Lord Macfadyen makes this very clear in paragraphs 19 and 20:

> '19. In my opinion the adjudicator fell into error in the first place by conflating his consideration of sections 110 and 111 of the 1996 Act. In my opinion Mr Howie [Counsel for Carillion] was correct in his submission that these sections have different effects and the notices which they contemplate have different purposes. Section 110(2) prescribes a provision which every construction contract must contain. Section 110(3) deals with the case of a construction contract that does not contain the provision required by section 110(2) by making applicable in that case the relevant provision of the Scheme, namely paragraph 9 of Part II. By one or the other of these routes every construction contract will require the giving of the sort of notice contemplated in section 110(2). But there the matter stops. Section 110 makes no provision as to the consequence of failure to give the notice it contemplates. For the purposes of the present case, the important point is that there is no provision that failure to give a section 110(2) notice has any effect on the right of the party who has so failed to dispute the claims of the other party. A section 110(2) notice may, if it complies with the requirements of section 111, serve as a section 111 notice (section 111(1)). But that does not alter the fact that failure to give a section 110(2) notice does not, in any way or to any extent, preclude dispute about the sum claimed. In so far, therefore, as the adjudicator lumped together the defenders' failure to give a section 110(2) notice with their failure to give a timeous section 111 notice, I am of opinion that he fell into error. He ought properly to have held that their failure to give a section 110(2) notice was irrelevant to the question of the scope for dispute about the pursuers' claims.
>
> 20. The more significant issue in the present case, in my opinion, is whether the defenders' failure to give a timeous notice under section 111 had the effect that there could be no dispute at all before the adjudicator as to whether the sums claimed by the pursuers were payable. The section provides that a party "may not withhold payment after the final date for payment of a sum due under the contract unless he has given an effective notice of intention to withhold payment". In my opinion the words "sum due under the contract" cannot be equiparated with the words "sum claimed". The section is not, in my opinion, concerned with every refusal on the part of one party to pay a sum claimed by the other. It is concerned, rather, with the situation where a sum is due under the contract, and the party by whom that sum is due seeks to withhold payment on some separate ground.'

There is a clear difference between not paying money claimed because the paying party values that work at a lesser value, and deducting monies from a valuation for reasons not related to the value of the work as completed. If the work is defective this does not require a section 111 notice, but if the contractor suffers an associated loss other than in the work itself, then it may not reduce the sum payable to the subcontractor without an effective notice of withholding as required by that section. Further, the contractor will need to be able to justify any such claim.

Even where an effective notice of intent to withhold payment is issued, it is submitted that the contractor may only issue such a notice in respect of sums to which it is entitled under the subcontract. Unless there is express provision to the contrary, such sums must be established and it is not sufficient for the contractor just to make 'an assertion of some liability in the subcontractor to pay him something'. Although decided well before the coming into force of the HGCRA, the decision of the Court of Appeal in *Dawnays Ltd* v. *F. G. Minter Ltd and Trollope & Colls Ltd* (1971) 1 BLR 16, must still be regarded as good law, to this extent.

Judge Humphrey Lloyd QC in his judgment in *KNS Industrial Services (Birmingham) Limited* v. *Sindall Limited* (2000) 75 Con LR 71, indirectly makes reference to the *Dawnays* v. *Minter* decision in relation to the payment provisions in the DOM/1 form of subcontract, at paragraph 17:

> '17. Clause 21.3 (read with clause 21.4 and the additional subcontract provisions in Appendix A) provides that on the valuation dates the contractor is to notify the subcontractor of (1) the amount of the interim payment due to the subcontractor as the value of work properly executed (and materials etc.) and how it is calculated; (2) the amount of any deduction for any payment due by the subcontractor and how it is calculated. Accordingly the amount notified as due by the contractor will be the contractor's view of the true value of work, taking into account any factors that affect the true value of the work, such as the state of completion and (the) proper execution of the work, including whether it was properly carried out or has proved to be defective, i.e. matters which might qualify in law as entitling a contractor to the defences of abatement or set-off (see for example the well-known restatement by Lord Diplock in *Dawnays* v. *Minter* ([1974] AC 689 at page 714)). Any amount that is due by the subcontractor (e.g. a contra-charge for labour, plant or materials supplied at the subcontractor's request) will also be notified as a deduction so as to arrive at the net amount due under clause 21.3.2. Clause 21.3.3, in giving effect to section 111, entitles the contractor not to pay that amount to the subcontractor in so far (as) there are grounds for withholding it or part of it or for making a deduction from it provided that an effective notice is given, i.e. one which complies with section 111 (and with any additional condition precedent that may be included in the contract). Section 111(1) makes it clear that a notice of the kind required by section 110 (i.e. that provided by clause 21.3.2.2) may suffice, but of course the contractor is entitled to give another notice (in order to ensure that there is an effective notice) or to give a notice where none had previously been given. The term withhold is thus used in section 111 to cover both the situation where in arriving (at) a valuation the contractor had not taken account of a countervailing factor as well as the situation where there is to be (a) reduction in or deduction from an amount that had been declared or thought to be due. In the former case the word "withhold" may not always be correct for one cannot withhold what is not due.'

The citation referred to by Judge Lloyd is that of *Gilbert-Ash* v. *Modern Engineering*, and this House of Lords decision re-established the common law right to set-off as a principle which had been eroded following the decision of Lord Denning in *Dawnays* v. *Minter*. It is this right which is incorporated into section 111 of the HGCRA.

In *Dawnays* v. *Minter*, Lord Denning MR made the following pronouncements, first at page 21:

> 'Every businessman knows the reason why interim certificates are issued and why they have to be honoured. It is so that the subcontractor can have money in hand to get on with his work and the further work he has to do. Take this very case. The subcontractor has had to expend his money on steelwork and labour. He is out of pocket. He probably has an overdraft at the bank. He cannot go on unless he is paid for what he does as he does it. An interim certificate is to be regarded virtually as cash, like a bill of exchange. It must be honoured. Payment must not be withheld on account of a cross-claim, whether good or bad – except so far as the contract specifically provides. Otherwise any main contractor could always get out of payment by making all sorts of unfounded cross-claims. All the more so in a case like the present, when the main contractors have actually received the money.'

Later on in the judgment Lord Denning refers to the contractor's claim, at page 22:

> 'Mr Knight [Counsel for Trollope & Colls] submits that clause 13 must be read together with clause 11(b), with the result that, whenever the main contractor asserts a money claim against a subcontractor, he is entitled to deduct the amount of that claim against the sum to which the subcontractor would otherwise be entitled under the interim certificate.'

Lord Denning answers this submission with the following dismissive words:

> 'That seems to me, with respect, an alarming submission. The result would be that the simple fact that the main contractor makes an assertion of some liability in the subcontractor to pay him something is sufficient to entitle the former to hold up payment. The practical result would be to render available to any main contractor a ready means of avoiding payment out to the subcontractor even of sums which the building owner had already paid the main contractor in respect of subcontract work covered by interim certificates.'

These extracts from the judgment of Lord Denning deny a right to set off sums without due cause. This principle was supported in *Gilbert-Ash* v. *Modern Engineering*, by the speech of Lord Salmon, where he said, at page 101:

> 'Paragraph 3 purports to confer much more on the contractors than the law allows. According to the natural meaning of its language, it would enable the contractors to suspend or withhold payment of very large sums of money due by them to the subcontractors in the event of the subcontractors committing some minor breach of contract causing only trivial damage in no way comparable to the amount owed to the subcontractors. This paragraph is, therefore, unenforceable since it provides for the extraction of a penalty.'

Where SBCSub/C is the subcontract form there is a requirement for agreement by the subcontractor to the loss and/or expense arising from the subcontractor's delay, claimed by the contractor, before such sums may be withheld. Clause 4.21.2 refers to any amount agreed by the parties as due in respect of any loss and/or expense thereby caused to the contractor. It is this agreed sum that may be deducted from any monies due or to become due to the subcontractor, or it shall be recoverable by the contractor from the subcontractor as a debt.

The right for the contractor to either deduct from sums due or to recover as a debt his loss and/or expense under this clause, is subject to the requirements of the sub-clause, as above. This means that until the procedures set out in sub-clause 1 have been complied with and a sum or sums agreed, no deduction may be made. In the event that no agreement is reached, either to an entitlement at all or to the sum to be allowed, the contractor will need to resolve the dispute under the dispute resolution provisions of the subcontract before a lawful notice of intent to withhold payment may be made or such sums withheld.

The common law right to set-off upheld as a general principle in *Gilbert-Ash* v. *Modern Engineering* does not override any express provisions of the subcontract, such as clause 4.21.2 of SBCSub/C. Lord Reid states this clearly, at page 75:

> 'It is now admitted, and in my view properly admitted, that at common law there is a right of set-off in such circumstances; but that right can be excluded by contract.'

And as stated by Judge Humphrey Lloyd QC in *KNS* v. *Sindall* quoted above:

> 'Clause 21.3.3, in giving effect to section 111, entitles the contractor not to pay that amount to the subcontractor in so far (as) there are grounds for withholding it or part of it or for making a deduction from it provided that an effective notice is given, i.e. one which complies with section 111 (and with any additional condition precedent that may be included in the contract).'

A requirement to agree, as stated in clause 4.21.2 of SBCSub/C, must be considered an additional condition precedent, such as referred to by Judge Humphrey Lloyd QC.

It will not normally be possible to use an effective notice of intention to withhold payment against an adjudicator's decision. If, for example, the matter referred to the adjudicator is the value of an interim payment certificate, then as stated above the adjudicator will be empowered to consider any matters raised in defence by the contractor concerning the valuation of the work, but not otherwise. The decision of the adjudicator will be binding and the only sum that may be withheld against that decision would be a valid notice of withholding issued in due time, having regard to the final date for payment for that interim payment, and then only if considered by the adjudicator and included within his decision.

This was considered in the case of *Levolux A.T. Limited* v. *Ferson Contractors Limited* [2002] BLR 341, where the form of subcontract was GC/Works. Judge David Wilcox said:

> '41. A party has no right to set off claims not dealt with by the adjudicator as a defence to the enforcement of the adjudicator's decision. See *VHE Construction plc* v. *RBSTB Trust Company Limited* [2000] BLR 187, judgment of Judge Hicks QC at pages 199 and 196; *Northern Developments (Cumbria) Limited* v. *J & J Nicol* [2000] BLR 158, judgment of Judge Bowsher QC at pages 158 and 164; *Solland International Limited* v. *Darayden Holdings Limited* TCC, judgment of Judge Seymour QC, 15 February 02.
>
> 42. The clause [section 111(1) of the HGCRA] clearly recognises the right to set off a claim against the enforcement of an adjudicator's award, when that claim is compliant with

section 111 as to the requirements through withholding notice and this was dealt with by the adjudicator. It does not create a fresh right of set-off.'

Repayment of sums overpaid

While it will not be the contractor's intention to overvalue the subcontractor's work or to overpay the subcontractor, at least two possible situations may arise leading to an overpayment. First, latent defects may become apparent which will lead to expense greater than the further sums due to the subcontractor either at the relevant time or at all, including any retention that might subsequently become due. Secondly, where the contractor fails to issue a timely and/or otherwise effective notice of intent to withhold payment, the subcontractor will be entitled to the full sum due at that time, and the contractor will have to seek repayment.

Few if any subcontracts contain any express provision for the repayment of sums overpaid, by the subcontractor to the contractor, and in this regard SBCSub/C is no exception. It is suggested that in the absence of express provisions, those of the Scheme must be implied into the subcontract. This would normally mean payment after application by the contractor, with a payment period as set out in the subcontract. Should this occur in the period after practical completion, and the payment mechanism relates further payments to the issue of the final certificate under the main contract, it will be difficult for the contractor to claim that a term in its subcontract is ineffective as a payment mechanism.

Sub/MPF04, however, at clause 25.4, expressly makes payment of the sum stated by a payment advice payable to either party at the end of the payment period.

Works carried out off site

The general arrangement in construction contracts is that payment is for completed construction work. Where payment is by periodic valuation, then an allowance is generally made in such valuation for goods delivered to site for incorporation into the works. Payment for goods held off site and for services such as design or the preparation of shop drawings may be discretionary on the part of the employer or employer's agent, and be subject to conditions such as the transfer of title and/or conditions of insurance.

The risk for the contractor/employer in paying for work carried out off site, before delivery and/or installation into the works, is that of possible damage, loss or misappropriation. The benefit to the contractor/employer is that it owns the goods and in the event of insolvency by the subcontractor, the consequential delay in finding a new source or negotiating with the liquidator will be minimised. The benefit for the subcontractor is the benefit of improved cash flow. The risks do not relate to the provision of design information. It will frequently be the case that design information will be required for coordination with other design activities, ahead of the time when it would be necessary for the subcontractor's work. For all these reasons, the subcontractor's design needs to be considered separately from goods and materials which are stored off site.

The payment provisions under the HGCRA make no distinction between work carried out under a construction contract on site and that undertaken off site.

Section 104 expressly makes the provision of design or surveying work in relation to construction operations a construction contract for the purposes of the HGCRA, while under section 105(2)(d) the manufacture or delivery to site of materials, plant and components are stated not to be construction operations. It is therefore uncertain as to whether, where a single subcontract includes design, detailing, manufacture and installation, the statutory provision for payment by instalments under section 109 of the HGCRA applies to all the pre-site/off-site activities. However in the absence of any express agreement to the contrary, it is suggested, that pre-site/off-site works in a subcontract for the provision and installation of components will all be subject to the same terms, since they are part of the same subcontract.

It is therefore necessary to set out in the subcontract conditions the provisions for payment for pre-site activities. Such provision might include an extended payment period or might set an earliest date for payment related to the programme for on-site work, with the intention of avoiding payment prior to delivery. Such a payment mechanism might, in addition, require provision of other arrangements, for example for insurance, labelling or special storage, as may be specified in the subcontract.

A casual incorporation by reference of a standard form and/or terms of the main contract is unlikely to be effective because the subcontract is likely to contain some payment terms, however incomplete or ineffective, when the terms of the Scheme will apply in addition. In other words, where there is no adequate mechanism, the payment will become due on the making of a claim by the subcontractor (paragraphs 3 and 4(b) of the Scheme). The rights and obligations for the carrying out of pre-site works with regards to programme and/or valuation are considered in the sections 'Programming off-site or pre-site works' in Chapter 4 and 'Works carried out off site' in Chapter 9.

The subcontractor's right or otherwise to be paid for off-site and/or pre-site works will affect its willingness to progress such activities at an early date, and/or if site installation is delayed it should lead to an entitlement to financing costs and a claim for these either as loss and/or expense or as damages.

Settlement

Where sums have been in dispute and an agreement or decision is reached, then a normal and necessary part of the settlement, whether by agreement or by the decision of a third party, is that the final date for payment is defined. Such agreement or decision need not have had any relation to the previously agreed terms for payment within the subcontract. Indeed, an adjudicator's decision will generally define a new payment period by stating that payment is to be made immediately or within seven days. The parties may agree for payment to be by instalments of stated sums on stated days. Action by the payee, such as terminating an action in adjudication or at law, may be dependent on receipt of cleared funds. In the absence of any express agreement, it is likely that the date of the agreement will become the payment due date, and the final date of payment will be the contract payment period after that.

Compliance with an adjudicator's decision will, in the absence of specific directions as to the fulfilment of the obligations decided by the adjudicator, be subject to the requirements of the subcontract. Where the Scheme applies, paragraph 21 of Part I states that this shall be 'immediately upon delivery of the decision':

> '21. In the absence of any directions by the adjudicator relating to the time for performance of his decision, the parties shall be required to comply with any decision of the adjudicator immediately on delivery of the decision to the parties in accordance with this paragraph.'

Settlement agreements are usually made in the form of a collateral contract or deed, the parties having agreed that in consideration for the payment of the agreed final account sum, this will be in settlement of all claims and counterclaims between them. It may be that there will still be certain matters remaining, such as repayment of retention if the date for release of retention has not been reached by the date of the agreement. Payment may of course be subject to a condition such as the issue of a guarantee, record drawings, health and safety file, or such other matter.

Settlement may be the subject of a unilateral offer by one party, frequently the paying party, when a cheque for a lesser sum than that requested by the payee is sent under cover of a letter stating that this is in full and final settlement. Like any other contractual offer it is for the offeree to accept or reject that offer. Acceptance may be by conduct if the payee banks or cashes the cheque. The payee may well see such payment as being a contribution or part payment against its outstanding account. Should the payee intend to take such payment as a part payment, it must make a counter-offer to that effect and at the very least allow a reasonable period for the paying party to demand return of its cheque.

Lord Denning MR, in the case of *D & C Builders* v. *Rees* [1965] 3 All ER 837, reviewed the law with regard to such payments and stated clearly, at page 838, that payment in such circumstances will not settle the debt:

> 'It is a daily occurrence that a merchant or tradesman, who is owed a sum of money, is asked to take less. The debtor says he is in difficulties. He offers a lesser sum in settlement, cash down. He says he cannot pay more. The creditor is considerate. He accepts the proffered sum and forgives the rest of the debt. The question arises; is the settlement binding on the creditor? The answer is that, in point of law, the creditor is not bound by the settlement. He can the next day sue the debtor for the balance, and get judgment.'

However, the situation may be complicated where there is a pre-existing assertion that the work to be paid for is defective or where some other consideration is offered as part of the settlement offer.

As mentioned above, when a cheque for the offered sum accompanies a compromise offer, the recipient of the cheque may be considered to have accepted the offer of settlement if and when he presents the cheque to the bank for payment. However, this will not necessarily be the case; it will depend on the associated facts and the actions of the parties. The compromise offer must clearly be an offer for which the cheque is sent in settlement. Manifestly, posting the cheque with the offer cannot be said to be the fulfilment of the compromise agreement since no agreement exists at the time of the offer and the sending of the cheque. The offeror does not at that time know whether his offer will or will not be accepted. It is therefore for the offeror to make very clear what the attached cheque may be used for. In

Bracken and Another v. *Billinghurst* [2003] EWHC 1333 (TCC), the offeror made it very clear, not only that the settlement offer would be deemed to have been accepted if payment was tendered, but also that if the recipient was not willing to accept the payment on those terms, then the cheque was to be returned. The recipient was given no option other than to accept the terms and the payment or reject the offer and return the money. The actual words used were:

> 'My client, however, is not willing to pay this sum of £6,000 but will come close to that figure. We are therefore instructed to forward immediately to you the cheque for the sum of £5,000 on the strict understanding that this sum is offered to you in full and final settlement of all issues between yourself, Mr Billinghurst and Advance Building Technology Limited in relation to all matters of dispute concerning 1 Hurst Road Hawley. The payment is tendered as a compromise settlement. The payment is tendered as an offer of settlement which will [be] deemed to have [been] accepted by you and therefore be contractually binding if it is presented to your bank and cleared for payment. If you are not willing to accept the payment on these terms, would you please return the payment and we will assume therefore that the dispute will have to continue.'

However, compromise offers are frequently not in such clear terms, allowing the recipient to reject the offer and take the benefit of the payment. For many companies a real difficulty is that incoming mail will frequently be opened and distributed, for necessary action, by a relatively junior member of the staff. There will likely be procedures within the company to ensure that incoming payments are banked as soon as possible, while correspondence will be passed to the appropriate authority for action. For this and many other reasons the banking of a cheque will not of itself be evidence of an acceptance, as Judge David Wilcox in *Bracken* v. *Billinghurst*, quoting from the judgment in *Stour Valley Builders* v. *Stuart*, said at paragraph 25:

> 'It is a question of fact, of course where there is documentation as in this case the construction of such documentation is a matter of law and will give rise to facts which are part of the material events which must then be judged objectively by the court. See *Stour Valley Builders* v. *Stuart* [1974] 2 Lloyds Rep 13, CA, where Lord Justice Lloyd said:
>
> "As with any other bilateral contract what matters is not what the creditor himself intends but what by his words and conduct he has led the other party as a reasonable person . . . to believe".'

In the case of *Joinery Plus Limited (in administration)* v. *Laing Limited* [2003] BLR 184, the part payment of an adjudicator's decision, where part of that decision was being challenged, was claimed to have been acceptance of a compromise offer. It was held that acceptance of a cheque did not amount to acceptance as claimed. Judge Thornton QC said:

> '91. It is clear from Judge Seymour's judgment, and in accordance with principle, that the mere payment of money and its acceptance is not sufficient to characterise a party's acts as amounting to an acceptance of, or an agreement to settle, an underlying obligation to pay or to settle a dispute. Judge Seymour was considering whether one of the parties had unequivocally accepted an adjudicator's decision in the circumstances of that case.
>
> 92. A party does not, merely by accepting a cheque, accept that the debt, obligation or dispute underlying the payment of that cheque has been discharged or settled. Whether

or not a discharge or settlement results will depend on the intention of the party accepting the cheque, as determined objectively from the surrounding circumstances of that acceptance. If the acceptance is intended to be qualified so that the payment is accepted generally on account of that party's entitlement to payment, and it is clear from the surrounding circumstance objectively determined that the acceptance of the cheque was qualified in that way, the accepting party will not be taken to have fully and finally accepted or approbated or settled the underlying obligation or the situation giving rise to that obligation.

93. This principle may be seen from the well-known and long established case of *Day* v. *McLea* (1899) 22 QBD 610. This case was approved and followed by the Court of Appeal in *Stour Valley Builders* v. *Stuart*, The Independent, 9 February 1993 and by Lord Justice Jacob in *Inland Revenue Commissioners* v. *Barbara Fry*, Lawtel, 30 November 2001. It is both surprising and regrettable that neither of the modern authorities have yet been reported since the principle established by *Day* v. *McLea* is not as well known to modern litigants as it should be. That principle is clearly explained by Lord Justice Lloyd in the *Stour Valley Builders* case:

"Cashing the cheque is always strong evidence of acceptance, especially if not accompanied by immediate rejection of the offer. Retention of the cheque without rejection is also strong evidence of acceptance depending on the length of delay. But neither of these factors are conclusive, and it would, I think, be artificial to draw a hard and fast line between the cases where the payment is accompanied by immediate rejection of the offer and cases where objection comes within a day or within a few days." '

Judge David Wilcox in *Bracken* v. *Billinghurst* summarised the legal principles and said:

'24. The offer "in full and final settlement" of the dispute is made at the time the cheque is sent. There must be clear evidence of actual or potential disputes at that time. The presentation of the cheque may amount to an acceptance of the offer giving rise to an accord. In *Day* v. *McLea* (1889) 22 QBD 610, Lord Justice Bowen said, at page 613:

"If a person sends a sum of money on the terms that it is to be taken if at all, in satisfaction of a larger claim; if the money is kept, it is a question of fact as to the terms upon which it is so kept. The accord and satisfaction imply an agreement to take the money in satisfaction of the claim in respect of which it is sent. If accord is a question of agreement, there must be either two minds agreeing or one of the two persons acting in such a way as to induce the other to think that the money is taken [as] satisfaction of the claim, and accordingly to act upon that view."

However, while the return of a cheque tendered with a compromise offer will be certain evidence of the rejection of that offer, there appears to be no requirement at law for such a return. Where the offer is positively rejected a reasonable period before the presentation of the cheque and the offeror has been made aware of the basis on which that cheque is being accepted, it is unlikely that the offeror will be able to show that the compromise offer has been accepted. Thus a party seeking to close out a dispute by the unilateral action of sending a compromise offer accompanied by a cheque for the appropriate sum, stands to have its offer rejected and its money retained against the alleged debt.

The courts will not uphold a settlement agreement obtained under duress. In the case of *D & C Builders* v. *Rees*, the building owner, on completion of the works, realised that the builder was in urgent need of the cash for the balance of the account. Consequently the building owner offered £300 as 'in completion of account', a shortfall of £182 on a total value of £732. The trial judge found there had not been a binding settlement. On appeal, Lord Justice Danckwerts said, at page 841:

> 'I agree also that, in the circumstances of the present case, there was no true accord. Mr and Mrs Rees really behaved very badly. They knew of the plaintiffs' financial difficulties and used their awkward situation to intimidate them. The plaintiffs did not wish to accept the sum of £300 in the discharge of a debt of £482, but were desperate to get some money. It would appear also that the defendant and his wife misled the plaintiffs as to their own financial position. Mr Rees, in his evidence, said: "In June (1964) I could have paid £700 odd. I could have settled the whole bill". There is no evidence that by August, or even by November, their financial situation had deteriorated so that they could not pay the £482.'

The appeal was dismissed.

Prior to the completion of the works, the subcontractor may exert pressure on the contractor to agree, at an early date and/or prior to the completion of the work, the final sum sought by the subcontractor. This was the situation in the case of *Carillion Construction Ltd* v. *Felix (UK) Ltd* [2001] BLR 1, which concerned a subcontract where the subcontract work had commenced well before the conclusion of a final agreement between the parties and the execution of the articles of agreement. Work on site had commenced in late July 1999 and had been scheduled to be completed in the following January. The agreement to the final documents was reached in January 2000 but the subcontractor was refusing to exchange the signed agreement and was seeking to agree the valuation of the variations before signing, and more significantly was refusing further performance.

Mr Justice Dyson set out in detail the events and nature of correspondence and meetings that took place during the critical period of January–March 2000. In brief, the subcontractor refused to make further deliveries without prior agreement to its final contract sum. This included various elements where there was dispute as to whether they were additional to or included in the scope of work subcontracted for. After protracted negotiations, the contractor agreed with the subcontractor a schedule of deliveries and associated payments, leading to the sum sought by the subcontractor. Immediately on delivery of the final panels, the contractor commenced proceedings seeking an order rescinding the settlement agreement for duress.

The judge set out the law relating to duress, at paragraphs 24 and 25:

> 'The ingredients of actionable duress are that there must be pressure, (a) whose practical effect is that there is compulsion on, or a lack of practical choice for, the victim, (b) which is illegitimate, and (c) which is a significant cause inducing the claimant to enter into the contract. In determining whether there has been illegitimate pressure, the court takes into account a range of factors. These include whether there has been an actual or threatened breach of contract; whether the person allegedly exerting the pressure has acted in good or bad faith; whether the victim had any realistic practical alternative but to submit to the pressure; whether the victim protested at the time; and whether he affirmed and sought to rely on the contract. These are all relevant factors. Illegitimate pressure must be distinguished from the rough and tumble of the pressures of normal commercial bargaining.
>
> Accordingly, Carillion must show that there was (a) pressure or a threat, (b) which was illegitimate, (c) the practical effect of which was that it had no practical choice but to enter

into the agreement, and (d) which was a significant cause inducing it to enter into the contract.'

The judge then examined the evidence of the threat by the subcontractor not to carry on with the work until the contractor had agreed to the final account sum demanded by the subcontractor. He concluded, at paragraph 31:

'But for the threat, it (the contractor) would not have agreed the final account on 13 March at all. It felt compelled to agree the account only because it was determined to secure the delivery of the cladding units. The figure of £3.2 m was the best Carillion could achieve in the circumstances.'

The judge then set out the benefits to the subcontractor of the early final account settlement and that there was no benefit at all to the contractor, and concluded, at paragraph 33:

'These factors support my finding that the threat was made, and that Carillion was induced by the threat to enter into the settlement agreement.'

After discussing the level of pressure put on the contractor and what, if any, practical choice the contractor had, the judge set aside the agreement.

It is common for a subcontractor to be requested to commence work on a building project where there is no formalised or concluded agreement. However, in these circumstances the subcontractor will normally have stated in its original offer what it seeks by way of payment and other arrangements. Where the contractor knowingly decides to proceed before concluding an agreement, including agreement as to price with the subcontractor, it is for the contractor to persuade the subcontractor to move its position. Where the contractor fails to move the subcontractor from its position, this is unlikely to be seen as the subcontractor putting the contractor under duress. However, should the subcontractor, as Felix did, seek to introduce further conditions to its considerable benefit, without giving any substantial benefit in return, and agreement to such conditions is made a condition for proceeding further with the subcontract works, then on the principles set out in *Felix*, such action is likely to be regarded as duress and therefore unlawful.

It also frequently happens that the contractor will put the subcontractor under duress by insisting after work, or work on site, has progressed for some time, that the subcontractor sign up to terms not previously agreed and making such acceptance a condition of further payment. In such circumstances it is likely that the subcontractor's acceptance could be revoked as having been obtained under duress.

Disputes arising out of settlement agreements will not be regarded as disputes in respect of a construction contract and will not be subject to the requirements of the HGCRA. There is therefore no right to adjudication over the effect of such an agreement. However, while it is likely that deciding whether or not there is a settlement agreement may well be considered a matter of fact in a dispute arising under the contract, deciding whether a settlement agreement has been justly entered into, for example under duress, will not be a matter arising out of the subcontract and consequently not referable to an adjudicator, nor would a dispute concerning the terms of the settlement itself.

The effect of settlement agreements was considered in the case of *Shepherd Construction Ltd* v. *Mecright Ltd* [2000] BLR 489, where the contractor sought a stay in adjudication proceedings on the grounds that, since there was a settlement agreement, there was no dispute under the contract that could be referred to an adjudicator. Judge Humphrey Lloyd QC said, at page 493:

> 'In my judgment where parties have reached an agreement which settles their disputes there can thereafter be no dispute about what had been the subject matter of the settlement capable of being referred to adjudication under the provision such as clause 38A.1 or otherwise for the purposes of the HGCRA. The prior disputes have gone and no longer exist.'

Further on, at paragraph 14, he makes the point that the effect of a settlement agreement is to cancel the relevant parts of the existing Sub-Contract Agreement and to create new ones:

> 'I should make it clear that in my judgment a dispute about a settlement agreement of this kind could not be a dispute under the subcontract since the effect of a settlement agreement is one which replaces the original agreement to the extent to which it applies. Here the agreement has the effect of replacing Shepherd's obligations to value and pay Mecright under the subcontract the value of the work. The only subsisting obligation to pay that apparently was not extinguished was the obligation to release retention as and when the time arose.'

Payments under several subcontracts

It can frequently happen that a subcontractor is working for the same contractor on a number of projects under different subcontracts at the same time, or indeed may have more than one subcontract relating to the same project. When payment is made, it may not be apparent what sums have been paid under which subcontract. Should a dispute as to payment arise, it may be very necessary to know how the sums paid relate to each project.

It will be for the contractor to decide how it intends the money it pays to be allocated against its various debts to the subcontractor. However, if the contractor makes no selection, the subcontractor may opt to allocate the money as it feels fit. This has been stated in the case of *Cory Bros & Co Ltd* v. *The Mecca* [1897] AC 286, by Lord Macnaughton who said, at page 293:

> 'When a debtor is making a payment to his creditor he may appropriate the money as he pleases, and the creditor must apply it accordingly. If the debtor does not make any appropriation at the time when he makes the payment the right of application devolves on the creditor. In 1816, when *Clayton*'s case was decided, there seems to have been authority for saying that the creditor was bound to make his election at once according to the rule of the civil law, or at any rate, within a reasonable time, whatever that expression in such a connection may be taken to mean. But it has long been held and it is now quite settled that the creditor has the right of election "up to the very last moment" and he is not bound to declare his election in express terms. He may declare it by bringing an action or in any other way that makes his meaning and intention plain.'

This will allow a subcontractor to allocate all payment away from a dispute where there is a high certainty of success, and take action on that subcontract to recover sums the subcontractor claims are underpaid or undervalued.

Pay-when-certified clauses

After the HGCRA made unenforceable general clauses which make payment conditional on payment by another, a number of contractors have introduced conditions limiting their obligation to pay subcontractors to monies certificated for payment under the main contract. The question therefore frequently arises as to whether such clauses are in fact effective.

For all the reasons set out below under the section on pay-when-paid, such clauses will, it is submitted, in any event be unenforceable where the reason for non-certification is due to the contractor's own fault and/or breach of the main contract conditions. Such faults would include failure to make an application for payment or failure to make such an application in due time or in the correct manner. It will also frequently be the case that a certificate under the main contract may either not be issued or may not be in sufficient detail to define exactly what is being certificated.

Many standard forms of subcontract have a clause similar to SBCSub/C clause 3.24, requiring the contractor to obtain the benefits of the main contract for the subcontractor. Such action is to be at the request of the subcontractor and subject to such rights as may lawfully be obtained by the contractor and to the extent that they are applicable to the subcontract works.

Even where such express provisions do not exist, it is likely that where there is a pay-when-certified clause incorporated within the subcontract, such a provision will be implied into the subcontract. Failure by the contractor to progress a subcontractor's application for payment, both at the due time and in the event of the subcontractor's application not being certified either in whole or at all, will be a breach of the contractor's obligations to the subcontractor.

Unlike a provision for pay-when-paid, which is effectively a provision for withholding payment against monies otherwise due, a pay-when-certified clause may be seen as a valuation procedure. Certainly in practice this is how it may be used, especially for the valuation of variations, where the contractor may be unwilling to value such variations not knowing if it will or will not be paid an equal or lesser sum under the main contract. In such cases, it is suggested, the pay when certified clause falls foul of the requirement of section 110(1) of the HGCRA for an adequate mechanism for determining what payment becomes due under the subcontract and when. The existence of this problem demonstrates the unsatisfactory arrangements, currently incorporated within construction contracts, for the valuation of variations.

In practice 'pay-when-certified' conditions may be either a 'certify if and when certified' clause or a 'pay what is certified' one. Depending on which applies, the amount to be paid to the subcontractor will either be the contractor's valuation, provided such work has been valued and certified under the main contract, or a valuation proportional to that made under the main contract. It is likely that only in the situation where the contractor makes its own timely valuation and subsequently issues an effective notice of intent to withhold on the basis that the work concerned has either not been valued under the provisions of the main contract or has been valued at a stated lesser sum, may such a condition of 'pay-when-certified' be effective. Such a provision is unlikely to be effective where the contractor has received no valuation at all, but may be effective where there is no valuation or a reduced valuation of the items for which it seeks to withhold payment. In any event it is suggested that it will be for the contractor to demonstrate that certification has been excluded, rather than the reverse.

One problem in trying to implement a pay-when-certified clause is that the contractor's statutory obligation under the HGCRA section 110, to issue a notice of payment five days after the payment due date under the subcontract, will remain effective whether the contractor itself receives a valuation under the main contract or not, or before the date for the issue of a notice of payment under the subcontract. The general disregard for the legal requirement to issue a notice of payment under section 110 of the HGCRA may mean that the contractor in any event will not know in detail, if at all, what work has or has not been included in any certificate issued to it under the provisions of the main contract, and consequently will not be able to define whether the subcontractor's work has been certified or not.

There is no reason why the payment mechanism under a subcontract may not define the payment due date as a certain period after certification under the main contract is to be made. This may of course lead to significant delays in payment, especially if combined with a significantly long payment period. However, where a subcontract defines the payment due date as a specified period after the payment due date in the main contract, the contractor will achieve its objective of pay-when-paid by making the payment period under the subcontract the same as that under the main contract.

Judge Humphrey Lloyd QC considered the lawfulness of payment terms relating the payment of the subcontractor to the certification under the main contract, in his judgment in *Jarvis Facilities Limited* v. *Alstom Signalling Limited (trading as Alstom Transport Information Solutions) No. 2* [2004] EWHC 1285 (TCC). In that case Alstom's subcontract provided a payment mechanism which was complex but in which the relevant clause was 2.6, as follows:

> '2.6 The Contractor shall pay to the Subcontractor the amount due on the certificate within seven days of the Railtrack certificate being issued in accordance with Annex F1 – the Project Cut-Off Dates (which contains the amount certified against the application aforementioned).'

The adjudicator, under an earlier adjudication, considered the term 'pay-when-certified' when he decided that this was not an effective payment mechanism because of the uncertainty as to the date of the 'Railtrack certificate being issued'. Had the Alstom term stopped there the judge might have agreed with the adjudicator, but the term went on to say 'in accordance with Annex F1'. This schedule gave the dates when the Railtrack certificates were to be issued. Provided the Railtrack certificates were issued on time, there was an effective payment mechanism. This was not therefore an open 'pay-when-certified' term, but a requirement to pay in parallel with the payment under the main contract provisions. The judge made this point in the following words:

> '22. Section 110 of the Act calls for an adequate mechanism to determine the final date for payment. I find myself at somewhat of a loss to understand why Schedule F does not comply with section 110 of the Act in terms of an adequate mechanism to determine when payment was due for the purposes of section 110(1). The subcontract was made by reference to the main contract, both formally and financially. Clause 1.3 of the Schedule F demonstrated a common approach as regards a "Neutral Cash Flow" (see also clause 2.3). The applications for payment and payments were linked to the Railtrack cycle. That was shown in Annex F1. That Annex set out the Wednesday in each month in the years 2000 and 2001 when "cost of work done" was required – see clause 2.4 of Schedule F. Clause 1.1 said that "payment of the Subcontract Price shall be . . . at intervals as defined in Annex F1 – Project Cut-Off Dates". The contractor is to issue certificates (see clause 2.5). That establishes when a payment is due for the purposes of the second limb of section 110(1)(a) and the requirements for the content of the certificate (to) comply with section 110(2). (I deal later with the first limb of section 110(1)(a).) Conventionally it seems that Alstom was to issue a certificate within 14 days of the receipt of an application (see for example, its letter of 13 June 2003). Clause 2.6 said that payment would be made within seven days of the Railtrack certificate being issued in accordance with Annex F1. There was therefore certainty as to the final date for payment – seven days of the Railtrack certificate. That satisfies section 110(1)(b). The fact that Railtrack, probably in breach of its contract with Alstrom, might fail to issue its certificate in accordance with Annex F1 does not mean that for the purposes of section 110(1)(b) there was no final date. The final date for payment remains seven days after the issue of the certificate. The fact that a date is set by reference to a future event does not render it any less a final date. Section 110(1) says, very clearly:
>
> > "The parties are free to agree how long the period is to be between the date on which a sum becomes due and the final date for payment."
>
> 'The event could be a stage, or milestone or completion, practical or substantial. It could be the result of action by a third party, such as a certificate under a superior contract or transaction, as is found in financing arrangements. Provided that the event is readily recognisable and will produce a date by reference to which the final date can be set, there is no reason why it cannot be used. Payment of a subcontractor by reference to the date of a main contract certificate accords with industry practice and, on this project, is not at all inconsistent with the aim of neutral cash flow (taking into account clause 2.1(b) of Schedule F). No difficulty could arise after the end of 2001 as the pattern set by the two years could easily be projected beyond the end of 2001 (and, evidently, was so projected). Put another way, if Railtrack did not issue a certificate on time Alstom could hardly use it as a defence since clause 2.6 is written on the assumption of due compliance. I therefore do not understand how it could be said that the date could be changed unilaterally. Accordingly in my judgment the adjudicator was wrong in his reading of the contract and in his decision that the Scheme applies to obtain a final date for payment.'

The essential part of this judgment is the sentence, 'If Railtrack did not issue a certificate on time Alstom could hardly use it as a defence since clause 2.6 is written on the assumption of due compliance.' A distinction can therefore be made between open pay-when-certified clauses and those that relate to a predetermined payment schedule under a separate contract.

Pay-when-paid

A major provision of the HGCRA is to render pay-when-paid conditions in subcontracts generally unenforceable. This is done through section 113, which states:

> '113 (1) A provision making payment under a construction contract conditional on the payer receiving payment from a third party person is ineffective, unless that third person, or any other person payment by whom is under the contract (directly or indirectly) a condition of payment by that third person, is insolvent.
> (2) For the purposes of this section a company becomes insolvent –
> (a) on the making of an administration order against it under Part II of the Insolvency Act 1986,
> (b) on the appointment of an administrative receiver or a receiver or manager of its property under Chapter I of Part III of that Act, or the appointment of a receiver under Chapter II of that Part,
> (c) on the passing of a resolution for voluntary winding-up without a declaration of solvency under section 89 of the Act, or
> (d) on the making of a winding-up order under part IV or V of that Act.
> (3) For the purposes of this section a partnership becomes insolvent –
> (a) on the making of a winding-up order against it under any provision of the Insolvency Act 1986 as applied by an order under section 420 of that Act, or
> (b) when sequestration is awarded on the estate of the partnership under section 12 of the Bankruptcy (Scotland) Act 1985 or the partnership grants a trust deed for its creditors.
> (4) For the purposes of this section an individual becomes insolvent:
> (a) on the making of a bankruptcy order against him under Part IX of the Insolvency Act 1986, or
> (b) On the sequestration of his estate under the Bankruptcy (Scotland) Act 1985 or when he grants a trust deed for his creditors.
> (5) A company, partnership or individual shall also be treated as insolvent on the occurrence of any event corresponding to those specified in subsection (2), (3) or (4) under the law of Northern Ireland or of a country outside the United Kingdom.
> (6) Where a provision is rendered ineffective by subsection (1), the parties are free to agree other terms for payment.
>
> In the absence of such agreement, the relevant provisions of the Scheme for Construction Contracts apply.'

The first point to note is that these provisions, like all those in the HGCRA, only apply to construction operations under construction contracts as defined in sections 104 and 105. They will not apply to a construction contract not satisfying the requirements of section 107. It is difficult to conceive of an oral contract with a pay-when-paid provision, since by their nature oral contracts rarely have defined terms beyond the acceptance of a simple offer.

Secondly, the Act makes express allowance in subsection (1) for the parties to agree that a pay-when-paid provision may apply when the third party has become insolvent. It has been argued that this subsection makes a general pay-when-paid clause effective when there is insolvency, whether the subcontract makes that a condition of the clause or not.

Even when there is an effective pay-when-paid clause it will not be sufficient for the paying party to say 'I have not been paid'; it will need to demonstrate that it has not been paid the monies for which it seeks to withhold payment and that it was not paid despite its having taken all appropriate actions to obtain payment. As Judge Humphrey Lloyd QC said in *Durabella Ltd* v. *J. Jarvis & Sons Ltd* (2001) 83 Con LR 145, at paragraph 17:

> 'A contractor cannot rely on a pay-when-paid clause if the reason for non-payment is its own breach of contract or default.'

In that case, the employer, for a breach for which no responsibility lay with the subcontractor, determined the contractor's contract.

It is likely that before a pay-when-paid clause would hold, the contractor would have first to demonstrate that it applied for payment for the work for which payment is being withheld, or that it had been carried out so close to the insolvency of the third party that application was not yet due. Again from *Durabella*, Judge Humphrey Lloyd said, at paragraph 18:

> 'If the clause is to be effective then the contractor impliedly undertakes that it will pursue all means available to obtain payment, or it will not be able to rely on the provision to defeat the subcontractor's claim.'

Where, as is common under main contracts, the value of the work completed is assessed by a consultant employed by the building owner, the paying party will have to show that the work was properly valued by that consultant. An undervaluation by the employer's agent will not be a failure by the building owner to pay, since the employer's obligation will normally be only to pay that which has been valued and certified by others. Consultant's valuations are not always broken down into sufficient detail to determine the precise work for which payment is being certified. This situation will certainly be the case where payment to the contractor is by payment graph or stage payments. Should the contractor seek to withhold monies on the grounds of non-payment to it, it will need to demonstrate that the monies due were in fact certified and then not paid by reason of insolvency.

It will be necessary to demonstrate that failure to receive payment was due to the third party's insolvency. If the third party, for example, does not pay having issued a valid notice of withholding under section 111 of the HGCRA, or is withholding LADs, or if payment is stated only to be due following some act by the contractor such as providing a tax certificate or submitting a VAT invoice and non-payment is due to such a failure, then the non-payment will be the result of a breach by the contractor and not a failure to pay by the employer. It is submitted that the fact that such a breach would, but for insolvency, only have delayed payment does not mean that the cause of the contractor not being paid is the insolvency of the third party.

Similarly, a breach of the payment provisions of a higher contract, for example failure to pay by the final date for payment or failure to issue a valid notice of payment or notice to withhold payment, will, it is suggested, be a failure to pay not because of the insolvency, but because of the breach of the head or main contract.

A further difficulty will arise if the payment arrangements between the contractor and subcontractor are different to those between the contractor and employer, as will frequently be the case. Main contracts generally provide for payment on a monthly or four-weekly basis; if the subcontractor is to be paid more frequently, say fortnightly or weekly, then to what extent can payment of intermediate valuations be said to be dependent on payment by the third party? It can be argued that only those payments that would have been made after the receipt of the money not paid as a result of the insolvency may be denied to the subcontractor.

Some confusion exists as to the meaning of section 113. It is suggested that this section, unlike others in the Act which are mandatory, is permissive. The section allows the inclusion of a pay-when-paid condition in a subcontract, that is only effective provided it is for the limited circumstances allowed. It does not, it is suggested, imply in a pay-when-paid clause upon the insolvency of a third party. What is perhaps less certain is whether it turns a general pay-when-paid clause into an effective clause on the insolvency of a third party.

It is frequently said that to be effective under the HGCRA, a pay-when-paid clause must be limited to the situation where:

> 'A third person, or any other person payment by whom is under the contract (directly or indirectly) a condition of payment by that third person, is insolvent.'

That, it is submitted, is not what section 113(1) says. It is not phrased that such a condition is ineffective unless limited to the situation where the third party is insolvent, but is phrased that the provision is ineffective unless the third party is insolvent.

However, section 113(6) says that where a pay-when-paid clause exists, it will be made ineffective by section 113(1). At that time the parties must either agree other terms or the Scheme will apply. Having substituted either other terms by agreement or the provisions of the Scheme by default, the pay-when-paid clause can be said no longer to exist. If this interpretation is correct it can only be reintroduced by agreement and then only in a form where its effectiveness is dependent on third party insolvency. For this reason it is suggested a general pay-when-paid clause does not become effective on the insolvency of a third party.

The concept that section 113(1) in any way implies in a pay-when-paid clause on insolvency of a third party is, it is suggested, totally misconceived. The 'unless' in this subsection relates to the existence of a provision for conditional payment in the subcontract. Without a pre-existing conditional payment clause, section 113 is not applicable to the subcontract and section 113(6) cannot be interpreted as meaning that a non-existent clause is to be included by reliance on the words: 'Where a provision is rendered ineffective by subsection (1)'; the rectification of the contract is either for the parties to agree 'other terms of payment' or for the provisions of the Scheme to apply by default. The Scheme makes no provision for pay-when-paid under any circumstance and therefore none can be implied into the subcontract.

Not all monies due to a subcontractor will, in any event, be allowable under the main contract. While it is common practice to try and make subcontract terms back-to-back with the main contract, there will be circumstances where it is considered commercially advantageous or necessary to enter into a subcontract under different conditions. For example, the contractor may assume the risk of increased costs across the project but may allow a subcontractor increases in price.

In the situation where the insolvent party is higher up the payment chain, such as exists where the main contractor becomes insolvent and a sub-subcontractor is not paid by reason of an effective pay-when-paid clause, and in the event that the principal subcontractor is re-engaged under a new contract, either directly to the employer or by a new main contractor and negotiates payment of money not paid to it by the insolvent contractor, it may then have an obligation to pay the sub-subcontractor any further sums received on a pro-rata basis. Contractually, while the principal subcontractor will be under new arrangements, with the rights for and/or liabilities under the previous contracts having been lost, the sub-subcontract may still be intact and the right to be paid survive. In this regard, many collateral warranties make provision for the employer to take over the subcontract on the default or insolvency of the main contractor.

There will normally be a desire for the building owner to recover some of its lost rights, for which consideration will be necessary. Because of the normal payment arrangements the building owner will not normally have paid for all the work as installed. The employer may often find that some compensation to subcontractors for their loss is necessary to achieve continuity and undivided responsibility for the works. Even though the principal subcontractor will receive payment under a new subcontract, with a different principal the original sub-subcontract will probably still be intact and the requirement to pay will remain, subject only to where the principal subcontractor has not been paid as a result of insolvency.

In practice, regardless of the legal position, it is likely that this will be a necessary requirement to persuade the sub-subcontractor to complete the works, which may involve supplying documents such as test certificates and as-built drawings, which will be expensive to produce without the sub-subcontractor's full cooperation. Contractually these new arrangements may be separate subcontracts with the rights for, and/or liabilities under, the previous subcontracts having been lost.

The right to withhold payment on the basis that money has not been paid to the contractor because of insolvency, and that there is an effective pay-when-paid clause, will only itself be effective if a timely notice giving such reasons as grounds for withholding is issued to the payee. In the absence of such a notice, the subcontractor is likely to be able to obtain payment and the contractor will have to seek recovery by way of a separate claim and action. Since such notice of withholding must be issued before the specified number of days (5 under SBCSub/C clause 4.10.3 or 7 under paragraph of Part II of the Scheme) before the final date for payment, this reduces still further the window of opportunity for imposing such a provision.

SBCSub/C has no clause equivalent to condition 32 in DOM/1, allowing for the contractor not to be obliged to make any further payment to the subcontractor of any amount which is due or may become due to the subcontractor, if the employer is insolvent.

Interest

Where the contractor does not pay on time, the subcontractor may become entitled to receive interest on the sum not paid for the period payment remains outstanding. Whether such a right exists and the rate of interest to be paid will be decided by the terms of the subcontract or by statute. Traditionally the law has not allowed interest on debts, but this policy has been significantly changed over the past years following the decision in *Minter* v. *Welsh Health*, albeit that this decision was phrased as an entitlement to recover financing costs as a loss and as part of the provisions under the subcontract for the valuation of variations, and not as interest on an outstanding debt. The enactment of the Late Payment Act has ended what had been described in the *Minter* case as 'the medieval abhorrence to usury'.

The House of Lords in the *Minter* case drew a significant distinction between financing charges and a right to interest on a debt. As Lord Justice Ackner said, at page 25:

> 'The claimants may have been involved by way of finance charges upon the amounts otherwise certified and paid or payable thereunder and/or by being stood out of their money (if established) for the following period only, viz. between the loss and/or expense being incurred and the making of the written application for reimbursement of the same as required by the said contract.'

As discussed below, there is frequently debate as to whether an undervaluation of the works attracts interest for the period the work is undervalued. It follows from the quotation above that it does. This view has been reinforced by the judgment of Lord Drummond Young in his decision in the case of *Karl Construction Ltd* v. *Palisade Properties plc* [2002] CA 199/01, where at paragraph 26 he states that the decision of an arbitrator and by implication that of an adjudicator as well, is 'equivalent to a certificate and payment will be due'. Further, he states that 'that amount may include interest' and that by the contract being enforced in this way the breach of contract does not give rise to damages because the subcontractor will not have suffered any loss:

> 'That decree, whether of the arbiter or the court, is equivalent to a certificate, and payment will be due by the employer once it has been pronounced in the same way as it would have been had the architect issued a certificate. On this basis an award of damages will not be necessary; the contractor will be found entitled to payment of what is due under the contract, which will be precisely the same amount as should have been certified. That amount may include interest. It follows that, if the contract is enforced in this way, there can be no breach of contract causing loss to the contractor; the provisions of the contract can always be enforced, using the mechanism of arbitration.'

SBCSub/C, at clause 4.10.5, allows for payment of interest on interim payments and clause 4.12.4 on the final payment at the interest rate, which is stated in the schedule of definitions to be:

> 'a rate 5% per annum above the official dealing rate of the Bank of England current at the date that a payment due under this Subcontract becomes overdue.'

It should be noted that clause 4.10.5 refers to 'the amount or any part of it, *due* to the Subcontractor'. For reasons set out below it is doubtful whether the definition of interest rate is effective in ousting the statutory rate. It is in any event considered that for such an interest rate to be effective, the payment mechanism must be fully complied with by the contractor. In this respect the issue, within five days of the payment due date, of a proper notice of payment defining the sum due is prequisite to the ousting of the statutory provisions, since it is that notice which seeks to define the sum due.

The purpose of the provision of interest under the standard forms of subcontract is to introduce such a provision into contracts where the statutory provisions do not apply and/or to reduce by 'agreement' the rate of interest set under the Late Payment Act. Since, where there is uncertainty or ambiguity, the conditions of the subcontract will be interpreted against the contractor, then in such circumstances the statutory rate will take precedence over any other rate. The effect of this is discussed below.

The Late Payment Act gives a right to interest, for debts in commercial contracts, for qualifying businesses. Section 1 of the Act implies the provisions of the Act into the subcontract:

> '1. – (1) It is an implied term in a contract to which this Act applies that any qualifying debt created by the contract carries simple interest subject to and in accordance with this Part.
> (2) Interest carried under that implied term (in this Act referred to as 'statutory interest') shall be treated, for the purposes of any rule of law or enactment (other than this Act) relating to interest on debts, in the same way as interest carried under an express contract term.
> (3) This Part has effect subject to Part II (which in certain circumstances permits contract terms to oust or vary the right to statutory interest that would otherwise be conferred by virtue of the term implied by subsection (1)).'

Since the right is implied into the subcontract, any dispute under a construction subcontract with regard to interest may be referred to adjudication. The reference in section 1(3) to the right to oust or vary the statutory interest allows the SBCSub/C subcontract to incorporate a rate of interest other than the current statutory rate of 8% over base rate, as decided by the Secretary of State under section 6. For further discussion as to the effect of this provision see the section 'Late Payment of Commercial Debts (Interest) Act 1998' in Chapter 19. Section 3 defines qualifying debts and section 4 defines the period for which statutory interest runs as:

> '4. – (1) Statutory interest runs in relation to a qualifying debt in accordance with this section (unless section 5 applies).
> (2) Statutory interest starts to run on the day after the relevant day for the debt, at the rate prevailing under section 6 at the end of the relevant day.
> (3) Where the supplier and the purchaser agree a date for payment of the debt (that is, the day on which the debt is to be created by the contract), that is the relevant day unless the debt relates to an obligation to make an advance payment.
> A date so agreed may be a fixed one or may depend on the happening of an event or the failure of an event to happen.
> (4) Where the debt relates to an obligation to make an advance payment, the relevant day is the day on which the debt is treated by section 11 as having been created.
> (5) In any other case, the relevant day is the last day of the period of 30 days beginning with –

(a) the day on which the obligation of the supplier to which the debt relates is performed; or
(b) the day on which the purchaser has notice of the amount of the debt or (where that amount is unascertained) the sum which the supplier claims is the amount of the debt,

whichever is the later.

(6) Where the debt is created by virtue of an obligation to pay a sum due in respect of a period of hire of goods, subsection (5)(a) has effect as if it referred to the last day of that period.

(7) Statutory interest ceases to run when the interest would cease to run if it were carried under an express contract term.

(8) In this section 'advance payment' has the same meaning as in section 11.'

Under The Late Payment of Commercial Debts (Rate of Interest) (No 3) Order 2002, the rate of interest will be adjusted on a six-monthly basis under section 4 of this order, which states:

'Rate of statutory interest

4. The rate of interest for the purposes of the Late Payment of Commercial Debts (Interest) Act 1998 shall be 8 per cent per annum over the official dealing rate in force.

For the purposes of this Order the official dealing rate to be used is that in force on 30th June or 31st December in any year. This rate will apply as the official dealing rate for the following six-month period, namely 1st July to 31st December or 1st January to 30th June respectively.'

In addition, under The Late Payment of Commercial Debts Regulations 2002 for debts due after 7 August 2002, there will be a lump sum payment as well as the interest itself. Section 2 amends the 1998 Act, and in particular subsection (4) amends section 5 of the Act, as follows:

'Compensation arising out of late payment

5A. – (1) Once statutory interest begins to run in relation to a qualifying debt, the supplier shall be entitled to a fixed sum (in addition to the statutory interest on the debt).

(2) That sum shall be –

(a) for a debt less than £1,000, the sum of £40;
(b) for a debt of £1,000 or more, but less than £10,000, the sum of £70;
(c) for 303a debt of £10,000 or more, the sum of £100.

(3) The obligation to pay an additional fixed sum under this section in respect of a qualifying debt shall be treated as part of the term implied by section 1(1) in the contract creating the debt.'

The effect of this revised section is more far-reaching than may be at first thought, since it is to be applied to every debt. Thus now whenever a payment is made just one day late or is undervalued, there is an immediate lump-sum payment due as a penalty. This makes all the more important the clear definition of the payment mechanism, especially the final date for payment and the definition of payment. Payment has in the past been considered and three possible definitions have been identified:

1. the date the paying party posts its cheque;
2. the date the payee receives the cheque; or
3. the date the payee has cleared funds.

It will be for the parties to decide when they regard payment to have been made but it is most likely that the date of receipt of a cheque will be considered as payment having been made.

It should be noted that section 4(5) equates with paragraphs 5 and 6 of Part I of the Scheme, to the extent that there are alternative start dates: (1) the day on which the obligation is performed or (2) the day on which the purchaser has notice of the amount of the debt.

Where the contractor issues a notice of payment and does not issue a valid notice of intent to withhold payment, then there is little doubt that the provisions of clauses 4.10.5 and 4.12.4 of SBCSub/C will apply. Complications will arise where no notice of payment is made. It has been argued that the reference in clauses 4.10.5 and 4.12.4 to 'the amount or any part thereof, due to the Subcontractor' means that interest is not due to the subcontractor if work is not valued or is undervalued or if a valid notice of intent to withhold payment has been issued, whether spurious or not. It is suggested that 'due' means the sum properly due by way of the net of the true valuation of the work and any justified set-off.

In the case of *Henry Boot Construction Ltd* v. *Alstom Combined Cycles Ltd* (2005) 101 Con LR 52, the Court of Appeal had to consider the date that sums become due under a construction contract where interim payments were to be made against a certificate to be issued by the engineer under the ICE standard form. In that case this issue was specifically being considered for the purpose of deciding the limitation of a right to action under a contract under hand.

Lord Justice Dyson said, at paragraph 76:

> '76. In my judgment, once £x is overdue for payment in month 1, the cause of action is complete in respect of that overdue payment, whether the overdue payment results from a failure to pay £x when it has been certified, or from the engineer's failure to certify £x. The fact that, where it has not been certified, £x can be certified in later certificates does not affect the position. The right to payment of £x and interest when £x is overdue for payment accrues when Boot is first entitled to those payments. The right to have £x included in successive certificates probably gives rise to successive causes of action. This is because each successive interim certificate revalues the whole of the work carried out, and any failure in relation to one interim certificate will leave unaffected Boot's right to challenge the valuation of the whole of the work executed in later certificates. Likewise, the right to have included in a final certificate a sum which happens to be for £x for work for which £x was claimed at the interim certificate stage gives rise to a distinct cause of action. But the claim to interest on £x from the date when the arbitrator holds that £x should first have been included in an interim certificate, even if repeated in later applications for interest when it is capitalised and recalculated, does not become a different cause of action when it is so repeated. On the other hand, a claim to interest on £x from a later date, on the footing that £x should have been included in a later certificate as part of the monthly revaluation of the work, would, if upheld by the arbitrator, give rise to a different cause of action accruing at that later date.'

It must be noted that in this case there was an express term in that contract which allowed for interest:

'(7) In the event of
(a) failure by the Engineer to certify or the Employer to make payment in accordance with . . .'.

This provision can be compared with the relevant words of SBCSub/C, which refer to 'if the contractor fails properly to pay the amount'. However, in making such a comparison regard must be had to the fact that in a subcontract the contractor is normally both the certifier and the payer. By reason of the extract from Lord Justice Dyson's speech above, it is clear that the word overdue can include both failure to pay the sum certified and sums not certified and/or undervalued. On this basis it is suggested that the phrase 'not properly paid' must also include all sums which on a true valuation of the work should have been paid, including any sums deducted from the valuation by way of set-off and/or contractual withholding or otherwise.

Where the subcontract, by reason either of the date of formation of the subcontract or by reason of the size of the business, satisfies the provisions of section 4(5)(b) of the Late Payment Act, then the provisions of that Act will apply and interest will become due if payment is not made within 30 days of the claim from the subcontractor, unless the subcontract provides otherwise. Since the effect of the Late Payment Act is not dependent on certification by the contractor, it is suggested that statutory interest will become due on adjudicator's decisions amending valuations against subcontractor's applications, where contractual interest is made subject to non-payment of sums certified by the contractor.

Some adjudicators and commentators have suggested that section 111(4) of the HGCRA should be interpreted, by reason of the words:

'the decision shall be construed as requiring payment not later than:
(a) seven days from the date of the decision'

as meaning that in the case of referral of a valid notice of intention to withhold payment, even if totally spurious, to adjudication, then the debt only becomes due seven days after the decision and only then will a right to interest commence. It is suggested that even if this is the correct interpretation of section 111(4) of the HGCRA, it is overridden by the effect of section 4(5)(b) of the Late Payment Act.

The Late Payment Act allows for the express terms of a subcontract to provide an alternative substantial remedy instead of the payment of statutory interest on debts paid late. The contract terms relating to late payment of qualifying debts is set out in Part II of the Act. The allowable circumstances where interest may be ousted or varied are set out at section 8 as follows:

'8. – (1) Any contract terms are void to the extent that they purport to exclude the right to statutory interest in relation to the debt, unless there is a substantial contractual remedy for late payment of the debt.
(2) Where the parties agree a contractual remedy for late payment of the debt that is a substantial remedy, statutory interest is not carried by the debt (unless they agree otherwise).
(3) The parties may not agree to vary the right to statutory interest in relation to the debt unless either the right to statutory interest as varied or the overall remedy for late payment of the debt is a substantial remedy.

(4) Any contract terms are void to the extent that they purport to –

(a) confer a contractual right to interest that is not a substantial remedy for late payment of the debt, or

(b) vary the right to statutory interest so as to provide for a right to statutory interest that is not a substantial remedy for late payment of the debt,

unless the overall remedy for late payment of the debt is a substantial remedy.

(5) Subject to this section, the parties are free to agree contract terms which deal with the consequences of late payment of the debt.'

The definition of interest rate in SBCSub/C allows for interest at the rate of only 5% over the base rate and no lump sum payment as required by section 4 of the Late Payment of Commercial Debts Regulations 2002. The Late Payment Act requires at section 8 that any change from the statutory interest shall be a substantial remedy. It can at least be argued, in the absence of any authority to the contrary, that a substantial remedy must be such as to give to the creditor at least the same or better recompense for the outstanding debt as it would receive under the statutory interest provision. Section 9 of the Late Payment Act gives reasons for considering an alternative remedy as not substantial. This will be discussed fully in Chapter 19, Statutes. However, the overriding consideration would appear to be whether the agreed remedy is effective in deterring late payment. If it has not been effective, then by reason of the definition in section 9(1)(a) of the Late Payment Act, the remedy is not substantial.

Sub/MPF04, at clause 27.2, makes a statement of agreement that the provision of an interest rate of 5% over base rate, as stated at clause 27.1, is a substantial remedy for the purposes of section 9(1) of the Late Payment of Commercial Debts (Interest) Act 1998.

Since section 9(1) of the Late Payment Act makes a remedy that does not deter late payment not a substantial remedy by reference to section 8.4.(a), clause 27.2, it is suggested, will be operable if the rate of interest of 5% stated in clause 27.1 deters late payment. Pre-agreement will not change the effect the agreed rate has on the contractor's attitude to prompt payment.

There has been a considerable reluctance for commentators and others to consider that the contractual rates as set out in the standard forms such as SBCSub/C are not effective in ousting the statutory provisions. Only one authority is known which gives any guidance in this respect. In the case of *Allen Wilson Shopfitters* v. *Mr Anthony Buckingham* [2005] EWHC 1165 (TC), Judge Peter Coulson QC found that a contract had been concluded by way of the acceptance of a letter of intent and that the terms of the JCT 98 Standard Building Contract Without Quantities had been incorporated in it. Notwithstanding that the judge found that the JCT terms applied, in his conclusion he directed that the parties calculate interest pursuant to the Late Payment of Commercial Debts Act:

'46. . . . I conclude that the adjudicator had necessary jurisdiction to reach the decision he did. I therefore grant summary judgment to the claimant for the following:

. . .

(b) Interest in a sum which I direct the parties to agree, pursuant to the Late Payment of Commercial Debts (Interest) Act 1998;'

Where either the subcontract and/or the subcontractor do not qualify under the Late Payment Act, then the subcontractor may be able to make a claim either for loss and/or expense or in damages. The difference is that an entitlement to loss and/or expense arises out of the provisions of the subcontract and will arise when certain stated categories of event occur. An entitlement to loss and/or expense may be dependent on the giving of notice at the time it becomes apparent that such loss has occurred or is likely to occur. This can be distinguished from damages at law, where the loss either flows naturally from a breach of the subcontract or was in the contemplation of the parties at the time the subcontract was entered into. Thus the late payment of retention may lead to interest by way of loss and/or expense if the delayed repayment is due to the disturbance of the regular progress of the work.

A significant difference between most main contracts and most subcontracts is that in the main contract, the person valuing and certifying the value of the work will be an independent person, such as the employer's quantity surveyor or architect, while the paying party is the employer. This contrasts with the normal subcontract situation where the contractor both values the work and is liable for payment. In the main contract situation the employer will not be in breach of the contract where the work is undervalued by the professional, such as the contract administrator, appointed to that task under the contract. In the subcontract, the contractor will normally be responsible for both the true valuation of the work and the payment of the sum so valued. A significant undervaluation by the contractor will be a breach of the subcontract, as will unjustified withholdings even where there is a valid notice of intent to withhold.

Financing costs

It is suggested that the *Minter* case, referred to in the previous section 'Interest', distinguishes between interest on an unpaid debt and financing costs as loss and/or expense, under the subcontract, following a variation and/or an entitlement to loss and/or expense for any other reason. At the time of the Court of Appeal decision in *Minter*, the courts were reluctant to grant interest on unpaid debts. However, such reluctance has been removed or reduced following the introduction of statutory interest under the Late Payment Act and the inclusion of interest provisions in many standard forms of contract.

Contractual interest under SBCSub/C only becomes due if the 'Contractor fails properly to pay the amount, or any part of it due to the Subcontractor under these Conditions by the final date for its payment'. However, interest under the Late Payment Act becomes due on the relevant day, which if not otherwise defined is as set out in section 4(5) of the Act:

> '4.(5) In any other case, the relevant day is the last day of the period of 30 days beginning with –
>
> (a) the day on which the obligation of the supplier to which the debt relates is performed; or
>
> (b) the day on which the purchaser has notice of the amount of the debt or (where that amount is unascertained) the sum which the supplier claims is the amount of the debt,
>
> whichever is the later.'

However, where the subcontractor has additional expenditure over and above the contract value of the works, it is likely to incur additional financing costs over and above those it has allowed for in its price. This was the substance of the *Minter* case. Lord Justice Stephenson identifies three distinct periods within the certification process, with regard to the compliance with a variation instruction, at page 14:

> '(a) between the loss and/or expense being incurred and the making of a written application for reimbursement of the same;
> (b) during the ascertainment of the amount of the same; and/or
> (c) between the time of such ascertainment and the issue of the certificate including the amount thereby ascertained.'

For most subcontracts the stages between the incurring of a cost and payment might more accurately be stated as follows:

1. from the incurring of cost to the application by the subcontractor;
2. from the application by the subcontractor to the certification by the contractor;
3. from the certification by the contractor to the final date for payment.

After this, interest either under the Act or under the subcontract will become due.

In the *Minter* case, the Court of Appeal unanimously decided that financing costs, as loss and/or expense, were due under (a) only, i.e. up to the date of application by the subcontractor. *Minter* can be distinguished from subcontracts such as SBCSub/C which, at clause 4.19.1, refers to 'has been or is likely to be affected' and requires an application to be made as soon as it became or should reasonably have become apparent, whereas in *Minter* the written application had to be made within 21 days of the date of the loss 'having been incurred', and there was no scope for pre-notification as in SBCSub/C. It was held that for a continuing loss such as financing costs, regular applications were required; such applications could account for periods (b) and (c), as defined by Lord Justice Stephenson.

The *Minter* case dealt with the payment of financing costs as loss and/or expense on additional work since the contract allowed for the payment of loss and/or expense. This was for 'authorised variations or the execution by the Contractor of work for which a provisional sum is included in the Contract Bills', where that loss and/or expense 'would not be reimbursed by payment in respect of a valuation made in accordance with the rules' or the conditions.

In the Court of Appeal case of *Rees & Kirby Ltd* v. *Swansea City Council* (1985) 30 BLR 1, the court interpreted their decision in *Minter*. First, they distinguished financing costs from interest on late payment of a debt as being the direct loss and/or expense consequent on the need to pay interest on borrowed money or to lose interest paid on capital invested, which has been caused by a matter for which the subcontract makes direct loss and/or expense recoverable. Lord Justice Robert Goff said, at page 11:

> 'It was also common ground that since, in the construction industry, cash flow is vital to contractors and delay in payment results in the ordinary course of things in the contractor being short of working capital, the loss of interest which he has to pay on capital which

he has borrowed and the loss of interest on capital which he is not free to invest does fall within the first rule in *Hadley* v. *Baxendale*, and so is direct loss and/or expense. Furthermore, it made no difference that the law has declined to award, by way of damages, interest on debts which have been paid late.'

And on the next page:

'As Lord Justice Ackner put it (at p.23 of the report), the contractors were seeking to claim not interest on a debt, but a debt which had, as one of its constituent parts, interest charges which had been incurred.'

Secondly, Lord Justice Goff, on the basis of the *Minter* case, found the contractor was entitled to financing costs from the time of the notification that loss and/or expense was being incurred in that manner to the time such loss was valued by the architect. Lord Justice Goff said, at page 23:

'In my judgment, written applications containing a sufficient reference to such loss or expense were made not only by the delivery of Messrs Ivor Russell's [the quantity surveyor instructed by Swansea City Council's] report, with its reference to the fact that the respondents had been out of funds, and that they had "a just contractual entitlement to settlement of the value we have assessed, plus interest", but also by letters claiming interest, sent at periodical intervals, on 5 December 1978, 13 March 1979 and 17 July 1979, and by their signature on the draft final account on 10 August 1979 reserving their right to interest. It follows, in my judgment, that the respondents have established a right to recover financing charges under the clauses in respect of the period from 11 February 1977 to 10 August 1979, on the principle in *Minter*'s case.'

On the face of it, it may be thought strange that where the subcontractor takes its time to quantify its claim for loss and/or expense it is entitled to financing costs for that period, provided it has identified that costs are being incurred as part of that loss and/or expense, while should the contractor refuse to value such loss or values it at a value less than that subsequently found to be due, it may be denied such entitlement. The issue here is that once the subcontractor makes its application for payment, the financing ceases to be an expense flowing directly from the relevant event under the contract. It will be for the subcontractor to establish a different head of claim, such as damages for undervaluation, in breach of the subcontract. In *Rees & Kirby* Lord Justice Goff decided that no such breach had been established.

In the case of *Amec Process & Energy Limited* v. *Stork Engineers & Contractors BV* [2002] No. 1997 ORE 659, Judge Thornton QC had to consider Amec's entitlement to financing charges on their undervalued work. Amec were in fact financed by their holding company, to which they paid interest on all money borrowed from them for financing the subsidiary business. The holding company compounded interest on a monthly basis. The judge found that:

'11. Amec financed the work from its own resources and from the use of inter-company loans and financing facilities provided by its holding company, Amec plc. These facilities were provided by the holding company at a commercial rate of interest compounded at monthly rests.'

The judge also found that although the contract contained an effective payment mechanism, Stork had never carried out the valuations as required by those provisions. The judge stated:

> '13. From the outset, therefore, Stork did not operate the detailed variation provisions of the contract in the manner provided for. This was because the number and extent of the variations that occurred . . . were so many in number and because of the time pressures to which Stork was subject.'

Judge Thornton QC considered the judgments in *Minter* v. *WHTSO* and *Rees & Kirby* v. *Swansea City Council*, which he summarised at paragraph 20:

> 'The relevant principles that they enshrine may be summarised as follows:
>
> 1. Where additional direct costs have been incurred during the course of the work which are recoverable as contractual entitlements under the contract, it is a matter for the parties and the terms of their contract whether the additional cost to the contractor of financing those costs between the dates that the costs were incurred and paid or for any part of that period [is recoverable]. Such a recovery, even for compound interest, is not intrinsically irrecoverable as being usurious or contrary to public policy.
> 2. The expression "direct loss and expense" confers an entitlement to ascertainable loss caused by financing the expenditure for such parts of the period between the loss and expense arising and payment that are not directly excluded from an entitlement to payment by the terms of the contract. However, if financing costs were not directly incurred as a result of the instructions or other causes which gave rise to an obligation on the employer to pay but were incurred, as a result of some other cause, these costs would not be recoverable as direct loss and expense.'

Thus an entitlement to financing costs is a contractual entitlement not by way of interest on a bad debt but by way of a foreseeable loss allowed for within the provisions of the subcontract.

Judge Thornton QC summarised the matters to be shown by the subcontractor, which in that case were:

> '23. In this case, Amec's claim must show that:
>
> 1. It incurred the financing charges in question.
> 2. The basis of the relevant contractual evaluation is authorised by the terms of the contract.
> 3. Compound interest is recoverable
> 4. The relevant period is one for which it may recover interest under the terms of the contract.'

The judge found that Amec were in this case entitled to financing charges for funding the additional work from the date the variations were carried out until payment. The reason for this decision can be seen at paragraph 25 where the judge found that because the contractor failed to define the basis of payment for work during the duration of that work, and since the contractor failed in this respect, the contract provided for payment on a reimbursable cost basis. Part of that cost was the financing costs.

It is apparent that only in exceptional circumstances will *Minter* provide for financing costs between the date of application by the subcontractor and valuation by the contractor or dispute resolution tribunal. Another approach for the recovery of such costs may be by way of statutory interest, since for work not pre-priced, statutory interest on the relevant day for the purposes of the Late Payment Act, section 4(5)(b) is:

> 'the day on which the purchaser has notice of the amount of the debt or (where that amount is unascertained) the sum which the supplier claims is the amount of the debt.'

On this basis statutory interest will commence when the relevant period after the date on which either the work was completed or the contractor had notice of the sum sought. Thus where an effective provision for contractual interest exists for late payment, it is suggested that statutory interest may apply for the period up to the final date for payment of a late certification by the contractor.

In most subcontracts, the contractor has either an express or implied duty to value the work in accordance with the payment mechanism. If and when the contractor fails in this obligation, either in whole or in part, then the subcontractor will suffer loss by way of financing charges. Where such loss is recoverable under the subcontract, by way of loss and/or expense or otherwise, such recovery will be a contractual right. Where the subcontract makes no express provision for the recovery of such loss, then the subcontractor is likely to have a right to recover such loss by way of damages. To the extent that the subcontractor can demonstrate either the rate of interest actually being paid by it for its bank overdraft or the rate of interest it is receiving for money on deposit, it will demonstrate its actual rate of loss. It will be arguable that where there is an established rate of interest for late payment either under the contract or by statute, then that is in fact an agreed pre-estimate of the subcontractor's loss, which takes into account not only the actual cost of borrowing but any restriction by its bank on further borrowing and therefore the subcontractor's ability to conduct and develop its business.

Payment of an adjudicator's decision

Where a dispute over the sums to be paid to the subcontractor is referred to an adjudicator, he or she will normally be asked to decide when the sums he or she decides are due are to be paid. As with other matters, the adjudicator will only be empowered to make such a decision if he has jurisdiction, either by way of the reference and/or the provisions of the subcontract.

Where the adjudication is under the Scheme, such jurisdiction arises from paragraph 20 of Part I, as follows:

> 'Adjudicator's decision
>
> 20. The adjudicator shall decide the matters in dispute. He may take into account any other matters which the parties to the dispute agree should be within the scope of the adjudication or which are matters under the contract which he considers are necessarily connected with the dispute. In particular, he may –
>
> (a) open up, revise and review any decision taken or any certificate given by any person referred to in the contract unless the contract states that the decision or certificate is final and conclusive,

(b) decide that any of the parties to the dispute is liable to make a payment under the contract (whether in sterling or some other currency) and, subject to section 111(4) of the Act, when that payment is due and the final date for payment,
(c) having regard to any term of the contract relating to the payment of interest decide the circumstances in which, and the rates at which, and the periods for which simple or compound rates of interest shall be paid.'

The standard form of subcontract SBCSub/C incorporates this provision by reason of clause 8.2, which incorporates the provisions of the Scheme for adjudication and thereby incorporates paragraph 20 above.

Judge Lloyd QC considered the effect of these provisions in the case of *David McLean Housing Contractors Ltd* v. *Swansea Housing Association Ltd* [2002] BLR 125. He stated, at paragraph 15:

> 'I adhere to the view that I previously expressed. The Scheme (and, so as far as I am aware, other standard forms of contract) does not confer on an adjudicator a right to adapt, vary or otherwise change a contract. Under the statutory Scheme an adjudicator has to decide a dispute under the contract (and in other schemes, disputes arising out of or in connection with the contract). It is a decision about the rights and liabilities of the contract which are questioned. Thus paragraph 20 of the Scheme expressly provides for the review of a certificate that has been issued (sub-para (a)) and for the adjudicator to decide a person "is liable to make payment *under the contract* . . . and, subject to section 111(4) of the Act, when that payment is due and the final date for payment". His decision does not create or modify a right or liability except, perhaps, in one respect.
>
> 1. I agree with Mr Harding [Counsel for David McLean] that since the Scheme (see paras 20(b) and 21) provides for the time for compliance with an adjudicator's decision to be set, it or the adjudicator's decision may alter the time within which, for example, a payment might otherwise have had to have been made, where an adjudicator decided that there had been an under-payment or under-certification. The purpose of this is clear. If an adjudicator were merely to decide that a different certificate should be issued or a different payment should be made the paying party could properly take the view that it would have the contractual period in which to honour the decision. Hence the statutory provisions make it clear that it has not to have that time. Indeed it may have had it already, and more, and that therefore a shorter period of time may be appropriate. Thus the Scheme permits the time within payment is to be made to be altered. Indeed if the decision does not set a time compliance is immediate which in my view shows that the decision does not affect or create a new cause of action. The Scheme is an implied term of the contract. As part of the contractual scheme it therefore modifies the ordinary contractual relationship.'

It appears from this judgment that at least where the matter referred to adjudication is the review of a certificate, then where the Scheme applies the adjudicator may decide a date by which payment must be made, but where the Scheme does not apply, he or she may not have jurisdiction to so decide and payment will be immediate, unless the contractual final date for payment has not yet been reached. This judgment makes it clear that the final date for payment relates to the date the money would have been required to be paid had the valuation been properly made. While this judgment does not expressly relate to monies withheld, it is suggested that a similar principle must apply.

It is suggested that the use of the phrase 'when that payment is due and the final date for payment' in paragraph 20(b) of the Scheme is misleading. Both 'payment due date' and 'final date for payment' are dates determined by the contractual payment mechanism. It is not believed that the adjudicator, by setting dates for payment of his decision, has or may change the date by which the subcontract required that money to have been paid. The adjudicator's decision under paragraph 20(b) is to determine the date by which payment, of any underpayment, is now to be made.

Section 111(4) uses different phraseology and states that 'the decision shall be constructed as *requiring payment* not later than'. The phrase 'requiring payment', it is suggested, can manifestly be distinguished from 'the final date for payment'.

What, it may be asked, is the significance of this distinction? Failure by the contractor to pay the sum due under the contract gives rise to a number of rights to the subcontractor:

1. a right to recover the sum due as a debt;
2. a right to interest or financing costs on the unpaid amount;
3. the right to suspend performance, after notice;
4. may deny the paying party the right to discount for 'cash' or prompt payment.

The effect of the issue of a notice of payment, satisfying section 110(2), or a valid notice of intent to withhold payment, satisfying section 111(2), is frequently considered to define and/or reduce the sum due. Therefore it is argued that there can at that time be no right to recover the underpayment as a debt, since no debt has been established. Similarly, provided that any balance between the sum stated on the notice of payment and the notice of intention to withhold payment is paid, there will be no right to suspend performance. It is suggested that this interpretation is not correct. The sum properly due is the sum properly due, whatever the contractor has initially valued the work at. While there may be an obligation for the contractor only to pay the net sum identified in the notice of payment and the notice of withholding, should that sum be less than the sum finally decided as due, the subcontractor will not have been paid the whole sum due on time and will, in fact, have been paid the balance late.

Both the issue of the notice of payment and the issue of a notice of intent to withhold payment may create a dispute which after being referred to adjudication may lead to the sum due at the time of the notice being amended, and the creation of an obligation to pay a further sum. It is considered that this should be distinguished from any suggestion that the adjudicator's decision creates an additional sum due at a later date. Whether that sum should be further adjusted for interest, discount, retention, LADs or other set-offs, VAT, etc., will be a matter of the contract or legislation. Unless the adjudicator has been asked to decide such issues under the reference, a successful referring party may find that it is still not entitled to its money. This was the result in *David McLean Housing Contractors Ltd* v. *Swansea Housing Association Ltd* [2002] BLR 125, where it was found that the respondent was entitled to deduct LADs from the sum decided by the adjudicator. In that case Judge Lloyd QC said, at paragraph 20:

> 'The contract, as I said, was amended. The defendant employer had issued a valid notice under clause 24.1. Clause 24.2.1 reads as follows:
>
> "Provided the Employer has issued a notice under clause 24.1 and provided, before the date when the Final Account and Final Statement . . . becomes conclusive as to the

balance due between the parties by agreement or by the operation of clause 30.5.5 or clause 30.5.8, the Employer has informed the Contractor in writing that he may require the payment of, or may hold withhold or deduct, liquidated and ascertained damages then the Employer may not later than five days before the final date for payment of the debt due under clause 30.6:

Either:

1. require in writing the Contractor to pay to the Employer liquidated and ascertained damages at the rate stated in Appendix 1 (or at such lesser rate as may be specified in writing by the Employer) for the period between the Completion Date and the date of Practical Completion and the Employer may recover the same as the debt;

or;

2. give a notice pursuant to clause 30.3.4 or clause 30.6.2 to the Contractor that he will deduct from monies due to the Contractor liquidated and ascertained damages at the rate stated in Appendix 1 (or at such lesser rate as may be specified in the notice) for the period between the Completion Date and the date of Practical Completion." '

The judgment in *David McLean* concerned a contract not subject to the HGCRA. It is suggested that it is for the paying party to plead, as its defence, any set-off against the subcontractor's claim as part of the adjudication proceedings. It will be for the adjudicator to decide the effect of such a defence, in his decision. Even if a set-off were allowable against an adjudicator's decision, it would be, subject to the proper issue of a withholding notice, the specified period prior to the final date for payment. In the absence of any provision to the contrary, the adjudicator's decision is to be paid immediately, in which case no time is available for the issue of an effective notice of withholding.

In the Scottish case of *A* v. *B* [2002] Scot CS 325, which concerned a dispute between a contractor and its roofing and cladding subcontractor, Lord Drummond Young had to consider two principal issues: first, could enforcement procedures be postponed by an express provision within the subcontract until a later date, such as the completion of the main contract works, and secondly, could the contractor deduct LADs from the sum awarded by the adjudicator in his decision? Lord Drummond Young found with regard to the first issue that there were two matters that might arise from an adjudicator's decision that might need referring to the court: first, the enforcement of the decision of the adjudicator and secondly, the judicial review as to the adjudicator's behaviour. Although he decided that the wording of the subcontract did not in fact restrict enforcement, in any event the wording of section 108 of the HGCRA would not deny enforcement. On the second issue, he held that while it had been open for the contractor to have pleaded that should the subcontractor's claim for an extension of time fail, the contractor was entitled to deduct LADs for the period of delay, the contractor had failed to make such a plea and could not now make such a deduction.

Lord Drummond Young deals with the enforcement issue as follows:

'7. In my opinion the pursuers are already entitled to enforce the adjudicator's decision of 14 May 2002; they are not obliged to wait until the actual completion date of the last phase of the main contract before doing so. That conclusion is based both on section 108(3) of the 1996 Act and on the construction of the parties' contract, read in the light of that section. The following principles are relevant.

8. First, a distinction must be drawn between two types of court proceedings that may arise out of an adjudication. The first of these is judicial review of the adjudicator's

decision. . . . The second . . . is an action to enforce an adjudicator's award. The nature of the latter type of action is described by Lord Macfadyen in *The Construction Centre Group Limited* v. *The Highland Council*, 2002 SLT 1274, in the following terms (at paragraph [9]);

"It is, in my view, important to appreciate the nature of this action. In it, the pursuers do not ask the courts to endorse the correctness of the adjudicator's decision on the merits of the dispute referred to him. Rather the pursuers merely ask the court to recognise that the parties have bound themselves contractually to implement the adjudicator's decision. The pursuers seek a decree from the court, not because they are in the right of the dispute, but because they are contractually entitled to require the defenders to implement the adjudicator's provisional determination of the dispute, whether it be right or wrong."

9. Secondly, when pronounced, an adjudicator's decision is legally binding on the parties. Such an award is no doubt provisional, in the sense that its effect can be undone by subsequent litigation or arbitration; until then, however, it creates binding legal rights and obligations, which must be given effect. That result is clear in my opinion from the terms of section 108(3) of the 1996 Act . . .

10. Thirdly, if an obligation arising out of an adjudicator's decision is binding, it must be capable of enforcement. The Latin maxim *ubi jus ibi remedium* reflects obvious common sense. In the words of *Erskine* (Inst. IV.i.l) [Scottish case]:

"It would import us little, that rights belonged to us, or that persons stood obliged to us, if there were no method by which we might make those rights effectual, and attain the enjoyment of our property, or compel those who stand bound to us to perform their obligations. If we were left at liberty to do ourselves justice by our own authority, on occasion of every difference with a neighbour, there would soon be an end of government. The judge or magistrate therefore must be applied to, by a proper action".

Consequently the party in whose favour an adjudicator's award is made must be entitled to enforce that award. That is in my opinion an inevitable consequence of section 108(3) of the Act . . .

I am accordingly of opinion that any provision of a contract that is incompatible with the binding nature of an adjudicator's award is legally ineffective.'

He dealt with the LADs issue in these words:

'16. The decision of an adjudicator is in my opinion in exactly the same position as a decree arbitral. Adjudication is based on a term of the parties' contract, in the form of either an express term or the terms of the Scheme for Construction Contracts, as applied by section 108(5) of the 1996 Act. The functions of an adjudicator are essentially similar to those of an arbiter. The major difference is that an adjudicator's decision is only provisional, and may be undone by subsequent arbitration or litigation. Nevertheless, in terms of section 108(3) the parties' contract must provide that decisions of the adjudicator are binding until the final determination of the dispute. In my view that gives such decisions the same force as the awards of an arbiter for the purpose of the application of the principle of retention. The result is that a plea of retention cannot be taken against the decision of an adjudicator once the decision is pronounced. The position is different, however, while the adjudication process is still continuing. At that point, the party against whom an award is sought may take any pleas or defences that are relevant to the claim made against him. Those include a plea of retention, if the conditions for the application of the principle exist. On this matter, I respectfully agree with the views expressed by Lord Macfadyen in *The Construction Centre Group Limited* v. *The Highland Council*:

"it seems to me to be axiomatic that the adjudicator must entertain any relevant defence on which the responding party wishes to rely in arguing that the sum is not due. In particular, it seems to me . . . to be clear that an employer who claims to be entitled to liquidate damages and seeks to retain a sum that would otherwise be due to the contractor against that claim, is in principle entitled to put that contention forward before the adjudicator. In my view, an adjudicator who held otherwise, and declined to permit the responding party to raise the issue of retention would be misdirecting himself."

In that case the pursuers had referred a claim to arbitration [adjudication], and the adjudicator made an award in their favour. The defenders did not assert any right of retention as a defence to the claim in the adjudication, but asserted such a right, based on a liquidate damages clause, when the pursuers sought to enforce the adjudicator's award. Lord Macfadyen held that the question of retention could have been raised in the course of the adjudication, but that the defenders had chosen not to do so. Had they done so, they would have been entitled to insist that the adjudicator entertain it. He continued:

"The fact that the defenders chose not to advance their retention argument before the adjudicator does not, in my view, entitle them to rely on it now for the purpose of depriving the adjudicator's award of the enforceability which the Act and the parties' contract conferred upon it . . ."

17. In my opinion that statement of the law is directly applicable to the present case. The defenders could, had they wished, have asserted that they were entitled to liquidate and ascertained damages if the adjudicator failed to award the pursuers the full extension of time that they claimed in the adjudication. They did not do so, however. They contested the extension of time claimed by the pursuers, and clearly enjoyed some success in reducing the extension claimed. Nevertheless, disputing entitlement to an extension of time is not the same thing as making a claim for liquidate and ascertained damages; a claim to such damages raises distinct issues, including the question of whether the sum of liquidate damages is a penalty.'

One difficulty that arises from this judgment is the concept that the responding party can plead a matter that is not the subject of the referral notice. Few adjudicators are willing to consider a counterclaim that has not previously been raised and is not the subject of a valid notice of withholding. However, by reference to the case of *The Construction Centre Group Limited* v. *The Highland Council* [2002] CA127/02 referred to above, two factors are of relevance. First, the set-off claimed is for LADs, which are a specific category of withholding, being pre-ascertained and flowing directly from a delay by the contractor in completing the Works. Secondly, in the case of *The Construction Centre*, the engineer's valuation had found that no monies were due from the employer to the contractor. As a consequence the judge, Lord Macfadyen, found that the employer had had neither the opportunity nor the obligation to issue a section 111 notice of withholding:

'20. I do not consider that in the circumstances of this case section 111 affects the matter. On the one hand, I am of the opinion that it is correct that the defenders had neither opportunity nor obligation to give a section 111 notice in advance of the adjudication. That is because in the absence of an engineer's certificate there was at that stage no "sum due under the contract" in respect of which a "due date for payment" might pass. The context in which the section provides for the giving of a notice of intention to withhold payment thus did not arise.'

The Construction Centre Group case relies on two circumstances which will rarely apply in subcontracts. First, the valuation was made by a third party and therefore the sum due at the time was that decided by that third party, whether that sum is right or wrong. Secondly, the monies which were sought to be withheld were pre-ascertained damages for delay in completing the Works, which are rarely a provision within subcontract conditions, where any reference to LADs is by way of reference to the LADs incorporated within the main contract conditions. In the subcontract situation the sum due will not be that certified by the contractor, even where a notice of payment as required under section 110(2) has been given. It is therefore considered that the circumstances that arose in both *A* v. *B* and *The Construction Group* v. *The Highland Council* cases will not arise in domestic subcontracts but may in non-domestic subcontract situations such as nominated subcontracts or Works contracts under management contracting forms of head contract, where the subcontractor's work is to be valued other than by the contractor.

Chapter 10
Completion

General principles

The intention of the parties in any subcontract is for the work to be complete by the end of the subcontract period. Completion means complete in every respect, to the standard and quality required by the subcontract. The provision of stage payment provisions makes the strict interpretation of completion more reasonable and acceptable than where payment of any kind was dependent on completion. The courts have, however, found that in such circumstances completion can be defined as 'substantially performed'. In *Hoenig* v. *Isaacs* it was stated that where the contract has been substantially performed, the employer must pay the contract sum and cross-claim for the defects. Lord Denning said, at page 180:

> 'On the true construction of the contract, entire performance was a condition precedent to payment. It was a lump sum contract, but that does not mean that entire performance was a condition to payment. When a contract provides for a specific sum to be paid on completion of specified work, the courts lean against a construction of the contract which would deprive the contractor of any payment at all simply because there are some defects or omissions. It is not every breach of that term which absolves the employer from his promise to pay the price, but only a breach which goes to the root of the contract, such as an abandonment of the work when it is only half done. Unless the breach went to the root of the matter, the employer cannot resist payment of the price. He must pay it and bring a cross-claim for the defects and omissions, or alternatively, set up a diminution of price.... It is, of course, always open to the parties by express words to make entire performance a condition precedent. A familiar instance is when the contract provides for progress payments to be made as the work proceeds, but for retention money to be held until completion. Then entire performance is usually a condition precedent to payment of the retention money, but not, of course, to the progress payments. The contractor is entitled to payment pro rata as the work proceeds, less a deduction for retention money. But he is not entitled to the retention money until the work is entirely finished, without defects or omissions. In the present case the contract provided for "net cash, as the work proceeds; and the balance on completion". If the balance could be regarded as retention money, then it might well be that the contractor ought to have done all work correctly, without defects or omissions, in order to be entitled to the balance. But I do not think the balance should be regarded as retention money. Retention money is usually only ten per cent, or fifteen per cent, whereas this balance was more than fifty per cent. I think this contract should be regarded as an ordinary lump sum contract. It was substantially performed. The contractor is entitled, therefore, to the contract price, less a deduction for defects.
>
> Even if the entire performance was a condition precedent, nevertheless the result would be the same, because I think the condition was waived. It is always open to a party to

waive a condition which is inserted for his benefit. What amounts to a waiver depends on the circumstances. If this was an entire contract, then, when the plaintiff tendered the work to the defendant as being a fulfilment of the contract, the defendant could have refused to accept it until the defects were made good, in which case he would not have been liable for the balance of the price until they were made good. But he did not refuse to accept the work. On the contrary, he entered into possession of the flat and used the furniture as his own, including the defective items. That was a clear waiver of the condition precedent.'

From the decision in *Hoenig* v. *Isaacs*, by taking possession of the works, in that case 'the flat and furniture', the purchaser waived the right not to pay on the grounds that the work is not entirely complete.

Most main contract forms will have provision for practical completion or partial possession where the employer takes possession of the works, even where a list of defects or snags is recorded as work still to be carried out and completed. The extent to which an employer can defer taking possession on the grounds that the defects identified are sufficiently serious that practical completion has not been reached has been considered by the courts but no effective definition has been provided.

NEC/Sub at clause 11.2(13) provides a useful definition of completion, which sets the level of defects that must have been rectified prior to completion as those that prevent the use of the building by either the contractor or employer.

'(13) Completion is when the *Subcontractor* has

- done all the work which the Subcontract Works Information states he is to do by the Subcontract Completion Date and
- corrected notified Defects which would have prevented the *Contractor* or the *Employer* from using the *subcontract works*.'

This definition however does not allow for the concept that completion of a subcontractor's work is not in practice a single event, as is generally envisaged. A building subcontract may comprise a number of activities for work of the same nature and completion of separate sections may be required progressively. Also, in some circumstances the contractor may require or prevent completion of a section or area of work until a later date. See the section 'Completion of the works' in Chapter 5 for further details on the concept. In particular, the concept of completion for the contractor to proceed with the work of following trades, and of completion of snags identified by the contract administrator prior to acceptance of the Works on behalf of the employer, are differing concepts of completion.

The effect of completion

The effect of achieving completion is first and foremost to put an end to any liability to damages for failing to complete by the due date or, if the works are late, further damages. However, in practical terms, either by express words of the subcontract or by implication, there will be the following additional consequences:

1. The contractor or employer becomes responsible for the works and as a consequence insurance by the subcontractor will cease.
2. Any stated period for defects liability will commence, including that for latent defects under the subcontract and liability at law.
3. Where retention has been withheld, normally half of those monies will be released.
4. Any further works will have to be carried out at the convenience of the contractor or employer and not at a time suitable to the subcontractor.
5. Any rights to instruct additional or varied work under the contract will normally die.

Completion of the subcontract works

For many subcontractors, completion of their works will be before, and in many cases well before, the completion of the main contract works. For others, completion of the main contract works will be concurrent, or virtually so, with the completion of the work by the subcontractor.

Standard forms of subcontract reflect the procedures of the main contract, with the concept of acceptance at the time the subcontract works as a whole are complete. Further, many subcontracts require the subcontractor to retain liability for its work, including damage to it by others, until either the completion of the Works as a whole or the whole of the subcontract works. There is, of course, no reason why the parties should not agree for the risks to remain with the subcontractor, even where it has no presence on site and/or no control over its completed work.

SBCSub/C sets out the arrangements for practical completion of the subcontract work and the subcontractor's liability for defects, at clause 2.20. Under this clause, the subcontractor is to notify the contractor in writing when it considers its works are practically complete and it is then for the contractor to dissent if it considers that practical completion has not been achieved. Practical completion, under SBCSub/C, includes sufficiently complying with the clauses relating to information for the health and safety file, which for subcontracts under SBCSub/D/C includes the provision of the as-built drawings. If the contractor is silent for 14 days after the receipt by it of the subcontractor's notice, then practical completion will be deemed to have taken place on the date notified by the subcontractor. Should the contractor dissent, it must set out its reasons for such dissent.

Clause 2.20.2 sets out the consequences if the contractor dissents, when practical completion will be the date notified by the contractor as such other date as is agreed between the parties or decided under the dispute resolution procedures. The fallback position is that practical completion of the subcontract works will be on the date of practical completion of the Works, as certified under the main contract.

ICSub/C has the same provisions at clauses 2.14.1 and 2.14.2. ShortSub, at clause 8.3, requires the contractor to notify the subcontractor in writing when the subcontract works are complete. There is no provision for an application by the subcontractor as in the other forms.

Sub/MPF04, at clauses 14.5 and 14.6, makes similar provisions to those in SBCSub/C, requiring the subcontractor to notify the contractor when it considers practical completion to have been achieved. Should the contractor not agree, then the contractor is to notify the subcontractor of the work that the contractor requires the subcontractor to carry out before the contractor will regard practical completion of the subcontract works to have been achieved. When the subcontractor considers that work to be complete, it is to notify the contractor and the contractor is to issue a statement recording the date of practical completion.

Clause 14.5 requires that in the event the contractor does not accept the subcontractor's claim that practical completion has been achieved, it notifies the subcontractor of the further work to be carried out. This requirement is entirely lost by reason of clause 14.6 if the contractor fails in its obligations under 14.5, since practical completion of the subcontract works becomes practical completion of the project. It is to be questioned if the reference to 'contractor' is not an error and should read 'subcontractor', which would make more sense.

In the absence of express terms under the subcontract, practical completion is likely to be on the taking of possession by the contractor/employer, by reason of the decision in *Hoenig* v. *Isaacs*. Lord Denning said, at page 181:

> 'Even if entire performance was a condition precedent, nevertheless the result would be the same, because I think the condition was waived. . . . the defendant could have refused to accept it until the defects were made good, in which case he would not have been liable for the balance of the price until they were made good. But he did not refuse to accept the work. On the contrary, he entered into possession of the flat and used the furniture as his own, including the defective items. That was a clear waiver of the condition precedent.'

The taking of possession of the subcontract works may be a difficult matter to decide and is likely to have differing meanings for different trades. A steel frame subcontractor will probably have a well-defined completion date, with a concurrent date of taking possession. For finishing trades, this will be much more difficult to determine and less certain. For other trades, especially where the provision of documentation, such as test certificates or guarantees, decides practical completion, then completion for the purposes of progressing the Works may and normally will be achieved at an earlier date.

For many trades in the subcontract situation, possession by the contractor may take place when the contractor instructs or allows the following trade to carry out work to the work of the subcontractor, such as the plastering of brick or block walls or the decoration of plasterwork. Once another covers up the subcontractor's work, serious problems can arise if it is found to be incomplete and/or defective. There will be an assumption, in the absence of evidence to the contrary, that at the time the contractor allowed the following trade to commence work, the contractor and/or the following subcontractor considered the work of the first subcontractor to be complete and satisfactory. In practice, many subcontract conditions make it an obligation of the following subcontractor to satisfy itself that the work of the preceding trade is satisfactory. The effect of such a condition must be limited to the second subcontractor having no grounds for considering the work of the first trade not to be suitable to receive its work.

In the case of *The Lord Mayor, Aldermen and Citizens of the City of Westminster* v. *J. Jarvis & Sons Ltd and Another* (1970) 7 BLR 64, the House of Lords had to consider when defective piling works, under a nominated subcontract, had been completed for the purposes of assessing an extension of time under the main contract. Two possibilities existed: either that the work was complete, although defective, when it was taken over by the contractor, or it was not complete until the rectification work had been carried out and completed some months later. Lord Wilberforce considered these alternatives, at page 78:

> 'That leaves two alternatives. The subcontractor contends that he is in delay whenever in any respect he fails to fulfil in time his contractual obligation. The employer contends that there is only delay if, by the subcontract date, the subcontractor fails to achieve such completion of his work that he cannot hand over to the contractor. Or, putting it negatively, that the subcontractor is not in delay so long as, by the subcontract date, he achieves such apparent completion that the contractor is able to take over, notwithstanding that the work so apparently completed may in reality be defective. This, on the employer's argument, may involve a breach of contract, but does not involve delay.'

Because of the consequences under these contractual arrangements there was much reluctance to decide the matter. In the end the decision was that the works were complete when accepted. However, if and when the subcontractor's work is subsequently found to be defective, such a discovery would establish a breach of the subcontract for which the subcontractor would be liable in damages. Lord Wilberforce put this in the following words, at page 79:

> 'It is not without its difficulties. Some of these were clearly demonstrated in the Court of Appeal: It is necessary to point to the fact that if the defects in the piles had been discovered before the subcontract completion date, and work had been at once put in hand to remedy them – thereby producing a similar period of delay in the completion of the main contract, the clause would, it seems, have applied, but it does not do so if the work was 'complete' (though defective) on that same date so that the contractor could take over. One must set against this the advantage that, if the subcontract work is apparently completed and handed over, and some defects appear very much later, but before the contract date, as they might in a large contract, this would not, on the employer's construction, be a case of delay, though it might be so on the subcontractor's. But even so the first type of difficulty is a very grave defect and a serious reflection on the clause: indeed, I cannot believe that the professional body, realising how defective the clause is, will allow it to remain in this present form. But in my opinion, though it is never agreeable to have to choose the lesser of two incongruities, we have to do so here and I find the employer's version qualifies for this not very flattering description.'

Lord Wilberforce's concern in this case was that because the subcontractor was nominated, liability for the delay to the contract would rest with either the employer or the subcontractor depending on the decision made by the court. In a normal domestic situation, the subcontractor would be liable in either case. The general importance of this decision is, where there is no question of defective work, is the work complete when accepted by the contractor? On the basis of this judgment the answer, in the absence of express conditions to the contrary, is yes.

Thus, where a subcontractor's work is defective and this is identified prior to practical completion of the subcontracted work, there will be an obligation for the subcontractor to rectify that defect which, if completed before the end of the subcontract period, will probably not result in a breach of the subcontract. However, if as a consequence of the defective work the subcontract completion is delayed, there will be a breach of the subcontract for the delayed completion, whereas if the defect is discovered after the completion of the subcontract work, the breach will be for the provision of defective work.

Completion of the subcontract works may include matters beyond the physical work on site and may also entail the provision of guarantees, or warranties, maintenance and operation manuals, test certificates, as-built drawings and other requirements for the health and safety file. For this reason it will normally be necessary to distinguish between practical completion under the subcontract and completion for the purposes of delay to the project and any associated loss and/or expense to the contractor. Further, where such matters are stated in the subcontract to be 'without further charge' or some similar phrase, then the contractor will not be entitled to make any deduction from the contract sum pending the provision of such documents.

The practical implications of linking the provision of documents for the health and safety file with physical completion of the work on site is that for many trades the provision of the information for the health and safety file, some time after the completion of the physical work, will not delay the Works as a whole or lead to the contractor suffering loss, whereas a delay in completion of the actual work on site will normally do so. Where the subcontractor is required to provide the information necessary for the health and safety file before practical completion, there will be an implied term that the subcontractor will have the time in which to complete its work on site in time to compile the necessary information. Thus the date when the last area of work must be released to the subcontractor by the contractor, if the contractor is not to delay completion, will be the sum of the time necessary to carry out the last element of work together with the time necessary to compile or complete the information necessary for the health and safety file.

Practical completion

The concept of practical completion is not one of law but of contract. In the absence of a practical completion clause completion will mean completion, which will generally mean acceptance by the contractor. However, where there is no express provision for practical completion, such a provision will, it is suggested, be incorporated by reference to a form of contract and/or main contract with such a provision or by custom and practice of the industry.

In the case of *Emson Easton Ltd (in receivership)* v. *E M E Developments Ltd* (1991) 55 BLR 114, Judge John Newey QC, considering the meaning of practical completion under the JCT main form of contract, found that completion and practical completion were the same thing for the purposes of the employer having a right under the contract to withhold monies and deduct costs, after the issuing by the architect of a certificate of practical completion, even when the contractor had become insolvent. The judge said, at page 122:

> 'In my opinion there is no room for "completion" as distinct from "practical completion". Because a building can seldom if ever be built precisely as required by drawings and specification, the contract realistically refers to "practical completion", and not "completion" but they mean the same.'

This judgment does not relate to the situation in SBCSub/C clause 2.20, which extends practical completion to incorporate activities in addition to the physical work on site. It is suggested that the effect of this judgment for subcontract works is that completion of the subcontract works will be achieved when all of the subcontracted work is available for the work of the next following trade.

Practical completion generally brings with it the same rights as completion, coupled with an obligation to carry out the remaining works and/or remedy defects, in the absence of an express provision, within a reasonable time. In this instance, what is reasonable will take into consideration both the seriousness with which the contractor and/or the employer regard the defects and the availability and/or the timing of access to the workplace for the carrying out of such further and/or remedial work.

The rights that can be incorporated by reference will be limited to those that do not conflict with the provisions within the subcontract from other sources, such as express provisions of the subcontract or those incorporated by statute. Of special significance may be the definition of the payment due date for the final payment. If the subcontract is silent, then paragraph 5 of Part II of the Scheme will make the final payment due on the expiry of –

> '(a) 30 days following completion of the work, or
> (b) the making of a claim by the payee
>
> Whichever is the later.'

Thus even if a concept of practical completion can be incorporated into the subcontract by reference, in the absence of any specific provision to the contrary, in a subcontract for work on a construction contract or for construction work the final payment will not be due until after completion, not practical completion, of the work. However it may be possible to rely on the decision in *Emson Easton Ltd* v. *E M E Developments Ltd* and consider they are the same thing. In the normal subcontract situation, and in conflict with the express terms of SBCSub/C, completion may in fact need to be considered as, in practice, having the following stages:

1. physical completion of each area so as to allow the progress of the next following trade;
2. completion of the subcontract work defined in the Sub-Contract Documents and as subsequently amended by instructions;
3. completion of pre-handover snags;
4. provision of all necessary documents – practical completion as defined by SBCSub/C;
5. completion of any latent defects at the end of the defects liability period.

As can be seen from the references to clause 2.20 of SBCSub/C above, practical completion can require more than just the physical or practical completion of the constructed work. Clause 2.20 specifically requires that the subcontractor shall have complied sufficiently with clause 3.20.4, and for SBCSub/D/C clause 2.24 as well, which are the clauses requiring the information for the health and safety file. Clause 3.20.4 of SBCSub/C requires that the subcontractor, within the reasonable time required by the contractor, is to provide the contractor with the information that is reasonably necessary for the contractor to comply with clause 3.25.3 of the main contract conditions (health and safety file).

Clause 2.24 of SBCSub/D/C requires that before practical completion of the subcontract works the subcontractor is to supply to the contractor, without further charge, the design documents and related information that is specified in the subcontract design documents or, as the contractor may reasonably require, to show or describe the Sub-Contractor's Designed Portion as built. This is expressly to be without effect to the subcontractor's obligations under clause 3.20.4 (health and safety file), concerning the maintenance and operation of that portion, including any installations forming part of it.

ICSub/C contains the same provisions as SBCSub/C clause 3.20.4, at clause 3.18.4, and ICSub/D/C, the same provisions as SBCSub/D/C clause 2.24, at clause 2.19.

Sub/MPF04 has no similar clause to that of SBCSub/C but achieves the same effect by reference to section 43 'Definitions and meanings', where practical completion is defined as:

> 'Practical Completion takes place when the Sub-Contract Works are complete for all practical purposes and, in particular:
>
> - the relevant Statutory Requirements have been complied with and any necessary consents or approvals obtained,
> - neither the existence nor the execution of any minor outstanding works would affect their use,
> - any stipulations identified by the Requirements as being essential for Practical Completion to take place have been satisfied and,
> - all as-built information and operating and maintenance information required by the Subcontract to be delivered at Practical Completion has been so delivered to the Contractor.'

These are manifestly obligations of the subcontractor and are obligations which, if not carried out in sufficient time, may in turn prevent the achievement of practical completion by the contractor of the main contract works. SBCSub/C makes no distinction between the completion of the subcontract work so as not to prevent the regular progress of the main contract works, and completion of the provision of information for the health and safety file. It will not generally be necessary to provide these documents prior to completion of the physical installation and/or construction work, or for the execution of the work of following trades. It follows therefore that there is no reason why failure to provide the documents required under the subcontract by the date of physical completion should leave the subcontractor with full responsibility for its work, when such work has in fact passed into the possession of the contractor.

It is suggested that it is in the best interests of both parties to record fully the dates on which all areas of the subcontract work are both made available to, and taken back from, the subcontractor. Regardless of the defined obligations set out in the subcontract, it is to be recommended that such records be made on a unit-by-unit or section-by-section basis, as appropriate to the specific project.

Clause 2.24 of SBCSub/D/C, quoted above, requires that before 'practical completion' the subcontractor is to supply such subcontractor's design documents and related information as may be specified, which show or describe the work as built. It is suggested that relating this requirement to practical completion of the subcontractor's work is unrealistic since by definition as-built drawings and information must inevitably require a period of time after practical completion of the works to finalise and submit. A much more realistic requirement would be to require the subcontractor to provide such drawings and information a stated period after practical completion of the subcontract works.

The provision of drawings and information required under clause 2.24 of SBCSub/D/C is stated to be 'without further charge to the Contractor'. This implies that the subcontract sum is for the execution of the physical work on site and that this provision is in addition to or outside the work for which the subcontractor is to receive a valuation and payment. Thus, subject only to any retention where the drawings and information under clause 2.24 are supplied, between the completion of the subcontract work and practical completion of the main contract there should be no reduction to the value of the subcontract work at the time of its completion to cover the provision of documents under clause 2.24.

Under SBCSub/C there would appear to be a distinction between 'Practical Completion' and completion of 'the Sub-Contract Works on site'. Clause 2.1 requires that 'The Sub-contractor shall carry out and complete the Sub-Contract Works in a proper and workmanlike manner, in compliance with the Sub-Contract Documents, the Health and Safety Plan and the Statutory Regulations and in conformity with directions given in accordance with clause 3.4 and all other reasonable requirements of the Contractor.' The subcontractor's contractual obligation as to programme is detailed at item 5 of the Sub-Contract Particulars in SBCSub/D/A; this makes reference to the period for carrying out the Sub-Contract Works on site but has no period allowed for the finalisation of the as-built drawings. Thus the articles of agreement make no provision for practical completion as described in the general conditions, which include the provision of health and safety documents under clauses 2.24 and 3.20.4 of SBCSub/C and SBCSub/D/C, respectively.

Sectional completion and partial possession

The subcontract may require parts of the subcontract works to be complete by express dates ahead of the main works. Like the concept of practical completion, sectional completion will arise out of the subcontract. Where the contractor requires sectional completion it must make it an express requirement within the subcontract, defining clearly the period for and/or the date for completion of each section.

It is considered unlikely that a requirement to complete the works on a sectional basis can be implied into the subcontract by reference alone, since, to be effective, the subcontractor must be aware of the requirements of the contractor with regard to the subcontract works and to have agreed to them.

Sectional completion can be distinguished from partial possession by the employer, under the main contract, of parts of the works for which no prior agreement exists other than for the giving of possession by a post-contract agreement with the main contractor. This is a right of the employer, not an obligation of the main contractor. Where the employer seeks agreement to partial possession, post-contract, it is suggested that the benefits and obligations of the main contract will be incorporated by reference, where appropriate, or may be implied into the subcontract.

In general, the achievement of sectional completion and/or partial possession will have the same effect as completion and/or practical completion, with regard to relief from liability and rights to payment etc. for the relevant part or section.

In practice, most subcontracted work is not required or considered or carried out as a stand-alone activity, but will be required to be carried out in sections, each section being made available to the subcontractor at differing times and to be taken back into the possession of the contractor progressively, prior to the end of the subcontract period. Further, for many finishing trades it will be necessary, for the integration of their work with that of others, to have several separate activities in each section of the works. Rarely do the provisions of a subcontract make any allowance or provision for these requirements.

A subcontract that only sets out a start on site date and a period for completion of the subcontract works, such as item 5 of the Sub-Contract Particulars in SBCSub/A, only has any effective meaning if all the subcontracted work can be, and is, made available to the subcontractor at the date for commencement on site by the subcontractor, and none of the work is required to be complete before the subcontract completion date. Even where the work is to be carried out in several stages, such as first and second fix and finals, and a programme showing the staggered start and finish for those stages is a subcontract document, the subcontractor's obligation is only defined as to be in stages and not to complete any section or area of the work by any definable date.

In the circumstances described above, without an express provision as to how the work is to be released to the subcontractor, it is suggested that there will be an implied term that the work will either be released progressively at a steady rate or that it will be released at a rate so as to be reasonable for the subcontractor to carry out its obligations under the subcontract. It is less certain that the subcontractor will have an implied reciprocal obligation to complete sections of the work on a progressive or regular basis, or in such a way that the contractor can reasonably complete the works as required by the main contract.

If, post-subcontract, the contractor seeks to require that the subcontractor complete the work in sections, ahead of the subcontract completion date, then in the light of the Court of Appeal judgment in *Lester Williams* v. *Roffey Brothers*, such a requirement is likely to be regarded as a variation to the subcontract for which the subcontractor may be entitled to an additional payment. In upholding the agreement that the contractor was to pay an additional sum for the subcontractor to complete on a stage-by-stage basis, the Court of Appeal found that the additional sum was valid consideration for a change to the requirements of the subcontract. Lord Justice Russell said, at page 91:

> 'True it was that the plaintiff did not undertake to do any work additional to that which he had originally undertaken to do but the terms upon which he was to carry out the work were varied and, in my judgment, that variation was supported by consideration which a pragmatic approach to the true relationship between the parties readily demonstrates.'

It is so much part of the normal expectations of the parties to a subcontract that the subcontractor's work may be used and worked on by following trades, that none of the standard forms of subcontract make any provision for the contractor to have beneficial use of the subcontractor's work prior to its completion. It follows that in the absence of any such express requirement, the contractor may only rely on some implied term for the subcontractor to cooperate or for the subcontractor to carry out its work reasonably in accordance with the progress of the main contract work. Despite there being no agreement for the subcontractor to progress at a certain rate or to make areas of work available to following trades at a specified time or period after commencement, contractors frequently seek to claim loss and/or expense on the basis that the subcontractor has delayed the regular progress of the works.

Where a subcontract makes adequate provision for the carrying out of the subcontract works in sections, there need to be procedures in place for the notification of the availability of and acceptance of each section, both at the start and the completion of the subcontractor's work.

Beneficial use or occupation

As mentioned above, it can be argued that the contractor accepts the subcontractor's work when he permits the following trade to fix to or cover up the subcontractor's work. For most trades this can be regarded as a passive use of the subcontractor's work. The plastering of a brick or block wall or the boxing in of service pipes cannot be regarded as use, in the active sense, of the subcontractor's work.

However, for many other trades the same is not true. The use of permanent lighting and/or heating systems is more than just a passive carrying out of the next operation, but a use of the permanent works ahead of acceptance. Since all such systems and/or surfaces start to wear and degrade with use, prolonged use may affect the warranty and/or appearance of the installation. Without express provision within the subcontract, the subcontractor will have no obligation to allow use of its system and/or work without the contractor first taking over the subcontract works.

In certain situations, especially on projects of long duration, there may be a need for a period of refurbishment before handover by the main contractor to the employer. This can involve replacement of air filters, re-roping of lifts and/or re-lamping of light fittings.

In the event that the contractor makes use of the subcontractor's work without express agreement and/or acceptance, there is likely, it is suggested, to be an implied term that at the time of beneficial use the subcontractor's work was considered by the contractor to be free of patent defects; or at least, it would be for the contractor to prove that any defects subsequently identified were either latent defects or manifestly defects that existed prior to the contractor taking beneficial use.

Even where the contractor has not sought to use the installed work but has covered it up or enclosed it by following work, further problems will arise if such work subsequently is found to be incomplete or defective; for example, where services are found, on testing, to require remedial work. Unless there is an obligation to have tested work by a certain stage or defined time or the contractor covers up the work in reliance on some statement or assurance by the subcontractor that its work is complete, tested and to specification, then the subcontractor will not be in breach of its obligation to complete its work by the date for completion of the subcontract works. It is suggested that a temporary non-conformity, not normally being a breach of the subcontract giving rise to damages, cannot leave the subcontractor liable for more than the costs of the rectification of its work. Costs to the contractor may include the requirements for other trades to take down and replace work or to carry out repair work directly resulting from the subcontractor's defect, such as might result from a water leak. By providing for formal sectional completions of the subcontractor's work on a stage-by-stage or area-by-area basis, the contractor will ensure that the subcontractor takes responsibility for latent, pre-completion defects.

Notification of completion

Many standard forms of subcontract have an express procedure for establishing the date of completion of the subcontract works (see reference to SBCSub/C clause 2.20, above). Many subcontracts make the date of completion of the main contract works practical completion of the subcontract works. The aim is not only to retain the retention until the release of retention under the main contract, but also to avoid having to accept the subcontract works before acceptance from the contractor by the contract administrator. Such a term produces a conflict within the subcontract provisions since there will on the one hand be a requirement for the subcontractor to complete by a certain date or within a certain period, and on the other hand, express denial of completion until completion of the main contract works.

The true interpretation of such conflict of terms will turn on the exact wording of the relevant clauses. However, a likely interpretation is that the Sub-Contract Works are to be substantially complete within the period or periods or on the date or dates stated for completion of the Sub-Contract Works, but that sundry items and rectification of minor works will not be complete and finally accepted until completion of the main contract works. For the reasons stated elsewhere in this chapter and in the section 'Completion of the works' in Chapter 5, this is the natural requirement of a contractor to have a two-stage completion and it is surprising that this need is not reflected in the standard forms of subcontract.

Many standard forms of subcontract, including SBCSub/C at clause 2.20, allow for notification by the subcontractor that its work has been completed, and if not denied by the contractor within a specified period, it is the date stated by the subcontractor that, for the provision of the subcontract, becomes the subcontract completion date. Where there is no express provision for either acceptance of the subcontract works and/or defining completion, it is likely that a procedure, such as that in SBCSub/C at clause 2.20, may be implied by reason of common trade custom and practice.

SBCSub/C clause 2.20 requires the subcontractor to notify the contractor when two conditions have been fulfilled:

1. The Sub-Contract Works are practically complete.
2. The subcontractor has complied sufficiently with clause 3.20.4 and in the case of SBCSub/D/C, clause 2.24 in addition.

It is suggested that, since in practice the finalisation of the information required under clauses 3.20.4 and 2.24 may not be achieved or indeed requested until some time after the subcontract works are physically complete and taken back into the contractor's possession, the subcontractor should make notification on the works being physically complete and again once clauses 3.20.4 and 2.24 are complied with.

Provided that notice is given, such as notification under clause 2.20 of SBCSub/C, it is likely that failure by the contractor to dissent within 14 days will, for all purposes of the subcontract, be evidence of the subcontract works being physically or practically complete, even for subcontracts where no such express term exists. It is probable that where the contractor takes possession on a section-by-section basis, such notices for each section would be similarly evidence of physical completion of those sections.

For many reasons, described elsewhere, the subcontract works may either have been delayed in their entirety or parts rescheduled to a later date, as either the design and/or detailed planning of the works is carried out. Irrespective of any express requirement in the subcontract, it will be prudent for the subcontractor, at the date set in the subcontract for completion of the subcontract works, to notify and record to the contractor the state of the subcontract works and the reasons, where applicable, for any non-completion.

Special cases

It should be noted that the normal definition of completion of a subcontractor's work does not apply equally readily to all specialist trades. A subcontract for the provision of temporary works may also involve, at the end of its requirement on site, the removal by the subcontractor of its work. It is suggested that completion of the subcontract works in such circumstances may be the completion of the installation, not its removal. For example, a scaffold to an existing building for the purpose of carrying out demolition, repairs or alteration works is complete when safely completed and available for use by the contractor. The period between erection and dismantling may depend on the contractor's progress and not be defined as an obligation for the subcontractor to remove it by a stated date, regardless of the contractor's progress. Alternatively, the subcontract could be arranged with two distinct requirements as to time, one period for the installation and a further discrete period for the dismantling and removal from site.

For many trades and activities, the carrying out of the work is entirely dependent for its progress on the work of others. Such a trade may be the provision of scaffolding for new construction work, which will be required to be installed at a rate of progress neither too far ahead nor too far behind that of the associated trades for which the scaffold is being provided. Many trades such as building services and joinery will have a phased installation, with each phase separated by work of other trades. In practical terms, each stage is effectively a separate task requiring formal acceptance and thereby creating a number of stage completions.

As mentioned earlier, the date for completion triggers certain other matters, such as release from an obligation for protection of and insurance for the work, and the release of retention. Such triggers may be totally inappropriate for trades where completion is achieved only when nothing remains on site. Such trades include demolition works or temporary works where the achievement of completion may be considered to be at the end of the dismantling process, rather than completion of erection or installation.

Failure to complete on time

Contractors will generally take a very simplistic view of a subcontractor's performance obligations as to time. Where the subcontract sets out a period for the subcontract works, a simplistic view is that the date for completion can simply be calculated by taking the date the subcontractor commenced work on site and adding to it the subcontract period. Another view would be to take the date on which the last area or element of the subcontract work was made available to the subcontractor and allow the contract period thereafter. It will frequently be the case that neither would give a just method or basis for calculating any loss to the contractor for the subcontractor's performance failure.

SBCSub/A at item 5 of the Sub-Contract Particulars refers to 'the period for the carrying out of the Sub-Contract Works on site after delivery and after the expiry of the period to commence work' and states a number of weeks. This has the effect of defining the completion date for the Sub-Contract Works by reference to the notice for commencement on site and not to the date the subcontractor actually starts on site.

Delivery relates to delivery including procurement and fabrication of materials before commencement on site. This leaves open the effect of the subcontractor not receiving all necessary drawings, details, approvals, etc. by the stated period prior to a commencement on site. It is suggested that such delivery period relates to the whole works. However, just as for work on site, as discussed above, it may be considered that the subcontract has an implied term making delivery for each section subject to the stated period of notice, in advance of the time that the subcontractor may reasonably be required to commence work on site, for each and every section and/or item of work.

The expression 'after the expiry of the period to commence work' is used in item 5.3 of the Sub-Contract Particulars, relating to sub-item 5.2 in this condition, which requires the contractor to give notice of a specified period prior to its requiring the subcontractor to commence work on site. It is suggested that a notice can only be an effective notice if made in a formal manner and will be limited to the work that at the expiry date of the notice is, in fact, available to the subcontractor. The formal notice may of course allow a progressive release of work, in which case it would effectively be a number of notices issued at the same time. In this respect SBCSub/A item 5.4 refers to 'the Sub-Contract Works/each Section'. While it is considered that this is intended to make allowance for the situation where the main contract works are being carried out on a sectional basis, it is also considered that this phrase may be construed as applying to the sectional release of the subcontractor's works themselves. As discussed elsewhere, for most subcontract trades, an obligation for the subcontractor to complete sections, stages or areas of its work in an identifiable period after notice may be the only really effective method of defining an obligation for a subcontractor's work.

For the above reasons, the period of notice under item 5.4 of the Sub-Contract Particulars can be regarded as the period of notice for each element of the subcontract works. This will not be the case where the subcontract incorporates a programme defining the period for the release of subsequent sections of the work in relation to the initial release date. The difficulty for the contractor is that the more release dates it establishes for the commencement of differing sections of the work, relative to the commencement of the subcontract works, the more it becomes contractually bound as to the timing and progress of the main contract work.

The requirement for notice by the contractor must be taken into account when considering the meaning and effect of clause 2.17, which has been discussed in the section 'Extensions of time' in Chapter 4.

SBCSub/C clause 2.21 requires the subcontractor, in the event of the subcontractor exceeding the period allowed under the subcontract for the completion of the subcontract works, if notified by the contractor within a reasonable time of the expiry of the period for the subcontract works, to pay or allow the contractor the amount of any direct loss and/or expense suffered or incurred by the contractor.

ICSub/C incorporates the same provisions at clause 2.15 and a similar provision is provided in Sub/MPF04 at clause 15.1.

It is suggested that before the contractor may issue a notice under clause 2.21, it must consider any revised period(s) under clause 2.18 and it will be necessary for the contractor first to have addressed any notice, particulars or estimate received by the contractor from the subcontractor under clause 2.17. Where the contractor has not revised the period(s) for the subcontract works, the reasonable time for issuing a notice of failure to complete under clause 2.21 will run from the expiry of the unextended subcontract period(s). In the event that the contractor fails to issue a notice within a reasonable time, which may be held to be a short period, then it may be held that the contractor has conceded that the failure to complete is not the subcontractor's fault or liability, and the contractor may in addition be considered to have waived any right to loss and/or expense, although such loss and/or expense may be held to run from the time the contractor issues such a notice.

After receipt of such notice, under clause 2.21, the subcontractor is to pay or allow to the contractor a sum equivalent to any direct loss and/or expense suffered or incurred by the contractor resulting from that failure. Such right will be subject to clause 4.21 and subject to prior agreement by the subcontractor.

SBCSub/C, at clause 7.4, makes allowance for the determination of the subcontract by the contractor in specific circumstances and by notice, before the date of practical completion of the subcontract works. No definition of practical completion under this clause is given, so it is likely that in making such a definition, regard must be had to clause 2.20, which requires the subcontractor to notify the contractor when in his opinion the Works are practically complete. For the purposes of this clause, it may be that complying with clause 3.20.4, the provision of information for the health and safety file, and in addition for subcontracts under SBCSub/D/C clause 2.24, the provision of drawings and information, in connection with performance specified work, will be considered separate from the Works being 'practically complete'.

It is suggested that there is a difference between 'practical completion' for the purposes of clause 7.4 and 'practically complete' for the purposes of clause 2.20. This interpretation is supported by the reference in clause 7.4.1.5 to failure by the subcontractor to comply with the requirements of clause 3.20 of the CDM Regulations and in particular under clause 3.20.4, the failure to provide information for the health and safety file.

The contention that under SBCSub/C there are in fact two completion situations is supported by the wording of item 5 of the Sub-Contract Particulars in SBCSub/A, which refers to the period 'for the carrying out of the Sub-Contract

Works on site'. If this interpretation is correct, i.e. that there are two stages of completion that the subcontractor is to achieve, then the articles of agreement only provide for one, namely the completion of the works on site, and there is no defined period or date for the achievement of practical completion if that includes the provision of the documents required under clauses 3.20.4 and 2.24.

Guarantees

It is not uncommon for the installers of specialist systems to be required to obtain, in the name of the employer, or to make provision for the employer to obtain, a guarantee or warranty as to the quality and durability of the installation. The outworking of these requirements is rarely fully considered pre-subcontract and, although not of direct concern to this work, is not fully considered by the arrangements in the main contract either.

The concept is simple enough. The manufacturer or supplier of the system in question, for example a roofing or glazing system, first vets or approves the specific construction details, subsequently monitors the installation and finally may carry out testing of the installed product. Once satisfied that the system has been installed satisfactorily, the manufacturer, in return for a small fee, provides its guarantee or warranty to the employer.

A number of difficulties and practicalities may arise in practice. A guarantee will only be issued when, and indeed would be of little benefit until, the specialist work has been completed. This will include any rectification work whether necessary for the function of the product or to improve on the appearance of the as installed work. Since the final inspection for the purpose of the provision of the guarantee can only be made when the whole project is complete, there will be a conflict if the provision of the guarantee is considered necessary for the achievement of practical completion. This is especially the case as many building contracts provide for the rectification of defects during a period, often twelve months, after practical completion.

It is frequently a requirement for the provision of the manufacturer's guarantee that the specialist subcontractor has been paid in full. In conflict with this requirement is likely to be the attitude of the contractor and the normal provisions of the subcontract. Even where there is no dispute as to the pricing of variations and other matters which adjust the subcontract sum, the contractor will consider that the subcontractor's obligations are not complete without the provision of the guarantee, and that an appropriate sum should be withheld pending the provision of the guarantee.

It can be argued that where, as is frequently the case in building subcontracts, there is a provision for the contractor to withhold retention on the subcontractor's work, this is effectively insurance for the rectification of any latent defects. If this is in fact the purpose of retention, then the subcontract should allow for the release of the retention in exchange for the provision of the guarantee or bond, since the purpose of the guarantee is to provide for the rectification of latent defects. As stated in the section 'Completion of the subcontract works' earlier in this chapter, the subcontract may require the provision of the guarantee to be 'at no cost to the contractor'. In such circumstances the contractor will have no defined price to withhold. Indeed, the consideration is frequently a nominal sum payable directly from the employer to the manufacturer under a collateral agreement, which may have as one of its requirements that the subcontractor has been paid fully for the work.

An arrangement, such as described above, would need to be reflected throughout the chain of contracts, otherwise one party (probably the employer) would have the benefit of both the manufacturer's guarantee and the retention through the main contract.

In many cases, the specialist system will be covered up by further work, either by the specialist subcontractor or by the work of others. It will be necessary to ensure that the work has satisfied all the requirements of the guarantee prior to covering up. In an ideal situation, this will require that the guarantee be provided before the carrying out of following work. Such an aim, because of the difficulties of releasing the guarantee before full payment is made, may be unachievable under the normal provisions of the standard forms of contract and subcontract.

Chapter 11
Breach of Contract

General principles

A breach of the subcontract can be said to have taken place when one party fails to carry out and complete its obligation or obligations to the other or where, by its act or failure to act, it prevents the other party fulfilling its obligations. The obligations may either be defined by express terms or conditions of the subcontract or may be implied either by statute or by law.

The consequence of the breach will depend on the seriousness of that breach and will, in most cases, lead to an adjustment of the obligations between the parties either by relieving or relaxing the obligations for performance and/or leading to compensation by way of damages. In extreme situations, there may be an entitlement for one party to consider that the other has repudiated the subcontract and for the former party to be entitled to regard its obligations to the other to be at an end.

In practice, the terms of the subcontract will make provision for the contractor to default in its obligations to the subcontractor and/or change its requirements under the subcontract, without being in breach of the subcontract. The subcontract terms must, at the same time, set out the means or method of compensating the subcontractor for that default or change. Even where there are no express provisions, normal trade custom and practice may well imply into the subcontract rights and remedies common to the industry.

The obligations of the subcontractor will be those set out in the terms of the subcontract and in the articles of agreement, including incorporated documents such as bills of quantities, specification, drawings, programme, etc. Generally, the subcontractor will have two obligations: first, as to the description and quality of the work, and secondly, as to the timing of the execution of that work.

The subcontractor's obligations for quality

The nature and quality of the work to be executed by the subcontractor will normally be set out in the contract drawings and specification and any variations issued during the course of the works, under the provisions of the subcontract or by agreement. In the absence of any specific requirements, it will normally be implied into the subcontract that the work is to be generally fit for its purpose and of good quality. The specific provisions of the subcontract may limit these general requirements. If the subcontractor is to provide and install materials selected or specified by others, then its obligation will be limited to providing goods or materials which are good of their type. The House of Lords, in the case of *Young &*

Marten Ltd v. *McManus Childs Ltd* (1968) 9 BLF
goods in a labour and materials subcontract, the
as to the suitability of the material or product
product supplied was to be of good quality. As I
ring to the decision of Mr Justice du Parcq in *M*
1 KB 46:

> 'That statement I have no doubt involves two wa
> frequently happens that for any relevant purpose
>
> Judged by this test the appellant was instructed
> inherently was of high grade quality and well fit
> warranty would apply unless the circumstances
> opinion, the fact that the appellant was instructed to supply in terms a Somerset No. 13 red tile is not sufficient to exclude such implied warranty, which must be judged as at the time of the contract. This is not a case analogous to those cases where the employer may have ordered the contractor to supply certain particular goods which are found to be inherently unsuitable for the purpose. Such cases may give rise to many difficult factual and legal problems. The truth is that the appellants through no fault of their own had the misfortune to supply a particular batch of tiles which failed to conform to the proper standard of quality and fitness. I can see no reason why they should be absolved from their implied obligations.'

In this case, had the Somerset No. 13 red tiles not been 'inherently of high quality', the contractor would, provided that the tiles as supplied were of the normal quality, not been liable for any defect, since it did not warrant the selected type of tiles. The subcontractor did, as found in this case, warrant that the tiles were good of their type.

In the case of *Rotherham Metropolitan Borough Council* v. *Frank Haslam Milan & Co Ltd and M. J. Gleeson (Northern) Ltd* (1996) 78 BLR 1, the Court of Appeal had to consider whether, in supplying steel slag as hardcore fill, the contractors had failed in their obligations to the employer. The court considered the matter from the point of view of fitness for purpose and from that of marketable quality. The court found for the contractor on both counts.

The specification, which had been prepared by the borough's engineer or architect, required the hardcore to be 'graded or crushed gravel, stone, rock fill, crushed concrete or slag or natural sand or a combination of any of these'. The steel slag used by the contractors, in fact and unknown to them, expanded when it was in wet conditions. This is what happened and the structural slab bearing on the fill was disrupted.

The speech of Lord Justice Ward, at page 38, makes it clear that where matters are specified by the employer, this negatives the idea that anything is left to the skill and judgment of the contractor:

> 'The reality in this case was that the architects and engineers assumed the responsibility for the whole design and specification of the work. They specified hardcore to include steel slag because the state of their actual knowledge was that slag was suitable for the purpose. The reality is that they were trusting their own knowledge. They did not need to rely on anyone else. In this case, in my judgment, the circumstances show in the words of Mr Justice Bayley in *Duncan* v. *Blundell* (1820) 3 Stark 6, 7:
>
> "Of course it is otherwise if the party employing him chose to supersede the workman's judgment by using his own."

[obscured]eks to escape that conclusion by reference to *Cammell Laird & Co Ltd* v. *Manganese* [obscured] *& Brass Co Ltd* [1934] AC 402 that "reliance partial but substantial and effective will [obscured]ng the implied condition into play", per Lord Wright, at page 428.

Adopting Lord McMillan's speech at page 419 I would pose the question in these terms: Did the employer by its detailed specification so tie the contractors' hands as to negative the idea that anything was left or intended to be left to their skill and judgment or was there an important margin within which the contractors were free and had scope to exercise their skill and judgment?

I was attracted to the submission that there was a field of choice for the contractors. There was, it was argued, a choice of the various forms of hardcore specified from gravel to sand, and a choice between blast furnace slag and steel slag, and, of course, a choice between weathered steel slag and unhydrated steel slag. That much is true. But in my judgment the freedom of choice that was afforded to the contractors was no more than a freedom to choose one (or more) of the specified forms of hardcore and the only skill and judgment (upon which the employer did not rely because it was of no direct concern to them) was in ensuring its availability at the best price. It was certainly not a choice dependent upon the skill and judgment of the contractor to find among the permitted alternatives that hardcore which possessed the quality of inertness.

If one reduces the issue in this case to one of colloquial simplicity, was Rotherham in effect saying to Haslam and Gleeson:

"This is what we want from this job, we don't mind which hardcore you choose but we rely upon you to choose one which is inert."

This may not be a textbook test but it provides a very emphatic answer –"No!" In my judgment it is impossible to impute such an intention to both parties and it is unreasonable to expect that reliance was being placed on the contractor's skill and judgment. A warranty of fitness for purpose cannot be implied and the learned judge was wrong to hold that it was.'

In this case, the contractor was to select material from a list of acceptable options. In many cases, the subcontractor will be required to use materials of a specific nature or 'equal and approved'. From the judgment in *Rotherham*, quoted above, it is suggested that in such a case where the contractor or the employer or its representative approves an alternative, it does so without reliance on the subcontractor. However, if the words 'and approved' are omitted and the subcontractor is permitted to use similar materials of its own volition, then should those alternative materials prove to be inferior to those specified and unfit for their purpose as a consequence, then the subcontractor will be liable to the contractor.

Having considered the contractor's duty as to fitness for purpose, in the *Rotherham* case Lord Justice Ward considered the nature of marketability. Quoting from *Hardwick Game Farm* v. *Suffolk Agricultural & Poultry Producers Association* [1969] 2 AC 31, he says on page 39:

'In *Hardwick Game Farm* Lord Reid spoke at page 75B of the warranty of merchantable quality being "of more general application" whereas the warranty for fitness has "a narrower scope"– page 79E.'

and on page 40:

'Lord Justice Roach has quoted from Lord Reid's speech in *Hardwick Game Farm* at page 77, I also draw attention to page 75C:

> "Merchantable can only mean commercially saleable. If the description is a familiar one it may be that in practice only one quality of goods answers that description – then that quality and only that quality is merchantable quality. Or it may be that various qualities of goods are commonly sold under that description – then it is not disputed that the lowest quality commonly so sold is what is meant by merchantable quality; it is commercially saleable under that description." '

Many specialist subcontractors take on a design development obligation. The subcontractor's obligation again will be to the extent set out in the subcontract provisions. Where the subcontractor is to design to a performance specification, then it is for the subcontractor to achieve and provide a design which satisfies that performance regardless of what is in fact needed or necessary. In such circumstances the adequacy or otherwise of the design leading to the specification of that performance level will not be to the subcontractor's liability.

The subcontractor's obligations by reason of the specification

The quality of the work a subcontractor is required to achieve will normally be that defined by the specification. The specification may be found in the descriptions within the schedule of work, in a pricing document or by reference to a separate document or documents. Such specification documents may be project specific, such as prepared by the consultant designer or the contractor, or industry standards such as the British Standards or ISO standards. Alternatively, or in addition, there may be a requirement to carry out the work in accordance with the manufacturer's requirements. Further, there may be a requirement, sometimes in addition to the above, for the work to be to the requirements of the architect, engineer or contractor and possibly all three.

While the aim in referring to a specification is to make sure that the employer's requirements are known to the subcontractor and that there is certainty as to what is to be provided by it, in practice a number of difficulties arise.

It is suggested that, in the absence of express words to the contrary, the use of a specification will not require the subcontractor to provide materials or incorporate components which have not been identified in the pricing schedule. The purpose of the specification will be to support, enhance and define the nature of the work to be carried out, but will not extend the scope of the works. It is suggested that this is still the case even when the subcontractor's offer is endorsed with words such as, 'our tender is in accordance with your specification'.

Many specifications used by design consultants are held within a computer database file, to be downloaded and edited for use on each individual project. As a consequence, such specifications may provide for materials or circumstances not required on the relevant project. More serious is where such a specification has not been edited to accommodate changes in building practice or to incorporate new materials or products, when these have been selected for the project.

A specification prepared by a design consultant may describe associated work which is an operation not normally the work of the trade to which the specification sections relate. Where a contractor seeks a price from a subcontractor, the subcontractor must make clear precisely what work is included in the subcontractor's offer. A typical area of uncertainty is a requirement for the cutting out and making good of holes and chases in walls and other structural and non-structural elements, to receive the work by the specialist subcontractor. This problem is more serious and requires thorough consideration where the subcontract enquiry is not issued by the contractor but by the employer's agent, with the intention that the successful tenderer will be named, nominated or novated to the contractor at a later stage. The inherent risk is that either work is priced by both contractor and subcontractor, or more likely by neither.

When general terms are used in a specification, it is suggested that any material satisfying that general term will be compliant. For example, the use of phrases such as timber, hardwood and oak are imprecise, since all can be limited by selection. The specification of natural products, so as to obtain the effect sought by the designer, is especially difficult. Different designers may require differing effects, such that the supply of material which is acceptable in one circumstance or to one designer will be unacceptable in another circumstance or to another designer. One designer may select a particular material for its range of shades or variety of texture, while another may seek a high degree of uniformity and consistency in appearance. Unless it is clearly stated in the enquiry documents and provided for in the accepted offer, a post-subcontract requirement to be selective in the choice of material used or as to its location in the final installation, will normally attract an additional price which may in certain circumstances be considerable.

A requirement to carry out the work in accordance with the manufacturer's recommendations or instructions will not normally mean providing additional components not identified in the pricing schedule or in the description of the work that is the subject of the accepted offer. However, this will be less certain where the product is an installation system, when it is likely that the subcontractor's offer will be deemed to be for the whole system, without each component being separately identified and/or quantified and priced. For example, where a natural or traditional tile is specified for use on a roof, the roofer will, it is suggested, not price for tilting fillets, flashings etc., unless expressly required and offering to do so. Whereas in using a tiling system, the roofer may be expected to have allowed for the necessary accessories listed by the manufacturer as comprising parts of its system. This can be made clear and certain by the use of express words in either the pricing schedule or enquiry letter.

If the specification calls for materials either to be to an approved sample or for samples to be submitted for the approval of the architect and/or contractor, several problems may arise. Selection invariably limits the range of materials to be used and will mean that where the product offered by the subcontractor does not achieve approval from the architect and/or the contractor, there will generally be an increase in cost and/or delay to the subcontract works. Who carries the risk is likely to depend on the time in the subcontract process when the selection is made. Where it is done pre-subcontract, any risks that flow from the selection will normally rest with the contractor. Where such selection is to be made post-subcontract, it will be advisable for the subcontractor to make it very clear in its offer exactly what product has been allowed for and, where appropriate, the extent of any allowance for selection it has incorporated within its pricing of the work. At the offer stage this may be expressed as a percentage of rejected material allowable without charge, or by the provision of a sample range.

Sample ranges create their own problems. First, only full-sized samples are truly representative of what will be supplied. Blocks of stone or planks of wood will vary across their surface. While a range of small samples may show the range of colour, texture, grain, etc. they will not show if the range will be manifest in an individual component or only between different components. Secondly, a range of samples can only represent supplies current at the time of submission. Many stone quarries have a range of colour within the stone beds. At any one time the rock will have specific qualities and appearance. Stone production is likely to be at a fairly constant rate. The availability of stone with specific qualities will depend on such stone being present in the area of quarry being worked. There will be no certainty as to the nature of the stone not yet quarried. Thus a subcontractor's ability to provide stone selected for certain qualities in the quantities required will not be assured, or at any rate assured of being available at the rate necessary for the work. Consequently there can be no certainty that a time schedule for work on site can be met.

There are no easy answers to this problem, other than for all the parties at each stage fully to understand the nature of the material being specified and/or offered for acceptance, and together to manage the selection and allocation process. This needs to be done in a spirit of cooperation and joint objectivity to achieve the best possible result from the material naturally available. Any other approach will lead to serious delays, cost increases and the inevitable disputes.

The subcontractor's obligations for time

The subcontractor's obligation as to the time for completion of its work is often stated as a specific period after notice. Some of the problems that may arise from such an arrangement are set out in Chapter 4, Programming of the Subcontract Work. SBCSub/C sets out the subcontractor's obligations at clause 2.3, which are that the subcontract works are to be carried out and completed in accordance with the programme details stated in the Sub-Contract Particulars (item 5) and reasonably in accordance with the progress of the main contract works. This requirement is subject to receipt by the subcontractor of the notice to commence work in accordance with those particulars, and subject to clauses 2.16 to 2.19.

This form makes provision neither for the work nor work within a section to be made available to the subcontractor either progressively or in stages. In practice, it will not often be the contractor's intention that all of the subcontractor's work will be available at the time of the subcontractor starting on site. Further, for many trades the subcontractor's work will be carried out in stages, interspersed by the work of other trades. Not only is there no provision within the SBCSub/C arrangements for staged or progressive release of the work to the subcontractor, there is no provision for the subcontractor to be required to make available its work on a progressive basis for the work of other trades.

The subcontractor's obligation as to time can be regarded as having two limbs. The first is an obligation to commence the work after a period of notice and complete in a stated period. The second is to carry out the work 'reasonably in accordance with the progress of the Works'.

The first can be regarded as requiring the subcontractor to be proactive. The subcontractor will be expected to plan and organise its work so as to complete within the agreed period. The subcontractor, however, will have the right to plan and organise its work as it chooses, subject only to any restrictive stipulations within the subcontract. In the second situation, the subcontractor is required to react to work becoming available to it. Thus, where a subcontract requires the subcontracted work to be completed within a stated time after a period of notice, should there be acts of prevention by the contractor, unless and until an amended agreement is made, either by the granting of an extension to the subcontract period or agreeing a programme for the remaining subcontract work, time will be at large and the subcontractor's obligation is to progress 'reasonably in accordance with the progress of the Works'.

SBCSub/C recognises the separate and different nature of these two possible obligations of the subcontractor by providing for separate recovery by the contractor of any loss and/or expense resulting from the subcontractor's breach. Clause 2.21 makes provision for where the subcontractor fails to complete on time, while clause 4.18 provides for where the subcontractor's default leads to disturbance of the regular progress of the Works. Both of these provisions are discussed in the next section, 'Loss and/or expense'.

These two limbs formalise what, it is suggested, would be the situation at law in any event. In a subcontract where there is a stated date for completion, should the contractor commit an act of prevention, time is likely to become at large. In such circumstances the subcontractor's obligation becomes one of completing the work in a reasonable time. What is a reasonable time will depend on all the circumstances. Such circumstances must include, among others, the making available of the work to the subcontractor by the contractor.

An obligation to perform in a reasonable time can therefore arise in three different ways: first, where there is no defined period within the subcontract or, it is suggested, where the contractor fails to trigger the mechanism for the commencement of the subcontract period, such as the issue of a notice to commence, where such is a condition of the subcontract, as is the case in the SBCSub forms; secondly, where the subcontract sets a specific time or period for the performance of the subcontract work but the subcontractor is prevented in his performance by events or actions for which the contractor is liable or where the parties have agreed as a condition of the subcontract that specified 'material' events will relieve the subcontractor of its obligations – such matters may include exceptionally inclement weather, strikes, fire, etc.; and thirdly, where the subcontractor is required to perform its work reasonably in accordance with the progress of the works on site.

The requirement to perform a contractual obligation in a reasonable time is not a situation unique to construction operations, and much of the decided law on this topic relates to the time taken in the unloading of ships. In such cases much debate has taken place as to the meaning of 'reasonable'. On the one hand, reasonable was taken to mean what was reasonable in normal circumstances, and on the other, it has been agreed that reasonable means reasonable in all the circumstances. It is the second, reasonable in all the circumstances, which it is considered represents the decided law.

In the case of *Pantland Hick* v. *Raymond & Reid* [1893] HL 22, the House of Lords reviewed the decided cases and came to a unanimous decision that a reasonable time should depend on the circumstances that actually exist. In this case the unloading time became prolonged by reason of a general strike at the port where the unloading was taking place. Lord Herschell LC said, at page 29:

'It appears to me that the appellant's contention would involve constant difficulty and dispute, and that the only sound principle is that the "reasonable time" should depend on the circumstances which actually exist. If the cargo has been taken with all reasonable despatch under those circumstances I think the obligation of the consignee has been fulfilled. When I say the circumstances which actually exist, I, of course, imply that those circumstances, in so far as they involve delay, have not been caused or contributed to by the consignee.'

Lord Watson stated the general principles leading to the implication of a reasonable time, at page 32:

'When the language of a contract does not expressly, or by necessary implication, fix any time for the performance of a contractual obligation, the law implies that it shall be performed within a reasonable time. The rule is of general application, and is not confined to contracts for the carriage of goods by sea. In the case of other contracts the condition of reasonable time has been frequently interpreted; and has invariably been held to mean that the party upon whom it is incumbent duly fulfils his obligation, notwithstanding protracted delay, so long as such delay is attributable to causes beyond his control, and he has neither acted negligently nor unreasonably.'

Lord Ashbourne, at page 34, combined these two concepts:

'The contract not naming any time for unloading, the obligation of the defendants was to unload within the time implied by law, that is, a reasonable time. What is the meaning of this expression "reasonable time"? It is obvious that "reasonable" cannot mean a definite and fixed time. It would not be "reasonable" if it was not sufficiently elastic to allow the consideration of circumstances, which all reason would require to be taken into account . . . If the consignee does all he can, is not his conduct reasonable? If by circumstances absolutely outside his control he can do nothing, is his inaction unreasonable? If it is reasonable to consider some circumstances outside his control in favour of the consignee, why are not all circumstances in the events which actually happen, and which he cannot control, also to be taken into account? In considering how to ascertain "reasonable time," must not the question come in, whether the consignee in the circumstances which eventuated acted unreasonably? If throughout the consignee acted reasonably, if he did all he could, if he omitted nothing that he should have done, why should all the circumstances be arbitrarily divided into ordinary and extraordinary for the purpose of putting a narrow and artificial meaning upon the words "reasonable time"?

and Lord Ashbourne concluded:

'I think when the parties have used a form of contract, which names no day, and leaves the discharge of the cargo to be made within the reasonable time implied by law, that reasonable time should be ascertained by a consideration of all circumstances which eventually happen, and which are outside the control of the consignee.'

What exactly will be regarded as circumstances 'caused or contributed to by the consignee', referred to by Lord Herschell in the extract from his judgment above, when applied to a building subcontract situation, is uncertain. It is suggested that such circumstances must be ones that result from negligent or irresponsible behaviour by the subcontractor and may not include all matters for which the subcontractor would normally be liable if time were fixed and not at large. Such risks must include delays in delivery of materials, plant breakdowns and labour failing to attend site. More controversial will be delay caused by the rectification of defective or incorrectly constructed work. For all such causes of delay the reasonableness test must be applied. Thus, for example, where the delay has been in part caused by the rectification of defective work, such delay will only be held to the subcontractor's account if the contractor can show that such defective work was unreasonable in all the circumstances. Such unreasonableness might be on grounds as to the nature of the defect or the number or extent of such defects.

The Court of Appeal in the case of *Shawton Engineering Limited* v. *(1) DGP International Limited (t/a Design Group Partnership) (2) DGP Limited (formerly known as DGP (Consulting Engineers) Limited* [2005] TCC 44/01, had to consider the meaning of completing within a reasonable time. In that case a sub-subcontractor design consultant had misunderstood the amount of work required of it; however, the work required was varied by the issue of a significant number of changes in the requirements. There was no effective method within the sub-subcontract provisions for establishing a new completion date. The judge had to consider whether the correct method of establishing a new completion date was to consider the work involved in actioning the variations themselves, and establish an extension of time on that basis, or to review the amount of work to be done as at the date a new completion date was being decided. In other words, was the sub-subcontractor, where it had become entitled to complete in a reasonable time, only to be granted an extension of time based on the additional work or for a new completion date based on the work left for it to do.

In addition, the court had to decide whether the contractor had been entitled to determine the sub-subcontract on the basis that the designer had been in breach by not completing within a reasonable time.

The judgment of Lord Justice May is lengthy and reviews in detail the judgment of the judge of the technical and construction court. The conclusion reached was that where time becomes at large and performance is to be within a reasonable time, that time is to be calculated and assessed on the basis of the work to be completed at the time of the assessment. Lord Justice May put it in these words:

> '72. The judge was, in my judgment, accordingly right to hold that Shawton had not established what was a reasonable time for completion. He was right to hold that DGP were not in breach for delay on 7 November 2000. He was right to hold, as he implicitly did, that on 7 November 2000 the reasonable time for completion was to be assessed afresh, mainly with reference to the outstanding work content including variations. That was not solely or mainly because Shawton had issued variation instructions, but because, until 7 November 2000, Shawton had not insisted on completion by any particular date or within any particular period.'

It had therefore been up to the contractor to make an assessment and seek to insist on completion by a particular date or within a particular period, and that date or period was to be reasonable having regard to the work to be completed at the date of that assessment.

Lord Justice May considered, at paragraph 76, the potential right of a contractor to determine a subcontract where time had become at large, without first having established a date for completion of the subcontract works. He defined the nature of the subcontractor's breach as having to be so serious as to substantially deprive the contractor of the whole benefit of the subcontract:

> 'I accept that, even if time is not of the essence, it is theoretically possible for a party to show that another party's delay is so profound as to be repudiatory. But what has to be shown is, not mere breach, but a breach of such gravity as to deprive the other party of substantially the whole benefit which it was the intention of the parties that they should obtain from the contract.'

It has been suggested that the obligation to progress the work 'reasonably in accordance with the progress of the Works', in addition to being an obligation to respond to work being made available to the subcontractor, requires the subcontractor to release work to the contractor to enable other trades to progress to their programmes. This contention was considered in the case of *Pigott Foundations Ltd* v. *Shepherd Construction Ltd* (1993) 67 BLR 49, where Judge Gilliland QC says, at page 61:

> 'The words "the progress of the Works" are in my judgment directed to requiring the subcontractor to carry out his subcontract works in such a manner as would not unreasonably interfere with the actual carrying out of any other works which can conveniently be carried out at the same time.'

He then goes on to state what he considers these words do not mean:

> 'The words do not however in my judgment require the subcontractor to plan his subcontract work so as to fit in with either any scheme of work of the main contractor or to finish any part of the subcontract works by a particular date so as to enable the main contractor to proceed with other parts of the work.'

SBCSub/C, at clause 2.17, requires the subcontractor to give written notice to the contractor if and whenever it becomes reasonably apparent that the commencement, progress or completion of the subcontract works is being or is likely to be delayed.

While this obligation may be reasonable where the subcontractor is carrying out the subcontract works in accordance with the programme detailed in item 5 of the Sub-Contract Particulars, when the subcontractor is in control of the timing of its work, it is suggested, this ceases to be a feasibility once the subcontractor is working in a responsive manner by way of its obligation, 'to carry out and complete the Sub-Contract Works . . . reasonably in accordance with the progress of the Works'. Further, the reference in clause 2.17 to 'commencement' being delayed is a strange requirement in a subcontract where the contractor is required to give notice of commencement. It is suggested that the subcontractor's obligation must relate to further delays following the notice to commence, and delays to activities for which no notice by the contractor is required.

Strangely, in SBCSub/C and ICSub/C, there is no equivalent requirement for the contractor to notify the subcontractor should it consider that either the commencement of the subcontract works is likely to be delayed, or the progress of the Works is being delayed by the subcontractor's breach of its obligations. This is in contrast to clause 14.4 of Sub/MPF04, where the contractor at all times is to ensure that the subcontractor is aware of the actual and projected progress of the project.

There are, in addition, provisions under clauses 4.19 and 4.21 of SBCSub/C for both the subcontractor and contractor to give notice to the other party, within a reasonable time, of the regular progress of the Works being affected or disturbed. Failure to give such notice may deny the claiming party the right to recover its loss and/or expense at least for that incurred after it became or should have become apparent that such loss was being or likely to be incurred, since failure to issue such a notice may be held as having prevented or at least not allowed the other party to rectify the situation. Similar provisions are made in Sub/MPF04, at clause 15.1, for delays caused by the subcontractor, and at clause 16.2 for delays to the subcontractor.

Loss and/or expense

SBCSub/C allows for the payment of direct loss and/or expense for matters affecting the regular progress of the works, as set out at clauses 4.19 to 4.21. The grounds for paying the subcontractor its direct loss and/or expense are set out in general terms at clause 4.19. These grounds provide that, if in the execution of the subcontract the subcontractor incurs or is likely to incur direct loss and/or expense for which he would not be reimbursed by a payment under any other provision in the conditions, because the regular progress of the subcontract works has been or is likely to be materially affected by any of the relevant subcontract matters, then the subcontractor may make written application to the contractor, and the amount agreed by the parties of the direct loss and/or expense caused to the subcontractor shall be taken into account in the calculation of the final subcontract sum. This is subject to the subcontractor satisfying the following three requirements:

1. The subcontractor makes his application as soon as it has become, or should reasonably have become, apparent to him that the regular progress has been or is likely to be affected
2. In support of his application the subcontractor submits to the contractor on request such information as is reasonably necessary to show that the regular progress has been or is likely to be affected
3. The subcontractor submits to the contractor such details of the loss and/or expense as the contractor reasonably requests to enable that direct loss and/or expense to be ascertained and agreed.

ICSub/C has the same provisions at clause 4.16. The provisions in ShortSub merely require that the subcontractor notify the contractor of any loss and/or expense 'as soon as it is reasonably practicable'.

Sub/MPF04 allows for the pricing of any loss and/or expense that will be incurred by the subcontractor as a consequence of a change to be priced as part of the valuation of the change under clause 23.6.4. Loss and /or expense due to other causes or a matter set out in clause 24.2 are not to be paid under this clause.

Clause 24.2 allows for the contractor to be liable to the subcontractor for loss and/or expense, subject to clause 24.1 which expressly states that no change or matter required by the subcontract as giving rise to a change shall entitle the subcontractor to loss and/or expense under clause 24. The only matters of loss and/or expense for which the contractor is to be liable are set out in clause 24.2, as three sub-clauses:

1. a breach or act of prevention on the part of the contractor, or by any of its subcontractors and suppliers or any other person under the control and direction of the contractor, other than the subcontractor and other than matters or actions that are expressly permitted by the subcontract and that are stated not to give rise to a change;
2. interference with the subcontractor's regular progress of the Sub-Contract Works by others on the site;
3. the valid exercise by the Subcontractor of its rights under section 112 of HGCRA 1996.

Clauses 24.3–24.7 detail the method and timing of notification, assessment and payment of loss and/or expense other than that related to a change. Both by the pre-pricing procedures for the valuation of a change under section 23, and the express words in clause 24.4, Sub/MPF04 clearly anticipates the potential for the valuation of loss and/or expense before such loss and/or expense is incurred. Under clause 24.4 the subcontractor is to provide its assessment of the loss and/or expense 'incurred or to be incurred' and the contractor is to make its assessment within 28 days. However, there is in addition provision under clause 24.6 for the subcontractor, within 28 days of practical completion, to submit a further assessment.

Sub-clause 4.19.1 of SBCSub/C, and indeed many other subcontract forms, require the subcontractor to give notice to the contractor as soon as it becomes apparent that loss and/or expense is either being incurred or is likely to be incurred by reason of disturbance to the regular progress of the subcontractor's work. The contractor, on becoming aware that the subcontractor is suffering loss, may take one of several actions:

1. The contractor may give instructions to the subcontractor so as to avoid or reduce any loss or, where loss has already been incurred, further loss.
2. The contractor may wish to take similar action to that of the subcontractor under the provisions of the main contract.
3. The contractor may wish to make records which it considers necessary to ensure that any loss and/or expense to be paid to the subcontractor represents the subcontractor's actual loss due to the relevant matters.

While failure by the subcontractor to issue its notice strictly in accordance with the express provisions of the subcontract may not be held to be fatal to such a claim, it will undoubtedly be a breach of the provisions of the subcontract, which will in turn give rise to a potential cross-claim from the contractor for its consequential losses. Such losses would include any that arose because the contractor could not take appropriate action, such as listed above. It is possible that a defence to such cross-claim was that the contractor was in any event fully aware of the situation. By reference to the Relevant Subcontract Matters listed at clause 4.20 of SBCSub/C, items 1–6 would normally fall into the category of matters known to the contractor, while items 7 and 8 will normally need notification by the subcontractor before the contractor has knowledge of the effect on the regular progress of the works.

Sub-clause 4.19.2 requires the subcontractor to support its application with such information as is reasonably necessary and is required by the contractor, and sub-clause 4.19.3 requires the subcontractor to provide such details as the contractor shall reasonably request, with regard to the subcontractor's loss and/or expense. Two possible situations may arise from these clauses:

1. The contractor, having been notified that there is or is likely to be loss and/or expense to the subcontractor under this clause, may require the subcontractor at that time to provide specific information or details and/or to keep specific records and/or agree such records on a regular basis with the contractor and/or another. This could be regarded as the pro-active approach.
2. The contractor, on receipt of a valuation from the subcontractor of its loss and/or expense account, may seek 'better and further particulars', which after the event may not be available. Whether such information may be considered reasonable for the contractor to request, if it did not at the relevant time advise the subcontractor of its requirements, will no doubt depend on the facts of each case. This approach would be regarded as retrospective and the subcontractor should not be denied a reasonable valuation of its loss because the contractor has failed to identify the level of information and/or details it will require to make or agree such valuation.

It must be noted that there is no requirement for the subcontractor to incur any loss or further loss after it has become apparent that the regular progress has been or is likely to be affected. Indeed, it is suggested that the subcontractor should take steps to prevent or minimise such further loss and/or expense and seek directions and/or instructions from the contractor, under clauses 3.2 and 3.3. Such directions and/or instructions may be required to be valued and paid for under the valuation provisions, and not therefore require proven actual loss.

It would appear, therefore, that there is at least an implied obligation for the contractor, on receipt of an application from the subcontractor notifying that the regular progress of the subcontract works has been affected, to advise the subcontractor what details it requires to 'enable . . . the direct loss and/or expense to be agreed'. Although there is no stated time-scale for the contractor's response to the subcontractor's notice advising that the regular progress of the works is being materially affected, it is suggested that where the contractor does not, within a reasonable time, identify the details that it will require in order to ascertain the loss and/or expense, it may be in difficulty if, at a later date, it denies an entitlement because of an absence of detailed information it then considers necessary.

It is suggested that the clear intention of clause 4.19 is that where the subcontract works become affected for one of several reasons defined in clause 4.20 (Relevant Subcontract Matters), the subcontractor must first notify the contractor and the contractor must issue its directions, which may include the keeping of records etc. It is submitted that once the directions of the contractor are requested, the subcontractor has no authority to continue to expend monies, in addition to the normal costs of the subcontract works, where such expenditure can be avoided, without the express instructions of the contractor. However, if the contractor fails to instruct as necessary, that will itself be a default by the contractor and may be a further matter affecting the regular progress of the works, which can be held to lead to further loss and/or expense to the subcontractor owing to the lack of direction from the contractor.

SBCSub/C, at clause 4.21, makes express provision for the recovery of direct loss and/or expense by the contractor from the subcontractor where the regular progress of the main contract works, or any part of them, is materially affected by any act, omission or default of the subcontractor. The contractor is, within a reasonable time of such material effect becoming apparent, to notify the subcontractor in writing giving reasonable particulars of the effects or likely effects on the regular progress and shall give such details of the resultant loss and/or expense as the subcontractor reasonably requests. Any amount agreed by the parties as due in respect of any loss and/or expense caused to the contractor may be deducted from any monies due or to become due to the subcontractor, or shall be recoverable by the contractor from the subcontractor as a debt.

ICSub/C has the same provisions at clause 4.18. ShortSub makes no express provision for such payment from the subcontractor to the contractor. Sub/MPF04 provides for the contractor to recover its loss and/or expense at section 15. These provisions are also subject to the contractor giving notice 'as soon as reasonably practical'.

As with clause 4.19, there is a requirement for the making of an application by the contractor to the subcontractor 'within a reasonable time of such material effect becoming apparent'. Again, there is express provision for payment 'of any amount agreed by the Parties of any loss and/or expense'. However, the contractor's right under this clause, unlike clause 4.19, is not 'subject to' notice. All the contractor is required to do is to notify the subcontractor, within a reasonable time, in writing. Since the sum that may be recoverable by the contractor as a debt is 'any amount agreed', it is unlikely that the contractor will be able to reach agreement where, by its failure to issue an early notification that it is incurring or likely to incur loss and/or expense, the subcontractor has not been able to monitor such loss and/or take action to limit such loss or further loss.

Several important points arise from the provisions of SBCSub/C clause 4.21. First, the right to recover loss and/or expense from the subcontractor is limited to where the regular progress of the works is materially delayed and then only when as a result of any act, omission or default by the subcontractor. It will therefore be necessary for the contractor to demonstrate that both of these situations exist and that the one has in fact caused the other, before being able to make a valid claim. It will not be sufficient to show that the works have been delayed, but in addition that they have been delayed by an act, failure or default by the subcontractor. Although not expressly stated in this sub-clause, it is submitted that it will be implied into this term that the act, omission or default will need to be shown to be in breach of the subcontractor's obligations under the subcontract.

Secondly, before the contractor may either deduct or recover any monies under this clause, it must first obtain agreement from the subcontractor. Since this condition expressly relates to 'any amount agreed by the Parties', it is submitted that this is an express condition and the issue of a notice of intent to withhold payment under the payment provisions will not override this requirement, or at least will impose an obligation first to seek and obtain agreement. It is to avoid this situation that the requirement for prior agreement is frequently deleted from the subcontract terms.

Thirdly, the contractor is required to make its application 'within a reasonable time of such material effect becoming apparent' by notice to the subcontractor in writing, giving reasonable particulars of the effects or likely effects on the regular progress of the Works and shall 'give such details of the resultant loss and/or expense as the Subcontractor reasonably requests'. One consequence of the contractor not complying with this requirement is that the subcontractor will not have the opportunity of verifying or disproving the contractor's assertions. Should the subcontractor's first knowledge of the contractor's claim for loss and/or expense be a notice of intent to withhold payment, then the contractor must be in breach of its obligations to make an application under this provision. Unlike clause 4.19, which makes reference to the loss having 'become, or should reasonably have become apparent', the words of clause 4.21 only refer to 'such material effect becoming apparent'. The extent to which a contractor will be able to claim loss under this clause at a late date, on the basis that it has only just become apparent that the subcontractor has caused the contractor loss and/or expense by reason of the subcontractor having affected the regular progress of the works, is likely to depend on the facts in each case.

Finally, the contractor is to submit to the subcontractor such details of such loss and/or expense as the subcontractor reasonably requests. It is submitted that this will include records of the progress of other trades, to identify whether there were other concurrent delays and their cause. Again, if the contractor is slow in notifying the subcontractor that it is being caused loss and/or expense, the subcontractor will have been denied the opportunity to request and obtain the details and/or records of such loss and/or expense that it may require to agree the contractor's loss. Further, should the contractor be either unable or unwilling to provide such details, then it may deny itself the right to such recovery by way of a breach of this obligation.

In addition to the general clause 4.21 in SBCSub/C, there is a specific clause 2.21 requiring the subcontractor to pay or allow to the contractor direct loss and/or expense due to the subcontractor's failure to complete the subcontract works within the period or revised period for it to do so. This provision is subject to the contractor notifying the subcontractor in writing within a reasonable time of the expiry of that period. Subject to that proviso, the subcontractor shall pay or allow to the contractor the amount of any direct loss and/or expense suffered or incurred by the contractor and caused by that failure.

It is suggested that the combined effect of the provisions under clauses 4.21 and 2.21 of SBCSub/C, when combined with the requirements for the carrying out of the subcontract work as defined in clause 2.1 and item 5 of the Sub-Contract Particulars in SBCSub/A, are unlikely to assist in a successful claim from the contractor. Chapter 4, Programming of the Sub-Contract Work, discusses more fully programming issues with regard to subcontractors, and identifies some of the programming considerations, which will differ from trade to trade, necessary for the effective control of the Works as a whole and individual subcontractors in particular.

Before a contractor may recover loss and/or expense from a subcontractor, it must be able to demonstrate a breach of the subcontract obligations by that subcontractor. It follows that a prerequisite is that the parties have agreed, in sufficient detail so as to be able subsequently to show a breach, those obligations and to have incorporated such agreement into the subcontract. The provision of item 5 of the

Sub-Contract Particulars is based on the presumption that failure to complete the Sub-Contract Works in a stated period will be the sole or at least principal reason that will lead to loss by the contractor. This presumption, it is submitted and as stated elsewhere, will rarely be justified and the contractor's potential for loss due to underperformance by the subcontractor will arise for other reasons.

For the majority of subcontracts, delay to the Works as a whole, and the consequent loss either directly or indirectly to the contractor from delays on the part of the subcontractor, will arise at a critical point in the subcontractor's progress some time before the expiry of the period for the subcontract works. Only in exceptional situations will the carrying out of some further work by the subcontractor, including snagging, clearing site and provision of information for the health and safety file after the expiry of the period for the subcontract works, cause significant if any loss to the contractor. For these reasons, unless the contractor sets out 'further details or arrangements that may qualify or clarify the above or otherwise relevant to the carrying out of the Sub-Contract Works', as provided for in SBCSub/A at item 5.4 of the Sub-Contract Particulars of that form of agreement or some similar provision, the contractor's right to loss and/or expense is likely to be limited to that actually incurred by the contractor by reason of the failure of the subcontractor to complete the subcontract works within the subcontract period.

In the absence of such further programme details or arrangements, it will be sufficient defence for the subcontractor to show that it was prevented from completing within the subcontract period by reason of the contractor's changed requirements and/or default. In many cases, such as those referred to in Chapter 4, Programming of the Sub-Contract Work, acts of prevention by the contractor will be very apparent.

As shown above, whether the claim, on the ground that the regular progress of the work has been affected or disturbed, is being made by the subcontractor against actions or inactions by the contractor or by the contractor for failure by the subcontractor, the procedures for claiming loss and/or expense are similar. Both require the issue of an application 'as soon as it becomes', and in the case of the subcontractor 'should reasonably have become apparent', that the party was suffering loss and/or expense or was likely to incur such for the reasons set out in the subcontract provisions.

It is often held, by way of resisting a subcontractor's claim for loss and/or expense, that it failed to issue a notice at the appropriate time. Failure to issue a notice of occurrence, especially where such a notice would anticipate loss, may be as a result of the party not wishing to antagonise the other and wishing to maintain amicable relations during the course of the works. For similar reasons contractors will not make any claim against the subcontractor until the Sub-Contract Works are complete or practically so. While both these approaches appear reasonable, they will not assist the progress of the Works. The need for the parties to a subcontract to be continually advising the other party as to its requirements is of fundamental importance to the efficient running of a project. It will also be easier for the parties to establish the information necessary to support such claims and thereby reduce the likelihood of a dispute as to the facts of such losses.

Where the contractor itself makes a claim against the subcontractor but has not itself issued a timely notice or notices of delay, it is likely that the contractor will, by its own actions, have waived the strict requirements of the subcontract, at any rate with regard to the giving of such notices, which should also apply to those notices due to have been issued by the subcontractor.

Establishing the obligation

As mentioned above, the right to recover damages from a subcontractor will only arise where it can be shown that the subcontractor has failed in its obligation. A contractor considering that it has suffered loss as a consequence of the subcontractor's default, will first need to demonstrate the subcontractor's obligation under the subcontract terms, and then detail the nature of the subcontractor's failure.

In most contractual arrangements the obligation is to carry out the work either within an express period or by a stated date. The subcontractor, subject only to any express limitations, will be free to plan and execute the work as it pleases. This principle, established in the judgment in *Wells* v. *Army & Navy Co-operative Society* (1903) Hudson's Building Contracts, 4th edn, Vol. 2, p. 346, was restated in the judgment of *Pigott Foundations* v. *Shepherd Construction Ltd* (1993) 67 BLR 49, referring to the provisions of the DOM/1 form of subcontract, at page 62:

> 'In my judgment clause 11.8 does not exclude or modify the general principle applicable to building and engineering contracts, that in the absence of any indication to the contrary, a contractor is entitled to plan and perform the work as he pleases provided that he finishes it by the time fixed in the contract.'

In practice, it is suggested, it is rarely the completion of the subcontract works, but a failure by the subcontractor to work so as not to delay the work of following trades, which leads to loss by the contractor. This was the background problem in *Piggott* v. *Shepherd*, where not only had the subcontractor failed to complete within the eight weeks required by the express terms of the subcontract, but in addition it had failed to construct the piles in a sequence that enabled the following trade, the construction of the pile caps, to commence either to time or when the piling had achieved a defined state of progress relative to the total piling work.

In that case, the general requirement for the subcontractor to 'plan and perform the work as he pleases provided that he finishes it by the time fixed in the contract', was not considered to impose on Pigott, the subcontractor, an obligation to construct the works in the order and sequence claimed or sought by the contractor, Shepherd.

It is rare, in construction work, that matters proceed without any problems and/or difficulties for which the subcontractor will be responsible. These will range from the non-availability of labour, plant or materials as required, to the construction of non-compliant work and its removal and/or rectification. Encountering difficulties and/or constructing non-compliant work are not in themselves breaches of the terms of the subcontract. They only become relevant matters if, as a consequence, the subcontractor fails in its obligation to complete its work to time.

Not only will the subcontractor not be in breach if it has difficulties in the performance of its obligation, provided it ultimately performs that obligation, but it will not have to perform that obligation at the maximum speed or immediately on sections of the work becoming available. In the case of *The Lord Mayor, Aldermen and Citizens of the City of Westminster* v. *J. Jarvis and Sons Ltd and Another* (1970) 7 BLR 64, albeit referring to the performance of a nominated subcontractor, Lord Wilberforce said, at page 78:

> 'One thing seems clear, that "delay" does not, as the appellant at one time contended, mean "sloth" or "dilatoriness" on the part of the subcontractor. There are at least two good reasons against this meaning; in the first place, it would put an impossible burden on the architect if he were required to form an opinion that the subcontractor had not worked as fast or diligently as he might have done and to measure the extent to which time could have been saved had he done so. This part of the contract would in practice, become unworkable. And, secondly, it is contractually irrelevant whether the subcontractor could have worked faster: what matters is whether he has done what he agreed to do in the contractual time. If he has, it does not matter, for the purposes of the contract, whether he achieved his target by leisurely methods: if he has not, it does not matter with what feverish energy he set about his work. This suggested test, then is both unworkable and irrelevant.'

This decision may appear to conflict, in part at least, with the provisions of SBCSub/C clause 7.4.1.2, which makes a failure by the subcontractor to proceed regularly and diligently with the subcontract works grounds for determination of the subcontract by the contractor. However, it is likely that these words do not mean 'with all possible speed' or some similar phrase. All they are likely to mean is that the subcontractor should progress at a reasonably steady rate and not in short bursts of high output interspersed by significant periods of inactivity, except if this is what is required of it. As will be mentioned in the section 'Determination on the grounds that progress is insufficient' in Chapter 12, the ability of the subcontractor to proceed regularly and diligently presupposes that the Works in general are progressing regularly and diligently; indeed, there is probably an implied obligation on a contractor, who is requiring a subcontractor to carry out the subcontract works in a regular manner, to ensure that the works as a whole are proceeding in a regular manner and if they are not the subcontractor may be relieved of its obligation to proceed regularly. For further discussion on the provisions for termination of the subcontractor's employment under SBCSub/C see Chapter 12, Determination of the subcontract.

Temporary disconformity

An obligation to provide work to a specified quality by a specified date raises the question that, if prior to the completion date work is constructed incorrectly, does any liability arise, even though the default may be rectified by the subcontract completion date?

The concept of temporary disconformity was considered in the House of Lords decision in *P & M Kaye Ltd* v. *Hosier & Dickinson* [1972] 1 All ER 144. In that case their Lordships were considering the clause in the RIBA form of main contract which made the architect's final certificate evidence 'that the Works have been properly carried out and completed in accordance with the terms of the contract'. Such a clause was capable of two meanings, as stated by Lord Pearson, at page 136f:

'In the wording of clause 30(7), "the said certificate shall be conclusive evidence . . . that the Works have been properly carried out and completed in accordance with the terms of this Contract", there is an ambiguity. What I will call "meaning no. 1" is that the whole series of building operations from beginning to end must be deemed to have been duly carried out and completed, so that any claim in respect of alleged past defects and their consequences is excluded. What I will call "meaning no. 2" is that everything that had to be done by way of building operations has now been done and properly done, all defects having been made good, so that the present state of the building is satisfactory but there is no exclusion of claims in respect of alleged past defects and their consequences.'

In that case there had been significant defects in the floor to the extent that it had been taken up and relaid by the contractor. In the same case Lord Diplock described the possibility of temporary disconformity, at page 139b and 139c:

'During the construction period it may, and generally will, occur that from time to time some part of the works done by the contractor does not initially conform with the terms of the contract either because it is not in accordance with the contract drawings or the contract bills or because the quality of the workmanship or materials is below the standard required by clause 6(1). The contract places upon the contractor the obligation to comply with any instructions of the architect to remedy any temporary disconformity with the requirements of the contract. If it is remedied no loss is sustained by the employer unless the time taken to remedy it results in practical completion being delayed beyond the date of completion designated in the contract. In this event the only loss caused is that the employer is kept out of the use of his building beyond the date on which it was agreed that it should be ready for use.'

Following the decision in *Kaye* v. *Hosier & Dickinson*, various attempts have been made to extend this principle to situations where defects were not rectified by the contractor or occurred so late and were so serious as to be certain not to be rectified by the completion date. In *Lintest Builders Ltd* v. *Roberts* (1980) 13 BLR 38, where the contract had been determined because the employer, Mr Roberts, had exhausted his funds, at that time there was defective work. The contractor claimed that the defects were merely temporary disconformities, which would have been rectified in the normal course of events and therefore should be paid for by Mr Roberts. Since the contractor was not intending to rectify the defects it was held by the Court of Appeal that the contractor had no right to be paid for the defective work.

Lord Justice Roskill did not support the view that the contractor was at liberty to carry out defective work 'ad lib' without any rights to the employer ensuing, but did not go so far as to suggest that such defective work constituted a breach of contract, which gave rise to any greater right to the employer than to instruct that the defective work be remedied. Lord Justice Roskill, referring to the claim on behalf of the contractor, said at page 44:

'He (the contractor) went so far as to say that there was no accrued right merely because defective work had been done and added – and I confess I find that submission rather surprising – that a builder in those circumstances can do defective work as often and as long and frequently as he liked provided that by the time the contract comes to an end and the defects period comes to an end he has remedied those defects. With respect, I do not think that is correct.

I think that the building owner acquired a right at the time the defective work was done.'

In the case of *Surrey Heath Borough Council* v. *Lovell Construction Ltd and Haden Young Ltd (Third Party)* (1988) 42 BLR 25, the building under construction suffered severe damage as a result of a fire. It was suggested by reference to *Kay* v. *Hosier & Dickinson* that there was a temporary disconformity. Judge James Fox-Andrews QC quickly dismissed that argument, at page 34:

> 'I would comment that the expression "temporary disconformity" does not immediately appear apt to describe a destruction of a building by fire when nearing completion.'

In the case of *William Tomkinson & Sons Limited* v. *The Parochial Church Council of St Michael-in-the-Hamlet with St Andrew Toxteth in the City and Diocese of Liverpool* [1990] OR 21/87, Judge Stannard had to consider the situation where the employer had carried out both repair and rectification works to the church prior to practical completion. The work carried out by the Church included both damage to the existing building as well damage to and/or defects in the work carried out by the contractor. The judge had to consider the rights of the parties under a JCT form of main contract where there was provision for the contractor to rectify defects that became apparent before the end of the defects liability period.

SBCSub/C, at clause 2.22, has a provision requiring the subcontractor to make good all defects, shrinkages and other faults in the Sub-Contract Works. This is to be done at the subcontractor's own cost and in accordance with any directions of the contractor.

Work not in accordance with the subcontract is dealt with at clause 3.11, which refers to non-compliant work and provides that the contractor may in addition to the other powers do one of the following:

1. Issue directions requiring the removal from the site or rectification of all or any of the non-compliant work provided that, unless those directions are issued following similar instructions under clause 3.18 of the main contract conditions, the contractor shall prior to the issue of such directions consult with the subcontractor and shall have regard to the subcontract code of practice set out in schedule 1.
2. After consultation with the subcontractor, issue such directions regarding a variation as are reasonably necessary as a consequence of any direction under clause 3.11.1 (but to the extent that such directions are reasonably necessary, no adjustment shall be taken into account in the calculation of the final subcontract sum and no extension of time shall be given).
3. Having due regard to the subcontract code of practice, issue such directions under clause 3.10 to open up for inspection or to test as are reasonable in all the circumstances to establish to the reasonable satisfaction of the contractor the likelihood or extent, as appropriate to the circumstances, of any further, similar non-compliance. To the extent that such directions are reasonable, whatever the results of the opening up, no adjustment shall be taken into account in the calculation of the final subcontract sum but clauses 2.18 and 2.19.2.3 shall apply unless the inspection or test shows that the work, materials or goods are non-compliant.

Workmanship is similarly covered at clause 3.13.1, which provides for where there is any failure to comply with clause 2.1 with regard to the carrying out of work in a proper and workmanlike manner and/or in accordance with the health and safety plan. Under such circumstances the contractor may, after consultation with the subcontractor, issue such directions (whether requiring a variation or otherwise) as are in consequence reasonably necessary.

By reference in clause 3.13.1 to clause 2.1, a workmanlike manner is defined as being 'in compliance with the Sub-Contract Documents, the Health and Safety Plan and the Statutory Requirements'.

These express provisions of SBCSub/C can be compared with the findings of Judge Stannard in *William Tomkinson*, in which he first found that damage caused by the contractor to the church structure and contents was a breach of contract covered by the provisions of the contract for the contractor to undertake rectification work at its own expense in the same manner as the rectification of work carried out by the contractor. It is possible, when this principle is applied to the provisions of SBCSub/C, that a subcontractor causing damage to the work of another may have an obligation to rectify such work itself, at its cost.

Secondly, Judge Stannard, quoting from several commentaries and authorities, held that at the time the damage occurred or the defective work was carried out, the contractor was in breach of the contract, and further that the contractor had a duty and therefore a right to carry out the repairs and/or rectification work at its expense. However, that right did not deny the employer the right to carry out such repairs and/or rectification work but not to recover its expense, but only such expense as would have been incurred by the contractor had it done the repairs and/or rectification work.

Judge Stannard said, at page 22:

> 'Thirdly, clause 2.5 is concerned with the mitigation of loss, in that it confers on the contractor a right to reduce the cost of remedial works by undertaking them himself. Effect is given to this last aspect of the clause if the damages recoverable by the employer for his outlay in correcting defects in the works are limited to such sum as represents the cost which the contractor would have incurred if he had been called on to remedy the defects.'

It is suggested that the provisions of SBCSub/C referred to above could have a similar effect. SBCSub/C, at clause 3.15, requires that the subcontractor indemnify the contractor against any liability and reimburse the contractor for any costs the contractor incurs as a consequence of:

1. compliance by the subcontractor with clauses 3.11, 3.12 and/or 3.13;
2. the operation of clause 3.14 (or of any equivalent provision) in any other subcontract let by the contractor in relation to the main contract Works (including any liability for direct loss and/or expense and/or as a result of any extension of time) in consequence of such non-compliance by the subcontractor.

However, such liability and cost is limited to the direct result of the operation of clauses 3.11, 3.12 and/or 3.13 (and/or of the corresponding provisions of the main contract) in respect of non-compliant work by the subcontractor and/or a failure by him to carry out work in accordance with this subcontract.

It may be that this clause is effective in preventing the subcontractor from having the benefit of a temporary non-conformity and is likely to be so where either the contractor issues instructions to the subcontractor under either clause 3.11 or 3.12 or where the workmanship is not in accordance with clause 2.1.

It must be questionable, in the event that a contractor carries out work, either to or enclosing or otherwise affecting the subcontractor's work, and the subcontractor's work subsequently proves to be defective, whether this is a direct cause of loss to the contractor, and whether the contractor will be able to recover such loss if it has carried out such further work without first obtaining an assurance from the subcontractor that its work is complete and/or ready for such further work. It is suggested therefore that where a subcontractor's obligation is to complete its work by a single date, then without express evidence that the subcontractor considers its work is complete any indemnity such as SBCSub/C clause 3.15 may be ineffective. It is considered that more is required by the subcontractor to bring an indemnity into existence than just performing some work.

'Temporary Disconformity' by Ellis Baker and Anthony Lavers, a paper given to the Society of Construction Law in April 2005, identifies three scenarios as possibly leading to differing degrees of liability. The first scenario, A, is where work is interrupted in an incomplete state. The second scenario, B, is defined as where there is bad quality workmanship which can be readily corrected. The third scenario, C, is where the defective work cannot be easily corrected without major demolition and which the contractor is taking no steps to rectify. The authors conclude that, in scenario A, provided the contractor completes the work within the time set in the contract then there will be no breach. In scenario B, provided rectification work is carried out so that the Works are completed by the due date, again there will be no breach; but in scenario C, since there is no apparent likelihood of the work being completed correctly and to time, there will be a breach which could lead to the employer being entitled to determine the contract.

While deciding the effect of a temporary disconformity under a main contract situation, where the employer has no right to the benefit of the Works until the date for their completion, may be relatively simple, the situation is much more complex under a subcontract, where the work of most trades becomes incorporated into the Works as a whole by the work of succeeding trades.

The situation is made much more uncertain since subcontracts are almost universally arranged on the basis of a period for the subcontract work or for completion by a specified date, with no arrangements for the taking over progressively of the work as completed although this is in practice what happens. Few trades will complete and hand over their entire work before the following trade commences its work. Demolition contractors and steel frame erectors may be frequent exceptions.

It is possible, and it is suggested to be good practice, to have formal procedures in place for the offer and acceptance of sections of each subcontractor's work prior to the commencement of the following operation. Where this is done it is likely that by offering its work for acceptance as complete, a subcontractor is in effect warranting that the work is complete and correctly constructed. Particular difficulties may exist where testing of an installation is dependent on a subsequent operation. An obvious example is the testing of hot water pipe-work, which until the source of hot water is operational can, at most, only be subjected to a pressure test and will not be subject to the movements resulting from heating up and/or pump pressure, where this is appropriate.

A major difference between the effect of a temporary disconformity under a main contract and the same disconformity under the subcontract is that, whereas the employer is unlikely to suffer actual loss or damage provided the disconformity is corrected in time for the Works to be completed on time, the contractor on the other hand is likely to suffer loss at least by having to accelerate following trades, but in addition may have to undo and/or rectify other work to allow rectification of the non-conforming work or work which may have become damaged as a consequence, such as by water damage from leaks.

It is fundamental to Lord Diplock's judgment in *P & M Kaye* v. *Hosier & Dickinson*, that 'no loss is sustained by the employer'. In the subcontractor condition there may and normally will be a loss to the contractor. Thus in the subcontract situation the question is less likely to be whether there is any liability, but the extent of that liability, which must depend on the facts and associated activities and behaviour of the contractor and/or its other subcontractors. Considering the scenarios described by Baker and Lavers, the following conclusions may be drawn.

In scenario A, where the work by trade 1 is incomplete at the time trade 2 carries out its work and as a consequence such work has to be undone and/or repaired, who will be liable for the costs of undoing and/or repair? Typical examples might be where an electrical subcontractor installs conduits and boxes in a wall prior to plastering. If after the plastering has been carried out it is realised that an additional conduit and box is required, then the wall will need to be chased and the missing conduit inserted and subsequently the plaster made good. Similarly, where a soil stack is inserted but it subsequently requires the fitting of a fire sleeve or branches for a sink or wash-basin waste, then in the event that the contractor boxes in those pipes before the fitting of the fire sleeve or branches, it will be necessary to remove that boxing-in to enable the plumber to complete its work.

In the case of the missing conduit and box this may result from a negligent omission by the electrical subcontractor, in which case the cost of plaster repairs may be to its liability, or on the other hand it may be as a consequence of incomplete information as to the location, which may be the subject of a site enquiry sheet, in which case the cost of the plaster repairs will be unlikely to be to the subcontractor's cost. In the case of the plumbing items as described, it is unlikely that these could in any event be to the plumber's cost unless it had 'signed-off' its work as being ready for boxing in. In either event, it may be the overzealous action of the following trade in carrying out its work without ensuring that the work was available to it. In such circumstances the liability must lie with the following subcontractor and/or the contractor.

The express terms of many subcontracts require a subcontractor to be satisfied that the work of the previous contractor is suitable to receive its work. It is suggested that this obligation is unlikely, in the absence of an express obligation to such effect, to extend to ensuring that the work of the previous subcontractor is complete, but may extend to the following subcontractor receiving authority to proceed. Such formalised close-up procedures often exist where an inspection by a third party is required under the main contract, such as the backfilling of drains and the installation of suspended ceilings.

In scenario B, where the work of trade 1 is defective or unsuitable but can readily be rectified, defects and/or unsuitable work may arise either as a result of bad workmanship or as a consequence of a failure to make suitable allowances in the design and/or specification, such as for construction tolerances and/or laps or joints in sheet materials. For some operations defects may not be immediately apparent and will only be discoverable after a period of time or as a result of outside influences, such as an unexpected frost or heat after a curing period.

The first consideration will be whether the work is in fact defective under the requirements of the subcontract drawings, specification and other documents, or needs alteration despite being conformant with the subcontract provisions. Where the subcontractor's work can be shown to be non-compliant, there will an expectation that it will be liable for the contractor's consequent losses. However, the extent of those losses will depend on the circumstances. It is likely that where there is a requirement under the subcontract for a specific action, such as the carrying out of a test prior to performing the next activity, and the contractor proceeds to the second stage prior to such test taking place, then the contractor's consequential loss may be limited to that resulting from any delay in the achievement of a satisfactory test result and not for any loss that might ensue from the failure of a test made subsequently.

However, where the subcontractor indicates that its work is complete, whether expressly or by its action, then it is likely that the contractor may assume it can proceed to the next activity and in the event that the earlier work is found to be defective, the contractor is likely to be able to recover both the cost of the delay and the costs of any resultant reworking. There must, in addition, be clarity of terms between the parties. Completion of first fix plumbing will involve the installation of primary pipe runs but will not necessarily include for testing, which may be required before boxing in of those pipes.

Scenario C may arise either between trades or within the work of a single subcontractor. For example, if a piling subcontractor's work is found to be defective during the construction of pile caps, then it would be for the contractor to stop further work until the problem had been solved and rectification work carried out. The subcontractor would undoubtedly be liable for the contractor's consequential losses. However, in the event that the contractor, on discovering the defective work, proceeds with the construction despite such knowledge, then it is unlikely that the subcontractor in default could be held liable for any additional loss the contractor might incur as a consequence of such action.

As is so often the case when considering the relationship between a contractor and its subcontractors, it is difficult to apply to all subcontractors the same interpretation of the law since each trade creates different circumstances, which lead to different considerations. Thus a plasterer who, having carried out work in a room or unit of a project, removes all its tools and material from that room or unit, may be assumed by its action to be notifying the contractor that its work is complete. However, the same may not be assumed of an electrician whose cables and wires may pass through areas totally disassociated with the purpose for which those cables or wires are required.

Contractor's right to rectify the subcontractor's defects

It is generally to the benefit of both parties that a subcontractor rectify any defect and/or damage to its work, whether or not it is due to the subcontractor's default and/or to its liability. Should the contractor employ others to carry out such rectification, issues are likely to arise both as to the ongoing warranty for the completed work and, in the event of a cross-charge from the contractor, the correct amount to be paid or allowed by the subcontractor against such claim.

For these reasons many subcontracts have express terms which require the subcontractor to rectify any defects that arise up to the end of the rectification period. SBCSub/C has such a provision at clause 2.22, which requires the subcontractor to make good at his own cost and in accordance with any direction of the contractor, all defects, shrinkages and other faults in the subcontract works or in any part of them due to materials or workmanship not in accordance with the subcontract.

ICSub/C has similar provisions, at clause 2.16. ShortSub makes express reference to the defects liability period at clause 8.4 and requires the contractor to notify the subcontractor of any defects that appear in the subcontract works during the defects liability period of the main contract works. The subcontractor is then, at its own expense, to make good such defects within a reasonable time from notification.

Sub/MPF04 has provision at clause 20.2 for issuing to the subcontractor instructions with regard to work, materials or goods not considered to be in accordance with the subcontract, when the contractor may:

1. instruct their removal from the site, either wholly or partially;
2. after consultation with the subcontractor, instruct that they may be used for the Sub-Contract Works, but subject to the subcontractor becoming liable to pay the contractor an appropriate amount calculated in accordance with the prices and principles set out in the pricing document and without the subcontractor having any entitlement to an adjustment to the period for completion and/or to the payment of loss and/or expense;
3. after consultation with the subcontractor, instruct such further works as are necessary as a consequence of the removal or use of the non-conforming work, materials or goods;
4. instruct such further opening up, testing or inspection as is reasonable in all the circumstances to establish to the reasonable satisfaction of the contractor that other similar work, materials or goods are in accordance with the subcontract.

Sub/MPF04, at clause 21.1, has provision for the contractor to issue instructions to the subcontractor for the rectification of defects during the rectification period, for any defect. The subcontractor is to comply with such instructions within a reasonable time and at no cost to the contractor, and in the event that it does not do so, the contractor may engage others in accordance with clause 3.3.

Where in the standard forms the phrase used is 'may instruct' or similar, there is no express provision for the contractor, as an alternative, being allowed to carry out the rectification work itself or employ others to do so, except where the subcontractor fails to carry out such rectification work itself.

In the case of *William Tomkinson* v. *The Parochial Church Council of St. Michael-in-the-Hamlet*, Judge Stannard had to consider the situation where a contractor, carrying out alteration and refurbishment work to a church, damaged parts of the building finishes and fixtures which were not part of the works. It was suggested by the Church that such items were not covered by the clause in the contract requiring the contractor to rectify defects, since these were not part of the contractor's work. It was held that the requirements of the relevant form of contract were wide enough to include rectification of the damage sustained. The judge said, on page 16:

> 'Mr James [Counsel for the Church Council] also submitted that clause 2.5 was not attracted because the defects under consideration were not in the works, but were in the existing structures. In my judgment this point was answered by Mr Field [Counsel for Holford] on behalf of the architects when he pointed out that the words "defects, excessive shrinkages or other faults" in clause 2.5 were not qualified by any reference to the works, and that in this respect clause 2.5 is to be contrasted with the preceding conditions of the JCT form, in all of which "the works" are expressly referred to. I agree with his submission that if clause 2.5 were restricted to defects in the works, it is to be expected that the draftsman would have adopted the same terminology as in the earlier clauses. Further, as to the alleged items of damage caused by the recurrent ingress of water, the contractors were under a contractual duty to protect the works until completion. This duty was implicit in their obligation to achieve completion, and in the terms of the specification set out above and clause 1.1 of the JCT Conditions, which require that "the (contractors) shall with due diligence and in a good and workmanlike manner carry out and complete the Works in accordance with the contractual documents . . .". If, as is alleged, these defects have been caused by shortcomings in workmanship or materials which were employed in protecting the works against weather, or by workmanship which left the works exposed to the elements, they were directly due to "materials or workmanship not in accordance with the contract" within the terms of clause 2.5 notwithstanding that these consequences were not manifested in the works themselves. In regard to further allegations that plasterwork has been damaged by hammering or incidentally when fixing a temporary cover to a rooflight, the contractors failed to deal properly with dry rot, columns were not brushed before they were painted, litter was not cleared away and a ceiling was damaged by dropping a bracket onto it, these are plainly allegations of "Workmanship not in accordance with the contract".'

The judge then considered whether the employer was, by the provision for the contractor to rectify defects, itself denied the right to rectify such defects. The judge considered the case of *P. & M. Kaye* v. *Hosier & Dickinson* and the judgment of Lord Diplock, and at pages 19–20 quoted from pages 139–140 of that judgment, which included the following words:

> 'I can read no such *necessary* implication into clause 15 or any other clause of the RIBA contract. Clause 15 imposes on the contractor a liability to mitigate the damage caused by his breach by making good the defects of construction at his own expense. It confers on him a corresponding right to do so. It is a necessary implication from this that the employer cannot, as he otherwise could, recover as damages from the contractor the difference between the value of the works if they had been constructed in conformity with the contract and their value in their defective condition, without first giving to the contractor the opportunity of making good the defects.'

However, Judge Stannard, at page 21, does not consider that such clauses deny the employer the right to recover damages where it carries out the rectification work:

> 'I accept that if a contractor remedies defects in works prior to the due date of practical completion, it may be concluded, as a matter of construction of the contract, that it was

not intended that damages should be recoverable in respect of temporary deficiencies in the works prior to their being corrected. However, it does not follow that where workmanship falls short of the standard required by the contract and the employer remedies it prior to practical completion, there is no breach of contract, or the employer is not entitled to recover as damages his outlay in remedying the defective works.'

The conclusion to be drawn from this judgment, which it is submitted is supported by the judgment of Judge Humphrey Lloyd QC in *Guinness plc* v. *CMD Property Developments Ltd (formerly Central Merchant Developments Ltd) and Others* (1995) 76 BLR 40, is that the execution of work which is defective is a breach of the subcontract for which the contractor has a potential right to damages. Where there is an express provision for the subcontractor to rectify such defective work, it has a corresponding right to do so. In the event that the contractor, without giving the subcontractor reasonable opportunity to rectify the defect, rectifies the defect or employs others to do so, then its right to damages is limited to the probable cost to the subcontractor had it carried out the work itself.

The reason for such damages being limited is because the remedy set out in the subcontract is for the subcontractor to rectify the defects. This is clear from the words of Lord Sutherland in the case of *Michael. A Johnson* v. *W. H. Brown Construction (Dundee) Ltd* [2000] BLR 243, where he said, at paragraph 5:

'The main problem for the reclaimer is that clause 16 appears to contain its own remedy. If there are defects discovered before the expiry of the defects liability period, the remedy is for the employer to give instructions to the contractor to remedy such defects. If they are duly remedied there is no further redress for the employer under clause 16.'

Differing obligations

Many standard forms of subcontract have provision for the adjustment of the period, as opposed to the duration, of the subcontract works and that the contractor will give a minimum period of notice as to the actual start on site date. It has been held in adjudication decisions that failure to start on the date requested is not a breach of the subcontract obligations which may lead to damages. It is considered that, while this is certainly often a correct interpretation of the subcontract terms, it is unlikely to provide the contractor with a workable subcontract. It should be a breach for the subcontractor not to commence at the required rate at the end of the period of notice, or at least within a very short period thereafter, since the ability of the contractor to give access to the follow-on subcontractors will almost certainly depend on the commencement of the first subcontractor's work on time and it proceeding at a rate so as to release work to the follow-on subcontractor at the planned interval.

However, it will rarely be a stated requirement that a given amount of work and/or stage is to be achieved by a set period after the instructed subcontract start date. It is possible, where the Sub-Contract Documents include an indicative programme for the entire works, that the contractor will be able to maintain that failure to make work available to another, after the planned lag period, will lead to a loss that flows naturally from a breach of the subcontract and/or what was in the contemplation of the parties at the time the subcontract was entered into. This will not be the situation if no such programme is incorporated in the subcontract, without some other express provision. The downside for the contractor in such a requirement is that any such obligation will be lost where the contractor fails to make the subcontractor's work available to time. There is a considerable dilemma for the contractor in setting up subcontract arrangements as to time, since an obligation imposed on the subcontractor will normally at the same time put a respective obligation on the contractor.

The concept of releasing work for following trades will manifest itself clearly where the project is for a series of distinct units, such as in the construction of housing estates or blocks of flats or even hotel bedrooms. Without an express requirement in the subcontract, the subcontractor may be entitled to a premium for completing progressively. This was the background problem in the case of *Lester Williams* v. *Roffey Brothers & Nicholls*.

The contractor, Williams, had agreed to pay its subcontractor, Roffey, an additional sum of £10,300 over and above the subcontract price. The question before the Court of Appeal was whether this undertaking could be upheld as an obligation, since in simple terms Roffey was to provide no additional work for that money and therefore, it was suggested, there was no consideration to make the undertaking to pay the additional money binding, as a contractual agreement.

All three Court of Appeal judges found consideration in differing ways, but the speech of Lord Justice Purchas is of significance, at page 92:

> 'Against this context the judge found that on 9 April 1986 a meeting took place between the plaintiff and a man called Hooper [an employee of Williams], on the one hand, and Mr Cottrel [the surveyor to Roffey] and Mr Roffey on the other hand. The arrangement was that the respondents would pay the plaintiff an extra £10,300 by way of increasing the lump sum for the total work. It was further agreed that the sum of £10,300 was to be paid at the rate of £575 per flat on the completion of each flat. This arrangement was beneficial to both sides. By completing one flat at a time rather than half completing all the flats the plaintiff was able to receive monies on account and the respondents were able to direct their other trades to do work in the completed flats which otherwise would have been held up until the plaintiff had completed his work.'

Interruption to the progress of subcontract works

Many subcontract arrangements expressly provide for the contractor's failure to provide continuity of work to the subcontractor. Such provisions may also require that the subcontractor still complete by the due date even where its work is interrupted. The effectiveness of such draconian requirements will depend on the actual wording of the clause and may be subject to a reasonableness test or for uncertainty, since it will be difficult for the contractor to demonstrate precisely what the subcontractor undertook with regard to time at the time the agreement was concluded.

In the absence of any effective condition to the contrary, the subcontractor will still be entitled to the time contracted to carry out its work, and where work is interrupted by the contractor, it may also be entitled to a further period of notice to restart, as set out in the subcontract, for each visit to site.

Obligations of the contractor

Where the contractor requires the subcontractor to complete its work within a fixed period, and in the absence of any agreement as to the release of the work on a progressive basis, the subcontractor may be entitled to regard the contractor as in default if the entire subcontract works are not available from day one.

While for many trades this may be an unreasonable expectation, for others the piecemeal release may prevent the subcontractor progressing its work at the necessary rate or may lead to it incurring additional costs. The subcontract provisions will rarely make much if any reference to the works to be carried out prior to the subcontractor's commencement on site. Again, in the absence of express provisions, the contractor's inability to provide full details of all its requirements on day one may lead to additional costs and/or delay.

A further matter for consideration in a subcontract such as SBCSub/C, which has provision for the giving of a notice by the contractor to the subcontractor for its start on site, is whether such a period of notice is to be given for each commencement on site or the release of each section of the works when released progressively, and if that is the situation, will the subcontract period run from the end of each period of notice. It is not considered that any universal conclusion can be given, and each subcontract will need to be considered on its facts. What can be said with some certainty is that in practice little thought is given, before concluding the Sub-Contract Agreement, by the contractor as to how it will require the subcontract works to be carried out. In such circumstances the contractor finds itself relying on the cooperation and goodwill of the subcontractor, more than on the agreed obligations within the subcontract provisions, for the successful completion of the works.

Valuation and payment by the contractor

Once the contractor requires the subcontractor to commence work, whether by way of pre-site work such as design and/or manufacture of components or by mobilising and commencing work on site, the primary duty and obligation of the contractor is to value and pay, strictly in accordance with the provisions of the subcontract, for the work carried out. Any failure by the contractor in this will be a breach of the subcontract, which in serious situations may be regarded as repudiatory. At clause 7.8.1.3, SBCSub/C makes 'failure (by the contractor) to make payment in accordance with the Subcontract' grounds for the subcontractor's determination under the contract. It is considered that the phrase 'payment in accordance with the Subcontract' must be interpreted as applying to the whole payment mechanism and procedures, including the proper valuation of the work, and is not limited to payment on time of a sum certified by the contractor.

The consequences of undervaluation of the subcontractor's work are that the subcontractor will require more capital to finance the work as a whole. It is generally acknowledged that only to a limited extent will work be financed from the subcontractor's own capital. Much of the financing will be by way of credit on the cost of labour and materials and by way of money borrowed from a bank. The overdraft limit will be set at an agreed sum depending on the subcontractor's circumstances.

The subcontractor's cash flow difficulties for any one project will be increased where, due to circumstances beyond its control, it suffers loss and/or expense, or where the value of the work is significantly increased by way of variations either to the quantity of the work to be carried out in the valuation period or in the nature of the work itself.

However the source of the difficulty may be expressed, should the subcontractor, for reasons not of its making, and not its liability, be required significantly to increase the capital necessary for it to finance the work, then manifestly the subcontract requirements will have changed in a material way and the subcontractor must give notice under SBCSub/C clause 2.17.1 or its equivalent, and seek to obtain the necessary instructions from the contractor.

While the only express provision under the SBCSub/C form of subcontract is for the contractor to grant an extension of time under clause 2.18, the parties would be free to agree any other remedy, such as for example more frequent payments to the subcontractor or other financial relief. In any event, extending the subcontract period will itself create an additional financial burden on the subcontractor by reason of its having to finance the retention on its work for a longer period.

In certain circumstances there may be an express or implied term of the subcontract as to the extent to which the subcontractor will be providing finance for the work. Where, for example, the subcontractor provided a cash-flow forecast, such a forecast could be held as defining the limit to which the subcontractor anticipated having to finance its work. In such circumstances, should the amount of such financing be significantly increased due to a default by the contractor, the subcontractor may be entitled to consider such increased requirement to be a breach of the contractor's obligations under the subcontract.

Omission of work to enable it to be done by another

In most construction subcontracts there will be either an express or implied term that the contractor may instruct variations to the subcontracted works. Such variations may either increase or decrease the amount of work to be carried out. Without very express provisions within the subcontract, the contractor may not omit work by way of a contractor's instruction and give the work to another.

The case of *Abbey Developments Limited* v. *PP Brickwork Limited* [2003] EWHC 1987 (Technology) has been quoted in the section 'Instruction to omit the remaining work' in Chapter 7. Another decided case relating to this issue is the Australian case of *Commissioner for Main Roads* v. *Reid & Stuart Pty Ltd and Another* (1974) 12 BLR 55. In that case the judge had to consider the situation where on a road works project additional top soil had to be imported and spread because there had been a shortfall in that recovered from the road works itself. The contract had made provision for such a shortfall by requiring the contractor to have tendered a rate per cubic yard for such additional material. When it came to the time for the additional top soil work to be carried out, the engineer omitted the importation of the additional soil by an instruction and arranged for the supply only by others.

The Court of Appeal upheld the decision at first instance that the omission and the giving of the work to others was a breach of the contract. Judge Gibbs expressed the breach as the employer making the work impossible for the contractor to perform, at page 58:

> 'In the absence of a requirement by the engineer under clause 18, it was a breach of the contract for the appellant to render it impossible for the respondent to do the work.'

Judge Stephen put it differently, first considering whether the engineer had powers to give the work to a third party. The judge said, at page 60:

> 'Were he (the engineer) legally entitled to do so it would, I think, run counter to a concept basic to the contract, namely that the contractor, as successful tenderer, should have the opportunity of performing the whole of the contract work.'

He went on to say, at page 61:

> 'Clause 18 is a common enough provision to be found in engineering contracts and permits of the omission from time to time by the proprietor of (a) portion of the contract works. What it clearly enough does not permit is the taking away of (a) portion of the contract work from the contractor so that the proprietor may have it performed by some other contractor – *Carr* v. *J. A. Berriman Pty Ltd* (1953) 89 CLR 327. Yet this was what the engineer sought to do in the present case in relation to spreading of topsoil.'

In practice, in addition to the possibility of the contractor seeking to omit the subcontractor's work and give it to a third party or do it itself, with the intention of making a saving on the price agreed with the first subcontractor, there will be other situations where such action will either be necessary and/or desirable.

One situation is where the design and/or specification under the main contract is changed in such a way that the appointed subcontractor is no longer competent to carry out the changed work. Manifestly two options exist: either the work is deleted from the first subcontractor's work and given to another subcontractor, who is a specialist in the varied work, or the first subcontractor must be instructed to perform the varied work by itself subletting it to a third party specialist. Inevitably the facts in each case will decide whether the unilateral omission of work from the first subcontractor will constitute a breach of the subcontract itself, and the method of remuneration and/or compensation to be paid.

A further situation that frequently occurs is when the period for the carrying out of the subcontract work is delayed and the subcontractor's resources are deployed to other commitments and it is therefore unable to provide either the required rate of performance or to perform at the time now required by the contractor. Alternatively, the main contract works having been delayed, the contractor wishes to recover the time lost by increasing the planned output for subsequent trades. It will not always be possible for individual subcontractors to respond to such requests, either economically or at all.

It is suggested that contractors need to incorporate well-constructed provisions within their subcontract terms, giving the contractor the express right to involve third party subcontractors under certain stated circumstances. Such provisions, as with the standard loss and/or expense provisions, must define the procedures for compensation to the subcontractor.

Chapter 12
Determination of the Subcontract

Introduction

Determination of a subcontract under English law is not a remedy generally available to the parties for a breach of the subcontract. The law requires that the parties, having reached an agreement and as a consequence entered into a subcontract, carry out their obligations to each other in accordance with the provisions of that agreement. Should one party or the other fail in their performance of the subcontract, either in whole or in part, then the remedy will be for the defaulting party to be liable in damages to the other for the value of the loss actually suffered by the non-defaulting or innocent party.

Under certain very special and highly unusual circumstances, the law will find that a contract ceases to exist by frustration or mistake. Such circumstances will be so unusual that they can, for the purposes of this book, be ignored. Generally, the only circumstance in which the law will allow one party to regard its obligations under the subcontract to be terminated is when the other party has expressly or by its actions led the other party to understand that the defaulting party has no intention of proceeding further with its obligations under the subcontract.

A fundamental breach of the subcontract terms, or the breach of a fundamental term, will be necessary before the non-defaulting party will be entitled to regard the defaulting party as having no intention of proceeding with the contract, and as a consequence it will be entitled to regard the subcontract as being at an end.

Even where a major breach of the subcontract and/or its terms occurs, the subcontract is not automatically determined. Where a fundamental breach or a breach of a fundamental term occurs, the non-defaulting party may either require further performance or accept the repudiation.

The law does not generally consider late payment or slow performance to be fundamental breaches of the subcontract. Nor will faulty construction, whether because it is not to specification or even fundamentally defective, be considered a fundamental breach. Such a breach will normally first require a refusal, either express or implied, by the subcontractor to rectify the defect or fault. Breaches must be serious enough that the other party could reasonably believe that the defaulting party had no intention either of fulfilling its obligations or of fulfilling them within a reasonable time. At any time prior to determination, the parties are free to agree new terms to resolve the situation that has arisen. Such new terms could include a revised price and/or period for payment, a new delivery or completion date or amendments to the specification for the materials or workmanship to be provided or achieved. Any of these supplementary agreements may be reached either before or after the event.

At the moment of determination of the subcontract, the subcontract itself does not come to an end; only the rights and obligations for further performance by the parties end. Such rights and obligations will include those incorporated into the subcontract by statute or other means. Thus the obligation to issue notices of payment and/or notices of intent to withhold payment will cease until the effect of the determination can be ascertained, as will the right to any interest on unpaid sums.

The obligation will be for the wronged party at the earliest opportunity to assess its damages and pay the balance, if any, of the sums due to the defaulting party. It is difficult to conceive circumstances where, when the subcontract is lawfully determined by the contractor, there could be a balance due to the subcontractor, since unless the contractor is suffering a loss by reason of the subcontractor's actions or failures, the contractor is unlikely to have any reason or cause to determine the subcontract. There may, however, be no obligation for the contractor to assess its loss until some other action has taken place, such as the issue of certificates of making good defects under the main contract.

Determination by agreement

The parties may at any time agree that the subcontract comes to an end. Such agreement may be for any reason and as a result of various circumstances. It will be for the parties to decide what, if any, payments by way of compensation will be made from one to the other as a consequence.

Where determination is by agreement and no terms are expressly agreed, then it will, as in any other commercial agreement, be for the tribunal or court to imply in terms so as to give business efficacy to the agreement. Typically, where the determination is at the request of the contractor, the subcontractor may be entitled to the value of the work completed together with the associated allowance for overheads and profit, any losses directly incurred as a result of the determination and, in some circumstances, the overheads contribution and profit it had anticipated making on the remainder of the subcontracted work, which can be considered to be a loss to it.

Conversely, where the termination is because of a failure by or at the request of the subcontractor, then the contractor may be entitled to withhold further payment until the extent of its losses arising from employing others to complete and/or rectify defective work is known.

Determination under the subcontract

Standard forms of Sub-Contract Agreements for use on construction contracts contain express provisions for the determination of the subcontract by either party in stated circumstances. The terms of the subcontract will normally state:

1. the circumstances under which a determination can be made, including any express procedures such as the giving of a warning notice and allowing a period for correction of a default;
2. the obligations of the non-defaulting party after determination.

In some subcontracts there may be a provision for determination at will. This is frequently a provision in government contracts but may also be used in provisional subcontracts formed under letters of intent, or for subcontracts where the project is being funded or financed on a progressive basis.

There are significant differences between the standard forms, both as to the circumstances in which determination is allowed and the giving of notice, which reflect the provisions of the main contract to which they relate or refer. Generally, in addition to determination when one party is in major breach of the subcontract conditions, there is provision for determination on the insolvency of the other party and/or on the determination of the employment of the contractor under the main contract.

In SBCSub/C these arrangements are set out at section 7. Specifically clauses 7.4–7.7 deal with determination of the employment of the subcontractor by the contractor, clauses 7.8, 7.10 and 7.11 with determination of the employment under the subcontract by the subcontractor and clauses 7.9 and 7.11 with determination of the contractor's employment under the main contract.

Determination by the contractor

Clause 7.4 of SBCSub/C and ICSub/C sets out the grounds on which the contractor may give notice to terminate the subcontract by reason of default by the subcontractor, before the date of practical completion of the subcontract works, by specifying one or more of the following defaults should the subcontractor:

1. without reasonable cause wholly or substantially suspend the carrying out of the Sub-Contract Works;
2. fail to proceed regularly and diligently with the Sub-Contract Works;
3. refuse or neglect to comply with a written direction from the contractor requiring him to remove any work, materials or goods not in accordance with this subcontract and by such refusal or neglect the main contract Works are materially affected;
4. fail to comply with clause 3.1 or 3.2;
5. fail to comply with clause 3.20.

Clause 7.2 of SBCSub/C and ICSub/C gives details of the nature and manner of service of notice generally under section 7 of the subcontract. Specifically such notices are not to be given unreasonably or vexatiously, are to be in writing and are to be by actual delivery or special or registered post.

ShortSub, at clause 14.1, has limited the grounds for the termination of the subcontract to those of 1, 2 and 4 of SBCSub/C clause 7.4.1. Clause 14.1 also provides for the giving of notice and a period of seven days for the ending of the default.

Sub/MPF04 provides for termination by the contractor at section 35. At clause 35.1 the termination procedures may be commenced by the contractor where the subcontractor commits a 'Material Breach' of the subcontract. A material breach is defined at section 43, 'Definitions and meanings', as:

> 'By the Subcontractor:
>
> - failure to proceed regularly and diligently with the performance of its obligations under the Subcontract;

- failure to comply with clause 3.1;
- failure to insure, as established in accordance with clause 31.3;
- suspension of the Sub-Contract Works or any part thereof, otherwise than in accordance with HGCRA 1996 or in the circumstances described in clause 37.1.

By either party:

- failure to make payment in the manner required by any payment advice issued under the Subcontract;
- any repudiatory breach of the Subcontract;
- breach of the CDM regulations;
- breach of the requirements of the (CIS) [Construction Industry Scheme].'

It is questionable whether this definition is at all helpful in defining the grounds for issuing a notice of determination. It is considered that a material breach may be more precisely defined as a breach of the subcontract, which in turn will prevent the contractor from carrying out its obligations to any third party or duty under any statute. Alternatively, it is the common law definition of being a fundamental breach or a breach of a fundamental term.

Sub/MPF04 allows for two periods for remedy of the breach: eight days where the contractor is under notice from the employer, and 14 days in all other circumstances.

The contractor my also terminate the subcontract at any time on the insolvency of the subcontractor.

In addition to the specific breaches by the subcontractor, the subcontract may be terminated by either party if the subcontract work is suspended for the stated period and on the grounds set out at clause 37.1. These are 'force majeure, the occurrence of any Special Peril or hostilities involving the United Kingdom, or the use or threat of terrorism as defined by the Terrorism Act 2000'.

The benefit to the contractor of having a period for the rectification of any breach by the subcontractor is that, unlike at common law where there must be a clear event by way of breach, which is either affirmed or accepted by the wronged party, there is an intermediate position of notice and request to rectify. In the situation of termination under the subcontract, should the contractor be wrong as to the existence of a breach that justifies termination, it can recover the situation without a breach of the subcontract. If the contractor is correct, rectification may in any event occur with limited further loss and at no risk to the contractor of an unlawful determination.

It is uncertain to what extent the provisions under the subcontract will deny a contractor its common law rights; however, in the case of SBCSub and ICSub, the provisions of the subcontract are at clause 7.3.1 expressly stated to be without prejudice to any other rights which the contractor may possess.

If, as is common in subcontract situations, there is a dispute as to the existence of or terms of the subcontract itself, should the contractor consider a subcontract exists and the subcontractor consider no subcontract exists, and on that basis the subcontractor wishes to cease its involvement in the project, will the contractor be bound by the terms of the subcontract it claims to exist and give notice before determination or may it rely on a common law determination? It is considered that, having offered terms requiring the giving of a notice before termination, the contractor is bound by that offer and may not rely on its common law rights just because the subcontractor asserts that no subcontract exists, unless the subcontract so provides.

Because of the serious regard the law has to the rights of a party to a subcontract to receive the benefit of that subcontract, and its reluctance to allow or support the termination of a subcontract, it is likely that a party seeking to terminate a subcontract under the terms of that subcontract will need to carry out the procedures strictly in accordance with the requirements of the subcontract, or risk being found to have terminated the subcontract unlawfully and consequently to have become the defaulting party rather than the wronged party.

The requirement, for example, to serve a written notice by registered post or recorded delivery is likely to be held as requiring the use of one of those methods for the service to be effective. In the Scottish case of *Muir Construction Ltd* v. *Hambly Ltd* [1990] SLT 830, Lord Prosser, in considering the effect of such a requirement, reviewed a number of previous cases and came to the conclusion that a notice served by hand to an employee of the employer had not satisfied the requirements of the contract for service of such a notice. He said, at page 833J:

> 'With some hesitation, I have come to the view that in the present case the notice was rendered invalid by the failure to send it by registered post or recorded delivery. Like Mr Justice Stephenson in *Goodwin* [*Goodwin & Sons Ltd* v. *Fawcett* (1965) 195 EG 27], I have no doubt that the contract must be construed in a commonsense business way. I am not however satisfied that there is anything contrary to commonsense, or anything inconsistent with a business approach, in concluding that precise words in a carefully structured provision are intended by the parties to have a precise effect in a carefully structured procedure. Like Lord Justice Ormrod in *Hill* [*J. M. Hill & Sons Ltd* v. *London Borough of Camden* (1982) 18 BLR 31], I think I would see it as sensible for a party determining employment to be allowed to do this by notice delivered by hand, as an alternative to sending it by registered post or recorded delivery. But I am none too confident that my feeling as to what might be sensible coincides with the views of those actually involved in building contracts, or that judicial impressions have much place in such an issue. More fundamentally, I think it important to distinguish between an overall view of what would be sensible, and what, when looking at specific words in an existing provision, is a sensible interpretation of those words.'

The phrase used in that contract, 'registered post or recorded delivery', is very much more prescriptive than the wording used in SBCSub/C, which does not for example refer to 'post' but to 'actual, special or recorded delivery', which must be taken to allow for delivery by any person or agency and to no expressly defined person or place.

It may be difficult to know when the contractor, by issuing a notice of termination, is acting with unreasonableness or vexation. When the contractor/subcontractor relationship has deteriorated to such an extent that termination of the subcontract is being considered by the contractor, there must be a distinct possibility that the contractor may act unreasonably or vexatiously. A likely indication as to whether this is the case will be the timing of the notice of default or notice of termination in relation to the progress of the subcontract works on site. Where such action takes place during the course of the subcontractor's work, there is likely to be a presumption that the notice is issued by reason of a genuine belief that there is a breach which gives rise to a right to terminate, and that the termination or notice of it will be necessary to ensure the proper progress of the works as a whole. Conversely, a notice of termination issued when the majority of the subcontractor's work is complete is likely to raise doubts that such notice has not been given unreasonably or vexatiously.

Clause 7.4.2 of SBCSub/C gives the subcontractor ten days to rectify its default(s), after which the contractor may terminate the subcontract, by notice, within a further ten days. The notice of termination is effective on receipt by the subcontractor. Under this condition, the right of the contractor to terminate the subcontract only holds for the period between 10 and 20 days after the notice of default. Should the contractor not issue a notice of termination, it must be assumed that either the contractor considers the subcontractor has rectified such defaults or the contractor has waived its right to terminate on that notice.

However, clause 7.4.3 allows the contractor to issue a notice of termination after the second period of ten days, stated in clause 7.4.2, if the contractor repeats the specified default. This right is defined as 'whether previously repeated or not' and may only be 'upon or within a reasonable time after such repetition'.

For conditions 7.4.2 and 7.4.3 to operate, the contractor wishing to issue a default notice must make very clear the nature of the default relied on. If the contractor is too general in its description of the default, it would be difficult to show that either the default has continued after the initial ten days' notice or, in the event of a notice under 7.4.3, that there is a repeat of the default or a continuance of the same default and not a default of a similar nature.

In identifying a default by the subcontractor, the contractor has to be able to identify precisely the obligation(s) under the terms of the subcontract that the subcontractor has failed to fulfil. In doing this, the contractor must express in clear terms how the subcontractor is defaulting on those obligations and how they are to be rectified. This is an onerous requirement and unless the contractor is able to present a well-defined case for the default, then it is unlikely that a termination of the subcontract will comply with the provisions of the subcontract and will be upheld at law. As mentioned elsewhere, it is common that the requirement for performance by the subcontractor as to time is frequently very imprecisely defined in the subcontract; where this is the case it will be much more difficult for a contractor to be able to demonstrate a default by the subcontractor than where the subcontractor's obligations are well defined. Where the contractor fails to issue an effective notice, it will be held to be in breach of the subcontract itself. Given the imprecise nature of the obligations under many subcontracts, and the fluid and varying nature of activities on construction projects, defining the obligation of the subcontractor at any specific moment in time may be extremely difficult.

Clause 7.4.1.1 requires the subcontractor to have wholly or substantially suspended the carrying out of the subcontract works. To demonstrate that the subcontractor has suspended the works, it will be necessary to show that work has been available on site for a significant period and the subcontractor has made no effort to carry it out despite notice and/or requests from the contractor. The word 'suspension' will itself require some definition. It is likely to involve more that just any interruption to, or pause in, the progress of the subcontractor's work, either on site or at all, and is likely to require a purposeful ceasing of work for no justifiable reason.

However, it may be argued that suspension only relates to the leaving of site when there is work available and not to an unreasonable period of delay in returning to site following the subcontractor's leaving site on a direction from the contractor or due to there being no work or insufficient work available to it. SBCSub/C, at clause 2.3, envisages a period of notice from the contractor for the subcontractor to commence work on site; this clause may be interpreted as a requirement for the contractor to issue a notice and for the subcontractor to have received such notice, each time the subcontractor is required to recommence work on site following a period when the subcontractor has left the site due to a lack of work being available to it. Thus, where the subcontractor is required, either by the programme or for any other reason, to leave site and demobilise, it will be for the contractor to issue its notice for the subcontractor's remobilisation. It may be an effective defence for a subcontractor to show that at the time the subcontractor suspended performance there was no work available for it, or that there was some other act of prevention such as no finalised details of the work required. SBCSub/C provides no express mechanism for the contractor to restart the subcontract works other than by giving a further notice to commence work on site.

Clause 7.4.1.2 relates to the subcontractor's failing to proceed regularly and diligently with the subcontract works. The phrase 'regularly and diligently', while being the phrase used in the SBC/Q main contract form at clause 8.4.1.2, is not something that can be measured with scientific accuracy and is highly subjective. In considering the meaning of this general requirement, regard must be taken of any specific requirements as to programme, whether in the Sub-Contract Agreement or resulting from post-subcontract directions of the contractor, and to the general right of the subcontractor to plan and organise its work as it chooses. The actual requirements of SBCSub/C for performance are set out at clause 2.3, which has two requirements: first, that the subcontractor carry out and complete the works in accordance with the provisions set out in the Sub-Contract Particulars, and secondly, that the subcontract works proceed in accordance with the progress of the main contract works. Neither of these requirements equates to one to progress regularly and diligently. It must therefore be questioned whether such a clause, where there is not an express requirement for such performance in the subcontract terms, can be enforced in any event.

Regular progress implies or suggests that the subcontractor is to work at a steady rate or achieve a constant output of work. It will be rare that the work for each trade is made available in such a manner that a unit of production will be able to operate efficiently at the planned rate. For example, if a gang of bricklayers comprises four bricklayers and two labourers and the programme requires a weekly output that does not equate to 20 bricklayer days, then it will not be possible for the gang to work both regularly and to programme. If the planned output is, say, 15 bricklayer days per week, then in practice the gang will be considered to have progressed regularly if its output averages the desired output. However, consideration may need to be made as to the period over which the average is taken. It will not be acceptable, and will therefore be contrary to the terms of the subcontract, if actual progress falls too far behind the work available on site.

There will, in addition, be an implied obligation on the part of the contractor to be making the work available to each trade by regular progress of the trade before. In practice, a similar situation will exist for other trades such that production will not be at a steady rate but will vary with the number of production units employed at any one time.

The concept of diligent working is less easily defined but, it is suggested, relates to the effective use of resources. Just having sufficient resources on site at the required time will not necessarily satisfy the obligation to proceed diligently. Those resources must be employed in work which is not only of the nature and quality required by the subcontract, but also in the sequence required for the regular progress of the Works as a whole.

It should be noted that the grounds for termination are that the subcontractor has failed in both regards, that of regular progress and that of diligent progress. For whatever reason, the ground for termination under clause 7.4.1.2 of SBCSub/C is not that the subcontractor has failed to progress either regularly or diligently, but regularly and diligently. It must therefore be a defence against termination, under this clause, that the subcontractor did not fail in both regards.

Even if a failure to proceed regularly and diligently is a justifiable ground for termination after notice, the contractor must be able to identify the contractual rate of progress that the subcontractor has failed to achieve and/or the nature of the failure to progress diligently, since it is only against a stated rate of output and/or other performance standards that it will be possible to ascertain whether or not the subcontractor has rectified his default at the end of the period of notice.

It is generally accepted that a precondition for a subcontractor's obligation to work regularly and diligently is that the main contract works have themselves been proceeding regularly and diligently or at least that part or aspect of the works that directly interfaces with the subcontractor's work. Thus, a contractor relying on this sub-clause must be able to demonstrate that the main contract works were and had been progressing regularly and diligently prior to the alleged breach by the subcontractor. Alternatively, a subcontractor may be able to show, by way of defence, that the main contract works had not been proceeding regularly and diligently and thereby were disrupting or preventing the progress of the subcontractor's work.

Clause 7.4.1.3 deals with the situation where the subcontractor 'refuses or neglects' to comply with a direction to remove work, materials or goods which are not in accordance with the subcontract, and then only when the works are materially affected. It must be noted that this is not a general requirement to keep the site either tidy or clear of unnecessary materials and/or rubbish. This sub-clause is solely related to work or materials not in accordance with the requirements of the subcontract, in other words not to the specification and/or drawings. Where the subcontractor considers and believes that the work, the subject of the contractor's direction, has been constructed to the requirements of the subcontract, for example a dispute as to quality, then clause 3.5.1 may apply and the subcontractor may make objection, in writing, requiring a variation before proceeding. Once the subcontractor has made a formal objection there can be no question of his refusing or neglecting to comply with the written directions. The effect may be to create a dispute between the parties which may need to be resolved urgently by means of the dispute resolution procedures such as adjudication by a third party.

Clause 7.4.1.4 gives the contractor the right to terminate where the subcontractor, without written notice and consent, assigns the subcontract to another and/or sublets the work. This leads to the bizarre situation where the contractor, having discovered that the subcontractor has either assigned or sublet all or part of the subcontract works without written consent, is given notice that unless it obtains written consent from the contractor, the contractor will terminate the subcontract.

The presumed intention of this clause, although not the actual provision, must be that where a contractor has discovered that the subcontract work has either been assigned or sublet to another without the contractor's written consent and in a manner, or to a sub-subcontractor, considered by the contractor to be unsuitable, then the contractor should be able to require that that other party be removed from site and replaced.

Clause 7.4.1.5 gives the contractor the right to terminate the subcontract, where the CDM Regulations apply, should the subcontractor fail to comply with those regulations. It is submitted that this does not give the contractor the rights it might reasonably expect. The reference is to the CDM Regulations rather than a more general reference to any breach of the health and safety legislation. It might be expected that this clause made reference to both clause 3.19, which relates to health and safety generally, as well as clause 3.20. Contractors may consider making a suitable amendment accordingly. However, since this clause is very restrictive, it is probably limited to failures such as not providing a health and safety plan before commencing work on site, and to cooperation with the principal contractor.

Subcontracts which, like SBCSub/C, provide for termination of the subcontract by the contractor where the subcontractor fails in one of a number of specified classes of default or breach, and after notice from the contractor continues in that default or breach, are much more difficult to apply in practice than they at first appear. For such a notice to be effective, it must first not just identify the class of default relied on but must state clearly the action required from the subcontractor. It is suggested, for example, that a notice referring to clause 7.4.1.2 of SBCSub/C, stating that the subcontractor has failed to proceed regularly and diligently with the subcontract works and that if the subcontractor does not rectify its default in seven days then the contractor will terminate the subcontract under the provisions of clause 7.4.2, will not be an effective notice. A notice such as this does not define the default nor put the subcontractor on notice as to what it has to achieve in sufficient detail that the subsequent performance can be quantitatively measured against the previous failure, so as not to give the contractor cause to terminate the subcontract.

Nor can a notice to achieve some target, the date for achievement of which has passed, be effective since it is manifestly not achievable. Thus a notice stating that completion was planned for a date which has passed and as a result of this failure the subcontractor is deemed to have failed to progress its work regularly and diligently, and as a consequence the subcontractor is put on notice to rectify its default within seven days or else the contractor will terminate the subcontract under a clause such as 7.4.2 of SBCSub/C, is unlikely to be effective as a notice.

The contractor must be able to identify a definable breach of the subcontract before an effective notice before termination can be issued. Where the alleged breach relates to failure to progress the works regularly and diligently, there must in the first instance have been a clear obligation as to the timing of the subcontractor's work against which the subcontractor's performance can be measured. If no such obligation exists, as is frequently the case, then the contractor will have extreme difficulty demonstrating a breach. It follows that if it is not possible to define the breach in precise measurable terms, it will not be possible to define the necessary action required from the subcontractor to rectify such breach.

A notice before termination must, it is suggested, define in measurable terms the requirements of the subcontractor. Without such a measurable requirement it will not be possible to say with certainty whether or not the subcontractor will have rectified the alleged breach. Manifestly, if there are clearly defined requirements for future performance stated in the notice and these are not achieved, then the contractor will consider that it is entitled to issue the further notice and terminate the subcontract.

However, even in the situation of failure by the subcontractor to achieve the required performance stated in the notice before termination, a subsequent termination will only be lawful if the requirements stated in the notice before termination accurately reflect the subcontractor's obligation under the subcontract. In the event that it can be shown that the performance stated as required was either not a condition of the subcontract or was in excess of that contractually due, then the termination of the subcontract will be unlawful.

Thus a failure accurately to define the performance required of the subcontractor at the time the subcontract is concluded will not only hinder the effective management of the subcontract works, but will also make it very difficult to terminate the subcontract on the grounds of non-performance.

Many subcontracts incorporate a clause entitling the contractor to terminate the subcontract on 'any breach by the subcontractor'. A clause of this type is highly biased in the contractor's favour and should be resisted by all potential subcontractors. Such a clause, unless accompanied by a provision defining the consequences of such termination, will not itself entitle the contractor to the costs it may incur as a consequence of terminating the subcontractor's employment, but will be limited to those costs that follow directly from the subcontractor's breach.

Thus, if the contractor has grounds for such a termination on the basis that the works are defective or progress is not at the rate contracted for, the contractor has two options under a subcontract with a termination clause on any breach. One option is to seek the necessary performance from the subcontractor and if the subcontractor fails in its obligation, to recover the contractor's consequential losses by way of damages. The other is to terminate the subcontract under the relevant provision, in which case the damages attributable to the subcontractor will be those that arose from its breach or breaches up to the date of termination.

As with any other provision for termination on a breach by the subcontractor, it will be for the contractor to demonstrate the breach. Alleged failures with regard to both the progress of the works and quality of workmanship are, in practice, difficult to demonstrate because of the difficulty of defining the performance requirements in the first place.

SBCSub/C, at clause 7.5, allows for termination on the insolvency of the subcontractor and, at clause 7.6, for termination in the event that 'the Subcontractor or any person employed by the Subcontractor or acting on his behalf' carries out any of the acts of corruption defined in the clause.

SBCSub/C, at clause 7.7, sets out the consequences of, and rights and obligations of the parties that arise on, the termination of the subcontract. The provisions of this clause are the same for both acts of default and acts of corruption and where the subcontractor has become insolvent. While some of the acts are permissive to the contractor, others are requirements of the subcontractor. It would seem inevitable in the case of the subcontractor's insolvency that a contractor would be unable to require the subcontractor to carry out its further obligations.

Clause 7.7.1 of SBCSub/C gives the contractor wide powers to take over the subcontract works, use the subcontractor's plant and materials and employ others to carry out and complete the subcontract works. This is made subject to obtaining any necessary third party consents, which presumably refers to the use of hired-in plant and equipment and/or goods supplied against purchase orders and for which payment has not been received by the supplier.

Even where the contractor believes it has paid for unfixed materials, the contractor may not be able to prove a good title to those goods. First, unless it has issued a notice of payment to the subcontractor, prior to the termination, which expressly states that the valuation includes a stated sum for the materials in question and can demonstrate that that valuation has been paid to the subcontractor in full, the contractor will not be able to show it has paid for those materials. The common attitude of contractors to the giving of a notice of payment means that rarely will such a notice be given, and even where such a notice is given it may not specify the basis on which the amount of the payment was calculated.

Secondly, a good title can only be passed to the contractor if the subcontractor itself has title in the goods; thus for goods delivered but not fixed, the contractor will have no right to them unless they have been paid for down the supply chain. This was clearly stated in the case of *Dawber Williamson Roofing Ltd* v. *Humberside County Council* (1979) 14 BLR 70, where a slate roofing subcontractor sought payment from the employer for slates delivered but not installed at the time the contractor went into liquidation. Mr Justice Mais stated, at page 78:

> 'Mr Lawrence [Counsel for Humberside CC] did say that it was obviously highly desirable that the contract should provide as to when the title should pass, and I agree wholeheartedly with him. He said unhappily that this was not so in this case, but he submitted that, despite that, the property did pass, and that there was in effect privity of contract. Mr Heppel [Counsel for DWR], on the other hand, submitted that there was no privity, and thus the clause in the main contract cannot be, and could not be, embodied in the contract between the plaintiffs and the main contractor, and he submitted that there was no provision in the contract between the plaintiffs and the main contractor as to the passing of title.
>
> I cannot accede to the submissions of Mr Lawrence, and, in my judgment, there was no privity of contract between the plaintiffs and the defendants. The main contractor did not buy or agree to buy the slates, and never had any title anyway. The plaintiffs remained the owners of the slates, and the title in them did not pass until the slates were fixed. The defendants failed to permit the plaintiffs to remove the slates. The defendants have not paid the plaintiffs for the slates, and have refused to do so. The slates have now been used, and the defendants have wrongfully converted them to their own use.'

Clause 7.7.2.1 of SBCSub/C allows the contractor to require the removal of the plant and surplus materials, either belonging to the subcontractor or to the subcontractor's persons, when required to do so by the contractor in writing. Inserted into this provision, in brackets, are the words 'but not before'. The intention of this insertion is presumably to deny the subcontractor or any of the subcontractor's persons any right to remove any plant or materials belonging to them from the site before the contractor issues a notice in writing. It is not understood why this clause is phrased in this way. It might have been expected that such a clause would have been expressed as preventing the subcontractor removing its plant or materials until so advised by the contractor, rather than requiring removal with a limitation provision.

Clause 7.7.2.2 of SBCSub/C makes provision for the assignment of any agreement for the supply of materials or goods or work, but makes no reference to plant or services. This is considered a strange omission. The taking over of temporary works such as standing scaffolds is an obvious example where a contractor will undoubtedly wish to continue with the existing arrangements. It is also not understood why this clause expressly provides for the assignment of the Sub-Contract Agreement, which may leave the contractor liable for any outstanding debts between the assignee and the subcontractor.

Surprisingly, clause 7.7.2.2 of SBCSub/C does not give the right to negotiate with any sub-subcontractor and/or supplier of the subcontractor for the purposes of completing the works and giving the contractor access to any records, such as 'as built' drawings, test results, etc. that the contractor may require.

Assignment being a three-way agreement, there can be no certainty, even if the contractor were to require the subcontractor to assign the sub-subcontract, that the sub-subcontractor will be willing to contract with the contractor on the same terms and conditions as it did with the subcontractor. In this respect it is surprising that there is no requirement for pre-agreement to assignment on the early termination of the subcontract at clause 3.3, the conditions for sub-letting by the subcontractor. The standard form of sub-subcontract, Sub/Sub, makes no express provision for the sub-subcontractor to agree in advance that in the event of the early termination of the subcontract it will, if required, allow its sub-subcontract to be assigned to the contractor.

Clauses 7.7.3 and 7.7.4 of SBCSub/C deal with the further payment of the subcontractor by the contractor. Clause 7.7.3 cancels the provisions of the subcontract for payment on the termination of the subcontract, and replaces them with the provisions of clause 7.7.4. Clause 7.7.4 puts the onus on the subcontractor to apply for further payment, which it may not do until the end of the rectification of defects period. Exactly how, in practice, the ex-subcontractor will be able to ascertain when that situation has been achieved is difficult to conceive. In any event the subcontractor will only be entitled to the balance between the outstanding value of its work and any direct loss and/or expense the contractor may have suffered as a result of the termination.

In practice, it will rarely be the case that the contractor will consider that any further sum is due to the subcontractor after its subcontract has been terminated. Although the contractor may recover the excess of its loss over the sum due to the subcontractor, there is no stated method or timing for the contractor to do this.

Sub/MPF04 at clause 35.4 lists four actions to be taken on the termination of the subcontract by the contractor:

1. The subcontractor shall not remove any materials, plant or equipment from the site unless expressly permitted to do so by the contractor.
2. The subcontractor shall provide to the contractor all design documents prepared in connection with the subcontract.
3. The contractor may make such other arrangements as it considers appropriate to complete the Sub-Contract Works.
4. The contractor shall not be obliged to make any further payment to the subcontractor other than in accordance with clause 35.

Clause 35.5 requires the contractor to issue a payment advice on the completion of the making good of defects under the main contract, which is to state the additional costs incurred by the contractor in undertaking the subcontract works when compared with the costs that would have been incurred had the employment of the subcontractor not been terminated, and any loss and/or damage suffered by the contractor and for which the subcontractor is liable, whether arising as a consequence of the termination or otherwise arising out of the subcontract.

The requirements of ShortSub on the subcontract being terminated by the contractor are set out, at clause 14.3, simply as requiring the subcontractor to leave the site immediately and allowing the contractor not to make further payment until the completion of the making good of defects.

Where there are no express provisions in a subcontract for the resultant rights and obligations on termination of the subcontract, it is likely that similar provisions to those of SBCSub/C will be implied. In other words, on termination the subcontractor will have no right to further payment until the work is complete and the contractor has been able to ascertain its damages that flow from the termination and the associated breach by the subcontractor.

Notwithstanding the express provisions of clauses 7.7.3 and 7.7.4 of SBCSub/C that the contractor is not to be bound, by any of the provisions of the subcontract, to make any further payment to the subcontractor until the making good of defects, an exception will be a lawful decision of an adjudicator that the contractor pay to the subcontractor a stated sum. The Court of Appeal, in the case of *Ferson Contractors Limited* v. *Levolux A. T. Limited* [2003] BLR 118, decided that such a clause would not defeat the intention of clause 108 of the HGCRA; further it would be contrary to the express provisions of the subcontract, which require that the decision of the adjudicator shall be binding on the parties until the dispute or difference is finally determined.

While *Ferson* v. *Levolux* concerned a contract under the GC/Works/Subcontract terms, there is little doubt that this decision is of general effect. Lord Justice Mantell said, at paragraph 30:

> 'The intended purpose of section 108 is plain. It is explained in those cases to which I have referred in an earlier part of this judgment. If Mr Collings [Counsel for Ferson] and Judge Thornton are right, that purpose would be defeated. The contract must be construed so as to give effect to the intention of Parliament rather than to defeat it. If that cannot be achieved by way of construction, then the offending clause must be struck down. I would suggest that it can be done without the need to strike out any particular clause and that is by the means adopted by Judge Wilcox. Clauses 29.8 and 29.9 must be read as not applying to monies due by reason of an adjudicator's decision.'

Lord Justice Longmore referred, at paragraph 33, to the express provision of that subcontract to comply, forthwith, with any decision of an adjudicator:

> 'Although the Parliamentary intention is clear, the parties themselves chose to underline it when they provided:
>
> "38A.9 Notwithstanding clause 38B, [the arbitration clause], the contractor and the subcontractor shall comply forthwith with any decision of the adjudicator; and shall submit to summary judgment/decree and enforcement in respect of all such decisions."
>
> The parties have thus agreed not merely that the adjudication is to be binding but also that they will comply with the adjudication notwithstanding the arbitration clause. For

> good measure, they have agreed they will submit to applications for summary judgment. If Ferson had a genuine point, there would then be a dispute which would have to be referred to arbitration but the parties have expressly agreed that course is not open to them once an adjudication has occurred. The clause thus prevents the party who has lost the adjudication from applying for a stay and, for good measure, requires him to submit to applications for summary judgment. The point of that must be not that the court should hear argument, at the stage of the application for summary judgment, about matters which (apart from the adjudication provision) should be referred to arbitration, but rather that summary judgment should be given without further ado.'

Although the wording in the SBCSub/C form, which directly incorporates the provisions of the Scheme, is different from the GC/works/Subcontract, it is submitted that its effect will be the same.

Determination by the subcontractor

The defaults by the contractor on which the subcontractor may rely when giving notice before termination to the contractor are set out in section 7.8 of SBCSub/C and ICSub/C. These are that if the contractor:

1. without reasonable cause wholly or substantially suspends the carrying out of the main contract Works; or
2. without reasonable cause fails to proceed with the main contract Works so that the reasonable progress of the Sub-Contract Works is seriously affected; or
3. fails to make payment in accordance with the subcontract; or
4. fails to comply with the CDM Regulations,

the provisions and requirements for the giving of notice are similar to those for the termination of the subcontract by the contractor.

ShortSub has no express provision for the determination of the subcontract by the subcontractor; thus an aggrieved subcontractor's rights are limited to its common law right to damages for any breach by the contractor and the right to accept repudiation by it, of the subcontract.

Sub/MPF04 provides for termination of the subcontract by the subcontractor after notice, where the contractor fails to rectify a material breach. Again, the list of matters comprising a material breach is set out at section 43, 'Definitions and meanings', as:

> 'By the Contractor:
>
> - failure to issue a payment advice in the manner required by the Subcontract
> - failure to insure, as established in accordance with clause 30.3.
>
> By either party:
>
> - failure to make payment in the manner required by any payment advice issued under the Subcontract;
> - any repudiatory breach of the Subcontract;
> - breach of the CDM Regulations;
> - breach of the requirements of CIS.'

Similar comments as to the benefit of this meaning, which are given in the section on termination by the contractor, also apply here. It is considered that a material breach may be more precisely defined as one which seriously interferes with the subcontractor's ability to perform its obligations under and in connection with the subcontract.

The period to be allowed to the contractor to rectify its breach is 19 days. There is no obvious reason why the contractor should be allowed 19 days when the subcontractor only has 8 or 14 depending on the nature of the breach.

The subcontractor may also terminate the subcontract at any time on the insolvency of the contractor.

In addition to the specific breaches by the contractor, the subcontract may be terminated by either party if the subcontract work is suspended for the stated period on the grounds set out at section 37. These are 'force majeure, the occurrence of any Specified Peril or hostilities involving the United Kingdom, or the use of threat of terrorism as defined by the Terrorism Act 2000'.

The use of the phrase, in clause 7.8.1.3, 'fails to make payment in accordance with the Subcontract' is wider than the provision in section 112 of the HGCRA which is restricted to the situation where 'a sum due under a construction contract is not paid in full by the final date for payment'. Condition 7.8.1.3 is not restricted to 'a payment', nor does it refer to the 'sum due', but it can be considered to relate to the entire payment mechanism. Thus a notice of default might be issued where the contractor fails to issue notices of payment within five days of the payment due date or any other payment default. While it is unlikely that a subcontractor would regard the failure to issue a notice of payment as sufficient grounds for termination, it might consider issuing such a notice and in the event of continued or repeated default, then that notice might give the subcontractor the right to terminate, at such later date, with no further notice. As with the provisions for the contractor to terminate, no notice may be given unreasonably or vexatiously. Clause 7.9 of SBCSub/C makes termination of the subcontractor's employment immediate on the termination of the contractor's employment under the main contract. Clause 7.10 of SBCSub/C sets out the arrangements where the contractor becomes insolvent. In that case the contractor is to notify the subcontractor and the subcontractor is to suspend work for a period of up to three weeks, during which time the subcontract may either be resumed or terminated.

The rights and obligations of the parties, on the termination of the subcontract by the subcontractor and for reasons of either the termination of the main contract or following the insolvency of the contractor, are set out in clause 7.11 of SBCSub/C.

On termination of the subcontract the subcontractor is with all reasonable despatch to remove or procure the removal from the site of its plant and equipment and all goods and materials not the property of the contractor. Although not specifically stated in this form of subcontract, this obligation must be subject to it being safe for the subcontractor to do so and that its work is left in a safe condition.

The practical implementation of this provision will be difficult in many circumstances, especially for those subcontractors where significant temporary works are involved which cannot or would not normally be removed until a much later stage in the progress of the subcontract works. In theory, it will be possible for the parties to agree that the contractor or, in the case of the contractor's insolvency or the termination of the contractor's employment, the employer takes over such temporary works and either purchases outright those materials and/or equipment or pays a reasonable hire for them during the further period they are required on site. Even if, under the circumstances of a major dispute likely to be associated with the termination of the subcontract, such agreement could be reached, then valuing the subcontractor's entitlement could still be difficult.

If, for example, an external cladding contractor has provided its own scaffolding or other access equipment, this may or may not be identifiable as a separate item in the subcontract offer. Its early removal may have resulted in a total cost to the subcontractor not directly proportional to the work actually completed. However, there is likely to be a benefit to the contractor or employer in leaving a standing scaffold and adding to and/or adapting it until all the subcontract works are complete, which will probably be much cheaper than having to erect a new scaffold from start.

In a similar manner to where the termination of the subcontract is due to the subcontractor's breach, the payment provisions of the subcontract are replaced by the payment provisions of clause 7.11.3. These provide that the subcontractor shall with reasonable dispatch prepare and submit to the contractor an account setting out the amounts referred to in items 1–4 below and, if applicable, item 5:

1. the total value of work properly executed at the date of termination of the subcontractor's employment, ascertained in accordance with these conditions as if the employment had not been terminated, together with any other amounts due to the subcontractor under these conditions;
2. any sums ascertained in respect of direct loss and/or expense under clause 4.19 (whether ascertained before or after the date of termination);
3. the reasonable cost of removal under clause 7.11.2;
4. the cost of materials or goods (including subcontract site materials) properly ordered for the Sub-Contract Works for which the subcontractor then has paid or is legally bound to pay;
5. any direct loss and/or damage caused to the subcontractor by the termination.

Clause 7.11.5 of SBCSub/C deals with the reconciliation of the amount then properly due to the subcontractor, which the contractor shall pay to the subcontractor within 28 days of the submission by the subcontractor of its account to the contractor, without the deduction of retention. It is to be noted that under clause 7.11.3 it is the subcontractor that makes the valuation. There is no express provision or requirement in this clause for the issue of a notice of payment or for the issue of a notice of intent to withhold payment. There is therefore no mechanism for ascertaining the amount properly due where, as is likely, the contractor disputes the subcontractor's account, other than by adjudication. It follows therefore that the clear intention of this clause is that the contractor pays the subcontractor's assessed valuation of its account in the first instance. This is considered to be unworkable in practice, since the contractor is unlikely to agree the sum sought or to make payment of the sum sought by the subcontractor.

Determination because of the contractor's failure to pay

Failure to pay in accordance with the payment provisions of the subcontract, for work properly carried out in accordance with the requirements of the subcontract, is unlikely to be regarded as a repudiatory breach since the law provides alternative remedies in the form of a right to suspend performance and the right to statutory compensation for late payment.

It is possible, as in SBCSub/C clause 7.8.1.3, to give a right to termination when payment is not in accordance with the subcontract. Where such a termination is sought, then the subcontractor must ensure that it follows precisely all the requirements of that subcontract, including allowing time for payment and the serving of notices as provided for under those terms.

However, the law will support determination on the grounds of a breach of the subcontract where, by its actions, the contractor shows that it has no intention of paying the further sums due to the subcontractor. It follows that the first requisite is for the subcontractor to establish that a sum is due to it, its value and that the contractor has not only not paid the sum due but is refusing to, and has no intention of, paying that sum. It will not be sufficient to claim that a sum is due and for such a claim to be disputed by the contractor. Such circumstances will not be a refusal on behalf of the contractor to pay, but a dispute over the right to be paid or the sum actually due. Again, either the subcontract and/or the law provide a remedy for resolving such a dispute.

In the case of *Weldmarc Site Services Limited* v. *Cubitt Building and Interiors Limited* [2002] HT-01–408, the subcontractor refused to deliver some steel staircases without receiving payment in advance. The issues were complex but Weldmarc claimed both that these stairs were additional to their subcontract and therefore they were entitled to an additional sum, and also that the terms of the subcontract required that they be paid before delivery. Cubitt disputed both these contentions and refused to pay until the work had been carried out. Cubitt gave notice that unless Weldmarc proceeded with these installations Cubitt would determine the subcontract by engaging others to complete the steel work, which they did and counterclaimed for the cost.

Judge Richard Harry QC found that Weldmarc, despite having been underpaid by Cubitt when it refused to return to site unless paid in advance, was itself in repudiatory breach of the subcontract. He said, at paragraph 64:

> 'Thus the parties proceeded on a common intention that monthly payments were a term of the contract. If monthly payments were a term of the contract, and Cubitt had retained the sum of £9597.55 against sums due, as alleged by Weldmarc, then Cubitt was in breach of the contract. Nevertheless, it is apparent that the sums claimed by Weldmarc on 20 June as a condition of resuming performance of the contract were substantially above any sums due at that date or on the dates demanded for the respective payments. In my judgment, that refusal to resume performance of the contract except on those conditions constituted a repudiation of the contract on the part of Weldmarc.'

In the *Weldmarc* case there appears to have been no contractual basis for Cubitt to have paid for the stairs before delivery and installation. However, even where the contractor is not properly paying the subcontractor under the terms of the subcontract, it will be necessary for the contractor, in addition, to have shown that it had no intention of fulfilling its obligations. In the case of *D. R. Bradley (Cable Jointing) Limited* v. *Jefco Mechanical Services Limited* [1988] OR 1986-D-959, the subcontractor ceased work on the grounds that it had not been paid. The judge found that as a result Bradley had been in breach of the subcontract. However, the judge further found that Jefco had affirmed the subcontract by giving Bradley seven days in which to recommence work. During that period the parties met to discuss payment. Jefco refused to pay the sum due in whole and such refusal was held by the judge to be sufficient cause for Bradley to claim that Jefco had repudiated the subcontract, albeit that Jefco were entitled to damages for the period of unlawful suspension prior to the ultimate repudiation by Jefco. Mr Recorder Rich put this in the following words, at page 18E:

> 'There remained a shortfall. It is also true that payments were barely if ever made within the 42-day period, which I have held was a reasonable time. Nevertheless, applying the test laid down by the Court of Appeal in *Decro-Wall International SA* v. *Practitioners in Marketing Ltd* [1971] 1 WLR 361, the failure to make full and prompt payment did not go to the root of the contract so as to amount to a repudiation by Jefco.
>
> Bradley complained and made threatening noises but took no sufficient steps to make prompt or full payment of the essence of the contract. Accordingly I hold that when they withdrew their labour on 12 December, they were guilty of a repudiatory breach of their contract with Jefco. However Mr Salveson's [the financial director of Jefco's] response on behalf of Jefco was the letter of 12 December (document 170) which I have already quoted. It gave Bradley seven days in which to return. That was not an acceptance of the repudiation but an election to continue the contract, continuing therefore Jefco's duty to comply with its obligations, whilst requiring Bradley to comply with theirs.
>
> I hold that it was after the date of this letter that a meeting took place between Mr Bradley and Mr Salveson. Although I agree with Mr Wynne-Griffiths [Counsel for Bradley] that it is easier to understand the chronology if the meeting preceded the withdrawal of labour and the letter, I see no ground for rejecting Mr Bradley's account of the order of events, whilst accepting, as I do, at least as to its broad effect, his account of the meeting. I hold therefore that within that 7-day period Mr Bradley went to see Mr Salveson to demand payment, claiming that some £31,000 was outstanding. I have held that the amount due was some £25,000. Mr Salveson offered £5,000 only and that was to be subject to a deduction of £2,000 as a set-off against another contract. That refusal to pay at that stage did in my judgment, go to the root of the contract. That in my judgment reasonably shattered Bradley's confidence in being paid, and it may be that Jefco's subsequent liquidation justified such conclusion. Bradley was therefore entitled to treat Jefco as having repudiated, and by their failure to return to work, they did so.'

Should the subcontractor's ability to continue the subcontract work be adversely affected because of the failure by the contractor to make payment as required by the terms of the subcontract, whether such failure is by way of undervaluation or late payment, the subcontractor has a choice of actions. These include: to continue with the work and seek a true valuation and or payment with interest via the dispute resolution process; the issue of a notice of breach of contract prior to determination by the subcontractor; or to proceed at a reduced performance level pending payment of the outstanding sum. This last option is likely to be the most ineffective since it will without doubt place the subcontractor in breach of the subcontract, despite being able to demonstrate a prior breach by the contractor. As stated above, the law only offers two options for a party affected by the repudiatory breach of the other: determination of the subcontract or affirmation by the wronged party.

While the standard forms of subcontract make the sum due to the subcontractor the sum valued by the contractor, it is possible for the subcontract to be set up on the basis that the contractor is to pay on receipt of an invoice from the subcontractor. In such a situation a failure to pay the sum as invoiced will be held to be a breach of the subcontract, which may be held to be a repudiatory breach and thereby allow the subcontractor to regard the subcontract as being at an end.

In the case of *Newton Woodhouse* v. *Trevor Toys Ltd* [1991] OBENF/1130/90 and OBENF/91/0684, the Court of Appeal upheld the judgment of the court of first instance and Lord Justice Glidewell said, at page 4E:

> 'The judge found that Newton Woodhouse contracted to erect the building for a fixed price for labour of £35,614, including overheads, together with the cost to them of materials and the hire of necessary plant. It was expressly agreed that payment would be made by the defendants in stages, against invoices submitted by Newton Woodhouse, and that in return for prompt payment the defendant would be entitled to deduct any trade discounts allowed to Newton Woodhouse on materials.'

And later, at page 5F:

> 'It followed from these conclusions that the judge decided that at 15 March 1982 Newton Woodhouse were entitled to be paid the amounts claimed in invoices 22 and 23, and Trevor Toys repudiated and breached the contract by failing to pay these sums. Newton Woodhouse were therefore entitled to cease work at that date.'

The decision in *Trevor Toys* must be considered on the basis of the terms of the contract and must not be considered to give a right for a subcontractor to seek to determine the subcontract on the basis that the contractor's valuations are low or inadequate. However, it may be that a failure of the contractor to issue a valuation and make payment, either under the express terms of a subcontract or where appropriate the Scheme and especially if repeated, would on *Trevor Toys* principles be held to be a repudiatory breach.

Determination on the grounds that progress is insufficient

The obligation for the subcontractor to carry out the work with sufficient speed is often expressed in the subcontract terms as a requirement for the subcontract work to proceed 'regularly and diligently'. Although this phrase is drawn from the main contract form, the concept is applied as justification for determination of subcontract works (see SBCSub/C clause 7.4.1.2). It is suggested that rarely will the requirements of the contractor from the subcontractor be for the subcontractor to proceed in that manner. In fact, it is likely that for very practical reasons the contractor will rarely be able to so arrange the work of all trades that the subcontractors are able to proceed at a steady rate. In practice, there will often be an express term that no guarantee or warranty to this effect is given, or to be implied, into the subcontract.

Depending on the terms of the subcontract, the subcontractor's obligations will either be to carry out its work within a stated period or to respond reasonably promptly to the contractor's requirements. The former is the most common arrangement and is discussed in more detail in Chapter 4, Programming of the Subcontract Works. Where no indication is given within the Sub-Contract Documents that the work will be released to the subcontractor in a certain way or at a certain rate, it may be argued that the entire works must be available to the subcontractor at the commencement of the subcontract period. However, an obligation for a subcontractor to carry out and complete its work in a fixed period will, almost certainly, carry with it an implied term that the contractor will make available the work to suit the reasonable requirements of the subcontractor, and more certainly in such time as to enable the subcontractor reasonably to fulfil its obligations. Both can be summarised as a requirement not to hinder or prevent the subcontractor from carrying out its obligations in accordance with the terms of the subcontract or executing the work in a regular and orderly manner, as decided in the case of *The London Borough of Merton* v. *Stanley Hugh Leach Ltd* (1985) 32 BLR 51, where Mr Justice Vinelott said, at page 79:

> 'I turn now to the terms which Leach seeks to imply. Implied terms (i) and (ii) are as follows:
>
> (i) [Merton] would not hinder or prevent Leach from carrying out their obligations in accordance with the terms of the contract [and from executing the works in a regular and orderly manner].
> (ii) [Merton] would take all steps reasonably necessary to enable [Leach] so to discharge their obligations [and to execute the works in a regular and orderly manner].
>
> The arbitrator held that these terms ought to be implied.
>
> In my judgment, the arbitrator was clearly right as regards the first of these terms. Lord Justice Vaughan Williams observed in *Barque Quilpué Ltd* v. *Brown* [1904] 2 KB 261 at 274:
>
> "There is an implied contract by each party that he will not do anything to prevent the other party from performing a contract or to delay him in performing it. I agree that generally such a term is by law imported into every contract."
>
> The implication of such a term seems to me to fall clearly within Lord Wilberforce's fourth category. The implied undertaking not to do anything to hinder the other party from performing his part of the contract may, of course, be qualified by a term express or to be implied from the contract and surrounding circumstances.'

While the *Merton* v. *Leach* case concerns the relationship between the employer and the main contractor, it is considered that the same principles apply, perhaps even to a greater extent, in the subcontract situation where, for many trades, the availability of their work is entirely dependent on the progress of others.

In the absence of any express provisions to the contrary, a subcontractor, at law if not in practice, is free to plan and perform the work as he pleases. This principle was established in the case of *Wells* v. *Army & Navy Co-operative Society*, when Mr Justice Wright said, at page 352:

> 'It seems to me clear that subject to certain restrictions as to uniform height of walls, the plaintiffs were to do the works in what order they pleased.'

This judgment was considered in conjunction with the provisions of DOM/1 clause 11.1, and was quoted in the judgment in *Pigott* v. *Shepherd*, at page 57G:

'The subcontractor shall carry out and complete the subcontract works in accordance with the details in the Appendix, part 4 and reasonably in accordance with the progress of the works . . . and to the operation of clause 11.'

This clause is substantially the same as clause 2.3 of SBCSub/C, which requires that the subcontract works shall be carried out and completed in accordance with the programme details stated in the Sub-Contract Particulars (item 5) and reasonably in accordance with the progress of the Main Contract Works, subject to receipt by the subcontractor of notice to commence work.

SBCSub/C, at clause 3.4, gives the contractor wide powers to issue directions described as 'any reasonable direction in writing'. A wide-ranging provision such as this must include giving to the contractor power to issue directions as to the timing and sequencing of the subcontractor's work. However, unless the contractor issues such directions, then the subcontractor's obligation remains to complete on the due date and/or reasonably in accordance with the progress of the main contract works. For subcontracts under the SBCSub/C terms, the right to terminate the subcontract after notice is stated at clause 7.4.1.2 to be for failing to proceed regularly and diligently with the subcontract works. It is not clear how the phrase 'reasonably in accordance with the progress of the main contract works' can be equated to the phrase 'regularly and diligently with the subcontract works', since the former implies or requires reactive performance to that of others and the latter requires a steady output independent of the progress of others.

It had been argued that clause 11 of DOM/1 required the subcontractor to carry out its work in such a way as to conform to the requirements of the contractor's programme and to facilitate the timely progress of following trades. Mr Justice Gilliland put a much less onerous interpretation on that clause, at page 61:

'In my judgment the obligation of the subcontractor under clause 11.1 of DOM/1 to carry out and complete the subcontract works "reasonably in accordance with the progress of the Works" does not upon its true construction require the subcontractor to comply with the main contractor's programme of works nor does it entitle the main contractor to claim that the subcontractor must finish or complete a particular part of the subcontract works by a particular date in order to enable the main contractor to proceed with other parts of the works. The words "the progress of the Works" are in my judgment directed to requiring the subcontractor to carry out his subcontract works in such a manner as would not unreasonably interfere with the actual carrying out of any other works which can conveniently be carried out at the same time. The words do not however in my judgment require the subcontractor to plan his subcontract work so as to fit in with either any scheme of work of the main contractor or to finish any part of the subcontract works by a particular date so as to enable the main contractor to proceed with other parts of the work.'

The judge then went on to consider the requirement for the subcontractor to proceed regularly and diligently with the subcontract works and by reference to *Keating on Building Contracts* (5th edn), quoted with approval as follows:

'It is suggested that this last obligation pre-supposes that the main contract Works are proceeding "regularly and diligently . . ." and that it is not to be construed as obliging the subcontractor to proceed without recompense in accordance with a main contract progress which is not in accordance with the agreed subcontract programme details.'

Mr Justice Gilliland then drew his conclusion that clause 11 is there to enable the subcontractor to complete in time, not to facilitate or further the contractor's rate of progress:

> 'The provisions of clause 11.8 of DOM/1 do not in my judgment assist the defendant in the present case. There is no contractual obligation upon the subcontractor to carry out his work in any particular order or at any rate of progress. The obligation under clause 11.8 is to prevent delay in the progress of the subcontract work and to prevent such delay resulting in the subcontract work overrunning beyond the period fixed for completion.'

Mr Justice Gilliland then goes on to restate the principle that the subcontractor is free to plan and organise its work as it pleases, within the confines of any specific requirements of the subcontract:

> 'Clause 11.8 is worded as a proviso and is introduced by the words "the operation of clause 11"; the same words appear in clause 11.1 and, there, must be a reference to the operation of the following provisions of clause 11, that is to the provisions which entitle the subcontractor to an extension of time if it becomes apparent that the "commencement, progress or completion of the subcontract works . . . is being or is likely to be delayed". In my judgment clause 11.8 does not exclude or modify the general principle applicable to building and engineering contracts, that in the absence of any indication to the contrary, a contractor is entitled to plan and perform the work as he pleases provided that he finishes it by the time fixed in the contract.'

From the above authorities, it is suggested that without very clear and express provisions within the terms of the subcontract, the contractor may not at a later date unilaterally impose a detailed programme requiring the performance of the subcontract works in a certain way or by a certain date, and subsequently claim that the subcontractor's non-compliance is a breach of the subcontract. Or, where there is an express provision for enabling the contractor to issue directions to that effect such as in SBCSub/C clause 3.4, the subcontractor will be able to claim such direction to be a variation for which it may be entitled to an adjustment to the subcontract rate or price.

In any event, should the subcontract works for whatever reason not be available so that the subcontractor can fulfil its obligations, then the contractor should at the very least advise the subcontractor in reasonable time of its changed requirements. In the event that the subcontract works cannot be completed within the subcontract period, by reason of any of its work not being available to the subcontractor, then, it is suggested, time will become at large and the subcontractor will only have had to carry out the works in a reasonable time having regard to all the circumstances, at least until either there is a new agreed completion date or the contractor issues its directions as to its revised requirements. If the contractor would seek to determine the subcontract on the grounds of non-performance by the subcontractor, then the contractor must be able to demonstrate that either it has made the work available in reasonable time so as to enable such performance by the subcontractor, and if not has made known in adequate time and/or instructed its reasonable changed requirements, and the subcontractor's response has been such as to be regarded as unreasonable in all the circumstances.

As described elsewhere, the subcontractor is not, without instructions from or agreement with the contractor, required to accelerate its work to recover time lost due to matters for which it does not carry the risk. It follows therefore where there is delay for reasons at the risk of others, the subcontractor must be instructed by the contractor, should the subcontractor be required to take action to recover lost time. Failure by the subcontractor to respond adequately to such an instruction will provide the contractor with such remedies as may be available to it under the subcontract.

Determination when time is of the essence

Subcontracts are sometimes written stating that time is of the essence. Frequently this term is used loosely implying no more than that the time clause is considered a condition and not just a term of the subcontract. By its true legal meaning, where time is stated to be of the essence the contractor will have the right to refuse performance should the subcontractor be late. This will not always be possible since a subcontractor is likely to have installed part of its work before the date for completion, and it will not therefore be possible at that time for the contractor to refuse performance. It is for this reason that a clause in a construction contract making time of the essence is generally considered to be ineffective or irrelevant.

However, there will be many subcontract situations where the majority of the work will be carried out off-site, with a relatively short, often very intensive, on-site installation process, in many cases only taking a few hours or days to install the results of many weeks' or months' design, procurement and off-site fabrication work. Such activities would include those generally defined under the heading of shopfitting but would also include entrance doors and screens, special gates and other architectural metal work and sundries such as clocks and signs. In all these cases, where performance by a fixed date is made the essence of the subcontract, it will be possible for the contractor to refuse performance if the delivery and/or installation are late. In such circumstances the contractor will have no liability to pay for work carried out.

Even though performance may be expressly stated in the subcontract to be of the essence, the contractor may in the event either accept performance late and/or waive its right to demand performance by the due date. Where the subcontractor's work is not available by the due date, the contractor may indicate that it will still accept the products and their installation late. However, at any time the contractor may give notice indicating that it requires performance by a new date. The contractor may set such date as is reasonable, but in this context reasonableness will be related to the amount of notice given and not to the amount of work remaining to be done at the time of the notice.

The case of *Charles Rickards Ltd* v. *Oppenheim* [1950] 1 All ER 420, while not being a construction case, decided the issue of the nature of notice necessary for the contractor to affirm its right for completion by a fixed date, and thereby making failure by the subcontractor a cause of action, giving the contractor the right to refuse performance. In this case, the employer ordered a Rolls Royce car through a dealer, who arranged for a specialist carriage maker to construct the body. The subcontractor failed to complete by the due date, but the employer indicated that he would allow further time for the subcontractor to complete. Eventually, having been advised that the subcontractor had two more weeks' work to do, the employer notified the supplier via the dealer that he must have the car no later than four weeks from the notice otherwise he would buy elsewhere, which he did. The relevant parts of the judgment of Lord Justice Denning are:

> 'It is clear on the findings of the judge that there was an initial stipulation making time of the essence of the contract between the plaintiffs and the defendant, namely, that it was to be completed "in six, or, at the most, seven months". Counsel for the plaintiffs did not seek to disturb that finding – indeed, he could not successfully have done so – but he said that the stipulation had been waived.
>
> I agree that that initial time was waived by reason of the requests for delivery which the defendant made after March 1948, and that, if delivery had been tendered in compliance with those requests, the defendant could not have refused to accept.
>
> If the defendant, as he did, led the plaintiffs to believe that he would not insist on the stipulation as to time, and that, if they carried out the work, he would accept it, and they did it, he could not afterwards set up the stipulation in regard to time against them. Whether it be called waiver or forbearance on his part, or an agreed variation or substituted performance, does not matter. It is a kind of estoppel. By his conduct he made a promise not to insist on his strict legal rights.
>
> Time and time again the defendant pressed for delivery, time and time again he was assured that he would have delivery, but he never got satisfaction, and eventually at the end of June he gave notice saying that, unless the car was delivered by 25 July he would not accept it. The question thus arises whether he was entitled to give such a notice, making time of the essence, and that is the question which counsel for the plaintiffs has argued before us. He agrees that, if this is a contract for the sale of goods, the defendant could give such a notice.
>
> The judge thought that the contract was one for the sale of goods, but in my view, it is unnecessary to determine whether it was a contract for the sale of goods or a contract for work and labour, because, whichever it was, the defendant was entitled to give a notice bringing the matter to a head. It would be most unreasonable if, having been lenient and having waived the initial expressed time, he should thereby have prevented himself from ever thereafter insisting on reasonably quick delivery. In my judgment, he was entitled to give a reasonable notice making time of the essence of the matter. Adequate protection to the suppliers is given by the requirement that the notice should be reasonable.
>
> In my opinion, however, the words of Lord Parker of Waddington in *Stickney* v. *Keeble* [[1915] AC 386] apply to such a case as the present, just as much as they do to a contract for sale of land. Lord Parker said:
>
> > "In considering whether the time so limited is a reasonable time the court will consider all the circumstances of the case. No doubt what remains to be done at the date of the notice is of importance, but it is by no means the only relevant fact. The fact that the purchaser has continually been pressing for completion, or has before given similar notices which he has waived, or that it is specially important to him to obtain early completion, are equally relevant facts."
>
> To that statement I would add, in the present case, the fact that the original contract made time of the essence.'

Determination by way of a term incorporated from another contract

As discussed above, it is likely that there will be express provisions under the subcontract for determination where the subcontractor persists in specific types of breach, for example breaches of health and safety requirements or persistent failure to fulfil the specification requirements. The subcontract may require that the contractor first issue a notice defining the breach or breaches that the contractor requires to be rectified, stating that failure to rectify the breach will give rise to an entitlement for the contractor to determine the subcontract. Where such a notice has been properly issued and, contrary to the provisions of the subcontract, not complied with, there will normally be a requirement for the contractor to give a further notice and only then will the subcontract be determined.

Even where there are no express provisions for determination, on the grounds listed above, the contractor may be able to claim such a right of determination either by reason of normal trade custom and practice or by reason of a term incorporated from the main contract or implied as a matter of business efficacy.

Where there is a requirement, stated in the provisions of the main contract, for the contractor to dismiss the subcontractor or determine the subcontract on an instruction given under the main contract, there may need to be express provision for such determination within the subcontract for the contractor to avoid being liable to the subcontractor for damages in the event of determination of the subcontract on those grounds. Such a situation arose in the Court of Appeal case of *Chandler Bros Ltd* v. *Boswell* [1936] CA 179. In that case the main contract, between Carnarvon County Council and Boswell, allowed for the engineer to require the contractor to remove the subcontractor from the works. Lord Justice Greer described this situation, at page 182:

> 'In the head contract there was a special clause to this effect (clause 54) under the head of "subcontractors," which said: "If the engineer shall at any time be dissatisfied with any subcontractor or other person similarly employed or engaged by the contractor upon or in connection with the works, the contractor shall, if required in writing by the engineer to do so, forthwith remove him or them and put an end to his or their employment or engagement." Now, it is quite clear that a contractor may undertake, as between himself and the head employer – in this case the Carnarvon County Council – something against which he may fail to guard himself in his contract with his subcontractor, and in order to find out the rights as between the contractor and subcontractor we have got to look at the subcontract. In the subcontract it seems to me perfectly clear that both parties, the contractor and the subcontractor, when they entered into their contract in December 1931, I think, had before them the contract which had been made, or was about to be made, between the head contractor and the county council, and they contemplated that they had clauses in it which would have the result of putting the contractor and the subcontractor into a difficult position, and under those circumstances, knowing that that might happen, or that there might be a decision by the engineer of the head contractor that a subcontractor should be ordered to cease work and to be no longer admitted on to the work to continue his job, the question we have to determine is whether in this subcontract, having regard to what I have called, and I think rightly called, an express omission of any reference to clause 54, we can imply a term that in the event of the engineer calling upon the contractor to dismiss a subcontractor, there is an implied term in the contract that in that event the contractor shall be entitled, as between him and the subcontractor, to put an end to the subcontract.'

The court were not prepared to imply into the subcontract a term to the effect that should the engineer, under the main contract, require the removal of the subcontractor, this gave the contractor a right to determine the subcontract as necessary and without penalty. It was stated that the parties had considered the main contract provisions at the time of the Sub-Contract Agreement, and failed to incorporate an express condition to the effect that they now sought to imply, and that the court could not imply a term which the parties had not considered necessary at the time of their agreement.

In addition to the specific terms incorporated into the subcontract by express agreement, there was a general agreement that the subcontract work would be carried out 'in accordance with the terms of the principal contract'. This phrase was taken to mean only 'that he was to provide work of the quality and with the dispatch which was stipulated for in the head contract'.

The effect of *Chandler*, it is submitted, is that for a provision under the main contract allowing for determination of the subcontract to be effective, there must be an express provision to that effect within the subcontract. This should define in clear words any limitation for compensation to the subcontractor for any loss the subcontractor may incur as a consequence of determination on these grounds. While a general obligation, such as that in SBCSub/C clause 2.5, may be held to give a contractor the right to determine a subcontract under such circumstances, it will not necessarily deny the subcontractor the right to be compensated for its associated loss.

Determination at common law

As stated above, the law does not readily accept that a breach of a subcontract condition is so serious as to warrant the parties believing that the contract is at an end. Such a situation would be where the contractor has good reason to believe that the subcontractor has abandoned the work and has no intention of continuing with it. One frequent ground for considering that the subcontractor can be held to have abandoned the contract is when this is evidenced by its removal of labour, plant and materials from the site. A failure to provide labour on a certain day or the removal from site of some plant or some material would not, in itself, demonstrate a repudiatory breach. However, unless there is express provision under the subcontract and/or express agreement, the engaging of another subcontractor by the contractor to carry out the work contracted to the first subcontractor would probably be evidence of repudiation by the contractor.

Similarly, a failure by the contractor to pay the subcontractor will not of itself be a repudiatory breach. In the case of *Decro-Wall* v. *Practitioners in Marketing Ltd*, the Court of Appeal had to consider whether the determination of a contract to market wall-covering materials had been lawful on the grounds that the marketing company had failed to pay the bills of exchange against which the product had been supplied. It was a fact that all but one bill had been paid late, but always before the next bill was due, and that on average payment was some eight days late. Further, the supplier had at various times throughout the contract extended by agreement the period for paying the bills. While the court found the marketing agents to have been in breach of the contract, that breach did not go to the root of the contract and could adequately be remedied by damages. Lord Justice Buckley said, at page 379:

> 'It is said on behalf of the plaintiff company that, if one party to a contract manifests an intention not to perform in accordance with the contract some part of his unperformed

obligations thereunder throughout the remainder of the subsistence of the contract, the other party is entitled to treat this as a repudiation of the contract. This, it is said, is so, however insubstantial the threatened departure from the due performance of the contract may be. I cannot accept this contention.'

After referring to a contract for the sale of timber where 'the suppliers were proposing to deliver goods which were not in accordance with the contract', he says:

> 'Each party to an agreement is entitled to performance of the contract according to its terms in every particular, and any breach, however slight, which causes damage to the other party will afford a cause of action for damages; but not every breach, even if its continuance is threatened throughout the contract or the remainder of its subsistence, will amount to a repudiation. To constitute repudiation, the threatened breach must be such as to deprive the injured party of a substantial part of the benefit to which he is entitled under the contract. The measure of the necessary degree of substantiality has been expressed in a variety of ways in the cases. It has been said that the breach must be of an essential term, or of a fundamental term of the contract, or that it must go to the root of the contract. Various tests have been suggested.
>
> I venture to put the test in my own words as follows:
>
> Will the consequences of the breach be such that it would be unfair to the injured party to hold him to the contract and leave him to his remedy in damages as and when a breach or breaches may occur? If this would be so, then a repudiation has taken place.'

On the basis of this definition, depending on the circumstances of the injured party, a similar breach may be repudiatory in one case but not in another. Such a circumstance might be the creditworthiness of a subcontractor, and a subcontractor with a low amount of working capital or overdraft facility at the bank might be able to regard a breach of the payment terms of the subcontract as being more serious than a subcontractor with substantial capital or a substantial overdraft facility at the bank.

For there to be a repudiatory breach on the grounds of non-payment or late payment, there will need to be a clear intention on behalf of the contractor not to make further payments and/or to pay in accordance with the agreement. In these circumstances, the subcontractor will have an alternative remedy by reason of section 112 of the HGCRA, which is to suspend performance after due notice. Care must be taken, however, to distinguish between an intention not to pay the sums agreed and a dispute as to the sums agreed and/or the correct basis of payment. If there is disagreement as to the payment due, that will not be grounds for determination and may not be grounds for suspending performance, but will be a dispute under the subcontract. Such dispute must be resolved by the procedures set out for that purpose in the subcontract or at law.

Both these actions frequently interface, in that the subcontractor believes that he has not been paid properly and/or at all and as a consequence either abandons the site or reduces his performance to a level well below that required. Many of the reported decisions appear to be in conflict. This is almost certainly because the courts have to find one party to have been in breach of the subcontract and to be held, by a specific action, to have repudiated the subcontract. This will be particularly difficult where, as so often happens, there is a progressively worsening relationship between the parties over a period of time. It is to be hoped that the provision under section 112 of the HGCRA, allowing for suspension of performance after notice while payment issues are resolved, will reduce such difficulties. It remains to be seen under what circumstances, if any, the courts may find that a suspension of performance after due notice constitutes a repudiatory breach.

In the case of *Sweatfield Limited (formerly JT Design Build Limited)* v. *Hathaway Roofing Limited* (31 January 1996, unreported), Sir Michael Ogden QC had to decide a dispute over a roofing subcontract. The contractor alleged that the subcontractor was not proceeding fast enough, while the subcontractor claimed that no additional operatives could be employed economically. The contractor then brought additional roofing labour to site but claimed they were only carrying out rectification work to the roof purlins fixed by others. In the event, the second roofing contractor commenced fixing guttering, part of Hathaway's subcontract work. Hathaway then withdrew its own labour. The judge heard much evidence as to the detailed circumstances, but it was evident that the second subcontractor was on site prior to Hathaway withdrawing its labour and was fixing Hathaway's materials; this appears to have been of particular relevance.

In the event, the judge picked up on a statement by Hathaway's contracts director to the effect that he could not take responsibility for safety issues where unknown operatives were working alongside his own but outside his control. The judge, at page 19, found the contractor had determined the subcontract by its actions:

> 'Of course it must be remembered that steel erectors and roofers are probably more at risk of serious accident than any other trade in the construction industry. I consider that what happened was a matter of great gravity. I conclude that JT was plainly in repudiatory breach of contract, that Hathaway was entitled to regard it as such and accepted it by removing its men from site and also by writing the letter to that effect on the same day.'

Whether the judge would have considered the employment of others a repudiatory breach if there had not been the safety aspect, is uncertain; however, he does make the point in the passage quoted above that the acceptance of the contractor's repudiation was achieved both by the removal of labour and by confirmation by the subcontractor the same day.

While there can be no doubt that the courts regard breaches of subcontract involving breaches of safety requirements and/or regulations as very serious, nevertheless a breach of such requirements by an employee or even employees is unlikely to be regarded as a repudiatory breach. The Court of Appeal, in the case of *IJS Contractors Limited* v. *Dew Construction Limited* (2000) 85 Con LR 48, had to consider the effect on the subcontract where a group of operatives in the subcontractor's employment were found smoking marijuana.

The work involved the refurbishment of a railway station and all operatives were required to undertake a Personal Track Safety Course (PTS) when they were made aware that the use of alcohol and drugs was forbidden on site.

The judge at first instance, who was quoted at paragraphs 22 and 23 by Lord Justice Aldous, had been required to consider whether there was an implied term as follows:

> 'The first question is whether it was an implied term of the agreement between the parties that the claimant would be bound by the regulations set out in the Track Safety Handbook 1995.'

The judge decided that it was in the following words:

'Having heard the evidence of Mr Rooker [the managing director of IJS], I am entirely satisfied that this was an implied term of the agreement. Indeed, in cross-examination he said:

"It was a prerequisite to our undertaking the work that the operatives would undertake a track safety course and be issued with certificates . . . It was mandatory: all had to undergo the PTS course . . . I was fully aware of clause 2.1 in [CB 12X] . . . All operatives had to be in possession of PTS cards at all times. If there was no card the operative could not work. We would not allow operatives on site without a card . . . There were absolute strict rules about working on railway stations . . . There has been some relaxation since but I am not privy to it . . . We had an internal induction, explaining the rules on drugs and alcohol and the measures we would take if they breached those rules. I accept [the incident] was potentially very serious and I dealt with it."

Thus no operative was to be allowed to work on site unless he held a Personal Track Safety Certificate and it is absolutely plain that the rules set out in the Track Safety Handbook 1995 form an integral part of the training process by which the award of the PTS is achieved. Every employee was aware that the use of alcohol and drugs was forbidden on site and every employee was aware of the consequences in the event that such rule was breached.

It follows that the incident which gave rise to the dismissal of the five employees of the claimant amounted to a breach of the agreement between the claimant and the defendant.'

However, the Court of Appeal overturned that judgment in the following words of Lord Justice Aldous, who said:

'24. With respect, it does not follow that the marijuana incident amounted to a breach for the reasons given by the judge. As I have pointed out, the reply admitted the implied term: also it was not in dispute that the employees had to be properly instructed and carry the appropriate card. It did not follow from that, that a breach of rules by an employee meant that there had been an incident which was contrary to any implied obligation upon IJS.

25. There is, in my view, difficulty in implying into a contract, absent evidence, a term that IJS had to ensure that their employees complied with all the instructions in the handbook. For example, instruction 2.1 was in these terms:

"2.1. Instructions concerning alcohol and drugs.

You must NOT report for work on railway premises if you are unfit through alcohol or drugs of abuse.

You must NOT consume alcohol or any drug of abuse while at work in railway premises.

You must NOT be in possession of any drug of abuse while at work on railway premises."

26. To ensure that that was complied with, IJS would have to search each employee every day so as to prevent them from reporting for work under the influence of alcohol. On the strict terms of that clause it would not be sufficient to send them home or dismiss them when they had reported. However, the workforce was one which was employed by IJS and Dew had to look to IJS to see that the appropriate safety precautions were complied with. It follows, in my view, it would seem necessary to imply some term requiring IJS to ensure compliance with at least some of the safety requirements.

27. For my part I am uneasy implying a term requiring IJS to ensure compliance by its employees with all the terms of the safety handbook, but will assume that there was such an obligation on IJS. Upon that assumption I come to the next issue, namely whether the

> breach, the smoking of marijuana on site, amounted to a fundamental breach such that Dew were entitled to terminate the contract forthwith.'

While it was accepted that a subcontractor is responsible for the actions of its employees, a single breach of safety regulations by an employee or group of employees contrary to instructions given to them by the subcontractor and without the knowledge of the subcontractor's management is not itself a breach of the subcontract by the subcontractor. Lord Justice Aldous stated the law, at paragraph 35:

> '35. Despite the terms of the letter, it does not appear to me that this single breach of an implied term to ensure employees would abide by the safety handbook was so serious a breach of the contract as to entitle the defendant to bring it to an end forthwith. The fact that an employee breaks his contract of employment, thereby entitling the employer to terminate that contract, does not mean that the contract upon which the employee was working would also be terminable. The two contracts are very different. The employee by breaking his contract demonstrates that he is unfit to be employed under his contractual terms. But it does not follow that the employer has been shown unfit to continue with the contract, particularly when appropriate disciplinary action was taken immediately. It cannot be said that IJS renounced its obligations under the contract or rendered them impossible of performance. The particular incident had been dealt with immediately with the result that the employees could no longer work on the project. Thus, the immediate likelihood of it recurring had been removed. Further, the consequences of the breach were not such as to deprive Dew of substantially the whole benefit of that which they had obtained from the contract. The incident had not and did not affect the working of the contract and the summary dismissal of the employees meant that the contract could continue, hopefully without further incidents.'

Wrongful and invalid determination

A determination can be said to be wrongful when the grounds for which it is claimed the subcontract was determined, either do not in fact exist or are not grounds on which a determination may be made. Alternatively, a wrongful determination can be said to be one where there is no just cause for the determining party to take such action. An invalid determination can be distinguished from a wrongful determination as being one where the party determining the subcontract does not follow the procedures set out in the subcontract for determination.

If there is no just cause, then the determination will be unlawful and the party against whom the determination has been made will have an action in damages against the determining party. Where the wronged party is the subcontractor, such damages will include the cost of determination and the losses to overheads and profit from the shortfall in earnings. Where the determination is against the subcontractor, then such costs will include the costs of engaging a new subcontractor, including the increased cost of the work and the contractor's administration costs, and the costs of any consequential delay.

Where the subcontract is determined on just grounds and the procedures of the subcontract are followed, then the parties will have the rights and obligations set out for them under the terms of the subcontract, and only such rights and obligations. Thus where SBCSub/C is the form of subcontract, a valid termination by the contractor will entitle it not to make any further payment to the subcontractor until 'after the completion of the Sub-Contract Works and the making good of defects'.

However, when the determining party fails to carry out the procedures as set out in the terms of the subcontract, the termination will be invalid even though the grounds may be just. Under such circumstances, it is suggested that the rights and obligations incorporated in those terms, for the consequences of a valid determination, will not hold. Thus a contractor who, for example, fails to give the proper notices to a subcontractor before termination, will not be able to rely on a further term allowing it not to make any further payment to the subcontractor until the completion of the subcontract works. The obligation will be to pay for the work done less any substantiated set-off, which will be subject to the subcontract payment mechanism, including the need to issue a valid notice of intent to withhold payment, and in addition any subcontractor's losses.

Suspension

Under many forms of subcontract, the contractor will have the right under specific circumstances to suspend the subcontract works. In other situations this may be included by incorporation of provisions within the main contract, or even as an implied term. There may be requirements for the contractor to issue a notice in writing, stating the period of suspension, and to affirm the suspension at regular intervals. In the event that the suspension exceeds a certain stated period, the subcontractor may by way of notice to the contractor be able to determine unilaterally its subcontract. Such procedures are provided for in SBCSub/C, at clause 7.8.1.1, where suspension is without reasonable cause. A reasonable cause will be where the suspension is under the provisions of the main contract as advised to the subcontractor.

SBCSub/C, at clauses 3.21 and 3.22, expressly allows the contractor to instruct the subcontractor to suspend work where the contractor suspends its performance under the main contract, pursuant to clause 4.14 of the main contract conditions. First, the contractor is to copy to the subcontractor any written notice the contractor may issue to the employer of the contractor's intention to suspend performance of its obligations under the main contract. Then, if the contractor does suspend performance of those obligations it may direct the subcontractor to cease the carrying out of the subcontract works and give such further directions as are necessary. If the contractor resumes its performance under the main contract, then the contractor is to issue such directions to the subcontractor as are necessary.

ICSub/C has similar provisions at clauses 3.19 and 3.20.

Suspension of the works by the contractor, for whatever reason, must be distinguished from suspension by the subcontractor on grounds that it has not been paid the sum due under the subcontract. This is a specific provision of the HGCRA, which at section 112 states:

> '112. (1) Where a sum due under a construction contract is not paid in full by the final date for payment and no effective notice to withhold payment has been given, the person to whom the sum is due has the right (without prejudice to any other right or remedy) to suspend performance of his obligations under the contract to the party by whom payment ought to have been made ("the party in default").

(2) The right may not be exercised without first giving to the party in default at least seven days' notice of intention to suspend performance, stating the ground or grounds on which it is intended to suspend performance.

(3) The right to suspend performance ceases when the party in default makes payment in full of the amount due.

(4) Any period during which performance is suspended in pursuance of the right conferred by this section shall be disregarded in computing for the purposes of any contractual time limit the time taken, by the party exercising the right or by a third party, to complete any work directly or indirectly affected by the exercise of the right.
Where the contractual time limit is set by reference to a date rather than a period, the date shall be adjusted accordingly.'

What constitutes the sum due under the subcontract has been discussed in Chapter 9, Payment, and Chapter 13, Damages. Suffice it to say that it is of vital importance for the subcontractor to be able to demonstrate that the sum due under the contract has not been paid, before suspending performance under this provision. A wrongful suspension, even if correctly carried out, will be a breach of the subcontract, which will almost certainly be considered to be sufficiently serious to be repudiatory. A wronged subcontractor thereby becomes the wrongdoer and liable to pay damages for any loss suffered by the contractor as a consequence of the suspension. If the contractor accepts such a breach by the subcontractor, then the subcontractor will probably forfeit its rights to further payment until such damages can be assessed.

Section 112(2) of the HGCRA requires the subcontractor to give seven days' notice prior to suspension. Some bespoke subcontracts make the subcontractor's right to suspend for non-payment subject to a longer period of notice. It is considered that such provisions do not defeat the subcontractor's right under section 112 of the HGCRA to give only seven days' notice.

Section 112(2) of the HGCRA does not expressly state that the giving of the notice is to be after the final date for payment; however, it may be held that such notice may not be given until the contractor is actually in default. One problem for the subcontractor is that there may not be a clearly defined payment mechanism as required by section 110(1), or such mechanism itself may be in dispute. In such situations the subcontractor will be on safer ground to seek a decision of an adjudicator as to the payment mechanism and the sum due, before seeking to suspend performance. In any event once such a decision has been obtained for one disputed payment, the payment mechanism will be established for the duration of that subcontract.

SBCSub/C, at clause 4.11, makes provision for the subcontractor to suspend performance on the basis of the contractor's failure to pay in full the sum required by the subcontract, by the final date for payment. Although this form uses the words 'required by' rather than the words of the HGCRA of 'amount due', it is not considered that this is of any significant effect. If it were held that the amount 'required' under SBCSub/C was less than the 'amount due', then clause 4.11 would not satisfy section 112 of the HGCRA and the provisions of the Scheme would apply. The subcontractor's right to suspend is subject to its having given to the contractor a notice in writing of its intention and the contractor having continued in its failure to pay in full.

ICSub/C has a similar provision at clause 4.13. ShortSub at clause 13 provides for suspension by the subcontractor on the grounds of failure by the contractor to pay the 'sum due'.

There is considerable confusion as to the meaning and effect of the phrases 'amount due' in the HGCRA and 'required by' used in SBCSub/C. It has been argued that where the payment mechanism is correctly carried out and the contractor, within five days of the payment due date, issues a notice of payment which fully complies with the requirements of section 110(2) of the HGCRA and any additional express requirements in the subcontract, then the amount stated in the notice of payment is the 'amount due' and/or 'required by' the subcontract. Alternatively it can be argued and is, it is suggested, generally accepted, that the phrases 'amount due' and 'required by' must relate to the true valuation, irrespective of that made by the contractor by way of an interim valuation.

Further, since the contractor has a duty, either express or implied, to value the works properly, then where it values the work at a lesser sum than the subcontractor's application and the contractor's valuation is shown to have been low or at least unreasonably low, then the contractor will have in been breach of the subcontract on that account, in any event.

In valuing the amount due, the works are to be valued as if correctly carried out and, if appropriate, reduced by a sum or sums to cover for the necessary rectification work. The value of such abatement is to be the cost to the subcontractor of rectifying the defective work and not the likely cost to the contractor in the event of such rectification work not being carried out by the subcontractor. In the case of *C. J. Elvin Building Services Limited* v. *(1) Peter Noble, (2) Alexa Noble* [2003] EWHC 837 (TCC), albeit in a case where the contractor suspended work because of the failure by the employer to pay under a contract where the HGCRA did not apply, the judge repeatedly made this point. He said, at paragraph 31:

> 'The value of the works and the amount payable to the claimant must take into account the extent to which work was complete or incomplete as the case may be and it should have taken into account the defects, if any, in the work and/or materials provided by the claimant. Since no issues were raised as to repudiation on the part of the claimant prior to suspension the appropriate method of valuation of defective works is the amount which it is probable that it would have cost the claimant to put right the defects.'

And at paragraph 57:

> 'I now turn to Part B of Mr Ellis' [the joint single expert's] Scott Schedules. I do this in the context, at this stage of the judgment, of the need to determine what it would have cost the claimant to put right established defects.'

This principle has been fully supported by other authorities and commentaries, which were set out at length in the judgment of Judge Stannard in *William Tomkinson* v. *The Parochial Church Council of St. Michael-in-the-Hamlet*. In that case the Church had rectified both damage caused by the contractor and the defective work executed by it. The judge held that the Church had every right to employ other contractors to rectify the damaged and/or defective work prior to practical completion and/or the expiry of the defects liability period, but that the value of the resultant damages was limited to the costs the contractor would have incurred had it been allowed to rectify those defects. The judge said, on page 22:

> '... Thirdly, clause 2.5 is concerned with the mitigation of loss, in that it confers on the contractor a right to reduce the cost of remedial works by undertaking them itself. Effect

> is given to this last aspect of the clause if the damages recoverable by the employer for his outlay in correcting defects in the works are limited to such sum as represents the cost which the contractor would have incurred if he had been called on to remedy the defects.'

This decision was quoted with approval by Judge Richard Havery QC in the case of *Guinness plc* v. *CMD Property Developments Ltd (formally Central Merchant Developments Ltd) and Others* [1995] 76 BLR 40 at page 56F.

It must be clearly understood that the limitation to the amount that may be recovered by a contractor from a subcontractor, to the likely cost of rectification by the subcontractor, is subject to the subcontractor having an express right under the subcontract to rectify such defective work. Such right will frequently be in the form of an express obligation on the subcontractor to rectify its defects.

Where the suspension is valid, the HGCRA at section 112(4) makes provision for the period of suspension to be allowed to the subcontractor or disregarded when 'computing the contractual time limit'. It is to be noted that this section also refers to a 'third party' in the words 'by the party exercising the right or by a third party'. Thus if the work of the subcontractor is delayed due to the suspension of work under these provisions by another subcontractor, then the first subcontractor will also be entitled to a similar extension of time.

The HGCRA makes no reference to damages that may result from suspension. It is likely that suspension by the subcontractor will cause delay and/or disruption to the contract works. Where the suspension has been invalid or wrongful, there can be little doubt that the contractor will have a right to recover from the subcontractor the losses incurred by the contractor both during the period of suspension and as a consequence of the invalid or wrongful suspension.

What is less certain is whether the subcontractor can claim and recover any damages resulting from a lawful and justified suspension of its work. While it can be argued that such costs arise from the default of the contractor, in failing to pay the subcontractor in accordance with the terms of the subcontract, and therefore that such losses are losses flowing naturally from that breach, it can also be argued that suspension is itself a remedy under the subcontract for that breach, and the one the subcontractor has chosen to adopt in preference to the other remedies available to it, such as the substantial compensation for late payment under the provisions of the Late Payment of Commercial Debts (Interest) Act 1998.

There can be little doubt, however, that a third party subcontractor, whose work is delayed or otherwise affected by the subcontractor's suspension, will be entitled to recover its losses from the contractor, either by way of loss and/or expense under SBCSub/C clause 4.19 or such similar provision under the subcontract, or otherwise by way of damages.

Chapter 13
Damages

General

'Damages' is the general term given for the compensation to be paid to a party which has suffered loss as a result of the breach of a subcontract. In assessing the value of such damages the aim of the court or tribunal is to place the wronged party, so far as money can, in the position it would have been in had there not been the breach. It must be noted that the aim is compensation for loss, not to punish or otherwise penalise the party in breach.

In assessing the liability to damages, the courts will only consider those losses that flow naturally from the breach and/or were reasonably in the contemplation of the parties at the time the subcontract was entered into. This principle was established in 1854, in the judgment in the Court of Exchequer known as *Hadley* v. *Baxendale* (1854) 9 Exch. 341, when Baron Alderson said:

> 'The damages which the other party ought to receive in respect of such breach of contract should be such as may fairly and reasonably be considered either arising naturally, i.e. according to the usual course of things, from such breach of contract itself, or such as may reasonably be supposed to have been in the contemplation of both parties, at the time they made the contract, as the probable result of the breach of it. Now, if the special circumstances under which the contract was actually made were communicated by the plaintiffs to the defendants, and thus known to both parties, the damages resulting from the breach of such a contract, which they would reasonably contemplate, would be the amount of injury which would ordinarily follow from a breach of contract under these special circumstances so known and communicated. But, on the other hand, if these special circumstances were wholly unknown to the party breaking the contract, he, at the most, could only be supposed to have had in his contemplation the amount of injury which would arise generally, and in the great multitude of cases not affected by any special circumstances, from such a breach of contract. For, had the special circumstances been known, the parties might have specially provided for the breach of contract by special terms as to the damage in that case; and of this advantage it would be very unjust to deprive them. Now the above principles are those by which we think the jury ought to be guided in estimating the damages arising out of any breach of contract.'

This judgment gives rise to the concept of reasonable foreseeability and the definition of consequential loss, discussed below.

It is not uncommon for a subcontractor to seek to limit its exposure to damages and this can be done in at least two ways. First, a limit may be put on the value of damages recoverable, expressed either as a defined sum or as a percentage of the subcontract sum. Such limitation may be for any breach or for all breaches under the subcontract. Again they may be limited to certain classes of breach such as delay. Secondly, the limit may be expressed by reason or cause and/or consequence. Thus it might exclude negligence of a plant operator under the control of the hirer, or exclude damage to the contents of a building where work is being carried out. Difficulties can be experienced with the meaning of words, for example the use of the phrase 'consequential loss' has been interpreted by the courts as not including losses that flow naturally from the breach.

In addition, the Unfair Contract Terms Act 1977 will restrict the effect of exclusion clauses. However, for subcontractors these limitations will themselves be restricted since a subcontractor will, it is suggested, always be acting 'in the course of business'. While the courts are likely to uphold terms which limit or restrict liability, they are generally unsympathetic to total exclusion clauses.

A claim for damages can, of course, flow in either direction. An employer may suffer a loss because the project is completed late or because there are defects in the work. A subcontractor's losses may flow from either the work not being available to schedule or not in the manner expected or as a result of not being paid on time. The contractor's losses against the employer will include losses paid to its subcontractors for matters that result from actions by the employer, in addition to any direct loss it may suffer. Its losses against its subcontractor will include charges against it from either or both the employer and other subcontractors in addition to its own direct losses, arising from an action or inaction by the subcontractor or breach of the terms of the subcontract.

Damages will normally be the repayment of proven actual loss and will not normally be paid on an assessment basis. However, where it can be shown that the party has suffered a loss, then it will be for the tribunal to make an assessment of that loss on the basis of the evidence available. It will not be for the tribunal to find that a loss has been suffered and fail to value it because the evidence as to value is limited and/or uncertain. Damages may not necessarily have required the actual payment to others; a party may be entitled to damages because it has not got what it bargained for, but settles for the loss in value rather than the cost of rectification. In certain circumstances it may be held that the entitlement is only for loss of amenity. Where relevant the contractor will seek recompense for damages payable to the employer for the above from the subcontractor.

Reasonable performance

Where the subcontract makes no express provision with regard to the performance of the subcontract, the law may imply into the subcontract that performance is to be reasonable. In determining what is reasonable the court is likely to consider such matters as:

1. normal trade custom and practice;
2. whether the parties normally trade under specific terms;
3. what is reasonable under the circumstances.

A common situation in building subcontracts will be that either there is no defined time in which the subcontract work is to be carried out, or for various reasons the time stated in the subcontract has become at large. In the case of

Pantland Hick v. *Raymond & Reid*, the House of Lords had to consider the meaning of reasonable time where there was no period stated in the contract for discharging the cargo from a ship. Lord Herschell LC said, at page 29:

> 'I would observe, in the first place, that there is of course no such thing as a reasonable time in the abstract. . . . It appears to me that the appellant's contention would involve constant difficulty and dispute, and that the only sound principle is that the "reasonable time" should depend on the circumstances which actually exist. If the cargo has been taken with all reasonable despatch under those circumstances I think the obligation of the consignee has been fulfilled. When I say the circumstances which actually exist, I, of course, imply that those circumstances, in so far as they involve delay, have not been caused or contributed to by the consignee.'

Thus a contractor seeking damages from a subcontractor on the grounds that the subcontractor's performance was dilatory, and where the subcontractor's obligation was to perform in a reasonable time, will have to demonstrate not what would under normal circumstances have been a reasonable time but that the time actually taken was unreasonable because of the subcontractor's failure to perform reasonably.

The burden of proof

It is a basic principle of law that 'he who asserts must prove'. In the case of the subcontractor's failure to complete by an agreed date, the contractor will only have to show that the subcontract works were not finished by the due date to be entitled to damages. It will then be for the contractor to prove its damages. The subcontractor may well have a defence that it was prevented by acts of the contractor from performing the works and that either the time for performance should be extended or that time has become at large.

Where the time for the subcontractor to carry out its work is not defined and its obligation is to complete in a reasonable time, then it will be for the contractor to demonstrate what was a reasonable time, as well as proving its loss. Again by reference to *Pantland Hick* v. *Raymond & Reid*, and the words of Lord Herschell LC, at page 28:

> 'The bills of lading in the present case contained no such stipulation (period of time for the unloading), and, therefore, in accordance with ordinary and well-known principles the obligation of the respondents was that they should take discharge of the cargo within a reasonable time. The question is, has the appellant proved that this reasonable time has been exceeded?'

When the matter of the claim is for money, either by way of loss and/or expense under the contract or damages, the claimant will have to prove to the civil law standard that by reason of the other party's default it has suffered a loss for the reason(s) claimed. Once it has established its right to be recompensed, it will not be put to strict proof as to the quantum, which may be assessed by the tribunal on the basis of the evidence available.

In the case of *Ascon Contracting Limited* v. *Alfred McAlpine Construction Isle of Man Limited* [1999] 1998-ORB-315, Judge Hicks QC made this point very forcefully:

> '42. Although he included it under this head Mr Darling [Counsel for McAlpine] also raised what is conceptually the wholly distinct objection that there was no direct factual evidence of any of the expenditure. It is true that technically the entries in Ascon's accounting system, and even the invoices and wage records which Mr Kirwood [the quantity surveyor expert for Ascon] did inspect, are hearsay evidence. It is, however, unheard of in my experience for strict proof, of the kind required (apart from statutory exceptions) in criminal courts, to be regarded as necessary in substantial construction industry litigation. The documents on which Mr Kirwood relied were disclosed on discovery and available to the quantity surveying expert witnesses on both sides on the understanding that they would form the basis of their evidence on the quantum of Ascon's claim, in the same way as no doubt McAlpine's documents did in relation to its counterclaim.'

Judge Thornton QC, in the case of *Norwest Holst Construction Ltd* v. *Co-operative Wholesale Society Ltd* [1998] All ER (D) 61, also made it very clear that once it had been established that a sum was due from one party to the other as a result of that party's loss, it was open for the assessor of that loss to quantify that loss or losses in general terms. Judge Thornton made his judgment in the following words, in paragraph 362:

> 'In this award, the arbitrator has made unimpeachable findings of fact that the heads of loss were incurred. It was then open for him to qualify those losses in general terms without CWS having to prove the precise number of hours worked by management or the precise number of units of electricity used up.'

Refer also to the section 'Loss and/or expense' in Chapter 8.

Where neither party is at fault

Either by express agreement between the parties or by implication at law, the parties may have to bear their own losses as consequence of events which have disrupted or delayed the progress of the subcontract works. Under these circumstances, the loss is said to 'lie where it falls'. Such circumstances will include the costs arising out of delays caused by third parties where time is at large, or events agreed by the parties to be neutral events, under the provisions of the subcontract. Again by reference to *Pantland Hick* v. *Raymond & Reid*, where the unloading was delayed by a strike of dock labourers, Lord Ashbourne, referring to an earlier case, quoted as follows, at page 35:

> ' "We are aware of no authority for saying that the law implies a contract to discharge in the usual time, . . . We think that the contract which the law implies is only that the merchant and shipowner should each use reasonable despatch in performing his part. . . . If this be so, the delay having happened without fault on either side, and neither having undertaken by contract, express or implied, that there should be no delay, the loss must remain where it falls." '

Many subcontracts provide express circumstances for which the contractor is to grant an extension of time. In SBCSub/C, these are set out generally at clause 2.18, and 'Relevant Events' are listed at clause 2.19. The subcontract may also define the circumstances for which the subcontractor will be entitled to its loss and/or expense. In SBCSub/C, this is done at clause 4.20. To the extent that these may differ, by allowing to the subcontractor additional time for reasons beyond the control of either party, such matters may be considered 'neutral' events for which the loss or losses will lie where they fall.

Damages from and to the subcontractor

Damages for delay in construction contracts are frequently calculated in advance as a genuine pre-estimate of loss, known as liquidated and ascertained damages (LADs). It is rare that a subcontract incorporates LADs, although the LADs applicable under the main contract will normally be stated so as to put into the contemplation of the subcontractor the potential liability under the main contract for its default. An entitlement to damages for delay, whether liquidated or not, will depend on the subcontractor being in culpable delay. Where the subcontractor suffers delay due to causes other than those at his risk, whether or not such matters are the subject of an express provision of the subcontract works, then if the contractor has not revised the completion date for the subcontract works, time is undefined and may become at large, when the contractor will have to demonstrate that the subcontract works took longer than was reasonable under all the circumstances, in order to be able to recover damages.

Even where the contractor is able to demonstrate that the delay for which it has been charged damages by the employer, whether liquidated or not, was due to the subcontractor, the contractor must demonstrate that it has not wrongfully been charged the damages and/or accepted unjustified damages from the employer. It is not sufficient for the contractor to show that it has paid or allowed certain sums, but that it had an obligation, either under its contract or at law, to pay the sums that are claimed.

Apportioning the liability of a subcontractor to damages for delay to the completion of the works is frequently confused by the Works being delayed by a large number of events arising from differing causes and being to the liability of differing parties. It will be necessary for a contractor to be able to demonstrate that a specific period of delay was due to an express default or defaults by a subcontractor before being able to recover damages from that subcontractor. It will normally only be necessary for a subcontractor to demonstrate that it was denied the ability to perform for any reason for which the contractor is liable for the subcontractor to be able to recover loss from the contractor. The fact is that the contractor is liable to the subcontractor for any default that affects the subcontractor, even though the default may be attributed to another subcontractor or combination of subcontractors and/or the employer. It will generally not be necessary for the subcontractor to identify the reason for the breach, but only the breach itself.

Since damages are compensation for a breach of the subcontract, compensation for events allowed for within the subcontract will not be damages in the true meaning of the word. The losses from such compensation or relevant events are frequently considered under the heading of 'loss and/or expense'. This term, used in many subcontracts, provides for compensation to the subcontractor in the same way as would have been the situation if no such arrangements were incorporated into the subcontract and the subcontractor had been entitled to general damages.

Thus a claim for loss and/or expense will also require the proving of actual loss arising from the compensation or relevant event(s) relied on.

There is no reason why the subcontractor may not avoid the difficulties of proving the actual loss arising directly from a compensation or relevant event by agreeing some other compensation arrangement, such as pre-ascertained prolongation costs, where in a similar way to LADs a genuine pre-estimate of loss is made ahead of the actual event, to be used where specific circumstances occur.

Another express and agreed compensation provision for breach of contract is to be found in conditions for the payment of interest at predetermined rates. Generally, the payment of interest under the express provisions of the subcontract is restricted to failure to pay the sum that should be certified by the final date for payment, as in the SBCSub/C terms and conditions at clause 4.10.5 'the amount due to the Subcontractor'. Where the subcontractor is kept out of its money for other reasons, it may be entitled to damages rather than interest. Damages will normally entail payment at the rate of interest actually payable by the payee for bank loans or to be received from the bank for sums on deposit; such loss and/or expense may be described as financing charges. Where there is an established interest rate within the subcontract provisions, even if only for different circumstances, it will at least be arguable that the rate is pre-ascertained for all purposes. In addition to the direct financing costs a subcontractor may suffer loss in other associated ways, including the costs of arranging financing and/or progressing the debt, and in addition the subcontractor may suffer loss of business opportunity due to restrictive limits on the finance available to it.

Settlement between contractor and employer

A settlement agreement between the contractor and the employer, giving rise to a liability from the contractor to the employer, generally will be evidence of the contractor's loss. Where that loss is the direct consequence of a breach of subcontract by the subcontractor, the contractor will generally be able to recover the settled loss from the subcontractor without further proof. However, this is subject to the contractor being able to demonstrate that the settlement was reasonable on the facts and not just a commercial deal. Where other factors such as the desire to work for the employer on future projects is a consideration, the contractor's case will be uncertain.

In the case of *Babcock Energy Ltd* v. *Lodge Sturtevant Ltd* (1994) 41 Con LR 45, Judge Humphrey Lloyd QC had to consider such a situation. Lodge had supplied and installed electrostatic precipitators as a sub-subcontractor to Babcock on a power station project in India. In the event it was found that the precipitators did not perform as required by the specification. After considerable investigation it was agreed that the easiest and most satisfactory remedy was to introduce a flue gas conditioning system, rather than install the significantly enlarged precipitators needed to satisfy the specification. However, flue gas conditioning, while being considerably cheaper and less disruptive in its installation, would cause the employer to incur increased production costs. Babcock settled with the employer for a calculated sum based on replacement precipitators of the correct size.

Judge Humphrey Lloyd QC summed up the situation, at page 103:

> 'Whether as a matter of commonsense or law, it seems to me quite clear that in a case such as this, where there are tiers of contractors and subcontractors and the defendant is one such sub-subcontractor it must be in his reasonable contemplation that liability for breach of the subcontract for damages may either wholly or in part be quantified in terms of the amount assessed or agreed to be due by a party in a superior tier. Obviously such a liability is not of a nature of a 'blank cheque': the amount of the settlement must be justified. But in law in circumstances such as these the possibility of a settlement between Babcock and GEC [the main contractor] and between GEC and Balco [the employer] must be within the contemplation of a party such as PSL: see the latter part of the extract from the judgment of Lord Justice Somervell in the *Biggin & Co Ltd* case to which I have referred. The amount of the settlement has been justified as equivalent to what PSL [former name of Lodge Sturtevant] contended would have been the extent of their liability so unless that settlement contains anything irrecoverable PSL are liable to Babcock for that amount.'

The quotation from Lord Justice Somervell's judgment in *Biggin & Co Ltd* v. *Permanite Ltd and Others* [1951] 2 KB 314, referred to above, appears at page 92:

> 'The law, in my opinion, encourages reasonable settlements, particularly where, as here, strict proof would be a very expensive matter. The question, in my opinion, is: What evidence is necessary to establish reasonableness? I think it relevant to prove that the settlement was made under legal advice. The client himself could do that, but I do not think that the advisers would normally be relevant or admissible witnesses. I say "normally", for it may be that in special cases that they might be. The plaintiff must, I think, lead evidence, which can be cross-examined, as to facts which the witnesses themselves prove and as to what would probably be proved if, as here, the arbitration had proceeded, so that the court can come to a conclusion whether the sum paid was reasonable. *The defendant may, by cross-examination, as was done here, seek to show that it was not reasonable, or call evidence which leads to the same conclusion. He might in some cases show that some vital matter had been overlooked.* In the present case, of course, [counsel for the defendant] relies, rightly, on the learned judge's finding with regard to the first head of damage, on the fact that the evidence showed that too much was bought, and so on, but if there is evidence at the end of the matter of the kind which I have indicated, on which the court can come to the conclusion that this was a reasonable settlement in the circumstances, then I think it should be the measure. *Parties, Lord Justice Bowen said . . . have been held to contemplate litigation in the sort of circumstances which have arisen here. It would, I think, be unfortunate if they were not also held to contemplate reasonable settlements in the type of circumstances which have arisen here.'* (emphasis by Judge Humphrey Lloyd).

In the case of *Bovis Lend Lease (formerly Bovis Construction Limited)* v. *R. D. Fire Protection Limited and (1) Huthco Limited (2) Baris UK Limited* (2003) 89 Con LR 169, Judge Thornton QC had to consider the effect of a settlement agreement between the contractor Bovis and the employer Braehead, on the entitlement of Bovis to set-off and/or abate and/or recover damages from its subcontractors. It was Bovis's case that since the settlement agreement was global, it was not possible to establish what amount of its loss was due to the subcontractors in question and therefore the correct procedure was to examine the liability of the subcontractors, as though no settlement agreement existed. If this were the correct approach it was distinctly foreseeable that a contractor could recover from its subcontractors, in total, more than it had lost by way of the settlement agreement with its employer.

The subcontractors, which in this case had carried out the fire protection work between them, claimed that since Bovis could not demonstrate that it had in fact paid or allowed to the employer any specific sum(s) or indeed any sum(s) at all, Bovis could not prove its case, which was that its loss had been caused by the default of the subcontractors. If this argument were correct, then by settling its account with the employer on a global basis and without a breakdown of how such settlement had been reached, Bovis would be unable to pass down the contractual chain any of that loss. Since it was policy for the courts to encourage parties in dispute to reach a settlement without recourse to the courts, if the subcontractors' arguments were correct it would undermine such policy and inevitably lead to a reluctance to settle with an employer while claims against subcontractors remained to be resolved.

Bovis's claim against the subcontractors was on several grounds, one of which was that the work as carried out was defective; however it had been accepted and there was no intention by the employer to carry out remedial work, nor any liability on Bovis to do anything further. This decision only concerns certain preliminary issues related to the correct approach to resolving the dispute and especially the effect of the settlement agreement. Judge Thornton QC gave a very full and lengthy decision in which he reviewed the authorities and the effect of the express provision within the subcontract, in which the subcontractor was to indemnify the contractor from the effect of any breaches by the subcontractor, including acts involving the contractor in a liability to the employer. The judge stated the nature and differences between a judgment of a claim from the employer, before and after settlement of any contractor's claim against the subcontractor:

> '68. If the main contractor's claim against the subcontractor is tried or settled before the employer's claim against the contractor has been tried or settled, the main contractor's claim can only be quantified by reference to the best estimate of the quantum of that claim. However, if the main contract claim is determined by way of a judgment before the claim against the subcontractor has been tried, the judgment sum against the main contractor will provide the basis of the claim over against the subcontractor since it would not be reasonable to quantify that claim on any other basis. This is an obvious and clear example of the displacement of the ordinary rule that damages in breach of contract cases are assessed as at the date of the breach.'

He then compared a settlement agreement with a judgment of the court and found that the Court of Appeal had found that there was no effective difference:

> '70. Lord Justice Somervell relied on the earlier decision of the Court of Appeal in *Hammond & Co* v. *Bussey* (1887) 20 QBD 79, CA, and in particular, on the judgment of Lord Justice Bowen, in reaching his decision. Lord Justice Singleton's judgment is in similar terms and Lord Justice Birkett agreed with both judgments. Lord Justice Somervell stated:
>
> "Parties, Lord Justice Bowen said, have been held to contemplate litigation in the sort of circumstances which have arisen here. It would, I think, be unfortunate if they were not also held to contemplate reasonable settlements in the type of circumstances which have arisen here."
>
> 71. Ordinarily, a main contractor who stands in the middle of disputes raised by an employer and resisted by a subcontractor as to the quality of a subcontractor's work-

manship and suchlike will wish to adopt any settlement terms it might reach with the employer and pass on to and recover from the subcontractor the sum agreed by way of settlement. Biggin's case shows that in principle a main contractor may adopt that course of action, subject to proof that the settlement was reasonable, that the subcontractor was in breach of the subcontract and that those breaches caused breaches of the main contract, the resulting claim and its subsequent settlement. This is because such a settlement is within the reasonable contemplation of the parties lower down the contractual chain of the kind in question in this case.'

Thus while a settlement agreement is usually solely about the net sum to be paid between the parties, the first matters to be established are that the subcontractor was in breach of its obligations and that those breaches caused breaches of the main contract which were the matters or part of the matters considered in the settlement agreement.

The judge then considered the relevance of the settlement agreement itself. Two possibilities were considered: first, that the settlement sets an upper limit on the claim against the subcontractor, and secondly, that it is primary evidence of what should be paid by the subcontractor to meet the contractor's claim. The judge analysed the case of *Biggin* v. *Permanite* in the light of a number of subsequent cases, and concluded:

'80. The following conclusions may be deduced from the *Biggin* line of authorities:

1. The *Biggin* principles may be relied on where the claim over arises as a claim under an indemnity or is a claim for damages for breach of contract. Equally, they are applicable both when liability to the third party had been admitted by the claimant and when it remained in issue and had been included as part of the issues being settled.
2. The principles are applicable because, ordinarily, a defendant is to be taken to have foreseen that a consequence of its breach of contract would be both that the claimant would be liable to the third party and that that liability might give rise to litigation and a compromise under which the claimant incurred financial loss.
3. The principles are not merely an aspect of a claimant's duty to mitigate its loss but also involve the law concerned with the measure of recoverable damages following a breach of contract. Thus, the settlement figure, that is the sum being paid by the claimant to settle the third party's claim, is to be taken to be the upper limit of what may be recovered from the defendant in relation to the loss caused to the claimant as reflected in the claimant's liability to the third party.
4. As a starting point, a claimant may recover the sum paid in settlement if it can establish that that settlement was reasonable and that it was reasonable to settle the third party's claim. It is only necessary for the claimant to prove, in general terms, that the settlement was reasonable. It is not necessary for it to establish in great detail the extent and quantum of the third party's claim.
5. The evidence that may be adduced to establish reasonableness will vary but it may include, if the claimant so elects, evidence of advice given by relevant professionals to the claimant which was relied on in deciding to settle. The material that is disclosable when a settlement is relied on may, on occasion, include documents recording such advice but any consequent waiver or overriding of privilege involved in such disclosure will depend on how the claimant elects to establish reasonableness and on what material it proposes to rely for that purpose.

6. The claimant must establish by normal methods of proof and to the normal standard of proof that the defendant was in breach of contract and that that breach caused the claims to be made. Equally, it must establish that the defendant's breaches of contract led to breaches by it of its contract with the third party.
7. To the extent that the settlement was unreasonable or unreasonably entered into, it is irrecoverable since that element of the settlement was not caused by the defendant's breach and, equally, it resulted from a failure by the defendant to mitigate its loss.
8. The *Biggin* principles are not confined to cases where there is only one party causing the loss being settled or where the claims being settled are confined to those based on the defendant's breaches of contract. It is necessary, however, for the court to determine what part of an overall settlement of a multi-party or multi-issue dispute is attributable to the relevant breaches of contract of the defendant, or to each separate defendant where a *Biggin* claim is being made against more than one defendant. Any allocation of part of a reasonable overall settlement must itself be reasonable.'

Thus where there is a settlement agreement between the employer and the contractor, it will still be for the contractor to demonstrate the subcontractor's breach and to show that that breach caused loss, and to justify the sum claimed. However, the effect of the settlement agreement will normally be to put an upper limit on the sum recoverable from the subcontractor. Where the allocation of loss under the settlement agreement is undefined, it will be for the court or tribunal, on the facts available, to make such allocation as it considers just and reasonable.

Judge Thornton QC identified a further difficulty in the use of the settlement agreement to determine liability of a subcontractor to the contractor, and set out the problem as follows:

'82. The first disputed question to be resolved is whether it is for Bovis to establish what sum, if any, was included in the settlement agreement or, if they wish to rely on the settlement agreement to limit or eliminate their liability to pay damages, (or) for RD Fire and Baris to establish this.

83. The answer to this procedural conundrum lies in determining the nature of the *Biggin* principles. If they are substantive rules of law relating to causation and to the quantification of damages, then it is for Bovis to establish its loss which means that it has the burden of proving what sum Braehead's claims were settled for when there has been a settlement of the claim it seeks to pass on. Conversely, if the *Biggin* principles are procedural rules relating to the evidence that is admissible to prove loss, then a claimant may elect to prove its loss in some other way without reference to the settlement. In such circumstances, it will be for the defendant to adduce evidence as to what was included in the settlement for the claims or causes of action in question if the defendant regards the settlement as relevant. A defendant who wished to rely on the settlement in this way would be seeking to show that the claimant, in seeking the full claim and not the reduced sum contained in a settlement agreement, had failed to mitigate its loss. It is clear that a defendant bears the burden of proof to establish that a claimant has failed to mitigate its loss.'

In other words, the settlement agreement may either be used by the contractor to prove its loss or by the subcontractor to limit its liability, when the burden of proof will be with the subcontractor.

A major problem for contractors seeking to pass on damages to their subcontractors after concluding settlement agreements with their employers, which is not addressed by cases such as *Biggin*, is how to allocate those damages where more than one subcontractor is involved. A first step is to be able to establish how the settlement sum was made up. Rarely will there be a single issue giving rise to the settled damages and similarly the causes may, in turn, arise from breaches by a number of subcontractors and/or suppliers. In reaching and agreeing a settlement sum, consideration may or may not have been given to the way such a sum was calculated.

Even where the contractor is able to demonstrate the basis of its original claim against the employer, it may or may not be possible to demonstrate which heads of claim were accepted in full, totally rejected or subject to compromise, in the final agreement. In fact, the settlement agreement may just be a lump sum for all heads of claim, without distinction, even though some will almost certainly have been more meritorious than others.

Provided the agreement is for stated sums against each head of claim, then the contractor can rely on the *Biggin* principles. Where this is not the case, it is suggested that the starting point will be the contractor's claim against the employer. A pro-rata allocation of the accepted sum against the sums claimed may place a limit on the sums recoverable. It will then be for the contractor to demonstrate the validity of its claim against each subcontractor in turn. The contractor may be able to go a long way to achieving this by relying on the state of negotiations prior to the settlement.

Delay to the subcontractor's work

When the subcontractor, in breach of the subcontract terms and conditions, causes delay, then it will be liable for the loss to the contractor that results. Although this is a fairly straightforward statement, understanding its implications in the wide range of subcontract situations to be found in construction contracts is far from straightforward. There is a presumption in commercial contracts in general, and construction subcontracts are no exception, that the subcontractor is free to plan and organise its work as it wishes, provided it completes its obligations by the due date. In the case of most construction contracts, especially those for new construction, the contractor is given possession of the site on the date specified in the contract, and is contracted to provide the finished building and hand back the site a certain time after. Failure to do so will put the contractor in breach of the contract. An employer expecting to have use of its building on a certain day will suffer a loss, either actual or potential, if the building is not completed on time. However, the employer does not suffer any loss if, along the way to completion the builder fails to achieve any target or targets it may have set itself. As a result, there is no penalty on the builder if it fails in its intention in such a way.

The situation with most subcontracts is that, in a similar way to general and main contracts, the subcontractor contracts to carry out the on-site works within an agreed period of time after a set period of notice. Where there is a significant amount of off-site work, such as the preparation of shop or detail drawings and the fabrication of components, the minimum period for off-site work prior to commencement on site will normally be stated. Subcontractors can be classified into a number of general categories, which include:

1. those with a considerable amount of detailing and off-site works, including structural steel, pre-cast concrete, architectural metalwork, specialist joinery and shop-fitting work;
2. items requiring the taking of site dimensions prior to manufacture, including stair balustrading and specialist joinery;
3. trades requiring free access to the entire site, such as piling, structural frame and cladding activities;
4. trades making more than one visit to each area, such as joinery, mechanical, electrical and plumbing services;
5. trades applying readily available materials, such as plaster and render work, asphalt and most roofing, brick and block work and painting;
6. trades relying on materials from distant places and/or on long delivery, including stone and tile work.

A casual consideration of those categories will show that while some trades are in control of their own progress, others cannot progress until work by the previous trade has progressed. For some trades the final completion of their work may not have as much, if any, effect on the progress of the works as a whole as its rapid progress early on in the subcontract period. A tiled roof, for example, will to a very large extent achieve a weathertight structure once the felt and battens are fixed, while the completion of the ridge tiles or sundry flashings may have no immediate effect whatsoever on the progress of internal finishing or other critical trades.

What can be deduced from an analysis of the work of differing trades, is that delay to the works as a whole will, in most cases, be caused by a trade or trades not progressing at the desired time and/or rate and will less frequently be solely because that trade failed to complete the entirety of its work by the due date. Indeed, it is suggested, in many cases non-completion to time may have an insignificant if any effect on the completion of the works, whereas failure to commence on the required day or to progress well in the early stages of the subcontract period will cause major delays and disruption to the progress and likely completion of the works.

As mentioned in the section 'The subcontractor's obligations for time' in Chapter 11, the contractor's right to damages for delay will depend on its having agreed with its subcontractor obligations as to the intended progress of the subcontract works, such that a delaying event is a breach of the subcontract. It is suggested that for many trades a start and finish date does not achieve this.

A further difficulty for the contractor relying on a subcontract where there is solely an obligation to start on site after a period of notice and complete a stated period thereafter, is that there is no clear indication as to how the work is to be released to the subcontractor. In the absence of any clear indication to the contrary, it could be held that at the end of the period of notice the entire subcontract work is to be available to the subcontractor. While it rarely will be, other than for trades in category 3 above, that this is the expectation of the parties. Nevertheless there will be an implied term that the work will be released to the subcontractor so as to enable it to fulfil its obligations without unreasonable hindrance and/or prevention. This will lead to major uncertainty and it will only require the subcontractor to claim, at an early stage, that the work is not being released to it as necessary, for it to have grounds to be released from the obligation to complete to time. Conversely, in the absence of an agreement for a specified rate of progress or for the achievement of set milestones, it is unlikely that the contractor will be able to claim damages if the work does not, at an early stage or at any other time, proceed as the contractor, retrospectively, considers should have been achieved.

Many subcontracts incorporate clauses which seek to make the subcontractor liable for any additional costs arising out of the coordination of its work and/or working in close proximity with other trades. Typical clauses might include the following examples:

> The subcontractor shall include within his lump sum price and rates, for liaison with, and simultaneously working alongside other trades.
>
> The subcontractor shall note that the nature of the works may not necessarily allow continuity of work, particularly in view of the complex interrelationship of the construction activities, and the multiplicity of specialist trades associated with hospital works. The subcontractor is to include for any associated disruption within his lump sum price and rates.

Or a general requirement to carry out the work, as directed or required by the contractor:

> The subcontract works are to be carried out regularly and diligently and in such order, and at such time or times and in such manner as the main contractor shall direct or require so as to ensure completion of the main contract works or any portion thereof under the main contract by the completion date or dates or such extended date or dates as may be allowed under the main contract.

Such clauses will have to be considered on their actual wording and in association with the remainder of the subcontract provisions. It may be that these clauses do no more than preserve the subcontract, should any of the listed matters arise. Where the subcontractor is to incorporate an allowance within its price or rates, then that may be limited to the amount and extent of such parallel and/or interrupted working as is in the contemplation of the parties at the time the subcontract is concluded, or is normal for projects of such nature and duration. It is considered that a very express condition would need to have been clearly accepted, for a subcontractor to be liable for its loss from any action or inaction by the contractor or its other subcontractors.

As stated in the section 'Acceleration' in Chapter 8, a specific instruction to accelerate will normally be required from the contractor before the subcontractor will be entitled to payment for the expense incurred in attempting to reduce and/or eliminate the effect of the contractor's delays or delaying events. However, it may be that general terms such as described above, requiring the subcontractor to 'ensure completion of the main contract works or any portion thereof under the main contract by the completion date or dates' will be an open requirement for the subcontractor to take such action as it believes is necessary to achieve such completion. In any event, a subcontractor will be well advised to inform and seek agreement from the contractor to any action proposed to be taken in compliance with such a requirement prior to taking such action and in such a notification to make reference to the specific requirement of the subcontract.

From the examples given above, the subcontractor may suffer additional expense either from the requirement to operate under circumstances of parallel or disrupted working greater than could reasonably have been anticipated at the time of tender and to be included for within the subcontract sum or rates, or alternatively by way of an express instruction or direction of the contractor. Where the subcontractor considers it is incurring additional expense, it is likely that it will have an obligation, whether express or implied, either to seek an instruction to carry out its work under changed conditions, and/or to advise the contractor that it is incurring loss and/or expense as a consequence of the regular progress of its work being disrupted. In either case, the subcontractor may not have a right to recover its additional costs without an express direction from the contractor to take the action(s) notified by the subcontractor.

It will not infrequently be the case that by the end of the subcontract period there will still be areas of the work not available to the subcontractor. In that case the contractor not the subcontractor will be in breach of its obligations. It is suggested that this is so even where there is other work of the subcontractor still outstanding and incomplete. In such circumstances, while the subcontractor will be relieved of its obligation to complete the works by the due date, the contractor may still have a claim against the subcontractor related to the other areas of the incomplete work. However, it will be for the contractor to demonstrate that each of those areas was released to the subcontractor at a reasonable time, having regard to the subcontract completion date and to the dates the other work was made available to the subcontractor, and that the subcontractor failed to respond in a reasonable manner following the release of such work to the subcontractor. Should the contractor be able to demonstrate that the subcontractor's delay is, on the balance of probabilities, due to the subcontractor's dilatory progress, then the contractor will have to demonstrate the loss that flowed directly and solely from that breach of the subcontract.

Under the common subcontract arrangements, as envisaged by item 5 of the Sub-Contract Particulars in SBCSub/A and similar Sub-Contract Agreements, the contractor will have considerable difficulty making an effective claim for damages, for delay by the subcontractor, where the contractor cannot show that the works were all available at the end of the period of notice or were released to the subcontractor progressively so as to enable the subcontractor, acting reasonably, to complete the works to time. Where there is no specific requirement for the subcontractor to complete parts of the work and/or specified operations by stated dates, the contractor will have considerable difficulty proving a claim for losses resulting from any delay during the progress of the subcontract works and prior to the subcontract completion date. There will be an assumption that failure by the subcontractor to complete to time will give rise to an entitlement to damages for any associated loss by the contractor. It will, however, be for the contractor to demonstrate such loss by showing the effect of the late completion of the subcontract work. Any such entitlement will be lost where the subcontractor can demonstrate an act of hindrance or prevention on the part of the contractor.

Damages when delay due to two or more subcontractors

It will rarely be the case on building projects that delays are due to only one cause or only caused by one party to the construction process and works. Where more than one party can be shown to have caused delay, then the loss and/or expense that arises, including any pre-ascertained damages claimable by the employer, will need to be allocatable between those who actually cause the loss.

As discussed in detail in Chapter 11, before a subcontractor can be held liable for a contractor's loss it must first be shown to have been in breach of its obligations as defined in the subcontract. There must have been no acts of prevention on the part of the contractor which have either on their own or concurrently with other matters caused the subcontract works to be delayed and as a consequence be the reason for the contractor to have suffered its loss.

No standard form of subcontract is known which makes any provision for the allocation of loss between subcontractors. As a consequence, a subcontractor whose work has been in culpable delay and who can identify another subcontractor whose work has delayed the project may resist a claim against it on the grounds that the works would have been completed late in any event and therefore its delay did not in fact cause the contractor any loss. Conversely, should a contractor seek to recover loss against two or more subcontractors for the same delay, it will, if successful against more than one subcontractor, recover more than its loss and as a consequence obtain 'unjustified enrichment'.

In reality the true cause of delay to a building project may be that too much was attempted in too short a time having regard to the state of design and procurement at the time the main contract was let. It is rarely the situation when a subcontract is let that the subcontractor knows precisely what is required of it either with regard to the detailed work or as to the timing of its execution. It is the normal practice for work to be released to each subcontractor as the work of the trade in front completes a section of its work. Once the Works fall behind programme, without a mechanism to restore control there is no standard under the subcontract against which to measure progress and identify breaches by the subcontractor which lead to a delay to the Works.

One possibility is that the overall delay will be allocatable on some proportional basis between all parties. In theory this may be considered a reasonable approach and may be considered to have legal backing by way of the judgment of Judge Hicks QC in the case of *Ascon* v. *Alfred McAlpine* where, after considering the effect of programmed float, he said:

> '93. I do not see why that analysis should not still hold good if the constituent delays more than use up the float, so that completion is late. Six subcontractors, each responsible for a week's delay, will have caused no loss if there is a six weeks' float. They are equally at fault, and equally share in the "benefit". If the float is only five weeks, so that completion is a week late, the same principle should operate; they are equally at fault, should equally share in the reduced "benefit" and therefore equally in responsibility for the one week's loss. The allocation should not be in the gift of the main contractor.'

In practice, there will be considerable difficulties in deciding which parties are liable and in what proportion.

In his judgment in *Ascon*, Judge Hicks QC analyses in considerable detail the effect of delays in a subcontractor's work and discusses the possible methods of assessing the contractor's consequential loss. The significant parts of the judgment are as follows:

> '80. . . . The importance of practical completion as between employer and main contractor, and therefore derivatively as between main contractor and finishing trades, is that it makes the date on which possession returns to the employer, with the opportunity of beneficial use of the property. The relevance of that date to causation of loss from delay is therefore direct, and usually decisive. The date of practical completion will also govern liability for liquidated damages. Here no subcontract liquidated damages clause is relied upon and Ascon was not a finishing trade; the question whether delay on its part caused loss of the kind claimed by McAlpine turns not on any nice question of how practical completion is to be understood or its date identified but on the factual issue whether, and if so how much, delayed working affected the progress of the following trades and thereby the completion of the main contract. If there is a planned overlap delay *before* the date for completion may be causative, while on the other hand delay *in* completion may be irrelevant, whatever the programme, if in truth no following trade is hindered.
>
> . . .
>
> 83. McAlpine's pleaded case is that Ascon was ten weeks and two days late in completion and therefore responsible for the whole of McAlpine's delay of ten weeks in completing the main contract. Presumably it would concede, although the question was not canvassed, that the period for which Ascon is responsible should be reduced by the amount of the extension to which I have now held it to be entitled. The more fundamental question, however, is whether McAlpine has established causation by Ascon's breach of any part of the main contract delay, and if so of how much.
>
> 84. The trade immediately following Ascon's was cladding, which was to be carried out by a subcontractor called Structal. McAlpine led evidence from Mr Lamont [a witness for McAlpine] that Structal's commencement date, programmed for 13 January 1997, was progressively postponed to 1 April because of Ascon's delays. There was no specific evidence of a continuous causative chain carrying this delay through to completion; that was tacitly left to be inferred, albeit with the assistance of an argument that, in effect, to fail to draw that inference would be to give Ascon the benefit of a "float" of five weeks built into McAlpine's main contract programme, which would be wrong.'

Judge Hicks then considered the judgments in both *Pigott* v. *Shepherd* and *GLC* v. *The Cleveland Bridge and Engineering Co Ltd* (1986) 34 BLR 72, and the passage in *Keating* (6th edn, p. 838) and continues:

> '89. I do not, therefore, consider that on the point of law either party wholly succeeds. McAlpine cannot rely on any obligation by Ascon to comply with the detail of the programme for the commencement or execution of the cladding work, but is not precluded from contending that Ascon was in breach of the obligation to carry out its works "reasonably in accordance with the progress of the [main contract] works" and that in consequence commencement of cladding was delayed. The factual issue how far it has established that contention remains, but I propose to defer consideration of it until I have dealt with the second stage of causation, from that point to completion of the main contract.'

The judge did not explain the effect of the phrase 'reasonably in accordance with the progress of the [main contract] works' on a subcontractor whose work is the only or majority of the work being progressed at the time, as was the situation in this case. It is suggested that in such situations this requirement is without relevance. Judge Hicks considered the effect of the contractor's programme float as follows:

> '91. Before addressing those factual issues I must deal with the point raised by McAlpine as to the effect of its main contract "float", which would in whole or in part pre-empt them. It does not seem to be in dispute that McAlpine's programme contained a "float" of five weeks in the sense, as I understand it, that had work started on time and had all sub-programmes for subcontract works and for elements to be carried out by McAlpine's own labour been fulfilled without slippage the main contract would have been completed five weeks early. McAlpine's argument seems to be that it is entitled to the "benefit" or "value" of this float and can therefore use it at its option to "cancel" or reduce delays for which it or other subcontractors would be responsible in preference to those chargeable to Ascon.
>
> 92. In my judgment that argument is misconceived. The float is certainly of value to the main contractor in the sense that delays of up to that total amount, however caused, can be accommodated without involving him in liability for liquidated damages to the employer or, if he calculates his own prolongation costs from the contractual completion date (as McAlpine has here) rather than from the earlier date which might have been achieved, in any such costs. He cannot however, while accepting that benefit as against the employer, claim against subcontractors as if it did not exist. That is self-evident if total delays as against sub-programmes do not exceed the float. The main contractor, not having suffered any loss of the above kinds, cannot recover from subcontractors the hypothetical loss he would have suffered had the float not existed, and that will be so whether the delay is wholly the fault of one subcontractor, or wholly that of the main contractor himself, or spread in varying degrees between several subcontractors and the main contractor. No doubt those different situations can be described, in a sense, as ones in which the "benefit" of the float has accrued to the defaulting party or parties, but no-one could suppose that the main contractor has, or should have, any power to alter the result so as to shift that "benefit". The issues in any claim against a subcontractor remain simply breach, loss and causation.
>
> 93. I do not see why that analysis should not still hold good if the constituent delays more than use up the float, so that completion is late. Six subcontractors, each responsible for a week's delay, will have caused no loss if there is a six weeks' float. They are equally at fault, and equally share in the "benefit". If the float is only five weeks, so that completion is a week late, the same principle should operate; they are equally at fault, should equally share in the reduced "benefit" and therefore equally in responsibility for the one week's loss. The allocation should not be in the gift of the main contractor.
>
> 94. I therefore reject McAlpine's "float" argument. I make it clear that I do so on the basis that it did not raise questions of concurrent liability or contribution; the contention was explicitly that the "benefit", and therefore the residual liability, fell to be allocated among the parties responsible for delay and that that allocation was entirely in the main contractor's gift as among subcontractors, or as between them and the main contractor where the latter's own delay was in question.
>
> 95. That brings me back to the factual issues of causation. The first is whether it is proper, in the absence of other evidence, to infer that the causes of delay at one stage have a

continuing effect so as to produce the same delay at a later stage. I believe that that is in principle a proper inference, but that the probability that it will be drawn, or drawn to its full extent, is likely to diminish with the passage of time and the complexity of intervening events. My reasons for regarding it as being proper, with qualifications, are first that such an inference, at least over short periods, is tacitly assumed in all negotiation, arbitration and litigation of delay claims, and secondly that it represents the "neutral" position, in the sense that if all other activities proceed according to programme (and if there were any direct evidence that they had not that would contradict the basis on which this issue is raised) that would be the result.'

From this judgment it is possible to list a number of principles to be considered when allocating or assessing damages or loss and expense due from a subcontractor to a contractor following delay to the completion of the subcontractor's work:

1. Practical completion marks the date on which possession passes back to the contractor and/or the contractor has beneficial use of the work.
2. Point 1. above may be affected where there is either overlap with another trade before the planned completion of the subcontracted work or there is no following trade planned to be dependent on the completion of that subcontractor's work.
3. The subcontractor's liability will normally be reduced by the amount of any extension of time to which it is entitled. The contractor will then have to establish causation that the remaining delay or part thereof actually delayed the main contract works. The Ascon decision did not consider the situation where there are agreed sectional or partial handovers planned prior to completion of the subcontract works and the extension of time events occur after the date for the early handovers. In such circumstances the extension of time would not reduce the effect of any delay to the early handovers.
4. Has the start of the following trade actually been delayed and if so has such delayed start been carried though to the completion? The judgment makes no reference to the contractor being able to demonstrate that the follow-on trade was appointed and available to start on the programmed date. There will always be the possibility of concurrent effect of two delays, one related to the progress of the first subcontractor and the other due to the appointment and/or availability to commence work by the second subcontractor.
5. Where there is delay due to more than one cause leading to a net delay to the project, then it is reasonable to allocate that delay on some apportioned basis.
6. Where float in the programme exists, then the assessment of the subcontractor's liability must take notice of the contractor's benefit from such float. If the contractor as a result of programmed float suffers no loss, then it may not claim a non-existent loss from its subcontractor.
7. It will generally be assumed that a delay early in the project will result in a corresponding delay to the completion of the project. This is stated to represent the neutral position. Two related matters not considered by this judgment are:
   7.1. What is the situation where the contractor recovers some of the subcontractor's delay and subsequently, due to other matters, that recovered time is either in whole or in part lost again. It is suggested that the subcontractor will not in such circumstances be able to claim no liability for the final delay because at some intervening date the project was not in delay.

7.2. Will the subcontractor be liable for delays not directly its liability such as where a following trade is not able to commence work immediately on its delayed access date, because in the interval its resources have been dissipated and not immediately replaceable on the work being available to it? It is suggested that the first subcontractor is likely to be held liable for such additional delays reasonably incurred.

Undervaluation of the work

Even where there is no express term that the contractor, in valuing the works, will value it accurately and promptly in accordance with the terms of the subcontract, there is likely to be an implied term to that effect. Except in a few cases, such as those of a nominated subcontractor and works contractors under a management contract, valuations of the subcontractor's work will be by the contractor as the paying party, unlike that of a main contractor where valuation will be by a third party. It cannot, therefore, be said that the contractor's obligation is limited to paying the sum certified. The contractor's obligation is the much greater one of valuing the works properly and accurately and paying the sum found to be due, on or before the date required.

Where the contractor fails to value the works properly and accurately, the contractor will be in breach of the subcontract. Such a breach is likely to give rise to an entitlement to damages and/or loss and/or expense. In extreme cases such a breach might be regarded as repudiatory, and indeed such is included as justification for determination of the subcontract by the subcontractor under the SBCSub/C terms, at clause 7.8.1.3.

A difficult situation may arise where the contractor fails to value the work at all, which, it is suggested, has a different effect to the situation where the work is valued at nothing. Where the contractor issues a notice of payment and if appropriate a notice of intent to withhold payment, then the contractor's valuation has been defined and must be paid or the contractor will be in breach. If, however, no such notices are issued and the subcontractor, by the final date for payment, receives a payment which is less than that sought, then it must be assumed that the sum paid is the sum considered by the contractor to be that due to the subcontractor. However, if no notices of payment and/or withholding are issued and no payment is made, two possibilities exist: either no payment is considered due or the contractor, in breach of its obligations, has neither valued the work nor paid the value. It is submitted that the contractor in these circumstances will not be able to resist a notice of intent to suspend performance unless it is able to demonstrate, to the reasonable satisfaction of a tribunal, that a proper notice of payment would have been valued at nil.

One remedy for a subcontractor who considers that the contractor has underpaid it is to seek the decision of an adjudicator. It might be expected that it was in the reasonable contemplation of the parties that, in the event of an undervaluation of the subcontractor's work, the subcontractor would resort to adjudication and as a consequence suffer loss; further, that the loss would flow naturally from such a breach of the subcontract. However, the HGCRA provides that it is only by agreement that the parties can recover their costs in adjudication and there is no right to the recovery of such costs as damages.

In the case of *Total M & E Services Limited* v. *ABB Building Technologies Limited (formerly ABB Steward Limited* (2002) 87 Con LR 154, Judge Wilcox had to consider a claim, made in an action to enforce an adjudicator's decision, whether the claimant could in addition obtain as damages its costs in the adjudication reference, and he decided they could not, at paragraph 24:

> 'Mr Harding [Counsel for Total] contends that the costs of the adjudication are recoverable as a damages claim. He submits that if a defendant fails to pay under a construction subcontract to which the Act applies it is foreseeable that the claimant would seek adjudication and properly incur costs, and therefore seek to recover them. Mr Coulson [Counsel for ABB] submits that since the statutory scheme does not make any provision for the adjudicator to award costs unless the parties agree otherwise the adjudicator has no jurisdiction to order that one party's adjudication costs should be paid by the other. There is no such agreement in the present case; indeed, the entire basis of the claimant's claim is that there was no such agreement. I agree with Mr Coulson that since the Act does not provide for the recovery of costs the claim is misconceived. Furthermore, this claim is put as a claim for damages for breach of contract arising out of ABB's failure to pay. Because the statutory scheme envisages both parties may go to adjudication and incur costs which they cannot under the Act recover from the other side, it follows that such costs cannot therefore arise as damages for breach.
>
> To permit such claim would be to subvert the statutory scheme under the Act.'

The measure of damages

As stated earlier, the damages recoverable consequent on a breach of the subcontract, by either party, are the monetary recompense for the effect of that breach and therefore held to be the actual loss incurred. Traditionally, this has been expressed as either the cost of rectification or the diminished value as a consequence of the default. Where matters are purely commercial, these two definitions provide effective alternative methods for the valuation of loss. Either the work will be defective and will cost an assessable sum to rectify, or because of the defect the resultant work will have less value than it would have had if the work had been correctly carried out.

However, outside strictly commercial situations it has been acknowledged by the courts that work not entirely to specification may have no effect on the market value of a building, but at the same time would cost a totally unreasonable amount to rectify. This has led to the acceptance of a third category, of loss of amenity. As Lord Mustill said in his judgment in the House of Lords case of *Ruxley Electronics & Construction Ltd* v. *Forsyth* (1995) 73 BLR 1, at page 14I:

> 'There are not two alternative measures of damage, at opposite poles, but only one; namely, the loss truly suffered by the promisee. In some cases the loss cannot be fairly measured except by reference to the full cost of repairing the deficiency in performance. In others, and in particular those where the contract is designed to fulfil a purely commercial purpose, the loss will very often consist only of the monetary detriment brought about by the breach of contract. But these remedies are not exhaustive, for the law must cater for those occasions where the value of the promise to the promisee exceeds the financial enhancement of his position which full performance will secure. This excess, often

referred to in the literature as the "consumer surplus", is usually incapable of precise valuation in terms of money, exactly because it represents a personal, subjective and non-monetary gain. Nevertheless where it exists the law should recognise it and compensate the promisee if the mis-performance takes it away.'

The law relating to damages for defective work in construction projects was reviewed in detail in the speech of Lord Jauncey of Tullichettle, in the *Ruxley* v. *Forsyth* case. That case arose where a private swimming pool was constructed so that the depth at the deepest part was less than the minimum specified. Mr Forsyth refused to pay, despite considerable rectification work having been carried out, including rebuilding once. The judge, at first instance, had awarded damages on the basis of loss of amenity but the Court of Appeal overturned this decision and awarded the full cost of reconstruction to the correct depth. The House of Lords, in turn, reversed the decision. Lord Jauncey of Tullichettle reviewed a number of cases, at page 8:

'The general principles applicable to the measure of damages for breach of contract are not in doubt. In a very well-known passage Baron Park in *Robinson* v. *Harman* [(1848) 1 Exch 850] said:

"The next question is: what damages is the plaintiff entitled to recover? The rule of the common law is, that where a party sustains a loss by reason of a breach of contract, he is, so far as money can do it, to be placed in the same situation with respect to damages, as if the contract had been performed,"

In *British Westinghouse Electric & Manufacturing Co Ltd* v. *Underground Electric Railways Co of London Ltd* [[1912] AC 673] Viscount Haldane LC said:

"Subject to these observations I think that there are certain broad principles which are quite well settled. The first is that, as far as possible, he who has proved a breach of a bargain to supply what he contracted to get is to be placed, as far as money can do it, in as good a situation as if the contract had been performed.

The fundamental basis is thus compensation for pecuniary loss naturally flowing from the breach; but this first principle is qualified by a second, which imposes on a plaintiff the duty of taking all reasonable steps to mitigate the loss consequent on the breach."

More recently, in what is generally accepted as the leading authority on the measure of damages for defective building work, Lord Cohen in *East Ham Corporation* v. *Bernard Sunley & Sons Ltd* [1966] AC 406:

"the learned editors of *Hudson's Building and Engineering Contracts* say that there are in fact three possible bases of assessing damages, namely, (a) the cost of reinstatement; (b) the difference in cost to the builder of the actual work done and work specified; or (c) the diminution in value of the work done to the breach of contract. They go on:

'There is no doubt that wherever it is reasonable for the employer to insist upon reinstatement the courts will treat the cost of reinstatement as the measure of damage.' " '

Having set out the authorities for basing damages for defective work on the basis of the cost of rectification, Lord Jauncey then reviews, at page 10, the authorities for basing such damages on a reasonable assessment:

'In *CR Taylor (Wholesale) Ltd* v. *Hepworths Ltd* [[1977] 1 WLR 659] Mr Justice May referred with approval to a statement in *McGregor on Damages* that in deciding between diminution in value and cost of reinstatement the appropriate test was the reasonableness of the plaintiff's desire to reinstate the property and remarked that the damages to be awarded were to be reasonable as between plaintiff and defendant. . . . In *McGregor on Damages*, after a reference to the cost of reinstatement being the normal measure of damages in a case of defective building, it is stated:

"If, however, the cost of remedying the defect is disproportionate to the end to be attained, the damages fall to be measured by the value of the building had it been built as required by the contract less its value as it stands."

A similar approach to reasonableness was adopted by Mr Justice Cardozo delivering the judgment of the majority of the Court of Appeals of New York in *Jacob and Youngs* v. *Kent* [(1921) 129 NE 889].

Damages are designed to compensate for an established loss and not to provide a gratuitous benefit to the aggrieved party from which it follows that the reasonableness of an award of damages is to be linked directly to the loss sustained. If it is unreasonable in a particular case to award the cost of reinstatement it must be because the loss sustained does not extend to the need to reinstate. A failure to achieve the precise contractual objective does not necessarily result in the loss which is occasioned by total failure . . . Further support for this view is to be found in the following passage in the judgment of Sir Robert Megarry V-C in *Tito* v. *Waddell* [(No. 2) [1977] Ch 106, 332C]:

"Per contra, if the plaintiff has suffered little or no monetary loss in the reduction of the value of his land, and he has no intention of applying any damages towards carrying out the work contracted for, or its equivalent, I cannot see why he should recover the cost of doing work which will never be done. It would be a mere pretence to say that this cost was a loss and so should be recoverable as damages." '

Lord Mustill, at page 14B, rejected totally any notion that the damages due should be valued at nought just because the value of the property would not be less by reason of the defective work, and made many valuable observations:

'It is a common feature of small building works performed on residential property that the cost of the work is not fully reflected by an increase in the market value of the house, and that comparatively minor deviations from specification or sound workmanship may have no direct financial effect at all. Yet the householder must surely be entitled to say that he chose to obtain from the builder a promise to produce a particular result because he wanted to make his house more comfortable, more convenient and more conformable to his own particular tastes: not because he had in mind that the work might increase the amount which he would receive if, contrary to expectation, he thought it expedient in the future to exchange his home for cash.'

And at page 15C:

'The law should recognise it (consumer surplus) and compensate the promisee if the misperformance takes it away. The lurid bathroom tiles, or the grotesque folly instanced in argument by my noble and learned friend Lord Keith of Kinkel, may be so discordant with general taste that in purely economic terms the builder may be said to do the employer a favour by failing to install them. But this is too narrow and materialistic a view of the transaction. Neither the contractor nor the court has the right to substitute for the employer's individual expectation of performance a criterion derived from what ordinary people would regard as sensible.'

It is apparent from the extracts quoted above that the basis of the measure of damages will depend, as so often, on the facts. The cost of reinstatement is likely to be the appropriate measure of damages where the work is materially defective, such as with a structural fault. Similarly, where the appearance of the building, whether externally or internally by the use of the wrong cladding, finishing and/or decorating materials or components, is manifestly altered by the default, then the employer will have the right to have those materials or that work removed and reinstated to the correct detail or specification. However, where the building work has been carried out using material other than that specified, but one which would have much the same properties as the specified material, then damages might be limited to any saving made by the subcontractor by using the substituted material. Where the defect is such as to detract from the value of the building, then the damages are likely to be based on the diminution in value.

Where remedial works have to be carried out, it is likely that further damages will be suffered by the building owner either by way of lost rental or the rental of other premises. Such losses will be compounded by the value of lost production, revenue and profit. It will be for the court or tribunal to decide in each case where such losses become too remote either to flow naturally from the breach or to have been in the contemplation of the parties at the time the subcontract was concluded.

Pre-ascertained damages

In most subcontract arrangements there will be no directly incorporated provisions for pre-ascertained damages or LADs, for late completion. Such damages may and frequently will be incorporated from the main contract by reference, so as to bring such risk into the contemplation of the parties at the time of their agreement. However, in certain circumstances, specifically where the contractor is under a private finance initiative or where the contractor is the developer such as in housing developments, directly incorporated LADs are common.

The law allows the parties to agree, in advance, the damages that are expected to arise from specific breaches of the subcontract. Such breaches regularly relate to failures to complete by the due date, when damages are assessed at a rate per day or week of delay. However, such pre-ascertained damages may relate to other matters, such as failure to pay on time, when an agreed rate of damages will be set which will normally be expressed as a percentage of the debt, on an annual rate basis, with or without an associated lump sum for administration. A further cause of loss that may be considered suitable for pre-loss assessment and agreement is that of prolongation. Such loss could be limited to head office and visiting staff or the full weekly cost including site costs. However, the law does not allow such sums or rates to be penalties but must be a 'genuine covenanted pre-estimate of damage'.

In the House of Lords decision, in *Dunlop Pneumatic Tyre Company Limited* v. *New Garage & Motor Company Limited* [1915] AC 79, Lord Dunedin summarised the law, at page 86:

> '1. Though the parties to a contract who use the words "penalty" or "liquidated damages" may prima facie be supposed to mean what they say, yet the expression used is not conclusive. The court must find out whether the payment stipulated is in truth a penalty or liquidated damages. This doctrine may be said to be found passim in nearly every case.

2. The essence of a penalty is a payment of money stipulated as *in terrorem* of the offending party; the essence of liquidated damages is a genuine covenanted pre-estimate of damage (*Clydebank Engineering & Shipbuilding Co* v. *Don Jose Ramos Yzquierdo y Castaneda* [[1905] AC 6]).
3. The question whether a sum stipulated is penalty or liquidated damages is a question of construction to be decided upon the terms and inherent circumstances of each particular contract, judged as at the time of the making of the contract, not as at the time of the breach (*Public Works Commissioners* v. *Hills* [[1906] AC 368] and *Webster* v. *Bosanquet* [[1912] AC 394]).
4. To assist this task of construction various tests have been suggested, which if applicable to the case under consideration may prove helpful, or even conclusive. Such are:
   (a) It will be held to be penalty if the sum stipulated for is extravagant and unconscionable in amount in comparison with the greatest loss that could conceivably be proved to have followed from the breach. (Illustration given by Lord Halsbury in *Clydebank* case).
   (b) It will be held to be a penalty if the breach consists only in not paying a sum of money, and the sum stipulated is a sum greater than the sum which ought to have been paid (*Kemble* v. *Farren* [(1829) 6 Bing 141]). This though one of the most ancient instances is truly a corollary to the last test. Whether it had its historical origin in the doctrine of the common law that when A promised to pay B a sum of money on a certain day and did not do so, B could only recover the sum with, in certain cases, interest, but could never recover further damages for non-timeous payment, or whether it was a survival of the time when equity reformed unconscionable bargains merely because they were unconscionable – a subject which much exercised Master of the Rolls Jessel in *Wallis* v. *Smith* [(1882) 21 Ch D 243] – is probably more interesting than material.
   (c) There is a presumption (but no more) that it is penalty when "a single lump sum is made payable by way of compensation, on the occurrence of one or more or all of several events, some of which may occasion serious and others but trifling damage" (Lord Watson in *Lord Elphinstone* v. *Monkland Iron and Coal Co.* [(1886) 11 App Cas 332]).

On the other hand:

   (d) It is no obstacle to the sum stipulated being a genuine pre-estimate of damage, that the consequences of the breach are such as to make precise pre-estimation almost an impossibility. On the contrary, that is just the situation when it is probable that pre-estimated damage was the true bargain between the parties (*Clydebank Case*, Lord Halsbury; *Webster* v. *Bosanquet*, Lord Mersey).'

In the *Dunlop* case, the contract stated the value of damages, as a pre-estimated sum, to be allowed for every item sold at a cut-price rate, regardless of the value of that item. Having reviewed the authorities, the House of Lords found that this was not a penalty. A major reason for reaching that decision was that the damages were for 'indirect and not direct damage'. It was recognised that such losses were difficult to prove, as Lord Dunedin said, at page 88:

'though damage as a whole from such a practice would be certain, yet damage from any one sale would be impossible to forecast. It is just, therefore, one of those cases where it seems quite reasonable for parties to contract that they should estimate at a certain figure, and provided that figure is not extravagant there would seem no reason to suspect that it is not truly a bargain to assess damages, but rather a penalty to be held *in terrorem*.'

The analysis of Lord Dunedin has been quoted in full in the recent Court of Appeal decision in *Jeancharm Limited t/a Beaver International* v. *Barnet Football Club Limited* (2003) 92 Con LR 26. In that case, the court had to consider both a lump sum charge, per item supplied late, together with a provision for imposing a high rate of interest for late payment. Lord Justice Peter Gibson set out rules for distinguishing a penalty provision, at paragraph 27:

> 'The principles that are relevant, in my judgment, for distinguishing a penalty provision, with which the courts will interfere from a valid contractual provision for a payment or payments in the event of default by a party are these:
>
> (1) the court looks at the substance of the matter, rather than the form of words, to determine what was the real intention of the parties;
> (2) the essence of a penalty is a requirement *in terrorem* of the party in default, as distinct from being a genuine pre-estimate of loss resulting from the default;
> (3) the question whether a provision for payment on default is a penalty is a question of construction of the contract, and that is assessed at the time of the contract and not at the time of the breach;
> (4) if the required payment is extravagant and unconscionable in amount in comparison with the greatest loss that could conceivably be established as a consequence of a default, it is a penalty.'

The court found that a rate of interest of 5% per week was, in that case, a penalty. It will be possible to defeat a claim based on LADs where the subcontractor can show that the damages claimed are in fact a penalty and not a genuine pre-estimate. In the situation where a contractor has allowed LADs to be withheld or recovered under the main contract, the contractor will not be able to recover those damages under the subcontract, even where it can demonstrate it has suffered that loss as a result of the subcontractor's default, should the subcontractor be able to demonstrate that the LADs were a penalty to the contractor from the employer.

A rate of damages based on the value of the subcontract works is unlikely to represent a genuine pre-estimate of loss, for example for delay, since there will be no direct relation between the loss and the basis of calculation. The loss to a contractor, due to a certain delay to the works, will be the same regardless of which subcontractor may be held responsible. Thus a term stating that the contractor will recover a stated percentage of the subcontract sum for each week's delay in completing the subcontract works, could be held to be a penalty even in the case of a subcontract of low total value, where the damages claimed would fall well short of the actual loss suffered by the contractor. However, where the risks consequent on delay may be disproportionate to the value of the subcontract works, it is open to the parties to cap the damages allowable and such cap might be expressed as a percentage of the subcontract sum either as a weekly rate or maximum total sum.

Since damages, whether pre-ascertained or not, are the losses that flow from the defined breach by the other party, it follows that LADs must relate to a breach that leads to that associated loss or losses. In a main contract situation it is generally accepted that a failure to complete the contract works by the due date will entitle the employer to withhold LADs. However, should the employer take possession prior to completion, then the LADs provision will die from the date that the employer takes possession. As discussed below, the subcontract situation may be more complex since there may be events critical to the Works that occur prior to the completion date for the subcontract works.

As discussed more fully in Chapters 4 and 10, the operations which are critical to the main contract works may be completed ahead of completion of the entire subcontract works. Further, there may be a number of activities to be carried out by a subcontractor, each of which, if delayed, will delay completion of the project, even if the subcontract works in total are complete within the subcontract period. By definition a non-critical activity will not delay the overall completion of the project, unless such delay becomes so extensive that the activity itself becomes critical. If LADs are to be effectively incorporated as a provision of the subcontract, such that they are a genuine pre-estimate of loss, then they must be related to completion of critical activities and not completion of the subcontract work as a whole. It is often not the completion of an activity by the subcontractor that is the delaying event, but the prevention of the progressing of the dependent activity by another trade or subcontractor. For a LADs clause in a subcontract to be effective it must be couched in terms such that the damages relate to the non-performance by the subcontractor of the matter or event that actually gives rise to those damages. Any condition which seeks to refer the pre-ascertained loss to any other matter or event will not, it is suggested, be effective.

The concept of pre-ascertained prolongation costs, which has been considered elsewhere, falls into this category of damage, being difficult to quantify with sufficient accuracy. It will normally be difficult to show the actual costs incurred as a direct result of a prolonged subcontract period, and such a pre-estimate will avoid the need to keep and/or agree records. It will, however, be essential to establish the losses to be covered by the pre-ascertained sum or rate. For example, such a rate may not include site-based plant or equipment, crew van, etc. but may be restricted to head office administration and visiting staff costs only.

Two further benefits may also exist. First, there will be a positive inducement to the contractor to agree the achievement of completion and to take possession of and responsibility for the subcontracted work. Secondly, it is suggested, it will be for the contractor to demonstrate that the prolongation costs are not allowable due to the subcontractor's failures. It is likely that, just as with LADs for delay where it would be for the contractor to demonstrate that it is to be relieved from the obligation to pay, similarly, it is suggested, it would be for the contractor to demonstrate that the prolongation was of the subcontractor's making. Where the subcontract works are substantially complete but minor amounts of its work are left down or incomplete, pending work by others, and such work is carried out after the general demobilisation of the subcontractor's team, then the costs of managing such late activities would need to be considered separately and not as a general prolongation of the subcontract work.

Failure to pay by the final date for payment

Unlike the undervaluation situation, failure to pay on time is subject to definite sanctions under the construction contract, either by express terms of the subcontract or implied by legislation, either by the HGCRA or the LPCDIA. Basically two sanctions exist: first, the right to compensation for the late payment, and secondly, after notice, the right to suspend performance.

The HGCRA at section 112 refers to:

> 'a sum due under a construction contract'.

The sum due is considered to be that which has been valued by the contractor or, if disputed and referred for decision under the provisions for dispute resolution, by an adjudicator, arbitrator or the courts. In the case of *Karl Construction Ltd* v. *Palisade Properties plc* [2002] CA 199/01, for example, Lord Drummond Young, referring to the provisions of the JCT main form of contract, and referring to an earlier case *Costain Building & Civil Engineering Ltd* v. *Scottish Rugby Union PLC* [[1993] SC 650], said, at paragraph 18:

> 'the provisions of clause 30 of the JCT Standard Form, dealing with certification and payment, make it clear that payment for work done is to follow the issuing of a certificate by the architect. Such a certificate must, if appropriate, take account of any variations that have been instructed under clause 13 and the amount of any direct loss and expense sustained by the contractor in consequence of any matter specified in clause 26. In those circumstances, in accordance with the ratio of *Costain*, there is no debt due until a certificate has been issued, and the issue of a certificate is a condition precedent to the contractor's right to demand payment. On that basis, inhibition on the dependence will only be competent if special circumstances are averred by the pursuers. The pursuers make no such averments, and accordingly on this ground I conclude that their inhibition on the dependence is incompetent so far as the action is for payment.'

Lord Drummond Young then considered the right to damages following the failure to issue a certificate by the architect, and at paragraph 24 decided that where the certification mechanism fails, the payee can resort to arbitration or litigation or under the HGCRA to adjudication:

> 'In my opinion, it is a general principle of building contract law that, if there is any failure in the certification mechanism, for example because the architect refuses to act and the employer does not compel him to do so, the party adversely affected can resort to arbitration, if the matter falls within the terms of an arbitration clause, and failing that litigation, in order to have the amount that should have been certified determined. If the contract says nothing about failure to issue certificates, the legal basis for that principle is an implied term, based on business efficacy;'

The reference to arbitration or litigation will also include the reference to adjudication, where relevant, which, it is suggested, will be the normal initial step.

Then, at paragraph 26, he concludes that damages will not be necessary:

> 'That decree, whether of the arbiter or the court, is equivalent to a certificate, and payment will be due by the employer once it has been pronounced in the same way as it would have been had the architect issued a certificate. On this basis an award of damages will not be necessary; the contractor will be found entitled to payment of what is due under the contract, which will be precisely the same amount as should have been certified. That amount may include interest. It follows that, if the contract is enforced in this way, there can be no breach of contract causing loss to the contractor; the provisions of the contract can always be enforced, using the mechanism of arbitration.'

It is to be noted that Lord Drummond Young does envisage that the amount may include interest.

As described above, it is considered that where a contractor undervalues the work but pays on time the sum stated in its valuation, the subcontractor may not have a right to issue a notice of intent to suspend performance. The same is unlikely to apply where the contractor has neither issued a valuation nor made any payment. In the case of undervaluation of the work, if there should be a manifest refusal by the contractor to value the work in the correct manner, as set out in the subcontract, then the existence of a right for the subcontractor to suspend or indeed, in extreme circumstances, determine the subcontract is likely.

Damages where no agreed subcontract sum

For various reasons and under differing circumstances, subcontractors not infrequently commence work on site prior to completing their negotiations and concluding a formal agreement. In such circumstances, as discussed in detail in Chapter 2, Contract, there may or may not be a subcontract in law and that subcontract may or may not be held to be a subcontract in writing, for the purposes of the HGCRA. One major reason why work commences prior to a concluded agreement is that the parties are still negotiating with regard to the on-site arrangements and the associated tender price adjustments.

As mentioned above, damages are intended to be a money compensation for a breach of the subcontract and to put the aggrieved party in the position it would have been had there been no breach. It follows that, while there may be general agreement as to the subcontract terms but there are still some obligations of the parties to each other to be agreed, it is unlikely that the aggrieved party will be able to demonstrate what its financial position would have been had an agreement been finalised prior to the breach. In the Court of Appeal case of *Crown House Engineering Ltd* v. *Amec Projects Ltd* (1989) 48 BLR 32, Amec, at page 48, are reported as having pleaded that the judge, at first instance, had 'erred in law in holding that it was not permissible to take into consideration in evaluating such sum the levels of cost incurred by the defendants as a result of the timing and manner in which the plaintiffs undertook their work'. Lord Judge Slade, at page 54, referring both to the decision in *British Steel Corporation* v. *Cleveland Bridge & Engineering Co Ltd* (1981) 24 BLR 94 and to *Goff and Jones on the Law of Restitution*, expressed doubts as to the law, but appears to have supported the view that such deductions are not allowable:

> 'I am not convinced that either learned work, or any of the other reported cases cited to us, affords a clear answer to the crucial question of law. On the assessment of a claim for services rendered based on a *quantum meruit*, may it in some circumstances (and, if so, what circumstances) be open to the defendant to assert that the value of such services fall to be reduced because of their tardy performance, or because the unsatisfactory manner of their performance has exposed him to extra expense or claims by third parties?
>
> In my judgment, this question of law is a difficult one, the answer to which is uncertain and may depend on the facts of particular cases. If, as the learned judge apparently considered, the answer to it is an unqualified "No never", I cannot help thinking that, at least in some circumstances, there would result injustice of a nature which the whole law of restitution is intended to avoid.'

It will no doubt depend on the facts in each case. It is suggested that where there is a concluded subcontract and the other matters, such as the time for the subcontracted work, are defined, then the contractor would probably be able to recover its losses, flowing naturally from the subcontractor's breach, against the value of the work carried out. However, to the extent that an agreement has not been reached on the subcontract sum, because, for example, the integration of the subcontractor's work with that of others had yet to be agreed, then it is unlikely that the contractor would be entitled to a loss due to the non-availability of the subcontractor's work to the requirements of other trades, when such requirements had not been defined or accepted. In the absence of an agreed price, it will always be possible for the subcontractor to claim that it would have made a suitable allowance for the consequences of the breach claimed by the contractor, in its finalised price. Such allowance would be to cover the cost of the necessary action to avoid or allow for the damages claimed by the contractor. Whether such an argument would be successful would depend on the facts and the state of negotiations at the time.

Timing of entitlement

As mentioned previously, the purpose of damages is to repay to the aggrieved party money for its loss. The loss generally takes place at the time of the breach. Unusually, in the subcontract situation, the effect of a breach may not and generally will not be known, at any rate in full, at the time of that breach. If a piling subcontractor overruns its subcontract period and, as a consequence, the following work is delayed, the contractor may or may not recover that lost time during the course of the remaining works. At the time of the subcontractor's breach, there may be an expectation that the contractor will incur both overrun costs and LADs chargeable by the employer. The contractor will seek to withhold such costs from the subcontractor's account, pending the final outcome of the project, notwithstanding that at that time it has in fact incurred no loss.

In other circumstances, it may be possible to ascertain the costs flowing from the breach before they are incurred. For example, when a subcontractor wrongfully rescinds the subcontract and the contractor engages another, then its loss can, in part at any rate, be ascertained as the difference between the new subcontract sum and the value of the same work at the first subcontractor's rates, together with the costs of arranging the new subcontract.

The subcontractor may, by its default, cause the contractor a loss by way of damage during the course of the subcontract works. The contractor, in such circumstances, will be able to calculate its loss and seek recovery by way of a notice of intent to withhold payment against the subcontractor's next valuation. Thus three situations may arise with regard to the timing of a contractor's claim for damages:

1. a defined loss, which has actually been incurred at the time of withholding;
2. a defined loss, which has been valued but not incurred at the time of withholding;
3. undefined losses, which may or are likely to be incurred by the contractor and for which only an anticipatory provision may be identified at the time of withholding.

Unless there are specific provisions in the subcontract restricting the contractor's right to withhold damages in anticipation of its loss, it is suggested that, provided the sums withheld can be shown to be a genuine estimate of the likely losses to be incurred, the contractor will have the right to withhold those sums. This contention is supported by the House of Lords decision in the case of *Gilbert-Ash (Northern) Ltd* v. *Modern Engineering (Bristol) Ltd* (1973) 1 BLR 73 where on page 82, Lord Morris of Borth-y-Gest said, referring to the bespoke form of subcontract used:

> 'In the third sentence it is provided:
>
> > "If the Subcontractor fails to comply with any of the conditions of this Subcontract, the Contractor reserves the right to suspend or withhold payment of any monies due or becoming due to the Subcontractor."
>
> If this was a valid provision any failure, however unimportant, would enable the contractor to withhold present or future payments. Such a heavily penalising provision ought not to be accorded any validity. Nor would it be appropriate to embark on a process of redrafting the sentence by reading in suggested words which it does not contain.'

Having shown that that part of the subcontract term was to be regarded as a penalty and therefore not enforceable, he turns to the fourth sentence and decides that since the subcontractor had agreed to allow any bona fide contra account to be deducted by the contractor, that was sufficient reason to allow such a term:

> 'The fourth sentence of the term is as follows:
>
> > "The Contractor also reserves the right to deduct from any payments certified as due to the Subcontractor and/or otherwise to recover the amount of any bona fide contra accounts and/or other claims which he, the Contractor, may have against the Subcontractor in connection with this or any other contract."
>
> It is on the interpretation of those words that the present case, in my view, depends. A 'certified' payment is clearly a liquidated sum. To have a process of deduction from such a sum there must clearly be some other stated sum. There could, for example, be some other liquidated sum. There could be some sum which could be regarded as a contra account. But there would have to be some sum. There could not be a deduction of something that lacked any kind of specification. But need the sum to be deducted be a liquidated sum or an ascertained sum in the sense of an agreed sum or of a sum assessed by a court? The wording of the provision does not so indicate. There may be a deduction of the amount of any bona fide claim which the contractor may have against the subcontractor. Such claim may be in connection with the contract which has occasioned the certified payments or in connection with any other contract. As applied to the facts now before us the position is that the appellants have claims against the respondents in connection with the subcontract. Those claims have both been particularised and quantified. Their amount is known. Whether or not they can be substantiated it is accepted that as 'claims' they have been made in good faith. Whether it is wise for a subcontractor to agree to a provision which may delay his receipt of money certified by an architect as being payable is not for us to decide. I can see no escape, however, from the conclusion that the respondents agreed with the appellants that there could be a deduction of the amount of any bona fide claims. Here there were such. I consider therefore that the learned judge was correct in deciding the preliminary issue as he did.'

The courts are equally prepared to uphold terms limiting the contractor's right to deduct sums not incurred at the time of the intended withholding. In the case of *Chatbrown Ltd* v. *Alfred McAlpine Construction (Southern) Ltd* (1986) 35 BLR 44, the judge had to consider the effect of the phrase 'actually been incurred' in the old Blue Form of non-domestic subcontract. In the Court of Appeal decision, Lord Justice Kerr first, at page 49, quotes clause 15(2) of that subcontract:

> ' "The contractor shall be entitled to set off against any money (including any retention money) otherwise due under this subcontract the amount of any claim for loss and/or expense which has actually been incurred by the contractor by reason of any breach of, or failure to observe the provisions of, this subcontract by the subcontractor, provided,
>
> (a) The amount of such set-off has been quantified in detail and with reasonable accuracy by the contractor; and
> (b) The contractor has given to the subcontractor notice in writing specifying his intention to set off the amount quantified in accordance with proviso (a) of this sub-clause and the grounds on which such set-off is claimed to be made. Such notice shall be given not less than seventeen days before the money from which the amount is to be set-off becomes due and payable to the subcontractor; provided that such written notice shall not be binding insofar as the contractor may amend it in preparing his pleadings for any arbitration pursuant to the notice of arbitration referred to in clause 16(1)(a)(i) of this subcontract." '

The judge then stated, at page 51:

> 'Faced with that issue of construction, the judge dealt with it as follows. He referred to the fact that the agreed completion date fixed for the main contract was 30 March 1986, which was still nearly fourteen months beyond the date when the notice was given. Presumably what he had in mind was that it might be said that the additional costs or expenses due to delays by a subcontractor in operating the site would in fact not be incurred until the main contractor would have completed work on site but for such delays. However that may be, having referred to the completion date for the main contract, he said:
>
> "The gravamen of McAlpine's case is that although moneys would be laid out at a future date resulting from the past delays of Chatbrown, nevertheless they constituted a present loss or expense at the time the notice was given."
>
> Pausing there, if one reads on, one sees that what he has in mind is that an alleged liability had already been incurred.
>
> He went on:
>
> " 'Incurred' in clause 15 it was contended meant 'incurred a liability for loss and/or expense'. The word 'actually' was, to use Mr Fernyhough's [Counsel for McAlpine's] expression, 'mere verbiage' . . . Adverbially it added nothing to the word 'incurred'.
>
> Mr Fernyhough contended that unless the subcontract entitled the contractor in circumstances such as those that existed in March 1985 to set off moneys of the nature included in Appendix 1, the subcontract would be unworkable. He contended that at 8 March 1985" – that is the date of the notice – "McAlpine had incurred a liability for matters set out in Appendix 1."
>
> He clearly meant that Chatbrown had incurred the liability, or that McAlpine were entitled to assert a liability on the part of Chatbrown.
>
> So that was the submission with which the judge had to deal, and he rejected that construction.'

In SBCSub/C, clause 4.21 allows for 'Any amount agreed by the Parties as due in respect of any loss and/or expense thereby caused to the Contractor may be deducted from any monies due or to become due to the Subcontractor or shall be recoverable by the Contractor from the Subcontractor as a debt'. It is submitted, at least for loss and/or expense arising from where 'the regular progress of the Works is materially affected by any act, omission or default of the Subcontractor, his servants or agents', that it is a precondition to the issue of a valid notice of intent to withhold payment under clause 4.10.3, that the contractor first obtains agreement from the subcontractor.

Damages for latent defects

It is normal practice in building subcontracts for there to be a term, either express or implied, that during a specified period after the completion of the Works the subcontractor will be required to attend to any defect that becomes manifest.

Where such defects become apparent, it will be both the duty of the subcontractor to attend to such defects and its right to be allowed to do so. Thus, were such a defect to become apparent and the employer and/or the contractor to engage another to rectify the works, the subcontractor would, at most, only be liable for the cost it would have incurred in carrying out the rectification works and not the cost actually incurred by the employer or contractor.

It is possible that in an emergency, where the defect is such that there is a serious risk to persons and/or property, the employer/contractor may be justified in not seeking out the subcontractor but taking such action as is necessary, at least to remove the risk. It is likely that the risk may only need to be perceived by the employer/contractor and need not in fact have existed.

However, once the period for making good defects under the subcontract passes, these rights and obligations die and the employer/contractor, on the discovery of a latent defect, may employ whomever it chooses to rectify the defect, and seek to recover the costs by way of damages.

In *Michael A. Johnson* v. *W. H. Brown Construction (Dundee) Ltd* [2000] BLR 243, Lord Sutherland, referring to an earlier decision of *Shanks* v. *Grey* [[1977] SLT Notes 26], concluded that if defects occurred after the defects liability period the only remedy would be in damages. He said, at paragraph 4:

> 'The decision in *Shanks*, however, does point to the fact that if defects were discovered after the expiry of the defects liability period, the employer's only remedy would be an action of damages for breach of contract, and he would have no remedy under the contract itself by instructing the contractors to remedy the defects.'

While there may be no right on the part of the employer/contractor, nor an obligation on the part of the subcontractor, under these circumstances to rectify any defect in its work, a subcontractor if asked may well both wish and be willing to rectify the defect at its own cost, rather than be faced with an action to recover costs over which it had no control.

A further issue occurs with regard to the investigation of defects arising during the defects liability period, as to which party is responsible for arranging and carrying out any necessary investigations. In *Michael Johnson* v. *Brown*, the commentary at page 243 makes the point that the judgment leaves the employer with only one option and that is to instruct the contractor to rectify the defect:

'Essentially, it decides that under the JCT Standard Form (with design) and by analogy the other JCT contracts, the employer's only direct remedy for putting right defects during the defects liability period is to instruct the contractor to put right the defects.'

The provisions of clause 3.11 (work not in accordance with the subcontract) of SBCSub/C are less restrictive in that they allow, but do not require, the contractor to instruct the subcontractor with regard to any work, materials or goods not in accordance with the subcontract.

The difficulty is that, at the time the defect becomes apparent, while there is a manifest defect such as water penetration or poor performance of a system, the cause may or may not be that which it appears to be. Damp patches may be caused other than by leaks, and decorative finishes can fail because of the nature of the underlying structure rather than defects in the quality of finishing materials or workmanship in their application or installation. It is suggested that where there is an apparent defect in a subcontractor's work and where the defects liability period has not expired, then the subcontractor has the obligation, at least under the JCT family of contracts, to investigate the apparent defect and either to rectify it, should it be due to the subcontractor's breach, or demonstrate that there is no defect in its work and that the cause of the problem is something else.

It is not considered that the subcontractor's obligations extend to identifying the cause, but are restricted to demonstrating that the subcontractor is not in default. It is likely, should the problem prove not to be a default of the subcontractor, that the subcontractor will be able to recover the costs of its abortive investigation, although on the basis of *Shanks* v. *Gray* these may be regarded as non-legal expenses in defending a claim and therefore to be extrajudicial, whether or not there is litigation.

Costs of an expert's report

When a dispute arises as to the quality of a subcontractor's work, it may not be within the expertise of the contractor to challenge in detail the quality, or to identify fully the extent of defects or non-conformities. It is, however, part of the subcontracting process that the work is inspected and either accepted or rejected, and if rejected that the basis or grounds of such rejection are given to the subcontractor.

In the Scottish appeal case of *Michael Johnston* v. *Brown*, the court had to consider the employer's right to damages where, after completion of the Works and during the defects liability period, leaks were found in the roof. The court found that the contractor had attended to the defective work as required under the contract. However, the employer had employed an architect to examine the work and produce a schedule of defects. The employer had sought to recover, by way of damages, the costs of the architect's fee for the investigation and report, the solicitor's fees for contractual advice, and the management costs. The appeal related to the architect's fees only.

The court considered the matter in two ways: first, what would the situation have been if there were no provision under the contract for the remedy of defects, and secondly, the nature of the contract term itself. The court found that the employer had no cause of action for damages for the cost of advice. Lord Sutherland said, at paragraphs 4 and 5:

> 'The decision in *Shanks*, however, does not point to the fact that if defects were discovered after the expiry of the defects liability period, the employer's only remedy would be an action of damages for breach of contract, and he would have no remedy under the contract itself by instructing the contractors to remedy the defects. If, in that situation, the employer instructed architects to ascertain the nature and extent of the defects, he would not be able to recover the costs of so doing in the subsequent action for damages. It might therefore be regarded as illogical that he should be entitled to recover such costs against the contractor if the defects were discovered during the defects liability period.
>
> However, this is not, in our view, a conclusive consideration. The main problem for the reclaimer is that clause 16 appears to contain its own remedy. If there are defects discovered before the expiry of the defects liability period, the remedy is for the employer to give instructions to the contractor to remedy such defects. If they are duly remedied there is no further redress for the employer under clause 16.'

The case of *Shanks* v. *Gray* concerned the costs of preparation for litigation in a claim following a road traffic accident. The report states:

> 'It has not been doubted in over a hundred years and is in line with the established law and practice that the legal and other expenses, such as the cost of medical or other reports and plans, incurred in vindicating or establishing or defending a claim cannot be claimed as damages. The legal expenses which are recoverable are the judicial expenses allowed by the court. All other expenses of vindicating or establishing or defending a claim are extra-judicial, whether or not there is a litigation.'

It is possible that there are differences at law between the situation where the claimant has access to professional help, such as exists in construction contracts involving an architect and/or an engineer and possibly other specialist professionals, and the situation where the injured party has no such pre-existing assistance. There may be a further distinction between the costs of establishing the nature of the damage or default and the valuation of the loss. The case of *Morgan* v. *Perry* [1974] 299 Estates Gazette 1737, concerned an action against a building surveyor for failing to identify defects in a house which were so serious that subsequently the house proved to be valueless. Investigating the cause and effect of the defects involved considerable work by a firm of consulting engineers. In awarding damages to the building owner, the judge said in the last paragraph:

> 'That did not mean, however, that *Philips* v. *Ward* [[1956] 1 WLR 471] went so far as to prevent the recovery by the plaintiff of some comparatively minor consequential loss. There appeared to be no logical reason why such consequential loss should not include the reasonable cost of discovering the true loss which the plaintiff had suffered. Since the proper basis was the differential value and not the cost of reinstatement or repairs, the plaintiff was not entitled to recover the cost of carrying out interim repair work, but was entitled to the balance of £1,200 claimed in respect of consequential loss.'

Although this report does not set out how the sum awarded as consequential loss was calculated or what it comprised, it is clear that this loss was considered to be comparatively minor and for the purposes of valuation of the direct loss only.

Recovery of preliminary costs

As mentioned earlier, a right to recover money as damages is a right to recover proven actual loss incurred or arising as a result of the breach of the subcontract as alleged. This is true whichever party seeks to recover a loss.

When the work of the subcontractor takes longer than contemplated, either or both the subcontractor and contractor will have resources committed to that project for an extended period, either at the time the subcontract is concluded or the Works as a whole are delayed. It is often, it is considered wrongly, assumed that the subcontractor and/or the contractor will have suffered a loss as a consequence. Whether the parties have or have not suffered a loss will be for them to demonstrate.

Both plant and staff will be allocated to a specific project and therefore it can be shown that such plant and/or staff have been charged to that project for the period of the delay. It can therefore be demonstrated that for the purposes of the company's internal accounting the project concerned has suffered a book loss. However, this will not show that the company or business has suffered a loss. If plant is owned by the company, there will be a need to demonstrate either that it cost money to keep it on the site rather than at the plant depot, or that because of its non-availability other plant had to be hired in for another project. In the latter case, the actual loss may be for a period of time relevant to that other project and not the delay on the relevant one.

Similarly, with staff costs, it will rarely be the case that senior staff are engaged on a short-term basis, although staff levels may be adjusted by the use of agency staff. It will be more certain that the costs to the project are in fact a loss to the company where, as with hired plant, agency staff are employed. Permanent staff will not be hired and fired on a project-by-project basis, and consequently it will be difficult to define the loss incurred through the further involvement of specific staff on a defined project.

In the case of *Weldmarc* v. *Cubitt*, Judge Richard Harvey QC had to consider the measure of damages to be awarded to Cubitt for the delay due to a breach of the subcontract by Weldmarc. He said, at paragraph 76:

> 'Cubitt counterclaims £18,455.91 as "direct costs" of the kind I have mentioned in paragraph 59 above. Cubitt claims that as a result of failure on the part of Weldmarc to start erection of the staircase on 18 June and of Weldmarc's repudiation of the contract, Cubitt incurred a delay of ten days in the completion of its contract with the main contractor. During that period Cubitt is said to have incurred additional costs in order to maintain site staff at the project. It is not said that extra staff were taken on by the company. No loss of opportunity to earn profit is shown. Staff costs may no doubt be attributable to this project for the extra ten days for management accounting purposes. But that does not establish that Cubitt suffered loss in consequence. The same applies to costs of operatives such as foreman, storekeeper and general labour attributable to preliminaries. Mr Shannon [Cubitt's project surveyor] said "Cubitt have also produced invoices for plant and car hire for the period 29 June 2001 to 15 July 2001". It is not in evidence that those sums would not otherwise have been incurred.'

Here the judge was not satisfied that the additional costs claimed would not have been incurred by Cubitt in any event, and therefore those costs could not be said to be a loss due to Weldmarc's breach, as opposed to, say, a loss due to a lack of other work at the time. His decision may have been influenced by the fact that Cubitt did not allocate its costs to each project on a strictly factual basis.

In the case of *Alfred McAlpine Homes North Ltd* v. *Property & Land Contractors Ltd* (1995) 76 BLR 59, Judge Humphrey Lloyd QC had to consider whether a claim for loss and/or expense for the contractor's small plant, which remained underutilised during a period of suspension of the works, could be upheld where that claim was for small plant owned by the contractor and charged at the RICS Schedule of Basic Plant Charge rates. The judge found that there had to be actual loss and that no loss could be demonstrated in the circumstances. He said, at page 92:

> 'From the award it appears that the arbitrator regarded the fact that PLC owned the small plant as "fortuitous", and, on the basis that, if the plant had not been owned by PLC, the plant would have been hired by PLC, dismissed arguments about valuing the claim in terms of depreciation. Instead the arbitrator valued the claim by reference to what he regarded as reasonable hire charges. His reasons included the fact that the RICS Schedule of Basic Plant Charges states that "the rates apply to plant and machinery already on site, whether hired or owned by the contractor". It seems therefore that the underlying question of law is whether under clause 26 of the JCT conditions an arbitrator can make an ascertainment other than on the basis of the amounts actually incurred or lost by the contractor.
>
> The sums at stake would not in themselves have led to an appeal. However the arguments on this part of the appeal raised points of some importance which may be of significance in other instances. In my judgment, the arbitrator interpreted clause 26.1 incorrectly as permitting him to award as an ascertainment of direct loss or expense an amount which might have been the loss or expense suffered or incurred by PLC had it not owned the plant but had hired it which was not in fact its actual loss or expense since it owned the plant and had not hired it. Clause 26 requires the architect or quantity surveyor (or arbitrator) to ascertain what was the actual direct loss or expense and not a notional or hypothetical amount, using the method adopted by the arbitrator. The arbitrator seems not to have applied the principle which he recognised in paragraph 22.2 of his first award as the approach to clause 26. Ascertainment on the basis of hire charges might not have been questioned if there had been a finding that PLC would have hired out this plant but there is no such finding – nor is it likely that small tools etc. would have been hired out, as opposed to being hired by PLC. Only if there had been such a finding could the arbitrator's award have been justified as representing, as Mr Darling [Counsel for Property and Land] submitted, the valuation of a lost opportunity. The RICS schedule although primarily intended for the valuation of variations might then have been relevant as a check against the rates claimed. Similarly if PLC had not owned the plant but had in fact hired it in there might have been little difficulty in principle in using the actual charges incurred by it. Where plant is owned by the contractor which would not have been hired or which was not able to be hired out the ascertainment of loss and expense must be on the basis of the true cost to the contractor and must not be hypothetical or notional amounts. An ascertainment needs to take account of the substantiated cost of capital and depreciation but will (or may) not include elements which are included in hire rates and which are calculated, for example, on the basis the plant will be remunerative for only some of time and other times be off hire. Hire rates are usually higher for that reason alone and also, because the rates are overall rates and therefore include elements which may not be incurred by a contractor owning the plant of hire, their application over a period of any consequence will almost certainly produce figures higher than the true loss or expense incurred by the contractor whose own plant is tied up for the same period.

The question of law implicit in this part of the appeal will therefore be answered as follows:

"That in ascertaining direct loss or expense under clause 26 of the JCT conditions in respect of plant owned by the contractor the actual loss or expense incurred by the contractor must be ascertained and not any hypothetical loss or expense that might have been incurred whether by way of assumed or typical hire charges or otherwise." '

While in *McAlpine* v. *Property & Land* the contractor's claim for its plant was disallowed, its claim for the salaries of its senior management was found to be allowable, by way of loss and/or expense, under the provisions of their contract. In the case of *Euro Pools PLC* v. *Clydeside Steel Fabrications Limited* (17 January 2003, unreported) similar claims were made, by Euro Pools, by way of damages under the Sale of Goods Act. In that case, defective tanks had to be repaired in situ and the purchaser, Euro Pools, had to divert resources both to supervise the repair work and to carry out necessary associated work to facilitate the repairs. The defenders, Clydeside Steel, averred that the staff costs were not a loss to Euro Pools since the staff would be employed in any event. Lord Drummond Young found that there had been a loss to the company by diverting staff, including the managing director, from their normal work. He said:

'[11] The first submission by counsel for the defenders was in essence that, if the pursuers were to recover damages for disruption of their business, such disruption must take the form of additional costs in completing other contracts; the mere performance of remedial work under contracts between the pursuers and third parties was not sufficient to establish any loss, unless it resulted in additional costs that would not have been incurred but for the defenders' breach of contract. In my opinion that proposition is incorrect, for two reasons. In the first place, it confuses two matters that are conceptually quite distinct, the loss sustained in consequence of a breach of contract and the quantification of that loss. The loss is physical or economic damage sustained in consequence of the breach of contract. That loss must be quantified, in the sense that a monetary equivalent must be ascertained, but that exercise is not the same as the identification of the loss. In the second place, it ignores the fact that, if employees are required to perform additional work in consequence of a breach of contract, that work in itself represents a loss to their employer. These points can perhaps be seen most clearly in the case of a managing director. A managing director will typically be paid a fixed salary, perhaps with profit related bonuses. If he requires to spend part of his time organising rectification work following a breach of contract, he will not be paid any more, and there is thus no direct cost to his employer as a result of the need to perform the rectification work. In reality, however there is still a loss to his employer as a result of the rectification work. A managing director will normally be expected to devote the whole of his working time and effort to the affairs of his employer; a term of that nature is commonly found in managing directors' service contracts, and in many cases would be implied even if it were not expressed. The tasks performed by a managing director are obviously very variable, but they will almost invariably include the strategic planning of the company's business and, at a strategic level, the search for new markets and the development of new products. Those tasks are critical to the development of the business; indeed, in modern conditions, they are usually critical to its very survival. In a smaller company such as the present pursuers, the managing director may perform much more varied tasks, including sales, the supervision of employees and general administration; those tasks too are vital to the survival of the company's business. If one of the company's suppliers commits a breach of contract, and in consequence the managing director requires to spend a significant amount of time supervising remedial measures, that time is lost to the other tasks that the managing director is obliged to perform. That in my opinion clearly represents a loss to the company. It may not leave the company out of pocket, in the sense of having to pay more to the managing director; nevertheless, the company will inevitably be deprived of part of the services that it would normally expect from the managing director. That might mean, for example, that the managing director was unable to devote as much time as would otherwise have been possible to the planning of an important marketing initiative, or the development of a new product, or to general administration of the company's affairs and the supervision of its employees. In any of these cases, the loss of time inevitably represents a loss of services provided by the managing director to the company. Thus in any such case the existence of a loss is established. Thereafter it is necessary to consider how that loss is quantified.'

Subsequently the judge held that there is no one way of determining loss under such circumstances and at paragraph 12, referring to various cases, he said:

'Those cases make it clear that there is no single method of assessing the loss sustained in consequence of a breach of contract; the loss must be determined pragmatically, having regard to the particular circumstances of the case.'

He elaborated on this aspect of law:

'13. In relation to quantification, the overriding principle is that the loss must be quantified in a manner that is objectively reasonable. What is reasonable in any particular case will, however, depend on the facts and circumstances of that case. Where employees are required to spend time on rectification works following a breach of contract, the employer will normally be entitled to recover the cost of the salary or wages paid to those employees for the time taken to perform the remedial work. In addition, the employer will normally be entitled in my opinion to recover some contribution to the general overheads of its business.'

The common factor in *McAlpine* v. *Property and Land* and *Euro Pools* is that both the innocent parties were small businesses where senior managers, at director level, had a direct involvement in the management of the work and whose time could and was allocatable to the specific project. While subcontractors are frequently such small businesses, it is not so common that actual time and detail of involvement of the senior management will be recorded and able to be allocated to specific events.

In any event, it is suggested, there will be a need to show that any such costs can be linked directly to the event for which the remedy is sought. Claims for vague causes such as prolongation of an entire subcontract are unlikely to be successful, both by reason of their nature and for lack of supporting records.

The case of *Babcock Energy* v. *Lodge Sturtevant*, like that of *Euro Pools*, concerned rectification of work found to be defective after handover to the employer. In this case electrostatic precipitators provided and installed in a power station's flues by Lodge Sturtevant proved to be ineffective in meeting the employer's specification for dust extraction from the flue gases. Babcock had contracted to provide boilers for a power station in India, and Lodge were their subcontractors. After the defect was identified much time was spent in considering possible remedial action. Not all of the proposals were specification compliant and one in particular, which required the treating of the flue gases to assist in the dust extraction, resulted in a continuing additional production cost. After much negotiation a settlement agreement was entered into between Babcock and the employer. Babcock sought to recover both the value of the settlement agreement and their extensive management and other staff costs in reaching such agreement. Judge Humphrey Lloyd QC had to decide whether those management and other staff costs were in fact losses to Babcock and found that they were, at page 104:

> 'The issue to be resolved was whether Babcock had actually suffered a loss of this magnitude. The defendant's case was that the personnel whose hours were recorded had not been shown to have been diverted from other profitable work or that additional staff had been taken on in view.
>
> The evidence was very general. Mr Lace's [Expert witness for Babcock's] evidence, which I accept, showed that at the relevant times the plaintiffs had plenty of work to do and as it is colloquially put, a full order book. Work had to be farmed out to another company. Agency staff had to be employed. Given the history of this contract it is not surprising that he also stated that if the staff concerned had not been involved in dealing with the problems created by the defendants, they would have been fully and gainfully employed elsewhere.'

The judge then compared this situation with that in the case of *Tate & Lyle Food Distribution Ltd and Silvertown Services Lighterage Limited* v. *Greater London Council and Port of London Authority* (22 May 1981, unreported), from which he quoted Mr Justice Forbes as saying:

> ' "I have no doubt that the expenditure of managerial time in remedying an actionable wrong done to a trading concern can properly form a subject matter of a head of special damage. In a case such as this it would be wholly unrealistic to assume that no such additional managerial time was in fact expended. I would also accept that it must be extremely difficult to quantify. But modern office arrangements provide for the recording of the time spent by managerial staff on particular projects. I do not believe that it would have been impossible for the plaintiffs in this case to have kept some record to show the extent to which their trading routine was disturbed by the necessity for continual dredging sessions . . . While I am satisfied that this head of damage can properly be claimed, I am not prepared to advance into an area of speculation when it comes to quantum. I feel bound to hold that the plaintiffs have failed to prove that any sum is due under this head." '

Judge Humphrey Lloyd then considered Lodge's case that only agency and additional staff were chargeable, but found that Babcock's own staff costs were recoverable because of the nature and magnitude of the problem and time spent in its resolution:

> 'However I reject the latter proposition as did Mr Justice Forbes in the *Tate & Lyle Food Distribution Ltd* case. Managers are of course employed to sort out problems as they arise. If however the magnitude of the problem is such that an untoward degree of time is being spent on it then their costs are recoverable. Looking at the hours recorded I am quite satisfied that that is the position in this case. The costs of course go beyond those of managers and represent staff costs that would not have been incurred but for the breach.'

While in *Tate & Lyle Food Distribution Ltd* the plaintiffs failed to demonstrate the value of their loss, in the *Babcock* case the judge was satisfied that they had kept proper accounts of their costs, which demonstrated their loss:

> 'The plaintiffs' relevant cost codes were M283 which was for the 'additional ESP [electrostatic precipitators] proposal' and M554 'ESP performance investigation'. Under the former time was charged by personnel in the drawing office, in sales, in dealing with proposals for ash handling, in the project engineering department at Crawley, in the electrical control systems department at Crawley, in the drawing office at Crawley, in the boiler design project section at Crawley, in the pipework system GRP unit at Crawley, in the product development unit at Crawley and in the research development and service departments. The bulk of the costs where incurred in relation to project engineering systems, boiler project design, and production development. Under cost code M554, the majority of the costs were incurred in relation to project engineering, boiler project design, and product development. I am satisfied that the plaintiffs did incur costs under the heads described above . . .'

The conclusion to be drawn from the *Babcock* case is that the costs of permanent staff may be considered a loss where it can be shown, first, that a significant amount of their time was devoted to the work resulting from the defendant's breach, and secondly, that the time has been accurately recorded and reasonably costed. This contrasts with the situation in *Weldmarc* v. *Cubitt* but conforms to the judgment in *Euro Pools*.

Contribution to head office overheads

It is frequently argued that when the subcontract works extend over a longer period than that anticipated at the time the subcontract was entered into, the subcontractor or contractor will suffer a loss because the head office establishment will have been required to provide support to the project for a longer period and this must have led to a loss to the subcontractor, by reason of these resources not being available to earn money elsewhere. Such losses are frequently referred to as 'un-recovered overheads'. Over the years various formulae have been devised to calculate such loss by relating the anticipated turnover of the subcontractor as a proportion of the company's turnover and extending the head office costs in that ratio for the extended period. The principal formulae are Hudson, Emden and Eichleay.

For most subcontractor companies the business of tendering, obtaining and carrying out work is a continuous process and little or no effect will arise where the completion of a project is delayed. Exceptions may arise with trades which employ a relatively few pieces of specialist equipment, because work cannot be carried out elsewhere while plant is still employed, albeit inefficiently, on the project in delay. Such trades may include piling, thrust boring, spray concrete, slipform or paving work.

For other trades, a loss is more likely where the commencement of work is significantly delayed. Such delays will cause loss especially to those trades where a significant proportion of the subcontract work is carried out in a factory or workshop. Structural steelwork, timber or concrete structural components, manufactured joinery and other shop fitting operations are examples of trades where a delay to the commencement of the work may lead to a considerable loss in overhead recovery, during the period that the factory is unproductive. However, in the delayed start situation, the ultimate loss may only be the financing of those costs during the period of delay, unless the subcontractor is able to demonstrate that other work was turned away as a result of the delay or prolongation.

It is now generally accepted that losses of this nature will only occur in a limited number of situations, such as those described above. One other factor in assessing the likelihood of there being a provable case will be the number of projects being carried out by a subcontractor at any one time. Where this is limited to a very small number, the effect on the business of major delays or prolongation of the work is more likely to lead to a provable loss.

In any event, when considering the effect of an extended duration of the subcontract period, the effect of any additional work will be of significance. Under the normal method for valuing additional work, an allowance will be made for overheads and profit. Thus frequently the overall result is that, by reason of the additional work, the subcontractor receives an additional contribution to its overheads, albeit that the rate of recovery may not have been as anticipated.

Recent judgments have taken a very robust view of this head of claim. Judge Thornton QC, in the case of *Norwest* v. *Co-operative*, after considering the recently decided case of *McAlpine* v. *Property & Land*, did not rule out the use of formulae but listed five tests that needed to be satisfied, before a subcontractor will have justified such entitlement. He said:

> '350. Thus, an Emden-style formula is sustainable and may be used as the basis of ascertaining a contractor's entitlement to payment for loss and/or expense in the following circumstances:
>
> 1. The loss in question must be proved to have occurred.
> 2. The delay in question must be shown to have caused the contractor to decline to take on other work which was available and which would have contributed to its overhead recovery. Alternatively, it must have caused a reduction in the overhead recovery in the relevant financial year or years which would have been earned but for that delay.
> 3. The delay must not have had associated with it a commensurate increase in turnover and recovery towards overheads.
> 4. The overheads must not have been ones which would have been incurred in any event without the contractor achieving turnover to pay for them.
> 5. There must have been no change in the market affecting the possibility of earning profit elsewhere and an alternative market must have been available. Furthermore there must have been no means for the contractor to deploy its resources elsewhere despite the delay. In other words, there must not have been a constraint in recovery of overheads elsewhere.

In the case of *McAlpine* v. *Property & Land*, Judge Humphrey Lloyd QC had to consider the recovery of head office costs, resulting from a delay to the works, as loss and/or expense under the contract. The situation was unusual since the contractor, Property and Land, was found to be a single contract company and the head office staff were effectively project staff and part of the project preliminaries. The judge said, at page 89:

> 'The arbitrator's findings show that PLC was in a special position. Not only was it a subsidiary of PLS but it was dedicated to carrying out such contracts as PLS thought it ought to do and was not therefore comparable to an ordinary contractor getting work from whatever sources that might be available. Further the arbitrator made findings that PLC's head

office personnel were directly involved in the supervision of the Shipton contract and would have been so involved in the work of any other succeeding contract which, but for the Shipton contract, it might have undertaken. It must follow therefore from the arbitrator's findings that he was satisfied that these costs would not have been reduced by the personnel being employed elsewhere on other work, as might well be the case with an ordinary contractor. In my judgment the arbitrator's conclusions of fact make it clear that PLC's fixed overheads were peculiar to it and were therefore directly incurred by it in the situation in which it found itself.'

What appears to have been important in this case was that these costs were found to be 'direct' costs to the contract, which was necessary for a claim for loss and/or expense under the term of the contract.

However, perhaps the most significant difference in this case was that the contractor's work was being stopped and as a consequence its workload and the associated earnings were being reduced. This situation can be distinguished from the situation where the work period is extended. Where work is taken away from a subcontractor without significant notice, there will normally flow direct losses as well as a loss in contribution to overheads and profit which the subcontractor would have been expected to earn from the work cancelled or omitted. Such direct losses may include those arising from the redeployment of labour and plant and/or cancellation or restocking of materials. Even where costs such as the dismantling and removal from site of temporary buildings and other temporary works and plant and equipment were to be expected in any event, those costs will now be disproportional to the value of work completed and the subcontractor will have suffered a proportional loss.

Exclusion and limitation clauses

A subcontractor will often seek to reduce its risk by way of limiting or excluding its liability for non-performance and/or breach of subcontract and/or negligence. Such exclusions and/or limitations may either be general and seek to include all wrongs however caused, or be limited to certain matters. Where the subcontractor seeks to limit its liability this may be for certain causes of action, such as failure by a specific supplier or specialist sub-subcontractor or where relevant by time or money.

Generally the courts are reluctant to uphold total exclusion clauses, considering that very clear evidence of agreement between the parties is necessary that it was their intention that the contractor would have no recourse to any sanction for any breach or default by the subcontractor. This is on the basis that in a commercial contract it will be anticipated that the subcontractor will be accountable for its non-performance in damages to the contractor. On the other hand, the courts are quite prepared to accept that there will be many commercial contracts where it is reasonable for the contracting party to limit its liability on the basis that the price for accepting the additional risk would be disproportionate to the price for the work.

An exception to the above general rule may be where a subcontractor provides a service on a non-commercial basis for no tangible or real consideration, such as free advice or information such as indicative prices or delivery periods.

In a contract for the delivery of steel pipes for use in an underwater pipeline, the supplier's liability was restricted both as to the time from the delivery in which defects might become apparent and its maximum liability expressed as a percentage of the purchase price. Such an agreement was upheld by the Court of Appeal as being commercially sensible. Lord Justice May, in the case of *BHP Petroleum Ltd and Others* v. *British Steel Plc and Dalmine SpA* [2000] 2 Lloyds Rep 277, quoted from the judgment of the House of Lords, at page 288:

> 'I would in particular refer to two passages in the opinions in the House of Lords in *Ailsa Craig Fishing Co Ltd* v. *Malvern Fishing Co Ltd* [1983] 1 Lloyds Rep 183; [1983] 1 WLR 964. At p.184, col. 1, p.966F, Lord Wilberforce said:
>
> "Whether a clause limiting liability is effective or not is a question of construction of that clause in the context of the contract as a whole. If it is to exclude liability for negligence, it must be most clearly and unambiguously expressed, and in such a contract as this, must be construed *contra proferentem*. I do not think that there is any doubt so far. But I venture to add one further qualification, or at least clarification: one must not strive to create ambiguities by strained construction, as I think that the appellants have striven to do. The relevant words must be given, if possible, their natural, plain meaning. Clauses of limitation are not regarded by the courts with the same hostility as clauses of exclusion: this is because they must be related to other contractual terms, in particular to the risks to which the defending party may be exposed, the remuneration which he receives, and possibly also the opportunity of the other party to insure."
>
> 72. Then at p.186, col. 2, p.970C, Lord Fraser of Tullybelton said:
>
> "There are later authorities which lay down very strict principles to be applied when considering the effect of clauses of exclusion or of indemnity: see particularly the Privy Council case of *Canada Steamship Lines Ltd* v. *The King* [1952] 1 Lloyds Rep 1, at p.8; [1952] AC 192, 208, where Lord Morton of Henryton, delivering the advice of the Board, summarised the principles in terms which have recently been applied by this House in *Smith* v. *UMB Chrysler (Scotland) Ltd* 1978 SC (HL) 1. In my opinion these principles are not applicable in their full rigour when considering the effects of clauses merely limiting liability. Such clauses will of course be read *contra proferentem* and must be clearly expressed, but there is no reason why they should be judged by the specially exacting standards which are applied to exclusion and indemnity clauses. The reason for imposing such standards on these clauses is the inherent improbability that the other party to a contract including such a clause intended to release the proferens from a liability that would otherwise fall upon him. But there is no such high degree of improbability that he would agree to a limitation of the liability of the proferens, especially when, as explained in condition 4(i) of the present contract, the potential losses that might be caused by the negligence of the proferens or its servants are so great in proportion to the sums that can reasonably be charged for the services contracted for. It is enough in the present case that the clause must be clear and unambiguous." '

While being reluctant to uphold clauses seeking to exclude all liability for matters that flow naturally from a breach by the subcontractor under the principle known as the first limb in *Hadley* v. *Baxendale*, the courts have accepted clauses seeking to exclude liability for matters beyond those that come under this class of breach. In this respect the courts have considered the meaning of the word 'consequential'. In the Court of Appeal case of *Croudace Construction Ltd* v. *Cawoods Concrete Products Ltd* (1978) 8 BLR 20, Lord Justice Megaw said, at page 27:

> 'It is clear that the word "consequential" can be used in various senses. It may be difficult to be sure in some contexts precisely what it does mean. But I think that the meaning given

to the word in *Millar's* case [*Millar's Machinery Co Ltd* v. *David Way & Son* (1934) 40 Com Cas 24] is applicable to the present case. It is binding on us in this case. Even if strictly it were not binding, we ought to follow it. That case was decided in the year 1934. It has stood, therefore, now for more than 43 years. So far as I know it has never been adversely commented upon. It is referred to in a number of textbooks, including some of those to which we were referred by Mr Neill [Counsel for Cawoods]; it is referred to in *Halsbury's Laws* (fourth edition), volume 12, under the title of 'Damages' at paragraph 1113.

It would, of course, not be binding on us if the terms of the clause which were there being construed were materially different from the clause with which we are concerned. I cannot see any material difference, though of course there are verbal differences. It is true also that the report in Commercial Cases of the Court of Appeal decision is not set out at length or in detail and does not, I think, purport to give the precise words used, but merely a summary of what was said, by Lords Justices Greer, Maugham and Roche, who constituted the court. Lord Justice Maugham is reported as saying:

"On the question of damages, the word 'consequential' had come to mean 'not direct', and the damages recovered by the defendants on the counterclaim arose directly from the plaintiffs' breach of contract under section 51(2) of the Sale of Goods Act, 1893."

I have already quoted what Lord Justice Roche said:

"Lord Justice Roche agreed that the damages recovered by the defendants on the counterclaim were not merely 'consequential', but resulted directly and naturally from the plaintiffs' breach of contract." '

It is of course highly likely that subcontractors incorporating the word consequential intended it to include all matters flowing from a breach by them; however, as can be seen, the courts will seek to avoid such an interpretation as a way of killing an exclusion clause. Subcontractors seeking to avoid or reduce the natural consequences of their failures in breach of the subcontract are advised to seek to establish a limit to their liability at a commercially sensible level, rather than trying to avoid all liability, when they are likely to fail to achieve any relief at all.

Even where it is held that an exclusion clause is effectively incorporated into the subcontract, it will only be effective where the subcontractor is fulfilling the requirements of its subcontract. In the event that the cause of loss to the contractor is a breach by the subcontractor that goes to the root of the subcontract, then it is unlikely that the subcontractor will be able to rely on an exclusion or limitation clause to prevent the contractor from recovering its loss flowing from the subcontractor's breach. In the Court of Appeal case of *J. Spurling Ltd* v. *Bradshaw* [1956] 1 WLR 461, which was a case where a contract for the storage of barrels of orange juice had an exclusion clause, Lord Justice Denning said, at page 465:

'These exempting clauses are nowadays all held to be subject to the overriding proviso that they only avail to exempt a party when he is carrying out his contract, not when he is deviating from it or is guilty of a breach which goes to the root of it. Just as a party who is guilty of a radical breach is disentitled from insisting on the further performance by the other, so also he is disentitled from relying on an exempting clause. For instance, if a carrier by land agrees to collect goods and deliver them forthwith, and in breach of that contract he leaves them unattended for an hour instead of carrying them to their destination, with the result that they are stolen, he is disentitled from relying on the exempting clause. That was decided in 1944 by this court in *Brontex Knitting Works Ltd* v. *St John's Grange* [[1944] WN 85], expressly approving the judgment of Mr Justice Lewis, or if a bailee by mistake sells the goods or stores them in the wrong place, he is not covered by the exempting clause: see the decision of Mr Justice McNair in *Woolmer* v. *Delmer Price Ltd* [[1955] 1 QB 291].

The essence of the contract by a warehouseman is that he will store the goods in the contractual place and deliver them on demand to the bailor or his order. If he stores them in a different place, or if he consumes or destroys them instead of storing them, or if he sells them, or delivers them without excuse to somebody else, he is guilty of a breach which goes to the root of the contract and he cannot rely on the exempting clause. But if he should happen to damage them by some momentary piece of inadvertence, then he is liable to rely on the exempting clause: because negligence by itself, without more, is not a breach which goes to the root of the contract (see *Swan Hunter and Wigham Richardson Ltd* v. *France Fenwick Tyne & Wear Co Ltd, The Albion* [[1953] 1 WLR 1026]), any more than non-payment by itself is such a breach: see *Mersey Steel & Iron Co* v. *Naylor, Benzon & Co* [(1884) 9 App Cas 434]. I would not like to say, however, that negligence can never go to the root of the contract. If a warehouseman were to handle the goods so roughly as to warrant the inference that he was reckless and indifferent to their safety, he would, I think, be guilty of a breach going to the root of the contract and could not rely on the exempting clause. He cannot be allowed to escape from his obligation by saying to himself: "I am not going to trouble about these goods because I am covered by an exempting clause."'

Exclusion clauses may be sought by either party to a subcontract. For example, a contractor may expressly exclude liability for damage by others to the subcontractor's work. Where, as regularly happens with subcontracted work prior to the completion of the subcontract, the contractor allows following trades to work on parts of the subcontractor's work that has or is considered to have reached the stage where the following trade can proceed, then it is considered that the contractor puts itself in the position of a bailee and must take reasonable steps to ensure the work in its charge is not damaged. In such circumstances the onus of proof will lie with the contractor to show that it may rely on any such exclusion clause. As Lord Justice Denning said in *Spurling* v. *Bradshaw*, at page 466:

'Another thing to remember about these exempting clauses is that in the ordinary way the burden is on the bailee to bring himself within the exception. A bailor, by pleading and presenting his case properly, can always put on the bailee the burden of proof. In the case of non-delivery, for instance, all he need plead is the contract and a failure to deliver on demand. That puts on the bailee the burden of proving either loss without his fault (which, of course, would be a complete answer at common law) or, if it was due to his fault, it was a fault from which he is excused by the exempting clause.'

Loss of overheads and profit

Where the subcontractor's employment is terminated prior to its completing the works for reasons other than for its default, whether such determination is under the express provisions of the subcontract, such as by way of a termination at will clause, or by way of a repudiatory breach by the contractor, the subcontractor will normally be entitled to either loss and/or expense under the provisions of the subcontract or general damages. It is well-established law that in principle that loss and/or expense will include the loss of overhead contribution and profit that the subcontractor can demonstrate it would have made had the subcontract not been terminated. Such cases include that of *Robertson* v. *Amey-Miller*, referred to in the section 'Loss and/or expense' in Chapter 8.

Manifestly there can be no certainty as to the amount the subcontractor would in fact have paid out for its labour, materials and plant had it completed the work, and therefore no certainty as to the level of gain, if any, that would be made by way of overheads and profit. The subcontractor will therefore have to rely on indicators as to the likely profit it would have made. There are at least three possible methods for such assessment:

1. the use of the subcontractor's tender allowance;
2. the value of work completed compared with the actual costs to date, adjusted as necessary for such items as advance design work and purchases and for under-utilised preliminary costs;
3. the rate of profit and overhead recovery made across the subcontractor's business in the most recent trading period, adjusted as necessary for any exceptional items or circumstances.

In making an assessment of the subcontractor's likely loss, regard may be had to circumstances that exist at the time of the termination but not to matters that arise thereafter. Thus, where a subcontractor, post-termination of the subcontract, ceases trading, it will not be an effective defence against a liability for damages to say that the subcontractor would not have made a profit in any event, provided the cause of the subcontractor's cessation of trading was not prior to the termination of its subcontract.

In the case of *Chiemgauer Membran Und Zeltbau GmbH (formerly Koch Hightex GmbH)* v. *The New Millennium Experience Company Limited (formerly Millennium Central Limited)* [2000] Ch-1998-K-No 3692, the court had to consider not only whether the phrase 'loss and expense' could include loss of profit but also whether, since the subcontractor became insolvent after the termination but within the anticipated subcontract period, it could be said to have been able to make a profit in any event. The judge found as a fact that the event leading to the subcontractor's insolvency happened after the termination and therefore did not affect the subcontractor's right to damages for loss of profit.

The judge reviewed a number of authorities and concluded, at paragraph 72:

> '(1) As a matter of construction, Koch's entitlement to "direct loss and/or damage" under clause 32(2)(vi) of the contract allows Koch to claim the profit (if any) which it can prove it would have made from the contract, had it not been terminated.
>
> (2) In assessing Koch's entitlement under clause 32(2)(c) of the contract, the court must assume that Koch would have been able, had the contract not been terminated, to perform the contract according to its terms.'

Chapter 14
Sub-subcontracts

Introduction

Much construction work that is subcontracted by the contractor may in turn be sublet to a sub-subcontractor. The operations comprising the scope of a sub-subcontract may not be for the entire work but be limited to the supply of labour only or labour and plant. In other cases the sub-subcontractor may be responsible for the supply of basic materials only, for example a bricklaying sub-subcontractor may be responsible for the supply of sand and cement but not for the bricks or blocks.

On some projects the same sub-subcontractor may be employed by a number of subcontractors; typical examples include access equipment such as scaffolding, or mastic pointing, where the work is required to be continuous between the works of a number of different subcontractors. In management contracting projects especially, the works packages, particularly those for internal finishes, may be let on an area rather than a trade basis with the result that several trades become common to different subcontract or works packages. This is again a situation where different subcontractors will, either from choice or by reason of an express provision of the subcontract, employ a common sub-subcontractor.

In general the problems of preparing a standard form of sub-subcontract that will satisfy the range of differing arrangements required of sub-subcontractors for differing trades and/or different levels of provision, e.g. labour and materials or labour only, are greater than in the normal trade subcontract. SubSub seeks to do this by producing a very short form of sub-subcontract, which is based on and substantially the same as ShortSub. However, as stated below, it is considered to fail in its objective for two principle reasons: first, by making the basis of the sub-subcontractor's obligation to carry out the work in a fixed period after notice, and secondly, being significantly biased against the weaker party, the sub-subcontractor.

Incorporation of terms

Many sub-subcontracts are let by a reverse offer in that instead of the sub-subcontractor being sent an enquiry which it prices and returns to the subcontractor, who then accepts that offer either as originally made or as amended by negotiation, the subcontractor sends the sub-subcontractor a schedule of rates or lump sum and invites the sub-subcontractor to accept that offer. Where this happens a sub-subcontract will come into being on the sub-subcontractor accepting the subcontractor's offer. The terms of such a sub-subcontract will be those of the offer, which will rarely include the terms of a standard form such as SubSub.

Under section 7 of SubSub the sub-subcontractor is deemed to know the provisions of the subcontract in so far as they apply to the sub-subcontract works, and to this end the sub-subcontractor is to be provided with a copy of the subcontract (omitting details of the subcontractor's pricing), if requested by the sub-subcontractor.

It is suggested that this clause is even more onerous than the similar provision within SBCSub/C since the number of documents becomes greater for a diminishing amount of work. As suggested elsewhere, with regards to subcontracts the onus should be on the employing party, in this case the subcontractor, to identify any specific provisions of the subcontract that apply to the sub-subcontract works. Similarly with clauses 7.2 and 7.3, which respectively require the sub-subcontractor to carry out and complete the sub-subcontract Works so as not to result in any breach of the contract by the subcontractor and to perform the obligation and to assume any liability of the subcontractor under the subcontract to the extent that they relate to the sub-subcontract works.

The subcontractor's obligations

It is a normal presumption that the subcontractor will obtain for the sub-subcontractor all the rights due to the subcontractor under its subcontract. In particular this will include the provision of welfare facilities as required by statute. SubSub at clause 6 makes no such express provision, the subcontractor's obligation being little more than to ensure that the sub-subcontractor has access to the site and that the subcontractor does not to hinder the sub-subcontractor in the carrying out of its obligations.

Contrary to the normal arrangements at clause 5.4 of the sub-subcontractor's obligations, the sub-subcontractor is required to provide everything required except those attendances set out in the sub-subcontract to be provided free by others.

SubSub, unlike SBCSub/A at item 6 of the Sub-Contract Particulars, does not have a set list of attendances to be provided. Therefore the sub-subcontractor must rely on the incorporation of the provisions within the subcontract, details of which will rarely be provided to the sub-subcontractor and which may be insufficient for its requirements.

It is to be expected that the subcontractor will direct the sub-subcontractor with regards to the sequence and timing of the sub-subcontract work. This should involve the regular issue of directions, whether by way of formal progress meetings or casual directions between site managers. SubSub limits the subcontractor's obligations as in no way to hinder or prevent the sub-subcontractor, which is a totally negative obligation rather than a positive requirement to facilitate the regular progress of the sub-subcontractor's work. At clause 5.1 the sub-subcontractor is to carry out the sub-subcontract works with due diligence, and at clause 8.2.1 of SubSub the sub-subcontractor is to proceed regularly and diligently with the sub-subcontract works. As with the subcontract works, this obligation is likely to be dependent on the subcontract works proceeding regularly.

The sub-subcontractor's obligations

As with any construction contract a sub-subcontractor's obligations will be those set out in the documents that form the sub-subcontract, subject to any further terms that may be implied into the sub-subcontract either by statute or by law.

Such obligations will define the extent of the sub-subcontract works, the nature and quality of work to be provided and achieved and any restrictions as to the timing of those works. It will rarely be the case that a sub-subcontractor will be allocated a definite period within which to carry out its work, independently of any other on-site activities. The normal situation will be a requirement to respond to the actual progress of other work on site. SubSub, however, conforms to the arrangements in other JCT standard forms and bases the requirement for commencement and completion of the sub-subcontract works on a fixed period of time subject to notice to start. Clause 8.1, however, has the express requirement that the sub-subcontractor is to commence its work on site within ten days of the notice to commence. At clause 8.2 the sub-subcontractor has stated obligations as follows:

1. to proceed with the sub-subcontract Works regularly and diligently and reasonably in accordance with the progress of the subcontract Works and the main contract Works;
2. to achieve practical completion of the sub-subcontract Works within the period for completion.

The subcontractor has express obligations under clause 8.3 to notify the sub-subcontractor in writing when it has determined that the sub-subcontract works have reached practical completion. In addition, at clause 8.4 the subcontractor is to notify to the sub-subcontractor the dates of practical completion and the expiry of the defects liability period under the main contract. SubSub, like ShortSub, has not been amended to reflect the phraseology of SBCSub/C and does not use the phrase 'rectification period'. Finally, at clause 8.5 the subcontractor is to notify the sub-subcontractor of any defects that appear in the sub-subcontractor's work during the defects liability period and the sub-subcontractor is to make good such defects within a reasonable time of notification.

One noticeable difference between SubSub and the similar provisions of SBCSub/C is the requirement to commence work on site within ten days of the instruction to commence. This makes failure to commence by the end of the period of notice a breach of the sub-subcontractor's obligations. If, however, the sub-subcontractor should commence its work on site within the ten-day period of notice, it is submitted that the date for practical completion of the sub-subcontractor's work will be set from the end of the period of notice and not the date of its commencement on site.

The sub-subcontractor is to achieve practical completion of the sub-subcontract Works within the period for completion. Nowhere is practical completion for this purpose defined in SubSub. By reason of clause 8.3 this appears to be for the subcontractor to determine, thereby making it judge in its own cause. Unlike under SBCSub/C, there is no provision for the sub-subcontractor to record the achievement of practical completion, although any such record will be likely to be strong evidence of such completion if not challenged at the time by the subcontractor.

Clause 5.3 of SubSub has the usual requirement for the sub-subcontractor to take reasonable steps to encourage employees to be registered cardholders under the Construction Skills Certification Scheme.

Instructions and variations

While without an express provision within an agreement there is no requirement for a sub-subcontractor or any other contracting party to accept an instruction to vary the work contracted for, in practice no sub-subcontractor will refuse to vary the work required by the subcontractor. In the absence of an express provision, such an acceptance of the request to vary the work will either be regarded as a supplementary agreement to vary the original sub-subcontract, or a separate sub-subcontract if the work is additional to the original requirements, especially if it is to be carried out outside the original sub-subcontract period.

SubSub makes provision for subcontractor's instructions at section 9 and for variations at section 10. The provisions for the issuing and carrying out of the subcontractor's instructions are as follows:

1. The subcontractor may issue written instructions which the sub-subcontractor shall forthwith carry out.
2. If instructions are given orally, they shall, within two working days, be confirmed in writing by the subcontractor.
3. Except as provided in clause 10, the sub-subcontractor shall not be entitled to any additional payment in respect of instructions from the subcontractor.
4. If within five days after receipt of a written notice from the subcontractor requiring compliance with an instruction, the sub-subcontractor does not comply, then the subcontractor may employ and pay other persons to carry out the work and all additional costs incurred shall be due to the subcontractor.

The provisions for the carrying out, valuation and payment of variations is as follows:

1. The sub-subcontractor shall carry out any reasonable variation of the sub-subcontract Works that is instructed in writing by the subcontractor ('variation').
2. Variations shall be valued by the subcontractor on a fair and reasonable basis, with reference to, where available and relevant, rates and prices in the pricing documents.
3. The sub-subcontractor shall be paid any direct loss and/or expense incurred by the sub-subcontractor due to the regular progress of the sub-subcontract Works being affected by compliance with any variation, provided that the sub-subcontractor notifies the subcontractor of such as soon as is reasonably practicable. The subcontractor shall determine the fair and reasonable amount of that direct loss and/or expense.
4. The sub-subcontractor shall not make any alteration to the sub-subcontract Works, other than pursuant to sub-clause 10.1.

These provisions are brief but give rise to a number of anomalies. First, clause 9.1 allows the subcontractor to issue instructions in writing but although there is no provision for the subcontractor to issue instructions in any other way, clause 9.2 requires oral instructions to be confirmed within two working days. Should the subcontractor default in this there is no express provision for the sub-subcontractor to confirm oral instructions. However, provisions similar to those in SBCSub/C are likely to be implied into the sub-subcontract in any case.

Clause 9.3 expressly denies the sub-subcontractor the right to any additional payment in respect of instructions that are not variations. This, coupled with clause 10.3, which only allows for payment of loss and/or expense due to the regular progress of the sub-subcontract Works being affected by compliance with a variation, but not any instruction or matter, would seek to make any other instruction or an act of prevention matters for which there is no recovery under the sub-subcontract. This is a blatantly unjust clause and no sub-subcontractor should sign up to this term. Thus any claim for loss due to matters other than a variation will be a claim in damages and not loss and/or expense.

The most obvious consequence of clause 9.3 is that the subcontractor could issue instructions requiring the sub-subcontractor to change location and/or the rate of production by requiring its operatives to stop and start their work at the indiscriminate whim of the subcontractor without any recompense to the sub-subcontractor under the terms of the sub-subcontract.

Perhaps even more onerous is clause 9.4, which allows the subcontractor to employ others to carry out the sub-subcontractor's work and to charge the sub-subcontractor all additional costs merely as a result of giving a five-day notice of the subcontractor's requiring compliance with an instruction. There is no qualification that such an instruction is to be reasonable or that the five-day notice shall in the circumstances be reasonable. This clause gives rise to the possibility of severe abuse and should be suitably amended or omitted by any potential sub-subcontractor.

Payment

Sub-subcontracts, like all construction contracts, require an effective payment mechanism, the most important parts of which are to establish the dates for payment as defined in the HGCRA and a payment period.

SubSub, like most of the JCT family of contracts, fails to define the payment due date, clause 12.1 making the first payment due not later than one month after commencement and monthly thereafter. This clause will override the 45-day rule, relating to sub-subcontracts of short duration, in section 109(1) of the HGCRA, but by not defining the first payment due date will allow the sub-subcontractor to define this by the making of an application under paragraph 4(b) of the Scheme.

Clause 12.3 makes the payment period 31 days, which is ten days longer than that set in SBCSub/C at clause 4.9.3. The other clauses under this section reflect the provisions of SBCSub/C.

Clause 12.5 gives the subcontractor the statutory right to issue a notice of intent to withhold payment against the sum due to the sub-subcontractor and to withhold such sum. There is no provision for prior agreement to the sums claimed, as due to the sub-subcontractor having caused loss and/or expense due to the progress of the subcontract works being materially affected by an act or omission of the sub-subcontractor as provided for in SBCSub/C at clause 4.21.2.

Determination

As stated in the section 'Introduction' in Chapter 12, determination under English law is not the normal remedy for a breach of contract. Thus it is necessary to make express provisions within the sub-subcontract for determination under express circumstances.

SubSub provides for termination either by the subcontractor under the provisions of clause 14.1 or on the termination of the subcontract under clause 14.3. There is no provision for the sub-subcontractor to terminate the sub-subcontract for reasons such as set out in SBCSub/C clause 7.8.

Disputes

Provided that the sub-subcontract fulfils the requirements of the HGCRA as being a construction contract in writing, there will be a right either express or implied to be able to refer any dispute to adjudication. In SubSub there is an express provision for adjudication at clause 16.2, which incorporates the provisions of the Scheme and allows for the referring party to select the adjudicator-naming body from the following list:

- Royal Institute of British Architects
- The Royal Institution of Chartered Surveyors
- Construction Confederation
- National Specialist Contractors Council Limited, or
- Chartered Institute of Arbitrators.

The only limitation is that the nominating body is to be selected by the referring party from one of a list of five. There is no pre-selection or default body.

SubSub at clause 16.1 makes provision for dispute resolution by mediation should the parties agree. Since the use of this procedure requires an agreement between the parties when they are in dispute, the successful application of the clause seems extremely unlikely. No reason can be conceived why either party should seek to use this dispute resolution process in place of the relative certainty of adjudication.

Extension of time

The purpose and effect of an extension of time provision in subcontract conditions have been considered in the section 'Extensions of time' in Chapter 4, and it is regarded as being of limited effect. It is considered that such a provision in a sub-subcontract is of even less necessity or value.

SubSub has such a clause (11.1), which is written in such general and imprecise terms as to limit still further any benefit of such a clause. This clause makes provision for where the sub-subcontractor is delayed in completing the sub-subcontract works within the period for completion, by the ordering of any variation of the sub-subcontract works or for other reasons beyond the control of the sub-subcontractor. Should such occur then the sub-subcontractor shall notify the subcontractor in writing and the subcontractor shall make such extension of time (if any) as is reasonable. The use of phrases such as 'other reasons beyond the control of the Sub-subcontractor' and 'such extension of time as is reasonable' could be given extremely wide and differing interpretations.

Chapter 15
Works Contracts under Management Contracting Arrangements

Introduction

Both construction management and management contracting under forms such as the JCT management contract seek to appoint the principal contractor as a member of the employer's professional team, on a par with its other consultants, rather than as tradesmen engaged as such. In each case the Works are divided into works packages, each of which is tendered after the appointment of the construction manager or management contractor. The preparation of tender documents and selection of the works contractors will be a joint operation between the employer's team of consultants, which will include the construction manager or management contractor.

A major difference between these arrangements is that in the construction management situation the works contractor's contract is direct with the employer, while in the management contractor situation the works contractor is contracted to the management contractor.

The works contract will define the roles, obligations and duties of the works contractor, which necessarily differ between works contracts under construction management and those under management contracting arrangements.

Management contracting

The main standard form of works contract for use under the JCT Management Contract is Works Contract/2. This seeks to allocate responsibility for the management of the Works and the valuation and payment of the works contractor between the management contractor and the architect and the quantity surveyor. Specifically, the valuation of the works contractor's work is the obligation of the quantity surveyor and/or the architect and provides for the works contractor, if and when dissatisfied with a decision or certificate from the architect, to challenge such a decision by direct action against the employer by way of the name borrowing provisions of clause 4.27.

The arrangements between employer and management contractor may vary but generally the management contractor will be required to provide specific on-site services and to act in the role of principal contractor for the purposes of the CDM Regulations. Such services may be provided by way of a separate works contractor's package or as part of the management contract itself. These services will vary from contract to contract but will range from the provision of site accommodation for the subcontractor, either individually or on a shared basis, to support services such as first aid and the setting out of the Works by way of establishing on-site lines and levels related to the site grid lines and floor levels. Other facilities may include access and hoisting and/or disposal of rubbish arising from the works.

While in general this will be no different to traditional subcontracting, the management contractor's ability to negotiate with the works contractor will be limited by the provisions of the management contract. If, for example, the management contractor is to provide all access scaffold, and a reinforced concrete frame contractor proposes a method of construction which combines a working platform within the shuttering system, the necessary access arrangements will be reduced and there may be an overall saving to the total project costs, but this will only be manifest if there is an adjustment downwards to the management contractor's price for scaffolding.

Works Contract/2 defines within the conditions which of the employer's team shall action the different matters within those conditions. So far as the works contractor is concerned the management contractor is neither wholly responsible nor entirely in control. In practice this has led to difficulties, which are not satisfactorily catered for within Works Contract/2. Just as in the traditional subcontract situation, under the works contract the management contractor retains control of the site and remains responsible to the employer for integrating the work on site and coordinating the production of the various works contractors, while the architect retains control of the design process and the valuation of the works as work proceeds.

Loss and/or expense by the works contractor

Works Contract/2 provides for the works contractor to recover loss and/or expense under several clauses, clause 4.45 for reason that the possession of the works has been delayed by the employer, or for any of the matters set out in clause 4.46 and clause 4.49, which relate to acts, omissions or defaults of the management contractor. Clause 4.45 requires the architect to ascertain the amount of direct loss and/or expense that is due to the works contractor as a result of any of the matters listed in clause 4.46. In making an ascertainment the architect is to consult with the management contractor. All such action is dependent on the works contractor making a written application to the management contractor, to the effect that the works contractor is incurring or is likely to incur direct loss and/or expense. Such loss and/or expense is to be in the execution of the works contract and to be such that it would not be reimbursed by payment under any of the other provisions of the works contract and by reason of the regular progress of the works being materially affected by one or more of the listed matters. Such entitlement is subject to the following:

1. The Works contractor's application shall be made as soon as it has become, or should reasonably have become, apparent to it that the regular progress of the Works or of any part thereof has been or was likely to be affected as aforesaid, and
2. The Works contractor, in order reasonably to enable the direct loss and/or expense to be ascertained, shall submit to the management contractor such

information in support of its application including details of such loss and/or expense as the management contractor is requested by the architect or the quantity surveyor to obtain from the Works contractor, and
3. The Works contractor has complied with clause 4.48.

The following are the matters referred to in clause 4.45:

1. the management contractor, or the Works contractor through the management contractor, not having received in due time necessary instructions (including those for or in regard to the expenditure of provisional sums), drawings, details or levels from the professional team for which the management contractor, or the Works contractor through the management contractor, specifically applied in writing provided that such application was made on a date which having regard to the completion date or the period or periods for completion of the Works was neither unreasonably distant from nor unreasonably close to the date on which it was necessary for the management contractor or the Works contractor to receive the same; or
2. the opening up for inspection of any work covered up or the testing of any of the work, materials or goods in accordance with clause 3.10 of the management contract conditions and/or clause 3.3.2 (including making good in consequence of such opening up or testing), unless the inspection or test showed that the work, materials or goods were not in accordance with the management contract or the Works contract as the case may be; or
3. any discrepancy in or divergence between the documents provided by the employer as referred to in article 7 of the management contract and/or any instructions or directions issued by the management contractor (save in so far as any such instruction or direction requires a variation) and/or the Works contract; or
4. instructions issued under clause 3.5 of the management contract conditions in regard to the postponement of any work to be executed under the provisions of the management contract or the Works contract; or
5. the execution of work not forming part of the project by the employer itself or by persons employed or otherwise engaged by the employer as referred to in clauses 3.23 to 3.25 of the management contract conditions or the failure to execute such work or the supply by the employer of materials and goods which the employer has agreed to provide for the project or the failure so to supply; or
6. failure of the employer to give in due time ingress to or egress from the site of the project or any part thereof through or over any land, buildings, way or passage adjoining or connected with the site and in the possession and control of the employer, in accordance with the management contract, after receipt by the architect of such notice, if any, as the management contractor is required to give or failure of the employer to give such ingress or egress as otherwise agreed between the architect and the management contractor;
7. instructions issued under clause 3.4 (other than where bills of quantities are included in the numbered documents, an instruction for the expenditure of a provisional sum for defined work or of a provisional sum for performance specified work) and clause 3.27 of the management contract conditions; or

8. the execution of work for which an approximate quantity is included in the numbered documents which is not a reasonably accurate forecast of the quantity of work required;
9. compliance or non-compliance by the employer with clause 5.18 of the management contract; (employer's obligation – planning supervisor – principal contractor where not the management contractor);
10. suspension by the Works contractor of the performance of its obligations under this Works contract to the management contractor pursuant to clause 4.28 provided the suspension was not frivolous or vexatious.

At clause 4.49 the situation is dealt with where the regular progress of the works contractor's work is affected by matters for which the management contractor has liability, and not the employer. Again, the works contractor is to give written notice to the management contractor within a reasonable time of such loss becoming apparent. The works contractor is to recover from the management contractor the agreed amount of direct loss and/or expense thereby caused. Again, this is subject to the works contractor fulfilling specified provisions:

1. The Works contractor's application shall be made as soon as it has become, or should reasonably have become, apparent to it that the regular progress of the works or of any part thereof has been or was likely to be affected as aforesaid; and
2. The Works contractor, in order to enable the direct loss and/or expense to be ascertained, shall submit to the management contractor such information in support of its application including details of the loss and/or expense as the management contractor may reasonably require from the Works contractor.

It is to be noted that there is no express, or it is suggested, any implied term requiring that the works contractor allocate its claim or claims for loss and/or expense against clauses 4.45 or 4.49. In particular, where the works contractor identifies that it has suffered loss and/or expense directly as a consequence of items listed in clause 4.46, it will no doubt relate its loss and/or expense to such matter or event. However, in many instances where the direct cause of loss and/or expense will be a failure under clause 4.49, the management contractor may well be able to justify its failure by reference to a failure by the consultant team or other cause for which neither the management contractor nor the works contractor bears the risk.

It may be that the delays to the works contractor's work are complex, resulting from a number of matters whose effect is concurrent. It is suggested that it is not for the works contractor to demonstrate whether the dominant cause was a matter under clause 4.45 or clause 4.49. While it is always necessary for the works contractor to show due cause for the loss and/or expense it seeks, it is suggested that where it is due to a number of issues, any of which justify the works contractor's claim for loss and/or expense, then it will have satisfied the requirements necessary to recover such loss and/or expense by identifying those causes without in addition relating its claim to any express clause in the works contract.

Notwithstanding there being no requirement for the works contractor to allocate its loss and/or expense specifically to causes listed under either clause 4.45 or 4.49, it must be noted that responsibility for valuing such loss rests with different persons for the different clauses. In the case of loss and/or expense under clause 4.45 the amount is to be ascertained by the architect, or if so instructed the quantity surveyor, in consultation with the management contractor, while for causes arising under clause 4.49 the loss and/or expense is a matter for agreement between the management contractor and the works contractor.

In practice, the works contractor is likely to be pushed from pillar to post because there is no single point of responsibility for the valuation of the works contractor's loss. As indicated elsewhere in this section, this will be particularly likely where the true cause of the loss is not a direct cause such as a delay by another works contractor, which itself has a claim for delay for reasons not at its risk and which may also involve claims under both of the clauses relating to loss and/or expense.

A further complication arises from the separation of the loss and/or expense provisions into two separate groups: while matters to be valued under clause 4.45 are to be included within interim certificates to be issued by the architect under the management contract, and to which the payment mechanism at clause 4.26 applies, there is no express payment mechanism for loss and/or expense arising out of an agreement under clause 4.49.

Loss and/or expense by the management contractor

Clause 2.11 of Works Contract/2 requires that the management contractor is to notify the works contractor of its failure to complete the works within the period or periods for completion. This is subject to the management contractor having given a decision on all outstanding notices by the works contractor under clause 2.2 and giving its notice to the works contractor in writing within a reasonable time of the expiry of that period or those periods.

Once the works contractor has been given a notice under clause 2.11, it is by reason of clause 2.12, subject to the requirements of clause 4.26.2, to pay or allow the management contractor for any direct loss and/or expense arising. Such loss shall include, but shall not be limited to, any liquidated damages which the management contractor is obliged to pay or allow to the employer by reason of the works contractor's failure.

In this respect it is submitted that a works contractor under Works Contract/2 is in no different a position from a similar trade engaged under a normal domestic subcontract. Any direct loss and/or expense to the management contractor may flow from a range of matters other than the completion by the date set in the works contract. In practice it is failure to allow following trades to commence to programme or at the stated period after the commencement of the works contractor's work, and/or for the subsequent trades to progress as planned, that will cause loss to the management contractor.

While clause 2.12 makes reference to clause 4.26.2, the giving of a notice of intention to withhold or deduct from the sums due to the works contractor, no specific reference is made to clause 4.50 which is the usual requirement for notification that loss and/or expense is being or is likely to be incurred and for agreement of the amount of that loss and/or expense before such sums may be deducted as a debt.

There is no provision for the employer to recover loss and/or expense directly from the works contractor. It appears inconsistent that the works contractor should be provided with the right to name borrow but no similar right is granted to the employer. Further, it is conceivable that the employer may suffer a loss by way of a works contractor's default for which the management contractor has no liability to the employer. Such loss may arise out of difficulties in design development or any matters which the management contractor has specifically excluded from its liability under its contract with the employer.

Extension of time

There is a frequently held misconception that delays to a works contractor's work will have a direct and similar delaying effect on the progress of the works as a whole. The effect of delay to any programme activity will only delay the work of others if there is another activity directly dependent on the delayed activity and they are on a programme chain that is critical at the time to the completion of the works. Not only must there be a critical activity dependent on the delayed activity but that further activity must itself not be in delay and must be in a position to follow on when the first activity reaches the necessary level of progress for that to happen.

As with any domestic subcontractor, delays to the works contractor may result from matters for which the design team and/or employer are liable or matters for which the management contractor is at risk. Not only will the management contractor normally have a duty to the employer to complete the works as a whole by a defined date, but the management contractor may well have sold its services under this contract procedure on the basis that its early involvement with the consultant team will reduce the risk of project over-run through the management contractor's planning and management or control of the design and procurement activities in conjunction with the consultant team.

It follows that in assessing the effects of any delaying event there needs to be consideration as to the effect on the works contractor's ability to complete within the period set for the works contract and separately on the effect on the completion of the project as a whole.

Works Contract/2, at clause 2.2, requires the works contractor in the usual manner to notify the management contractor whenever it becomes apparent that the commencement, progress or completion of the works is likely to be delayed, and to identify both the particulars of the delay and the works contractor's estimate of the likely delay. Such notice is to be in writing and to give the material circumstances including, in so far as the works contractor is able, the cause or causes of delay, and to identify in such notice any matter which in the works contractor's opinion comes within clause 2.3.1.1.

In addition, for each and every matter the works contractor shall, if practicable at that time or otherwise in writing as soon as possible after such notice, give particulars of the expected effects. In addition, the works contractor is to estimate the extent, if any, of the expected delay in the completion of the Works or any part thereof beyond the expiry of the period or periods stated in Works Contract/1, section 2, item 1, or any revision of such period or periods which results, whether or not concurrently with delay resulting from any other matter which comes within clause 2.3.1.1.

The works contractor is also to give such further written notices to the management contractor as may be reasonably necessary or as the management contractor may reasonably require for keeping up-to-date the particulars and estimate referred to in clause 2.2.2.1 and 2.2.2.2, including any material change in such particulars or estimate.

On receipt of a notice under clause 2.2, the management contractor is required under the provisions of clause 2.3 to consider whether the work of the works contractor is likely to be delayed and whether any of such delay is down to a relevant event or an act of omission or default of the management contractor. Having made his assessment the management contractor is to notify the architect and then give an extension of time to the works contractor. The anomaly is that subsequent to the management contractor's making its assessment and giving an extension of time, the architect may express dissent from the management contractor's assessment and extension of time award, in which case the management contractor must notify the works contractor. On the other hand, if the management contractor does not consider an extension of time should be given, it again is to notify the works contractor, and again after having notified the architect.

The works contract is silent on the effect of a dissent by the architect. However, the likely effect is for the management contractor to make his assessment and to notify the architect of his assessment and to delay the giving of an extension of time to the works contractor until the management contractor receives a comment from the architect on the assessment of the intended extension of time award. Clause 2.4 of the works contract provides a timetable for the management contractor to give his decision to the works contractor. This gives a maximum period of 12 weeks but in any event not later than the existing completion date for the works contractor's work.

The management contract at clause 2.14 requires the architect to express any dissent 'before the Management Contractor is required to notify the Works Contractor of his decision in accordance with the provisions of clauses 2.3 and 2.4 of the Works Contract Conditions'. Clause 2.14 of the management contract does not define the effect of a dissent by the architect on the management contractor's assessment. It is strange that the employer, having opted for the management contracting arrangements so as to have the benefit of the management contractor's management skills and in particular those of planning and control of the works, provides for the involvement of the architect in this decision at all.

It is likely that, by following the procedures set out in clause 2.4 of the works contract and 2.14 of the management contract, the redefining of the works contract period is likely to be protracted, which will be contrary to the needs of the management contractor, who is required to coordinate the work of all the works contractors, issuing instructions as necessary to reprogramme the works. In this regard the right of the management contractor not to give a decision on the award of an extension of time for up to 12 weeks from the date of notification is both unreasonable and ineffective for the purposes of managing the Works. The effect of clause 2.14 of the management contract could be to render a prompt assessment by the management contractor to no avail because of a tardy assessment by the architect.

Dispute resolution

One consequence of the employer's team having two leaders is that while changes to both the work and the manner and/or timing of such work may lead to the works contractor being entitled to have either or both its contract sum and/or period for the work adjusted, such rights may require the decision of either the architect and/or the management contractor. At first glance the works contract at section 9 provides for the settlement of disputes by one of section 9A adjudication, 9B arbitration or 9C legal proceedings. Section 9A.1 gives either party the right to refer any dispute to adjudication, which complies with the provisions of the HGCRA.

There is no requirement for disputes concerning decisions of the architect to be treated in a different manner to disputes resulting from decisions by the management contractor. While the management contractor will be in a position to defend its actions, it will almost certainly need to involve the architect and where necessary other members of the professional team in other disputes.

To a limited extent this problem is reduced by the provision for name borrowing at clause 4.27. This clause is restricted to the situation where the works contractor feels aggrieved in regard to any amount certified by the architect under clause 4.2 of the management contract conditions and included in a direction in respect of the works as referred to in clause 8.3.2 of the management contract conditions, or by his failure so to certify or direct. Under these circumstances, subject to clause 1.11, the management contractor is to allow the works contractor to use the management contractor's name and if necessary to join with the works contractor in the relevant procedures applicable under the management contract to the resolution of disputes or differences at the instigation of the works contractor in respect of the said matters complained of by the works contractor.

In practice, however, this clause is not entirely satisfactory. First, it only relates to 'any amount certified by the Architect' and does not refer generally to decisions of the architect and consequently does not specifically include decisions as to the quality of the work, architect's instructions or other matters. Secondly, and perhaps most importantly, this clause requires that the management contractor 'shall allow the works contractor to use the management contractor's name'.

In considering the provisions of clause 4.27 it must be borne in mind that the management contractor's obligations under clause 4.17 are to ensure that the employer operates the valuation procedures and that the architect directs the management contractor as to the amount in respect of the works contractor which is included in the amount stated as due in interim certificates issued under clauses 4.2.2, 4.2.3 and 4.2.4 of the management contract conditions. Under clause 4.26 the management contractor is to notify the works contractor not later than five days after the issue of the interim certificate, of the sum due, and to make payment at the end of the payment period, which is stated to be 17 days from the date of issue of each interim certificate.

It follows that the management contractor's obligation to the works contractor is to pay the sum certified by the architect at the end of the payment period. Only in the event of a failure by the management contractor to make payment would the works contractor have a cause of action against the management contractor. It further follows that without the express right of action against the employer, the works contractor would be without any rights should it consider the architect's valuation of its work wrong.

Clause 4.27 makes reference to clause 1.11, which provides for the management contractor to obtain for the works contractor the rights and benefits under the main contract so far as are applicable to the works contract. Any such action taken by the management contractor is to be at the works contractor's cost and the management contractor may require both an indemnity and security for costs.

It may be that there are provisions other than dispute resolution within the management contract which the works contractor might request the management contractor to obtain for it, for which the provisions of clause 1.11 are appropriate, but it is suggested name borrowing under clause 4.27 is not one.

Two issues frequently arise when applying clause 4.27. First, it is frequently suggested that the works contractor must first obtain permission from the management contractor before it may use these procedures. However, the wording of clause 4.27 is very clear: 'the Management Contractor *shall* allow the Works Contractor to use the Management Contractor's name'. While it may be considered correct to inform the management contractor in advance of the intention to us its name, there are no grounds for the management contractor to prevent or deny its use.

Secondly, it is frequently suggested that prior to taking action under clause 4.27 by reason of the reference to clause 1.11, the works contractor must provide to the management contractor both 'such indemnity and security as the Management Contractor may reasonably require'. It is submitted that under clause 1.11 such indemnity and security is only to be provided when the management contractor has been requested to take action to obtain the rights and benefits referred to. The normal course of action will be for the works contractor to commence its action for a revised valuation of its work without any need or involvement of the management contractor, who will therefore have no rights under clause 1.11.

The problems set out above do not arise in the construction management situation since the works contractor's contract, being directly with the employer, automatically contains the right to act against the employer. In certain circumstances the employer as a consequence seeks to recover from the construction manager sums for which the employer may have become liable to the works contractor, considered to be the result of a failure by the construction manager.

Name borrowing

As mentioned above, name borrowing is a form of action which gives the works contractor the ability to act directly against the employer, subject to the conditions of the works contract. This right is of particular importance in disputes where the management contractor, under the works contract, is only required to execute the decisions of a party to the management contract. There will in such circumstances be no dispute under the works contract and therefore no possibility of the works contractor having its grievance rectified under the works contract.

The provisions for name borrowing under both clauses 4.27 and 1.11 of Works Contract/2 were considered in detail by Judge Humphrey Lloyd QC in the case of *Belgravia Property Company Limited* v. *(1) S & R (London) Limited (2) Taylor Woodrow Management Limited* [2001] BLR 424. While this case considered these clauses in relation to arbitration, it is considered the same requirements would apply where the dispute is referred to adjudication, although the simpler procedures in adjudication make some of the judge's observations highly improbable in adjudication.

The judge was required to consider and reply to five questions as follows:

1. whether the first respondent is entitled to bring these arbitration proceedings against the applicant, there being no arbitration agreement in existence between them;
2. whether, if the first respondent is entitled to bring these or any arbitration proceedings against the applicant, it is obliged (under the terms of the Works contract made between it and the second respondent) to do so in the name of the second respondent;
3. whether, if the first respondent is obliged to bring these or any other arbitration proceedings in the name of the second respondent, the second respondent must consent to the use of its name by the first respondent;
4. whether, if it is a necessary pre-condition to the use of the second respondent's name by the first respondent that the second respondent shall have consented to the same, the second respondent is entitled to impose any condition upon the first respondent in respect of the grant of any such consent;
5. in particular, whether the second respondent is entitled, under the terms of the Works contract, to require that the first respondent should indemnify the second respondent and/or provide security in respect of any liability which the second respondent may incur as a result of the first respondent being permitted to use its name.

In this case, the works contractor, S & R, had commenced an arbitration against the employer, Belgravia, in its own name but allegedly under clause 4.27 of the Works contract. The management contractor, Taylor Woodrow Management Limited, had sought a stay to the arbitration pending the receipt by it of an indemnity from S & R under clause 1.11. The judge considered the cases referred to him, all of which referred to nominated subcontracts under earlier forms of both main contract and nominated subcontract. The judge commenced his judgment with a warning against taking too much cognisance of earlier cases dealing with nominated subcontracts. He said:

> '17. In order to understand the reference to clause 1.11 in clause 4.27 it is necessary to look at the contractual arrangements for payment. In doing so it must be assumed, in the case of the 1987 Works Contract, that the selection of the scheme to be found in clause 4.27 took into account not just the cases on the old "green" form of nominated subcontract which was intended for use with the 1963 editions of the JCT forms but also the changes made to the JCT forms in 1980 and thereafter.'

In his judgment Judge Humphrey Lloyd QC spent much time debating the difference in effect of clauses 1.11 and 4.27. Since under clause 1.11 the management contractor is to obtain for the works contractor 'any rights or benefits of the provisions of the Management Contract' these must surely include the right to arbitrate or adjudicate 'any amount certified by the Architect', which is the specific provision of clause 4.27. It is to be noted that clause 4.27 also makes reference to clause 8.3.2, which is the provision for nominated subcontractors to the works contractor. Clause 4.27 therefore seeks to facilitate the possibility of the nominated subcontractor being able to borrow the management contractor's name.

It follows therefore that any other remedy that the works contractor may seek against decisions of the architect or actions of the employer must be sought under clause 1.11 by requiring the management contractor to take action. How such action is in practice managed will be for agreement at the time between the management and works contractors. Judge Lloyd said, at paragraph 21:

> 'The wide terms of clause 1.11 would however permit it to be used to enable a Works contractor to use the management contractor's right to commence an arbitration against the employer, e.g. about the amounts to be included in certificates or to make a claim for damages for breach of the management contract where the sole damages were suffered by the Works contractor.'

The judge went on to say:

> '22. In contrast clause 4.27 is specific and limited. It is concerned only with complaints about what ought to have been included in certificates issued under the management contract (and thus also about failure to issue a certificate or to give direction). The arbitration or litigation contemplated must be against the employer since the commercial purpose is to obtain a revision of the amount certified and its replacement with a larger amount or, possibly, the same or similar amount arrived at in a different way (where, for example, the dispute is about the rate applicable). The clause refers to "proceedings . . . at the instigation of . . .". (The forms of subcontract the subject of earlier cases said "proceedings . . . by [the nominated subcontractor]".) Since the amount in dispute is that due under the management contract and since the parties to that contract and its arbitration agreement are the employer and the management contractor, the only person who could give notice of arbitration to the employer for the purposes of article 8 and clause 9.1 is the management contractor and the only person who had a cause of action under that contract in respect of the amounts claimed that might be justiciable in litigation would be the management contractor.'

A further significant difference between clauses 1.11 and 4.27 relates to which party instigates the action. Under clause 1.11 the management contractor is to take action – "the Management Contractor will" – at the request of the works contractor. However, under clause 4.27 the management contractor is to allow – "the Management Contractor shall allow" – the works contractor to use the management contractor's name in the relevant procedures under the management contract.

Judge Lloyd considered the effect of the provisions of clause 4.27:

> '23. Thus clause 4.27 authorises the Works contractor "to use the Management Contractor's name" for otherwise the Works contractor would have to make a request to the management contractor under clause 1.11 to give notice of arbitration and thereafter to conduct an arbitration to obtain from the employer whatever the Works contractor desired. The mechanism is commonplace. . . . The permission given by clause 4.27 is apparently unqualified but, if it were considered by itself, it must be subject to some obvious and commonsense implied conditions. To that extent Mr Sears [Counsel for Belgravia] is right in his submission that "allow" signifies something more than a synonym for "permit" or "authorise" (although not otherwise). For example, the Works contractor must inform the management contractor of its intention to start proceedings in the name of the management contractor and the management contractor is obliged to inform the Works contractor of any notices or other correspondence that it receives. . . . Similarly the Works contractor will have to ensure that the management is promptly informed of everything that is done or received in the name of the management contractor. The management contractor may therefore have to devote significant time and expense to monitoring the

> conduct of the arbitration in case it affects its interests. Although the management contractor has no right to withdraw the permission and authority granted by clause 4.27 the management contractor, in my judgment, has the right to do so if the arbitration or its conduct were, for example, to reflect badly on its reputation (in the eyes of the employer since an arbitration is a private matter). The Works contractor's right to use the name is a device to enable it to make its claim against the employer and it is not a right to use or abuse the name in any other way.'

What is clear from the words of paragraph 23 of this judgment is that while under clause 4.27 the works contractor has a contractual right to seek a revision of the architect's certificate, it must at the same time keep the management contractor informed of what it is doing.

The comments about the reputation of the management contractor in the eyes of the employer are unlikely to be an issue where the dispute relates only to the valuation of the works as completed, whether just the contracted work or the valuation of variations as well. However, such issues may arise where the works contractor is seeking loss and/or expense or compensation for late payment and/or to recover discounts. However, Judge Lloyd made some further comments, at paragraph 24, including:

> '. . . The employer is entitled to defend the claim on any grounds available to it (e.g. that the work is now defective as a result of some neglect by the management contractor) or to bring a counterclaim to reduce the amount that might be payable to the management contractor on grounds which do not concern the Works contractor so the management contractor has to be aware of them. Clause 4.27 does not prevent the employer counterclaiming. So far as it is involved in an arbitration under clause 9 of the management contract and has the right to raise any defence or counterclaim.'

Later the judge said, in the same paragraph:

> 'Furthermore clause 4.27 confers a benefit on the parties to the Works contract. It enables the management contractor to avoid two arbitrations.'

Since clause 4.27, unlike 1.11, relates only to matters where the management contractor merely passes on the decision of the architect, it must be questioned whether there could ever be two arbitrations since, provided the management contractor has notified the works contractor of the amount certified by the architect, there is no breach by the management contractor and no cause of action against it.

The situation may be very different for disputes where clause 1.11 is invoked by the works contractor; then, should the management contractor fail to obtain for the works contractor the right of action against the employer, the works contractor may bring an action against the management contractor. It will then be for the management contractor either to defend the substantial issues in dispute or defend its action in failing to obtain for the works contractor the right of action against the employer.

Judge Humphrey Lloyd considered the problem of enforcing an award under a clause 4.27 action:

> '25. There may also be circumstances, especially where the proceedings continue at or after the end of the process of certification, when the employer if held liable will wish to know

to whom any amount awarded should be paid. Clause 4.27 apparently only authorises the Works contractor to instigate proceedings. It would be an empty right, if limited to giving notice of arbitration, so the authority to instigate proceedings must presumably extend to pursuing those proceedings. The clause does not however authorise the Works contractor to enforce the award since the award has to be in the form of a new or revised certificate the effect of which will then be dealt with in accordance with the terms of the two contracts . . .'

The situation may be simpler after the HGCRA since the management contractor will not have a right not to pay on the grounds that it has not itself been paid. It is considered that an adjudicator's decision or arbitrator's award will replace the architect's certificate and that a new certificate from the architect itself is not actually necessary before the works contractor has a right to payment from the management contractor.

In arbitration, and less frequently in adjudication proceedings, the parties' costs will be awarded against the losing party together with the arbitrator's or adjudicator's fees. To prevent the management contractor suffering any loss should the works contractor be unsuccessful and default in payment, there is provision within clause 1.11 for the management contractor to obtain from the works contractor an indemnity for such loss. Since the action is in the name of the management contractor it will in the first instance be liable for such costs and fees. This point was made by Judge Humphrey Lloyd QC, in paragraph 30:

'. . . The "action taken by the Management Contractor" as described in clause 1.11 will, for the purpose of clause 4.27, be that taken by the Works contractor in so far as it does or may render the management contractor liable to others (primarily the employer and the arbitrator). The most obvious example would be where the Works contractor was unable to pay the management contractor the costs of the arbitration were it to be unsuccessful and an award of costs was made in favour of the employer against the management contractor.'

However, it is considered that in an adjudication, where the parties bear their own costs and the liability for the adjudicator's fees will be between the employer and the works contractor, since the contract with the adjudicator will not involve the management contractor, but the parties in dispute, the situation referred to by the judge does not arise.

Judge Humphrey Lloyd QC then considered a number of practical difficulties with the process of arbitration, and by implication adjudication, which may require the management contractor to join with the works contractor. While these comments were made with reference to the provisions of clause 4.27, it is considered they will be more likely under other heads of dispute and consequently more relevant to matters under clause 1.11.

Judge Humphrey Lloyd QC said:

'26. On the other hand the arbitrator may at any stage of the proceedings order or award costs to the employer. An order might be made for discovery by the management contractor. The employer has no right to recover costs or to obtain sight or knowledge of the management contractor's documents from the Works contractor. So as to be sure that the management contractor is a real and not a nominal "off stage" party for these and other purposes, it may be necessary, e.g. by direction of the arbitrator to meet the employer's request, for the management contractor to be required to join with the Works contractor (for the same reasons that an assignor may have to be joined in litigation). The employer will thus be certain that its rights and liabilities to the management contractor in respect of the subject matter of the claim by the Works contractor have been settled conclusively and that they are not liable to be set aside on a technicality, such as an assertion that the management contractor was not actually aware of what was being done in its name. The reference to joining the management contractor is also apt so as to enable the Works contractor to comply with obligations as to discovery etc.'

At paragraph 27 he emphasises that the action is between the employer and the management contractor:

'. . . There is therefore no possibility of reading clause 4.27 as entitling the arbitrator to make orders against the Works contractor directly: they must be made against the management contractor. The Works contractor will then be obliged to secure compliance with them if it is to avoid sanctions against the management contractor which might prevent or prejudice the further pursuit of its claims. In the same way the employer cannot be ordered to do anything more than it would be obliged to do had the arbitration been an ordinary one under clause 9 of the management contract, e.g. disclosing documents to a third party such as the Works contractor. (The Works contractor and the employer may of course arrive at a special agreement whereby the Works contractor obtains access to and use of the documents in return for the usual undertaking to preserve confidentially.) There are therefore considerable practical limitations in an arbitration under clause 4.27.'

Judge Humphrey Lloyd QC reiterated, in paragraph 29, the point that clause 1.11 is primarily for disputes which cannot be actioned under clause 4.27:

'. . . even if one were to ignore section 9, such interpretation does not make sense. Clause 4.27 only permits certain types of claims to be made. If the subcontractor has other claims, and provided that the management contractor has no adverse interest and is otherwise agreeable, they may be pursued by arbitration via clause 1.11 (as was done in the proceedings that led to the *Birse* case [*Co-operative Wholesale Society Ltd* v. *Birse Construction Ltd* (1996) 46 Con LR 110]). The employer's defence to claims made via clause 4.27 may cause the subcontractor to reconsider their basis and to resort to clause 1.11. The two routes cannot therefore be exclusive, although in practice it is highly unlikely that an arbitration commenced via clause 1.11 would be succeeded by one pursued via clause 4.27, where the subject matter was essentially the same.'

An important practical issue in name borrowing is that the works contractor must seek the appointment of an adjudicator or arbitrator from the nominating body applicable to the management contract. These details should be set out in Works Contract/1, section 1. It is to be noted that unless the form is amended by the parties, the appointment of an adjudicator will be the same as for the management contract, while the appointment of an arbitrator has differing default provisions. No reason for this arrangement can be conceived. However, Works Contract/1 section 2, at items 4 and 5, has provision for the tendering contractor to make its own proposals as to the nominating body.

In summary it appears that the works contractor may commence an action under clause 4.27 for those restricted issues, subject to notifying the management contractor and complying with any requirements for an indemnity that the management contractor may be entitled to and require. Any other action will require the management contractor to obtain for the works contractor the benefit of the provisions of the management contract. In either case the tribunal must have regard to the fact that the action is between the employer and the management contractor and that the works contractor is not in fact a party to the action despite having commenced it and running the reference.

Adjudication/arbitration under the name borrowing provisions

The name borrowing provisions in Works Contract/2 under clause 4.27 are, by way of the reference to clause 1.11, that the management contractor is entitled to 'such indemnity and security as the management contractor may reasonably require'. What is reasonable for the management contractor to require must relate to the risk, if any, it is put to, not so much by reason of the works contractor borrowing its name as to its being required by the works contractor to take action. It is suggested that in practice, and especially in a reference to adjudication when using the management contractor's name, no such action will be required.

First, and this is frequently overlooked when a management contractor seeks either or both an indemnity or security from the works contractor, the right to such indemnity and security is not available to the management contractor on the works contractor using the management contractor's name, but on the management contractor being required as a result of a request from the works contractor to take action itself. It therefore must be the case that until the works contractor requests the management contractor to take action, there will be no right for the management contractor to have either an indemnity or security.

In adjudication under the JCT management form of contract each party will be liable for its own costs and therefore the management contractor will not be at risk of having to meet the employer's costs in the event of default or insolvency by the works contractor, as would be the case in an action in either litigation or arbitration.

The contract with the adjudicator or arbitrator will not be under the name borrowing arrangements. It will be the works contractor who has made the application and who will be liable to the adjudicator or arbitrator for their fees, which may, in the event of the works contractor being successful, be payable by the employer. The management contractor does not become a party to the contract with the adjudicator or arbitrator.

Rights of third parties

It may be considered strange that the JCT forms of contract generally have taken advantage of the provisions of the RTPA to deny third parties the rights that the RTPA provides. Why the JCT has decided on this course of action is not known. By amendment 2 of the Works Contract/2 and the new clause 1.25, the rights of third parties provided by the RTPA have been removed.

Considering the much closer link between employer and works contractor that is intended under this form of contract, in comparison with the more usual domestic subcontract arrangements, this dogmatic approach to removing the benefits provided under the statute is unexplainable. Indeed, considering the right of the works contractor to use the name of the management contractor against the employer, it might be anticipated that a similar provision might be given for the employer to use the name of the management contractor against the works contractor.

In addition, clause 1.9 eliminates any liability on the works contractor in respect of any default on the part of any third party within the exhaustive list specified and excludes the creation of any privity of contract between those parties. This clause will include any liability for delay to the works contractor's work due to any such act or omission. The elimination of the creation of any privity of contract between the works contractor(s) and the employer again prevents any action by the employer against the works contractor.

Chapter 16
The Legal Approach

Introduction

Many, if not most, disputes in construction contracts, and especially subcontracts, are settled by negotiation and not by legal entitlement through a formal dispute resolution process. The subcontractor, as the claimant, decides its bottom line, which will normally start with its known costs and add its anticipated margin and then a bit for negotiation. The contractor as the paying party, on the other hand, will start with the amount it anticipates recovering from the employer, with an allowance for its overheads and profit. Provided agreement can be reached to both parties' reasonable satisfaction, it will not matter what are the rights or wrongs of each party's arguments or indeed their strict contractual or legal entitlement. Different principles will apply if and when the dispute goes to a legal tribunal such as adjudication, arbitration or to law.

It will no longer be relevant to declare that a certain sum is the minimum that can be accepted or to assert that the work has cost such an amount to achieve. It will be necessary to demonstrate a right to the sum claimed, either as an entitlement under the contract or as a right at law. A subcontractor can look to its entitlement first from the words of the subcontract it entered into with the contractor, and where this is unclear or insufficient can seek an entitlement under the law. In extreme cases, the subcontractor may be able to receive a judgment in what is called restitution, when the court will make an award on the principle of what it considers is right for the subcontractor to be paid.

To a very large extent, the subcontracting parties can make any arrangements they wish, and conversely only those things that they have agreed between themselves will be allowable and enforceable under the subcontract. For example, only by agreement can a subcontract be varied; there is not an automatic right for an employer or contractor to issue variations and even when there is such a right, it will be limited to those matters allowed for under the subcontract. Similarly, how such changes are to be paid for will be decided by the terms of the subcontract. Thus whether the subcontractor is to be paid on the basis of bill rates or material and labour costs or by reference to commercial rates or indeed any other basis, will be decided by the terms of the subcontract. Only where the subcontract is deficient or ambiguous will the law intervene to decide the necessary terms of the subcontract.

In an ideal situation, the terms and conditions of a subcontract will be fully agreed before work starts, but it is frequently the case that full agreement has not been reached prior to commencement. When this happens either the parties agree all matters at some time subsequently, when the agreement will probably apply retrospectively, or it will be left for a tribunal to decide the nature and terms of the subcontract. In the case of subcontracts to which the HGCRA applies, terms for payment and dispute resolution, etc. will, where necessary, be implied into the subcontract in the form set out in the Scheme. It is possible for parties, having entered into a subcontract in one form, subsequently by agreement change those arrangements. Thus if the subcontract makes no provision for change, it is possible for changes to be made, carried out and paid for by agreement. Similarly, if the subcontract has no provision for early use or occupation of part of the building before completion, such use or occupation may be granted by agreement. If such action were taken, without agreement or without agreeing the method and level of compensation to be paid to the subcontractor, then the law would determine such compensation. One method of doing this is to consider what is the normal trade custom and practice within the industry under such circumstances.

The way the law decides matters where there is no contractual or statutory provision is by considering previously decided and reported cases. The speeches of judges over the years have been recorded. Many cases are published in the law reports, such as the Building Law Reports; other cases are 'unreported' and have transcripts available and are also cited in argument. In each case, the judge has set out the facts he has found and then stated the law he has applied to reach his judgment. Lawyers will argue each new case by reference to the reported cases, claiming either that the new situation is the same or similar to that referred to and therefore the same conclusion must be drawn, or conversely that since the circumstances are different an opposite decision must be made.

Construction of the subcontract

One consequence of this analysis of cases is to consider very carefully the language used. Often, in English, different words have similar meanings and will be used as alternatives, to give interest to the prose. For the lawyer they may be similar but they will each be different, so where in a subcontract different words are used in different places, the lawyer will consider that a different meaning was intended. The English legal system has a hierarchy of courts: County Court, High Court, Court of Appeal and the House of Lords. A decision in a higher court may overturn or vary a decision of a lower court; thus a decision quoted from the House of Lords will have greater authority and weight in argument than those of the Court of Appeal or High Court.

Those terms that are to be found in writing in the words of the subcontract are called 'express terms', while those added by the law are called 'implied terms'. The courts are reluctant to interfere in the free bargaining of contracting parties and have established two tests for the implication of a term:

1. the business efficacy test;
2. the officious bystander test.

The Court of Appeal case known as *The Moorcock* (1889) 14 PD 64, CA, concerned a steamship of that name, which was being unloaded at a jetty on the river Thames. As the tide went out the ship came to rest on the riverbed, where she settled; she was straddling a piece of hard ground and was damaged as a consequence. In deciding the case, Lord Justice Bowen said:

'The question which arises here is whether, when a contract is made to let the use of this jetty to a ship which can only use it, as is known by both parties, by taking the ground, there is any implied warranty on the part of the wharfingers (owners of the jetty), and if so, what is the extent of the warranty.

An implied warranty, or, as it is called a covenant in law, as distinguished from an express contract or express warranty, really is in every instance founded on the presumed intention of the parties, and upon reason. It is the implication which the law draws from what must obviously have been the intention of the parties, an implication which the law draws with the object of giving efficacy to the transaction and preventing such a failure of consideration as cannot have been within the contemplation of either of the parties; and I believe if one were to take all the instances, which are many, of implied warranties and covenants in law, it will be seen that in all of these causes the law is raising an implication from the presumed intention of the parties with the object of giving to the transaction such efficacy as both parties must have intended it should have. In business transactions, what the law desires to effect by the implication is to give such *business efficacy* to the transaction as must have been intended by both parties; not to impose on one side all perils of the transaction, or to emancipate one side from all the burdens, but to make each party promise in law as much, at all events, as it must have been in the contemplation of both parties that he should be responsible for.'

The judgment in *The Moorcock* was quoted as the main authority for implied terms, until the judgment of Lord Justice MacKinnon, in 1939, in *Shirlaw* v. *Southern Foundries (1926) Ltd* [1939] 2 KB 206, when he said, at page 227:

'I recognise that the right or duty of a court to find the existence of an implied term or implied terms in a written contract is a matter to be exercised with care; and a court is too often invited to do so upon vague and uncertain grounds. Too often also such an invitation is backed by the citation of a sentence or two from the judgment of Lord Justice Bowen in *The Moorcock*. They are sentences from an extempore judgment as sound and sensible as all utterances of that great judge; but I fancy that he would have been rather surprised if he could have foreseen that these general remarks of his would come to be a favourite citation of supposed principle of law, and I even think that he might sympathise with the occasional impatience of his successors when *The Moorcock* is so often flushed for them in that guise.

For my part, I think that there is a test that may be at least as useful as such generalities. If I may quote from an essay which I wrote some years ago, I then said; "Prima facie that which in any contract is left to be implied and need not be expressed is something so obvious that it goes without saying; so that, if, while the parties were making their bargain, an officious bystander were to suggest some express provision for it in their agreement, they would testily suppress him with a common 'Oh, of course!'"

At least it is true, I think, that if a term were never implied by a judge unless it could pass that test, he could not be held to be wrong.'

The law does not concern itself with what the parties intended or what they believed they had agreed, but only what they actually subcontracted to do. In construction subcontracts there is frequently a lengthy period between an initial enquiry and the finalisation of an agreement. During that intervening period there may be extensive discussions as to the nature of the final agreement. If these discussions do not get recorded or the record does not get incorporated into the subcontract, as a named document, those agreements will, if subsequently disputed, rarely be of any effect.

A subcontract will frequently be contained in a number of documents; these may include the enquiry from the contractor, the subcontractor's tender, subsequent correspondence, the form of subcontract and the contractor's acceptance letter or articles of agreement. Such an extensive number of documents, some of which will be in a standard form and general applications while others will refer specifically to the particular project, main contract and/or the scope of the specific subcontract, will almost certainly contain conflicting provisions. It is therefore necessary to establish the order of precedence of the documents in the event of such conflict. SBCSub/C does this, at clause 1.3, and requires the subcontract to be read as a whole and sets out the following provisions:

1. If there is any inconsistency between the Sub-Contract Agreement and these conditions, the Sub-Contract Agreement shall prevail.
2. If there is any inconsistency between the Sub-Contract Documents (other than the numbered documents) and the numbered documents (excluding the schedule of modifications (if any)), those Sub-Contract Documents shall prevail.
3. If there is any inconsistency between the subcontract documents and the main contract conditions, the Sub-Contract Documents shall prevail.
4. Nothing in any descriptive schedule or similar document issued in connection with and for use in carrying out the Sub-Contract Works shall impose any obligation beyond those imposed by the Sub-Contract Documents.
5. Nothing in the Sub-Contract Documents shall be construed as imposing any liability on the subcontractor in respect of any act, omission or default on the part of the employer, the employer's persons, the contractor or the contractor's persons.

In addition, in the absence of express provisions to the contrary, the law will regard written/typewritten text as overriding printed text, and specifically written as overriding general conditions. By reference to condition 1.3 of SBCSub/C, the Sub-Contract Agreement takes precedence over the conditions, and the Sub-Contract Documents over the numbered documents. With such a clause, only an alteration to the signed agreement or actual alteration and signing of the printed text of the subcontract conditions may be held to be an effective method of changing the text of the printed contract. However, SBCSub/C does allow for a schedule of modifications which is excluded from the general provision that the Sub-Contract Documents take precedence over the numbered documents.

A third factor in determining the arrangements between subcontracting parties is the law established by Parliament through the statutes. Statute law may operate either by compelling the parties to behave in a certain way, such as when tax law requires a paying party to make certain deductions or pay additional sums, to be paid as tax by the benefiting party, or a statute may create an obligation to make certain arrangements between the parties. In the latter case, should the parties fail to comply, the law will normally imply terms into the subcontract. Such an example is Part II of the HGCRA. In that case a complete set of supplementary conditions known as the Scheme for Construction Contracts (England and Wales) Regulations 1998 (the Scheme), will be incorporated as necessary into relevant construction subcontracts.

It perhaps goes without saying that the law will not uphold an unlawful or illegal act. So neither the courts nor a legal tribunal will support a subcontract in breach of the law, however carefully prepared. In construction work this may, for example, include unlawfully carrying out work to listed buildings or buildings in conservation areas or agreements not to charge VAT. It will be for the subcontractor to ascertain for itself that the work being contracted for is not outside the law.

Obligations of the parties

In a simple subcontract there will be an agreement for the subcontractor to provide the goods and services as described, and an agreement for the contractor to pay the stated price once the work is complete. Generally there will also be an obligation to complete the work within a stated period or by a stated date. There is no need for a subcontract to make further provisions but, because of the nature of building construction work, standard forms make express provisions for many other matters such as the right to instruct variations to the work, the valuation of the work, progress payments, the consequence of delays and disruption, the right to suspend performance and the right to terminate the subcontract.

In deciding the actual obligations of the parties, reference will be made to the documents the parties have agreed form the subcontract as a whole, but not to exchanges between the parties either before or after the formation of the subcontract, unless expressly incorporated within it. In general terms it will not be for either party to rely on obligations said to be 'good practice' or 'normally required' or some such other reason if not specifically incorporated into the agreement. Conversely, where the Sub-Contract Documents place an obligation on one of the parties, that party will not subsequently be able to resist such an obligation on the grounds that they 'never agree to such terms' or 'that they are not a normal' requirement of a subcontract of this nature.

Chapter 17
Dispute Resolution

Nature of a dispute

Disputes or differences of opinion can arise at any time and/or stage during the period in which the subcontract works are being carried out. Major fields of dispute may include:

1. uncertainty as to the terms or requirements of the subcontract;
2. uncertainty as to the extent and/or content of the subcontract work;
3. whether instructions of the contractor are variation instructions;
4. the quality of work contracted for and that actually constructed and/or provided;
5. valuation of variations;
6. the timing of the work;
7. the right to suspend performance or to determine the subcontract.

Some of the above may have serious implications for the ongoing performance of the subcontract. For example, a dispute over the quality of work will decide whether or not the work is to be removed and replaced. Once the work has been removed or improved, it will be difficult for a third party to make any decision on the matter when there is no longer any possibility of physical examination. If the work is to be removed and replaced there will be a cost and a time implication, which will be attributed to the subcontractor if the work is not to specification or to the contractor if it is. The decision and/or requirements of a third party, such as an architect and/or employer, may be involved.

Should the work be found to be to specification, then the employer's agent may seek to have it removed and the specification amended to their requirements and by doing so accepts the resultant delay involved and the obligation for additional costs. Alternatively, where the work is not to specification, the employer may be prepared for the work to remain and abate the cost on the basis of 'loss of amenity' or that the subcontractor will carry out agreed remedial works which may not entirely conform to the original specification.

Similarly, where variations are instructed and the subcontractor considers there is an entitlement to a significant change in the subcontract rates and/or sum, early notification and/or agreement may lead the contractor or employer to issue further instructions as part of its control procedures, to contain increases in cost.

What is a dispute?

In simple terms a dispute arises where one party to the subcontract makes a claim on the other, which after a reasonable period for consideration by the other party is not accepted. It must be noted that it is not the making of the claim which creates the dispute but the adverse response to it by the other side. An adverse response may be either active by a rejection of the claim or passive by a failure to respond within a reasonable time. It is considered that only by acceptance of the claim will a dispute be avoided.

Many disputes are resolved by agreement, either by one side accepting the arguments of the other or by compromise. For a compromise solution to be binding on the parties, there needs to some formal mechanism amending the subcontract, which may need to have evidence of consideration or to be by way of a deed. In practice, each dispute will not be formalised in this manner but may be left until the settlement of a final account for the entire subcontracted works.

It must be noted that claims may be made or originated by either side to the subcontract during the normal progress of the subcontract works. For example, a notice of non-conformity, issued by the contractor to the subcontractor, may create a dispute if not accepted by the subcontractor either in writing or by its action in removing and replacing its substandard work, and will be just as much a dispute as a failure to agree a subcontractor's application for payment for additional or varied work.

In a similar manner either party may make a money claim against the other. An abatement and/or set-off by a contractor against a subcontractor, whether presented in a detailed format or not, will be a claim by it against the subcontractor and if the costs are not accepted by the subcontractor, will comprise a dispute, in just the same way that a subcontractor's application for payment will, if reduced or rejected, create a dispute.

When does a claim become a dispute?

What constitutes a dispute between two parties, and particularly what constitutes a dispute that can be referred to statutory adjudication under the HGCRA, was considered in considerable detail in the judgment of Judge Seymour QC in *Edmund Nuttall Limited* v. *RG Carter Limited* [2002] BLR 312. The judge reviewed a number of previous judgments.

At paragraph 27:

> 'Lord Denning MR in *Monmouthshire County Council* v. *Costello & Kemple Ltd* (1965) 5 BLR 83 at page 89 as to what constituted a "dispute or difference" fit to be referred for determination in accordance with the disputes resolution procedure incorporated in the form of contract in issue in that case. Lord Denning said:
>
> "The first point is this: was there any dispute or difference arising between the contractors and the engineer? It is accepted that, in order that a dispute or difference can arise on this contract, there must in the first place be a claim by the contractors. Until that claim is rejected you cannot say that there is a dispute or difference. There must be a claim and a rejection in order to constitute a dispute or difference." '

At paragraph 29:

'A question which Judge Thornton QC had to consider in *Fastrack Contractors Ltd* v. *Morrison Construction Ltd* [[2000] BLR 168] was what was the meaning of the word "dispute" in the context of adjudication. That, of course, is exactly the issue which arises before me in this case. The passage upon which Mr Pennicott and Mr Bowdery [Counsels for Nuttall and Carter] relied in the judgment of Judge Thornton QC was as follows:

> "23. In some cases, a referring party might decide to cut out of the reference some of the pre-existing matters in dispute and to confine the referred dispute to something less than the totality of the matters then in dispute. So long as that exercise does not transform the pre-existing dispute into a different dispute, such a pruning exercise is clearly permissible. However, a party cannot unilaterally tag onto the existing range of matters in dispute a further list of matters not yet in dispute and then seek to argue that the resulting 'dispute' is substantially the same as the pre-existing dispute . . ." '

At paragraph 30:

'Judge Humphrey Lloyd QC handed down on 15 June 2001, *Sindall Ltd* v. *Solland* [15 June 2001, unreported]. In the course of his judgment in that case Judge Lloyd QC said:

> "For there to be a dispute for the purposes of exercising the statutory right to adjudication it must be clear that a point has emerged from the process of discussion or negotiation has ended and that there is something which needs to be decided.
>
> . . . Instead Sindall asked MEA [Contract Administrator and Quantity Surveyor] to look at a mass of information to which MEA had not been previously referred or specifically referred. Even if MEA had not said that it needed more time it would not have been required to provide an answer within seven days. A person in the position of the contract administrator must be given sufficient time to make up its mind before one can fairly draw the inference that the absence of a useful reply means that there is a dispute." '

Judge Seymour then summarised the previous cases and defined a dispute, at paragraph 36:

'In my judgment, both the definitions in the Shorter Oxford Dictionary and the decisions to which I have been referred in which the question of what constitutes a "dispute" has been considered have the common feature that for there to be a "dispute" there must have been an opportunity for the protagonists each to consider the position adopted by the other and to formulate arguments of a reasoned kind. It may be that it can be said that there is a "dispute" in a case in which a party which has been afforded an opportunity to evaluate rationally the position of an opposite party has either chosen not to avail himself of that opportunity or has refused to communicate the results of his evaluation. However, where a party has had an opportunity to consider the position of the opposite party and to formulate arguments in relation to that position, what constitutes a "dispute" between the parties is not only a "claim" which has been rejected, if that is what the dispute is about, but the whole package of arguments advanced and facts relied upon by each side. No doubt, for the purposes of a reference to adjudication under the 1996 Act or equivalent contractual provision, a party can redefine its arguments and abandon points not thought to be meritorious without altering fundamentally the nature of the "dispute" between them. However, what a party cannot do, in my judgment, is abandon wholesale facts previously relied upon or arguments previously advanced and contend that because the "claim" remains the same as that made previously, the "dispute" is the same.'

Despite the considerable review of the law by Judge Seymour QC in the case of *Nuttall* v. *Carter* there continues to be considerable scope for debate as to whether a dispute, referred to an adjudicator, is in fact a dispute that existed at the time the notice of adjudication was issued. Parties who seek to kill the adjudication process have relied on this decision and postulated that, if the referral includes any new argument or changed detail of the claim, then by reason of *Nuttall* v. *Carter* it was not a dispute which had crystallised and could be referred to adjudication.

In the case of *Dean & Dyball* v. *Kenneth Grubb Associates* (2003) 100 Con LR 92, Judge Seymour QC sought to enlarge on his judgment in *Nuttal* v. *Carter*:

> '42 . . . In my judgment whether, at the point of giving of notice of adjudication, there is a crystallised dispute in respect of the matter sought to be referred when the quantum of the sum claimed and the alleged composition of that sum has been altered from the sum previously claimed is a question of fact and degree the answer to which depends upon what, on the facts, the dispute between the parties was actually about. If liability in respect of a claim is not, at the point of giving notice of adjudication, in dispute, so that the only dispute is about quantum, it is likely to be difficult to say that there is a dispute about a formulation of quantum which is new at that stage and which the responding party has not had an opportunity to consider. However, as it seems to me, the situation is different if liability itself is in dispute and the party alleged to be liable has not accepted that it is bound to pay to the other party any sum whatever. In such a case there is a crystallised dispute as to liability – it is denied – and a crystallised dispute as to the obligation of the party to be liable to make a payment to the other party – the responding party's position is that it will pay nothing. Those crystallised disputes do not cease to be crystallised simply because the quantum of the claim is altered.'

The judge appears to be seeking to separate the dispute from the remedy sought. It is quite likely that in many disputes the referring party, while collecting together and arranging documents and calculations for presentation to an adjudicator, may make a number of corrections and/or adjustments to the sum or sums sought. So long as such adjustments are limited to minor matters and/or corrections of arithmetical errors, it is unlikely that such changes will be held to create a new dispute. It is the introduction of new heads of claim or the restyling of a claim using entirely or substantially new argument, facts or methods of calculation, that is likely to replace the existing dispute with a new claim.

Claim or counterclaim

In formal disputes, it is often the situation that both parties have claims against the other. For example, on the one hand the subcontractor may consider that the contractor has undervalued the work that it has completed, either by the contractor giving no specified reason for its valuation or by the contractor's valuation of specific items being less than that claimed by the subcontractor, while at the same time the contractor may have a claim against the subcontractor either for specific services carried out by the contractor, which it holds should be paid for by the subcontractor, or by way of payment of the loss incurred by the contractor arising from a breach of the subcontract by the subcontractor.

In a formal dispute resolution process, such as arbitration and/or litigation, there is provision for both claims being considered at the same time although they will be regarded as separate actions. In each case, i.e. for both claim and counterclaim, there will be a formal statement of case setting out the facts and arguments in support of the claim being made. The defendant, for both claim and counterclaim, will be allowed to respond, admitting or denying each point made in the statement of case and adding any further facts or legal arguments it wishes to be considered. Finally, each claimant will have the right to reply to the response.

Even where less formal methods of dispute resolution are to be attempted, it will be highly beneficial for the parties to set out the facts and law that they believe justify the claim they make. Disputed claims frequently are presented without justification by way of stating the facts relied on and/or the reasoning behind the claim. It must be remembered that the basic assumption is that once the subcontract work is completed, the subcontractor is to be paid the subcontract sum in whole but no more. It is for the parties to demonstrate that that sum should be changed by adjustment, either upwards by reason of a proven claim or downwards by way of a proven counterclaim.

Where the contractor has withheld payment from the subcontractor's account, the subcontractor may have to commence an action to obtain payment. In such circumstances the subcontractor as instigator of the action is technically the claimant, although the matter in dispute is the defendant's counterclaim. This can lead to some difficulties in disputes referred to adjudication, where there is no provision to consider more than one dispute in one reference.

Where one party has a claim against another, the second party will generally be allowed to raise any defence to that claim and be entitled to have all such defences heard and considered. However, an exception will be where a dispute as to the valuation of an interim account is submitted for the decision of an adjudicator, when the contractor will only be able to plead a cross-claim where it has issued a valid notice of withholding as required by section 111 of HGCRA. A distinction will be made between a claim relating to the value of the subcontract work, such as that the work was defective, and a claim associated with but directly related to the subcontract work, such as for loss to the contractor due to delay.

However, the limitations stated above will not apply where the subcontractor's claim relates to a final account or other final determination, when the responding party will have the normal right to raise any defence irrespective of whether a valid notice of withholding or other defence has been issued by it. This was the subject of the judgment of Judge Peter Coulson QC in the case of *William Verry (Glazing Systems)* v. *Furlong Homes Ltd* [2005] EWHC 138 (TCC). From paragraph 25 onwards the judge considers, whether if the claim made by the respondent Verry was a new claim, the adjudicator could take it into account. After reviewing a significant number of authorities he concluded that the adjudicator could consider such a claim and said:

> '35. I take on board all of the principles in the authorities which I have sited above. Plainly I have to take a robust view in relation to what is comprised in the word dispute and I also have to approach this case on its particular facts. Taking on board the facts as set out above and the principles derived from these authorities, I consider that even if, which I do not accept, that section D could be classified as a new claim for an extension of time, it formed part of the dispute which was referred by Furlong to adjudication.
>
> 36. Of course in coming to that view I am considerably strengthened in my analysis of the point that swayed the adjudicator in the first place. Namely that Verry were responding

to this claim, they were not the referring party; they did not start the adjudication. They had to defend themselves as best they could against the suggestion that their only entitlement to an extension of time was to 2 February 2004 and that liquidated damages should be deducted for the period of delay thereafter. They were not to be taken, in my judgment, as having agreed that in some way they could only defend themselves with half a shield relying on some matters of fact but not others. It seems to me that that would be an absurd result. In my judgment Verry were entitled to take whatever points they liked to defend themselves against this assertion and the adjudicator was obliged to consider all such points.

37. It seems to me that this is a classic example of what Judge Thornton QC was talking about in *Fastrack Contractors* v. *Morrison* where he said:

"The scheme [that is the Adjudication Scheme] gives the adjudicator two powers: to take (the) initiative in ascertaining the facts in the law . . . and to resign if the dispute varies significantly from the dispute referred to him . . . These powers show that it is possible that a dispute that has been validly referred to adjudication can in some circumstances, as the details unfold during the adjudication, become enlarged and change its nature and extent. If this happens it is conceivable that at least some of the matters and issues referred . . . which are not previously encompassed within a pre-existing dispute could legitimately become incorporated within the dispute that is being referred."

In similar vein it seems to me that what happened here was also the sort of thing envisaged by Judge Lloyd in the other case involving Sindall namely, *KNS* v. *Sindall* 75 LR Con 71, which is to the effect that:

"A party to a dispute who identifies the dispute in simple general terms has to accept that any ground that exists which might justify the action complained of is comprehended within the dispute for which adjudication is sought."

38. It seems to me that this is precisely the position here. Having made a general reference to the adjudicator to decide the question of extension of time, Furlong cannot now complain because in seeking to defend themselves against that point Verry have raised a variety of matters which on Furlong's approach are new. Furlong sought a resolution of that entitlement to an extension of time and they claimed positively that such an extension of time did not extend beyond 2 February 2004. Verry can defend themselves against that assertion and if that means referring to matters which were not part of a previously made formal claim then there is nothing to stop Verry from doing so.'

Presenting a claim

It is for the party making the claim to prove its case and not for the other side to show that the claimant has no entitlement. This principle applies equally whether the claimant is the subcontractor seeking payment for additional work or the contractor seeking to withhold money for some breach or default by the subcontractor.

It will be necessary to treat each head of claim separately. For each item claimed, it will be necessary to show a cause of action, which may be either an instruction from or an act of prevention on the part of the contractor, or, for a claim by the contractor, it will need to identify each failure by the subcontractor on which it relies. In many cases it will be necessary to demonstrate what was required of the other party under the subcontract terms and conditions, and that this obligation was not fulfilled. If there is no express provision within the subcontract terms and conditions, then it will be necessary to present a case that some term should be implied into the subcontract.

Once the claimant has shown a cause of action, it must demonstrate that it has suffered a loss or expense and that that loss or expense flows from the cause of action relied on. For some heads of claim, such as an instruction to carry out additional or varied work, the subcontract may have express provisions to be applied for valuation of the claim.

In most subcontract situations, there will be the alternative of two basic forms of claim: those involving instructed variations and those arising from breaches and/or failures by the other party. It must be understood that these two circumstances, generally, lead to a very different method of valuation and requirement for proof.

In the case of instructed variations these will be valued in accordance with the valuation rules set out in the subcontract and will normally require the claimant to show that the work was carried out, that it involved the quantity of additional work claimed and/or hours involved and the appropriate rate to be applied. The aim is that the subcontractor is rewarded for its additional work at the same rate as for the work carried out under the subcontract as originally envisaged, or at a similar rate of benefit to the subcontractor. In addition, it is possible that in any event the subcontractor is not to be the loser by way of complying with a variation instruction.

Loss and expense are valued in a similar way to damages where a breach of the subcontract is involved. The intention is to place the disadvantaged party in the same position, so far as money can, as it would have been had the matter complained of not occurred. This involves the claimant showing actual loss. Since this is a matter of recompense there can be no question of a profit mark-up. This principle must not be confused with a right to loss of profit, in certain circumstances, such as where work is wrongfully removed from the subcontractor and given to others.

In many situations the nature of any claim will be clear cut; for example, an instruction to a subcontractor to do some additional brickwork, plastering or painting will normally be valued by calculating the additional work involved and applying the bill rates. If there was no bill of quantities then the rate used in the subcontract can either be calculated from the subcontract sum or demonstrated from the subcontractor's tender calculations. It may be that the rates should be adjusted for the changed circumstances in which the work was performed, such as for small quantities and/or confined conditions and/or changed time of year at the time of return visits. Where the change is excessive, it may be correct to value the work on a time and materials basis applying the percentage uplift used and/or defined in the subcontract.

Other matters will be a clear loss, such as where work is stopped and labour and/or plant has to stand for periods of delay by the contractor. However, a much more common circumstance is where the subcontractor, for various reasons, is required to progress its work in a less productive and efficient manner to that anticipated by it at the time the subcontract was entered into. Or work may be postponed for a significant period which may lead to changes in working conditions and/or labour and/or material availability and/or cost. In these circumstances, the subcontractor may be able to claim either that there was an instructed variation to its work, for example to carry it out in winter rather than summer or that it suffered loss and expense because its commencement date was delayed.

It is for the claimant to state its case how it will. It will generally be to the subcontractor's advantage to plead an instruction to vary the work since, as explained above, this does not require proven actual loss but the application of reasonable rates and percentage mark-up based on the rates within the original subcontract sum.

There is rarely the same choice when making a counterclaim since such will rarely follow an instruction. An exception will be where the counterclaim involves charges from other third parties, such as other trades who have been instructed to vary their works following a breach or default by the subcontractor. In these circumstances, the claimant contractor will have to demonstrate that the sums paid to the other subcontractor were those it was entitled to under the provisions of that subcontract. It will not be sufficient to state or even demonstrate that it has in fact paid the other subcontractor the sums claimed.

It is apparent that many contractors and subcontractors commence the preparation of their claim by listing a number of heads of claim and then put a valuation to each head of claim as listed. It is suggested that such a procedure is fundamentally wrong and unlikely to lead to a successful conclusion. The starting point for a claim is, it is suggested, an analysis of the loss suffered in carrying out the work. The contractor or subcontractor must first compare its actual costs against its budget, and consider any additional costs for which there was no budget. In addition, it may look for areas of the work which cost more than similar work carried out in other areas of the project or at another time within the project duration.

Having analysed where it has incurred additional costs or lost money, the subcontractor/contractor must ascertain the true cause of that loss. Depending on the circumstances there may be several causes. Such causes may be consecutive when the loss is, in reality, a series of different losses or concurrent ones. Where the causes are concurrent there may be a dominant cause in which case the entire loss may be attributable to that cause, or they may all have been contributory in which case the total loss must be allocated between those causes.

Concurrent delays are rarely caused by events both of which are concurrent with the lost time. It is therefore necessary to distinguish between the concepts of concurrent event and concurrent effect. If work is delayed because of one or more events which arise at different times, the earlier event is likely to be considered the cause of the delay to the works. Where there has occurred a subsequent event which might also have delayed the work, it may not in fact have had any delaying effect because of the pre-existing delaying event. A variation instruction which requires the procurement of materials, especially if on a long lead period, may have a delaying effect up until the date of delivery, which must be considered in addition to the direct time of carrying out the instructed work. It follows that a subsequent cause of delay may have no overall effect if the work it concerns is carried out either prior to the delivery of the materials that are the subject of the earlier event or at the same time.

As well as demonstrating the cause(s) of its loss the subcontractor/contractor must identify the grounds for holding the other party liable for that loss. It is this link between cause and effect that is referred to as causation. Whether the claimant is able to prove actual loss or relies on a less precise claim such as a global or reasonable valuation basis for its claim, it will in any event have to demonstrate the causal effect. To do this it must show that it actually suffered a loss and that that loss was for the cause as claimed. Provided such a causal link is established, the courts are prepared to make assessments of the loss suffered where it is considered not possible or realistic to prove the actual loss flowing from each cause of action.

There will be circumstances where the subcontractor will be able to demonstrate that a number of events have caused it loss, but may find it difficult or impossible to allocate its loss accurately to each or any event. In such cases the subcontractor will have to make its claim on a global basis, relating a specific loss to a range of events. Where the courts are convinced that the loss claimed has been caused by the delaying events as claimed, and for which the subcontractor is not liable, then global claims will be allowed. Such claims run the risk of being defeated where the defending party is able to demonstrate that there is a contributory cause for which the other party is liable. For a fuller discussion, see the section 'Global claims' in Chapter 8.

In presenting its claim a subcontractor, or in the case of a counterclaim the contractor, will generally be well advised where possible to break down its claim into discrete issues and to avoid, so far as possible, claims for large sums relating to generalised heads of claim. This will be considerably helped if the subcontractor or contractor prepares and presents its claims, whether for the cost of variations, loss and/or expense or damages, at the earliest opportunity as part of its periodic valuation and application for payment procedure. The prompt presentation of a claim should make it easier for the other party to verify or deny the claim as made and/or indicated, at a time when action can be taken to keep additional records and/or obtain further evidence or substantiation of those claims.

Contractor's difficulties – combining disputes under different contracts

A dispute between a subcontractor and the contractor will rarely relate to a simple issue, but even where there are a number of issues it may be beneficial to have a certain major issue or issues considered independently. In many disputes the matter or matters at issue may relate to either the actions by a third party or a third party's response to a claim from one of the parties.

A common dispute may concern the quality of a subcontractor's work, the valuation of a change in the subcontractor's work, the effect on the regular progress of the subcontractor's work and the associated right to an extension of time, either for the subcontractor or for the subcontractor and contractor. In addition, the contractor may also seek to pass on charges from one subcontractor to another.

In these circumstances the risk to the contractor is that it may reach a settlement with, or obtain a tribunal's decision in connection with, the subcontractor's claim or other matter which is less advantageous to the contractor than the settlement or decision it obtains in its dispute with the other party. The contractor either has to sit on the fence and pass the claim and responses backwards and forwards acting solely as a post box, or it must take one side or the other. A further option is to resist both the subcontractor and the other party. Thus, if there is a quality issue the contractor may assert before the subcontractor that the work is substandard and abate the value of the work done by the subcontractor to compensate, while at the same time protesting to the employer that the work is to the specified requirement.

Some subcontracts have provision for the contractor to require the subcontractor to join its dispute against the contractor with the contractor's dispute against the employer, with the aim of achieving a consistent decision. While on the face of it this appears to be a straightforward and sensible method of resolving three-way disputes, it will in practice be more complex than immediately apparent. First, there will be a need to obtain the agreement of the other party and the tribunal to proceed in this way. Should this hurdle be overcome it is likely that the terms of the main contract will differ from those of the subcontract. What is a variation to the subcontract may or may not be a variation under the main contract, and in most cases the rates against which the variation is to be valued will differ between subcontract and main contract.

Where a variation leads to a delay to part of the works, the subcontract works may only require a minor amount of resequencing with little or no delay to completion. However, if the variation affects a critical part of the works, the project as a whole may be considerably delayed. Conversely, a change or default in the timely issue of construction details may considerably and adversely affect the subcontractor's off-site and/or pre-site works, but not delay installation and construction work on site.

Any procedure for the joining together of actions between a number of parties to a dispute must define a dispute as not arising until such dispute has been referred up the contractual chain and considered and responded to at each level. In the CECA/6th form of subcontract, clause 18(2)(b) states:

> '18(2)(b) Where in the opinion of the Contractor such a submission gives rise to a matter of dissatisfaction under the Main Contract, the Contractor shall so notify the Subcontractor in writing as soon as possible. In that event, the Contractor shall pursue the matter of dissatisfaction under the Main Contract promptly and shall keep the Subcontractor fully informed in writing of progress. The Subcontractor shall promptly provide such information and attend such meetings in connection with the matter of dissatisfaction as the Contractor may request. The Contractor and the Subcontractor agree that no such submission shall constitute nor be said to give rise to a dispute under the Subcontract unless and until the Contractor has had the time and opportunity to refer the matter of dissatisfaction to the Engineer under the Main Contract and either the Engineer has given his decision or the time for the giving of a decision by the Engineer has expired.'

Clause 18 of the CECA/6th requires the contractor promptly to pursue the matter in dispute under the main contract. What would be considered prompt, and what time lag would constitute not being prompt, will be a matter of fact to be decided in each case by the tribunal. However, it is suggested that where, for example, a subcontractor seeks a variation for some matter and seeks a valuation at a certain sum, the contractor must either agree with the subcontractor and value and pay for the work accordingly or reject it outright, or state that this is a matter to be referred under the main contract and make the appropriate application at the next application date. Should the main contractor delay in seeking a decision under the main contract, it is likely that the contractor will lose its rights under such a clause.

Should the subcontractor who fails to receive a satisfactory response to its application to the contractor wish to refer the matter for decision under the adjudication provisions of the subcontract, it is possible that the contractor will be able to rely on the provisions of a clause such as clause 18(2)(b) of CECA/6th, provided that the contractor has pursued the matter under the main contract promptly. It is considered that, since the parties have agreed to this procedure, it must be assumed that a dispute cannot be said to have crystallised until sufficient time has elapsed for the agreed procedure to run its course. Should the contractor not obtain agreement to the subcontractor's claim within the period allowed under the main contract, then there will be a dispute which either the contractor must action under the main contract, or in the absence of such action it is suggested it will be open to the subcontractor to commence a unilateral action in adjudication against the contractor.

Thus such clauses can be considered as a two-edged sword, empowering the contractor to seek acceptance of the claim under the main contract prior to agreeing to the subcontractor's claim as presented, but at the same time requiring the contractor to pursue the matter should it be disputed or be prepared for a single party action between it and the subcontractor.

It may be considered that to the extent that the contractor considers the matter justifies pursuing against the employer, the contractor may not defer paying the subcontractor by reason of the provisions of section 113 of the HGCRA as being a conditional payment. Should the contractor not pay to the subcontractor the value sought under the provisions of the main contract, the subcontractor should commence an action against the contractor on this basis.

The House of Lords, in the case of *Lafarge Redland Aggregates Limited* v. *Shephard Hill Civil Engineering Limited* [2000] BLR 385, had to consider the effect of a similar subcontract condition. The initial dispute was as to whether the subcontractor was barred from commencing arbitration proceedings while the contractor continued negotiations under the main contract. In that subcontract the operative words were:

> '18(2) If any dispute arises in connection with the main contract and the contractor is of the opinion that such dispute touches or concerns the subcontract works, then provided that an arbitrator has not already been agreed or appointed in pursuance of the proceeding sub-clause, the contractor may by notice in writing to the subcontractor require that any such dispute under this subcontract shall be dealt with jointly with the dispute under the main contract in accordance with the provisions of clause 66 thereof. In connection with such joint dispute the subcontractor shall be bound in like manner as the contractor by any decision of the engineer or any award by an arbitrator.'

Lord Hope, on page 392, found that while clause 18(2) bound the subcontractor 'in like manner as the contractor by any decision of the engineer or any award by an arbitrator' it was not binding in the case of a negotiated settlement:

> 'The purpose of clause 18(2) is to avoid the risk of inconsistent findings on matters which arise in connection with the main contract and touch on or concern the works under the subcontract. The risk of negotiations with the employer resulting in an agreement between the contractor and the employer which is unacceptable to the subcontractor is not within the mischief that clause 18(2) seeks to avoid. Negotiation is the antithesis of submitting the dispute for the decision of the engineer or an award by an arbitrator.'

Lord Hope then made reference to the merits of a negotiated settlement in avoiding the costs of litigation, but concluded that the contractor may not act solely in its own interests but must have regard to those of the subcontractor. He said, on page 393:

> 'But a contractor who seeks to take advantage of the power under clause 18(2) is not entitled to have regard only to its own interests in selecting a means of resolving its dispute with the employer. It must have regard also to the interests of the subcontractor, which is being deprived of its power to make use of the procedure set out in clause 18(1).'

Lord Hope then considered an earlier decision in *Erith Contractors Ltd* v. *Costain Civil Engineering* [1994] A & DRLJ 123, with which he agreed, and said:

> 'I agree. Clause 18(2) of the subcontract does not give the contractor the right to deprive the subcontractor of the benefit of the procedure in clause 18(1) while he attempts to settle the main contract dispute by negotiation with the employer. There is nothing in either clause 66 or in clause 18(2) to prevent the contractor from attempting to settle the dispute under the main contract by negotiation once it has initiated the procedure that clause 18(2) contemplates. But any delay which is attributable to the negotiation process (must) be left out of account when consideration is being given to the question whether the contractor has fulfilled its obligation to the subcontractor to have the dispute which has arisen under the subcontract resolved within a reasonable time under clause 66.'

He concluded by finding an implied term to the effect that the procedure of joint action, if required by the contractor to be followed, must be commenced in a reasonable time. He said:

> 'I would therefore hold that it is an implied condition of the exercise of the power under clause 18(2) that the contractor intends to invoke the procedure under clause 66 of the main contract. This means that it is no answer for the contractor, if challenged on the ground of its failure to invoke that procedure within a reasonable time, to attempt to explain the delay by referring to time which has elapsed due to negotiations entered into with the employer with a view to rendering that procedure unnecessary.'

The House of Lords then considered in detail the nature of the arbitration procedures which would follow if such joint action were invoked. Their lordships were divided in this regard and thereby highlight the complications of drafting such contractual provisions.

Lord Hope expressed the opinion that there cannot be such a thing as a tripartite arbitration. Lord Hope's interpretation of the subcontract was that there was to be joint arbitration under the main contract, the result of which would be binding on the subcontractor by way of clause 18(2). He said, at page 395:

> 'The absence of any machinery in clause 18(2) for the reference of the dispute between the contractor and the subcontractor to an arbitrator agreed or appointed under the subcontract means that clause 18(2) has been drafted on the assumption that there will be only one arbitration and only one arbitrator – that is to say, the arbitrator agreed under clause 66 of the main contract. The assumption is not that the arbitrator will make an award against the subcontractor – he could not do that unless the subcontractor was a party to his appointment as arbitrator – but that the arbitrator's award against the contractor will be binding on the subcontractor under the contractual arrangement between the contractor and the subcontractor which is set out in clause 18(2) of the subcontract.'

He went on to find that the mechanism required that the contractor in the arbitration against the employer must represent the interests of the subcontractor. He also saw a possibility that the interests of the contractor and subcontractor may not be the same. It must certainly be anticipated that they may at least have differing slants or angles on the same arguments. Lord Hope said, on page 396:

> 'Clause 18(2) assumes that once an arbitrator has been appointed the contractor will deal with all the issues which the subcontractor wishes to raise in the course of the presentation of its case to the arbitrator. The subcontractor has no right to appear as a party to the arbitration between the employer and the contractor. So here again the contractor must keep the subcontractor informed about progress and must take all reasonable steps to present the subcontractor's case to the arbitrator. This will involve providing the subcontractor with a reasonable opportunity to supply the contractor with the necessary evidence so that the contractor may then place that evidence before the arbitrator.'

The contrary view was held by Lord Cook of Thorndon, who considered that the subcontract clause did contemplate a tripartite arbitration. Further he found that the disputes, although with a common theme, 'remain different disputes between different parties though linked in their subject matter as they touch or concern the subcontract works'.

Lord Cook later gave a warning that the procedures set out in the subcontract may fail for various reasons since the employer is not bound by the subcontract and may not be willing to concur in a joint dealing with the disputes. He concluded that if the procedure failed for any reason, not the subcontractor's fault, 'the subcontractor will be able to fall back on arbitration under clause 18(1)'.

The only conclusion that can be drawn from these judgments is that while it will be highly desirable for major issues to be decided as between all parties, the practicalities of devising suitable arrangements and drafting the necessary contractual provisions are considerable. Further, unless such matters can be speedily resolved the subcontractor must be the one to suffer by being held out of its money for the duration.

These arrangements can be compared and contrasted with the name borrowing provisions under Works Contract/2, where the initiative rests with the subcontractor and where the employer has agreed to the procedure by requiring the management contractor to adopt the specified form of works contract. For further details see the section 'Name borrowing' in Chapter 15.

Methods of dispute resolution

The many methods of dispute resolution are outlined below, ranging from the informal progressively to the fully formal. Many of these methods have variants and hybrids especially at the informal end.

Discussion between the parties is the best and cheapest way to resolve differences and is for the parties to arrange and carry out between themselves. There is, however, a misconception that an agreement can be achieved by the parties just stating what they want and that rational fair-minded people can somehow reach a compromise without any further effort. This is a misconception which leads almost inevitably to the parties becoming more and more intractable, since no party is likely to give ground until satisfied that the other party has some justification for its claim. Once one party accepts that the other party has a just cause, then there is a distinct possibility of the parties dealing out the value of the claim.

Expert determination is where the parties engage a third party to decide any dispute between the parties. In traditional main contracts this has in part been the role of the architect or engineer; however, since there is not normally a person in such a role in subcontracts, such an expert must be selected and agreed between the parties. To ensure impartiality the parties may invite a professional body or trade association to make an appointment for them. The role of the expert will be defined by the parties by agreement, but will generally allow the expert access to documents and/or the personnel involved and to determine the facts and the law and reach a decision. Again, the effect of this decision will be what it has been agreed it is to be. However, unless such a decision is to be binding on the parties, it is unlikely to be of much benefit to them.

Mediation again is a process, by agreement between the parties, where the appointed mediator does not himself make decisions as to the resolution of the dispute, but seeks to assist the parties to reach a compromise agreement between them. It is the objective of the process that both parties sign up to the agreed solution. The probability of success will, like discussion between the parties without the assistance of a mediator, be dependent on a prior acceptance that there is some just entitlement and that what is disputed is the extent or value of that entitlement.

Adjudication is a process provided for either expressly in the subcontract or by implication through the HGCRA and the Scheme. It provides for the giving of a notice that the dispute is to be referred to an adjudicator for his or her decision. Following the issue of the notice or simultaneously, the referring party seeks the appointment of an adjudicator. On the appointment of the adjudicator or by the stated time, the referring party presents its case to him/her for consideration and decision. The other party has the right to respond, after which the adjudicator controls the proceedings to achieve the objective of determining the facts and the law and making a decision, within 28 days of the referral of the dispute to him or her.

The HGCRA makes the decision of the adjudicator binding on the parties, unless and until it is varied either by agreement, arbitration or litigation. The decision of the adjudicator will be enforced, unless that decision was outside his or her jurisdiction or the adjudicator acted with potential bias. In practice, assuming the adjudicator has been properly appointed and the matter referred to him or her is in dispute, then the adjudicator's decision even if wrong will be binding on the parties and enforceable at law.

Arbitration has been described as private justice. The right to have a dispute decided by arbitration arises from an express agreement within the subcontract or by a separate agreement. Since the right to arbitration denies the right of the parties to the courts, the courts will require clear evidence of a pre-existing agreement to arbitration before staying an action in the courts for determination by arbitration.

The arbitration procedures are established by the terms of the contract and by the Arbitration Act 1996. Model forms of arbitration procedures exist and they are incorporated into most standard forms of subcontract, but are not universally used. Arbitration may be conducted on a documents-only basis, which will greatly limit costs, but arbitration usually will involve a preliminary meeting, exchange of pleadings, followed by a full hearing where the cases for each side are made before the arbitrator and supported by witnesses.

Litigation is the process of taking a cause of action to the courts. Depending on the nature of the action it will be tried in the Small Claims Court or in the County or High Court. This is the most formal of procedures and is covered by the Civil Procedure Rules. Being a court action, except in the case of actions in the Small Claims Court, the parties will normally need to be represented by a practising barrister, who will have been briefed by a solicitor.

Chapter 18
Adjudication

Introduction

It is likely that most subcontractors' disputes which cannot be resolved by discussion between the parties will, at least in the first instance, be referred to adjudication. Although the process itself is straightforward, there is more to obtaining a satisfactory decision than seeking the appointment of an adjudicator and passing to him or her a bundle of documents or exchange of letters.

It is often said that the adjudication process favours the claimant. While this is not accepted, two factors may however have led to this misconception. First, the referring party is likely to be convinced that right is on his side and is unlikely to commence the process unless it feels sure of success. Secondly, in any dispute the claimant is likely to have better prepared its case. Unlike in arbitration and litigation, the respondent has very little time to analyse in detail the dispute and to prepare and submit sophisticated arguments, once the reference commences. Therefore where, as is often the case, the respondent has not fully considered the claim made by the other party at the time such claim was first made, the respondent will have placed itself at considerable disadvantage should a reference to adjudication be made.

The process

The adjudication process commences when one party to a dispute decides that it is unlikely that more is to be achieved by discussion and negotiation, and that the issue is important enough to warrant the time and expense of adjudication. It will then be for the referring party to define the dispute to be referred. The dispute may involve matters claimed by either or both parties.

The next stage is to issue a notice to the other party that it is the intention of the referring party to refer the dispute to adjudication. This notice gives the other party a warning of the impending reference and enables it to appoint and/or assemble its professional/legal consultants. Alternatively, it provides a last opportunity for the other party to give ground to a sufficient extent that an agreed settlement can be reached.

At the same time as, or very shortly after, the referring party issues its notice of adjudication, it must seek the appointment of an adjudicator, or where there is a named adjudicator, check his or her availability and ability and willingness to act, and request that he or she does so.

Once the adjudicator is appointed, the dispute can be formally referred to him or her and in any event must be referred within the stated period, normally seven days, following the notice of adjudication. The receipt by the adjudicator of the notice of referral commences the 28-day period within which the adjudicator must reach and publish his or her decision, unless that period is extended under the provisions of the subcontract.

The other party will be allowed a period of time in which to respond, which may either be set by the express terms of the subcontract or set by the adjudicator. Further exchanges may follow, either in response to questions and/or requirements of the adjudicator or in reply to submissions of the other party.

Finally, the adjudicator will publish his decision. The decision will be limited to answering the questions put to the adjudicator and if these have been badly phrased may lead to an unsatisfactory decision. At that time the adjudicator will seek payment, which he or she may require before releasing the decision to the parties.

Nature of adjudication

For all construction contracts, as defined in section 104 of the HGCRA, the right to refer a dispute to adjudication is derived from section 108:

> '108. – (1) A party to a construction contract has the right to refer a dispute arising under the contract for adjudication under a procedure complying with this section. For this purpose "dispute" includes any difference.
>
> (2) The contract shall –
>
> (a) enable a party to give notice at any time of his intention to refer a dispute to adjudication;
> (b) provide a timetable with the object of securing the appointment of the adjudicator and referral of the dispute to him within 7 days of such notice;
> (c) require the adjudicator to reach a decision within 28 days of referral or such longer period as is agreed by the parties after the dispute has been referred;
> (d) allow the adjudicator to extend the period of 28 days by up to 14 days, with the consent of the party by whom the dispute was referred;
> (e) impose a duty on the adjudicator to act impartially; and
> (f) enable the adjudicator to take the initiative in ascertaining the facts and the law.
>
> (3) The contract shall provide that the decision of the adjudicator is binding until the dispute is finally determined by legal proceedings, by arbitration (if the contract provides for arbitration or the parties otherwise agree to arbitration) or by agreement.
>
> The parties may agree to accept the decision of the adjudicator as finally determining the dispute.
>
> (4) The contract shall also provide that the adjudicator is not liable for anything done or omitted in the discharge or purported discharge of his functions as adjudicator unless the act or omission is in bad faith, and that any employee or agent of the adjudicator is similarly protected from liability.
>
> (5) If the contract does not comply with the requirements of subsections (1) to (4), the adjudication provisions of the Scheme for Construction Contracts apply.'

It can be seen that the HGCRA requires the subcontract to incorporate eight provisions with regard to adjudication, which are:

1. Allow for any dispute arising under the contract to be referred to adjudication at any time.
2. Provide a timetable with the object of securing the appointment of the adjudicator within seven days.
3. Require the adjudicator to reach a decision within 28 days of referral or such longer period as is agreed by the parties after the dispute has been referred.
4. Allow the adjudicator to extend the period by up to 14 days, with the consent of the referring party.
5. Impose a duty for the adjudicator to act impartially.
6. Enable the adjudicator to take the initiative in ascertaining the facts and the law.
7. Provide that the decision of the adjudicator is binding until the dispute is finally decided by legal proceedings, by arbitration or by agreement.
8. Provide that the adjudicator is not liable for anything done or omitted in the discharge of his functions as adjudicator.

Where a subcontract fails to satisfy any or all of the above requirements, any express provisions for adjudication are replaced by the provisions of Part I of the Scheme. It will be seen that the provisions of the Scheme may incorporate express provisions contained within the defective subcontract conditions. For example, if there is an identified adjudicator nominating body, that body will be the agreed body by reason of paragraph 2 of the Scheme.

SBCSub/C and ICSub/C provide for adjudication at clause 8.2 and by express words incorporate the Scheme. At sub-clause 1 the adjudicator is to be the person stated in the Sub-Contract Particulars (item 14) or in the absence of a named adjudicator a person nominated by the stated nominating body.

Sub-clause 2 refers to a dispute or difference relating to clause 3.11.3 (clause 3.10 in the case of ICSub/C) and seeks the appointment of an adjudicator with the appropriate expertise and experience but provides for the adjudicator, should he or she not have the appropriate expertise and experience, to appoint an independent expert to advise them. The reference to clause 3.11.3 (clause 3.10 in the case of ICSub/C) is to the contractor's right to issue directions that the subcontract work be opened up and inspected or tested.

Both ShortSub and Sub/MPF04 refer to the provisions of the Scheme.

Since adjudication is a process provided for under the terms of the subcontract, whether expressed in the subcontract terms or implied by law, without express agreement it can only apply to matters concerned with that contract. The fact that the law only requires a provision for adjudication in subcontracts that satisfy the definition of a contract in writing as set out in the HGCRA is no reason why any other subcontract should not incorporate such provisions. Any such provisions would, of course, only incorporate the express terms of the subcontract whether they complied with the eight requirements set out above or not, and then only where it was clear that the parties had agreed to refer disputes to adjudication.

Matters that may be referred

Unless the parties have agreed otherwise, only a dispute under the contract may be referred to adjudication. Thus the matter must first be a dispute and secondly a dispute under the contract. It will therefore be a defence if the other party is able to show that the matter referred to the adjudicator was not in dispute or was not a matter under the contract.

Further, the referring party, by its notice of intention to refer a dispute to adjudication as provided for at section 108(2)(a) of the HGCRA, defines the scope of the reference. Although this is only required, by paragraph 1(2) of Part I of the Scheme, to be a brief description of the dispute, it must be in suitable words for the other party to anticipate the nature of the referral. It not infrequently happens that the referring party by specifying the precise answer required from the adjudicator, limits his or her jurisdiction to a 'yes' or 'no' answer.

Judge Thornton QC, in the case of *Fastrack* v. *Morrison*, described the various ways of defining the dispute and the effect on the adjudicator's jurisdiction, as follows:

> '26. Thus, whether or not the reference is wholly or partly lacking in jurisdiction will depend on the nature and extent of the dispute that has purportedly been referred to adjudication by the referring party. A particular dispute may be correctly characterised as being in this form: "what sum is due for a particular interim payment?" or "what sum is due for a particular item of work?" or "what sum is due at the Final Account stage?" without any particular or finalised sum being included as part of that claim. Alternatively, the dispute may be correctly characterised as being one concerning the question of whether or not a particular specified sum is due. In the first type of dispute, it would not necessarily follow, if a larger sum had been included in the notice of adjudication than the sum previously claimed in the relevant application, that no dispute had yet arisen.'

It will also be necessary for the referring party to request that the adjudicator decides the action to be taken by the other party, for example, that one party is to pay the outstanding sum within a stated period. It is very easy to overlook the need for the adjudicator to decide the consequences on the parties of his or her decision as to the disputed matter.

As stated above, either party may refer a dispute to adjudication at any time. This means that the referring party need not be, and frequently will not be, the claimant. Alternatively the reference may comprise claims by both parties. Where the claim being referred is that of the other party, it will not be possible for that other party to plead that the claim is not in dispute because it was only a "preliminary presentation" of its claim or initial assessment of its loss. Further, the referring party may select only some of the issues in dispute between the parties at any one time. For example, the referring party may select a few significant items out of an extensive variation account. Finally, the referring party may request the adjudicator to make either an interim or final decision with regard to the issue referred.

The parties to the adjudication

In any formal dispute resolution process, and adjudication is no exception, it is important to be sure who the parties actually in dispute are. Many organisations have sister organisations with similar but discrete names, while other organisations regularly trade under a different or abridged version of their true company name. Again, where a sole proprietor owns a company, it is not uncommon for the business owner's name to appear on the contract documents as the party to the subcontract.

It will be an effective challenge to a referral to adjudication to establish that the party referred to in the notice of adjudication and/or referral is not a party to the subcontract. In the case of *A. J. Brenton t/a Manton Electrical Components* v. *Jack Palmer* [2001] HT 00/436, the matter of the identity of the parties to the contract was raised before the adjudicator, who decided as a matter of fact that Mr Jack Palmer and not his company, Lords of Princetown Limited, was a party to the contract. It was claimed not only that Mr Palmer was not a party to the contract, but also that the adjudicator had no jurisdiction to make such a decision.

In upholding the decision of the adjudicator the judge, Judge Harvey held that deciding the identity of the parties was a matter of fact and consequently within the jurisdiction of the adjudicator and as a result was not a matter the judge could review. His words at page 4 were:

> 'In this case the adjudicator purported to make a decision which he was empowered by the Act to make. If he made any error in it which in fact results in his having no jurisdiction, it was an error of fact which it was certainly within his jurisdiction to determine.
>
> In those circumstances, in my judgment, the award is binding to the extent of course that any adjudicator's award, i.e. provisionally, and subject to any further proceedings that might be taken to set it aside and so on. That being so, it seems to me quite clear that Mr Palmer can have no defence to this application.'

This is one of a number of decisions which lead to the conclusion that where the adjudicator expresses his or her decision as a matter of fact, it will be upheld without question whereas if he or she expresses the same opinion as being a view as to jurisdiction, it will be subject to the scrutiny of the courts.

The referral

Although, as stated above, only an existing dispute may be referred to adjudication, the referring party may make minor adjustments to its claim and more generally may adjust the presentation. Up to the time the referring party decides to seek a decision from an adjudicator, it will have been negotiating with someone with intimate knowledge of the project concerned, the work carried out and the circumstances of the dispute.

The adjudication process will frequently be on a documents-only basis and only occasionally will the adjudicator have any intimate knowledge of the construction project and/or the details of the specific subcontract. Further, the adjudicator will not be acquainted with the circumstances leading to the claim and/or the method of valuation of any monies sought.

For this reason, it is necessary to ensure that the description of the matters disputed is made in full. For example, it will normally be sufficient when making a claim to refer to work by a brief description such as "remove redundant piping at high level in such a location". Such a description will give no indication to an adjudicator the level of difficulty and or quantity of work involved. Redundant piping may be a few feet of small diameter copper or plastic pipe or be many metres of large diameter cast iron pipe. Similarly, at "high level" may mean working off a pair of stepladders or involve a mobile tower or scaffold of several lifts. Another factor that will affect the adjudicator's appreciation of the work involved will be a description of the location where the work took place, i.e. at ground floor or several floors up a building and whether accessible by stairs, lift or external scaffold. The referring party must state whether the sum claimed includes for the removal and disposal and/or making good where redundant hangers or fixings have been. In this example, an adjudicator is more likely to accept without much questioning a much larger valuation for the work where he or she knows and understands the full circumstances and nature of the work. Again, a significant claim may be more convincing when it is divided into its various constituents, each supported with evidence and calculations of its own.

The referral should be set out in a logical sequence with clear cross-referencing both to other parts of the text and to the documents submitted, as supporting evidence. A typical notice of referral will be divided into major sections, for example:

1. An introduction, which will state the work subcontracted, what has been done, how the dispute has arisen and how the adjudicator has jurisdiction to decide the matter.
2. The details of the subcontract and the provisions which will be relied on by the referring party.
3. Any statute or case law that the referring party wishes to refer to the adjudicator.
4. The detail of each part of the reference. Where what is being referred is the sum due from the contractor to the subcontractor, the reference may include a number of variations, claims and counterclaims. It may be suitable to deal with each as a separate appendix, just referring in the general narrative to each, by its description, appendix number and sum claimed.
5. There must also, either at the beginning or at the end, be a statement of what the adjudicator is asked to decide.

While the above method of presentation is by no means obligatory, it is suitable for most references. It will be for the referring party to decide how best to present its case. Where the referring party is seeking a decision on a claim by the other party, such as when denying a right to withhold payment or other set-off, then it will need to present its defence to that claim as well as the claim itself. Frequently, such a defence will involve demonstrating the defects in the other party's case, such as by stating that it has not shown that the subcontractor had the duty claimed to have been breached, and most commonly that the loss claimed is not proven actual loss but is only a "guesstimate".

It is suggested that, where the referring party is contesting the claim of the other party, it is not restricted to a reliance on facts and argument previously presented, as it is when presenting its own case. It will be free to use such material as it believes is relevant and as will be allowed to the other party when it is responding to the notice of referral. For this reason, it is believed it is necessary for an adjudicator, in references involving what is often described as a reverse adjudication, to allow to the referring party the right of response to the other party's submitted statement of case.

Notice of adjudication

Once the referring party is satisfied that it has set out its case in the most satisfactory manner in a draft notice of referral, and is satisfied that this incorporates all the matters and arguments that are in dispute, then the next stage is to issue to the other party a notice of intention to refer the dispute to adjudication.

The timing of this notice may be important having regard to other events. First, it may be beneficial for the referring party to delay the commencement of an action in adjudication until after the defects liability period, so that the release of retention can be incorporated into the sum due to the subcontractor. Secondly, it is necessary to ensure that its key personnel will be available if and when needed in the process. Thus the timing of the notice must be related to the taking of annual leave and other commitments.

It is not recommended purposely to time the notice so as to be of inconvenience to the other party. While the adjudicator is to make his or her decision on the basis of the facts and the law only, he or she is likely to be sympathetic to a party where the other party has, very obviously, set out to gain an unfair advantage by the timing of the notice of referral. For this reason, it is not advised to issue a notice of intention to refer a dispute to adjudication to time adversely with public holidays, especially Christmas and New Year.

The purpose of the notice of intention to refer a dispute to adjudication, is to give notice to the other party that the referring party wishes the dispute to be resolved and considers that adjudication is now the only effective way to achieve this, and to advise the nominating body and the potential adjudicator of the nature of the dispute the adjudicator is required to decide, so as to enable both parties to ensure that the nominated adjudicator has the appropriate expertise and experience to decide the dispute being referred. This was set out in the decision of Judge Humphrey Lloyd QC in the decision in *KNS* v. *Sindall*, at paragraph 19:

> 'A notice of intention to refer a dispute to adjudication has two purposes: it notifies the other party to the contract of the dispute, which, although it is to be "briefly identified in the notice", must nevertheless be identified and it also tells the person agreed to be the adjudicator of the existence of dispute (and thus, for example, of the possibility of having to make time available) or it informs the person or body responsible for making an appointment of the nature of the dispute. This may be important so as to avoid a conflict of interest or in the process of selection of a suitably qualified person and it may also affect the execution of a JCT Adjudication Agreement.'

The notice must therefore make it clear, where there may be any doubt, what dispute is being referred. The Scheme describes this at paragraph 1(3) of Part I as follows:

> '(3) The notice of adjudication shall set out briefly –
>
> (a) the nature and a brief description of the dispute and of the parties involved,
> (b) details of where and when the dispute has arisen,
> (c) the nature of the redress which is sought, and
> (d) the names and addresses of the parties to the contract (including, where appropriate, the addresses which the parties have specified for the giving of notices).'

And by reference to clause 8.2 of SBCSub/C, this arrangement is incorporated into this form of subcontract by the words 'the Scheme shall apply'.

Disputes have arisen as to the effective date of issue of a notice of adjudication and the resultant commencement of the seven-day period referred to in section 108(2)(b) of the HGCRA, for the issue of the referral. The phrase 'within 7 days of such notice' has been held to mean seven days from the date of the notice and not of its receipt by or service on the responding party. Similarly the 28-day period for an adjudicator to publish his or her decision starts from the date of the referral and not the receipt of the referral. However, while the 28-day period has been held to be mandatory to the extent that an adjudicator reaching his decision out of time has been held to lack jurisdiction, the issue of the referral within the seven-day period has been held not to be mandatory.

Both these issues were considered by Judge Thornton QC in the case of *William Verry Ltd* v. *North West London Communal Mikvah* [2004] BLR 308, when he said:

> '26. Verry seeks to meet these contentions as follows. It firstly contends that the critical 7-day period runs from the date on which NWLCM received the notice of adjudication and not from its date of issue. The referral notice was therefore served in time. However, section 108 and Clause 41A.4.1 both refer to "7 days of the notice" and not to "7 days of the service of the notice". Thus, the 7-day period runs from the date the adjudication notice was issued and not from the date of service where that date is a later date.'

The responding party's objective in seeking to establish the date of the notice of adjudication is to show that the referral itself is issued late and to claim that as a consequence the process either may not proceed or that it should not have proceeded, and that as a consequence any decision of an adjudicator is flawed and unenforceable. In practice it will normally be to the benefit of the responding party for the referral to be delayed, thereby giving the responding party additional time to prepare its response.

In the *William Verry* case, referred to above, the judge had to consider the situation where the adjudicator had issued directions for the submission of the referral seven days from the day the responding party would have received the notice of adjudication. In that case Judge Thornton QC found that the adjudicator could issue such directions and that the referring party had effectively complied with such directions. He said:

> '27. Verry secondly contends that the requirement imposed by section 108(1)(b) is not mandatory. That section merely requires that the contractual procedure should allow a referring party such as Verry to serve a referral notice within 7 days if that party wants to serve it within that short timescale. If the contract does allow for service within that timescale, the requirement(s) of that section are not infringed if the contract also allows the adjudicator to extend that timescale and the referring party then elects to make use of that extended period and serve the referral notice outside the 7-day period provided for.
>
> 28. In my judgment, Verry's contention is correct. The language of section 108(1)(b) is not rigid. It requires that the contractual timescale should have the objective of securing the referral of the dispute to the adjudicator within 7 days of the adjudication notice. Thus, the statute is setting a minimum requirement for the contract. The contract must allow a referring party, if it chooses, to issue a referral notice within the prescribed 7-day timescale. However, there is nothing in the language of section 108(1)(b) to preclude the contract

from being drafted so as to provide additionally a machinery that enables the adjudicator to extend that timescale and enable the referring party to refer the dispute outside it if it chooses to. In other words, the language of section 108(1)(b) requires contractual machinery that enables the referring party to refer the dispute within 7 days of the adjudication notice but it does not prohibit a machinery which additionally enables the referring party to refer the dispute outside that timescale if (it) elects to take longer in making the reference.

29. The language of section 108(1)(b) is in stark contrast to the language of section 108(1)(c) which provides that the contract shall:

"require the adjudicator to reach a decision within 28 days of referral or such longer period as is agreed by the parties after the dispute has been referred."

Thus, the contract cannot allow the adjudicator to take longer than 28 days to reach a decision in circumstances where the parties do not agree to an extension. Had it been intended that section 108(1)(b) should prohibit the service of a referral notice outside the 7-day timescale even if the referring party chooses to serve late, that section would have been drafted in the same way as section 108(1)(c) and would have required the referral notice to be served within 7 days of the adjudication notice.'

While not expressly stated by Judge Thornton, the inference in the *Verry* judgment is that the service of the referral more than seven days after the issue of the notice of adjudication is not fatal to the process. The issue of the referral, however, does define the date for the adjudicator to issue his or her decision. This judgment has been upheld by the majority decision of the second division, inner house, Court of Session in the Scottish case of *Ritchie Brothers (PWC) Limited* v. *David Philip (Commercials) Limited* [2005] BLR 384.

Appointment of an adjudicator

In some cases the parties agree in advance, as part of the subcontract terms and conditions, to a named person to act as adjudicator in the event of a dispute. Alternatively, the parties may define who is to appoint the adjudicator and the procedure for doing so. The HGCRA requires that this procedure 'provide a timetable with the objective of securing the appointment of the adjudicator and referral of the dispute to him/her within 7 days of such notice'. Since this seven days includes both the activity of appointment and the activity of referral, it follows that the procedures must have the objective of the appointment of the adjudicator in a shorter period than seven days, to allow for the subsequent serving of the referral documents.

In the absence of a conforming pre-agreed method of nomination and appointment of an adjudicator, the referring party is free to seek a nomination from any adjudicator nominating body. Most standard forms of subcontract will allow for the pre-selection of an adjudicator nominating body and, in the event of there being no pre-selection, define the default body to be used in such circumstances. SBCSub/A allows the use of any of the five selected adjudication nomination bodies where no pre-selection has been made.

While it is suggested that, generally, the adjudicator nomination procedures may not be incorporated into the subcontract by reference to either the terms of the main contract or to those of a standard form, where there is uncertainty the referring party may be best advised to select the nominating body that might be so incorporated. This should avoid any jurisdictional challenge on this issue, with the associated waste of both time and expense.

Where the adjudicator is named in the subcontract, it will be for the referring party to contact the adjudicator to establish that he or she considers the matter to be referred to be within his or her competence, and possibly more importantly that he or she is available and willing to decide the reference. If the adjudicator is unable or unwilling to act for any reason, then the referring party must again refer to the procedures defined in the subcontract and act accordingly. On agreement to act, the adjudicator will notify the parties of his or her fee rates and method of payment, and set out and define the procedures and timetable for the adjudication. In doing this the adjudicator must conform to any procedures set out in the subcontract itself, and in the absence of express procedures in accordance with the Scheme paragraph 13:

> '13. The adjudicator may take the initiative in ascertaining the facts and the law necessary to determine the dispute, and shall decide on the procedure to be followed in the adjudication. In particular he may –
>
> (a) request any party to the contract to supply him with such documents as he may require including, if he so directs, any written statement from any party to the contract supporting or supplementing the referral notice and any other documents given under paragraph 7(2),
> (b) decide the language or languages to be used in the adjudication and whether a translation of any document is to be provided and if so by whom,
> (c) meet and question any of the parties to the contract and their representatives,
> (d) subject to obtaining any necessary consent from a third party or parties, make such site visits and inspections as he considers appropriate, whether accompanied by the parties or not,
> (e) subject to obtaining any necessary consent from a third party or parties, carry out tests or experiments,
> (f) obtain and consider such representations and submissions as he requires, and, provided he has notified the parties of his intention, appoint experts, assessors or legal advisers,
> (g) give directions as to the timetable for the adjudication, any deadlines, or limits as to the length of written documents or oral representations to be complied with, and
> (h) issue other directions relating to the conduct of the adjudication.'

Jurisdiction of the adjudicator

It is well established that the adjudicator may not decide his or her jurisdiction unless there is express agreement between the parties to the dispute that the adjudicator has that power. However, where a challenge is raised to the adjudicator's jurisdiction, it is the adjudicator's duty to consider the objections and reach a decision as to whether or not to withdraw or to continue.

This was considered by Lord Macfadyen in the case of *Homer Burgess Limited* v. *Chirex (Annan) Limited* [2000] BLR 124, when he said, at page 133:

> 'It was not disputed by the defenders in the debate before me that in that situation it was necessary for the adjudicator, when the parties were in dispute before him as to whether

some of the works constituted construction operations, to apply his mind to that question, and to reach a view on it. That was a necessary preliminary step which he required to take before making his decision on the matters referred for adjudication. If he had decided that the works in question fell wholly within the scope of construction works, he would have proceeded to a decision on the whole matter. If he had decided that none of the works were construction works, he would have declined to proceed with the adjudication. If he had decided (as he in fact did) that some of the works were, and some were not, construction operations, he would have proceeded to a decision on the aspects of the dispute relating to those that were. In that sense it was inevitable that the adjudicator should address the question of the scope of his jurisdiction, and come to a conclusion on that matter. The legitimacy of his doing so is in my view clearly supported by part of the passage which I have quoted above from the speech of Lord Reid in *Anisminic* [*Anisminic Ltd* v. *Foreign Compensation Commission* [1969] 2 AC 147].'

The most common grounds for a challenge to an adjudicator's jurisdiction are:

1. that the contract is not a construction contract or not a construction contract in writing, for the purposes of the HGCRA;
2. that the matter referred to the adjudicator is not a dispute that may at that time be referred to adjudication;
3. that the adjudicator's appointment is itself invalid.

There are a number of issues that may be regarded as jurisdictional issues because, depending on how such issues are decided, the adjudicator may or may not have jurisdiction to consider the other matters referred to him. Such matters will include the nature and terms of the contract as well as the nature of the parties to that contract. In the case of *Tim Butler Contractors Limited* v. *Merewood Homes Limited* [2000] TCC 10/00, the adjudicator was required to decide what interim payment was due to be paid. A challenge was made to his jurisdiction on the basis that the contract was not one of 45 days' duration. Judge Gillard QC held that this was a matter of fact concerning the contract, which the adjudicator had jurisdiction to decide. He said:

'31. In my judgment the question as to whether a construction contract came into existence which entitled the claimant to staged payments is a dispute as to the terms of the contract and not a dispute which goes to jurisdiction of the adjudicator. The position is that the adjudicator had to decide what were contractually agreed terms and in particular whether the programme was a term of the contract. The adjudicator said it was not. The adjudicator arrived at this conclusion having regard to the documents before him. He considered the correspondence and in particular the acceptance by the defendant. It is clear that the programme was supplied but there is no express term referred to in the documents as to the period of the works.

32. It was squarely before the adjudicator whether there were contract terms regarding payment and the effect of paragraph 9 of the terms and conditions. It seems to me that even if he were wrong in interpreting section 110, he was entitled to reach the conclusion as to the existence of the provision for staged payments. Undoubtedly there was provision by paragraph 9 of the terms and conditions. Some of the provisions of part 2 of the Scheme apply to a relevant construction contract and some to construction contracts generally.'

The response to the referral

The response to the referral is the opportunity for the other party to present its case, by way of facts and/or arguments of law, and to show that either the facts as stated by the referring party are untrue or that the conclusions reached by the referring party in the referral document are wrong and/or incorrect. The respondent may well do this in two stages: first, by responding item by item or paragraph by paragraph to counteract or destroy the statements of the referring party, and secondly by building up its own argument for the adjudicator to consider, and decide that the facts were different to those stated in the referral and/or reach a different conclusion as to the liabilities and obligations of the parties.

While certain commentators have suggested, it is considered wrongly, that the judgment in some decided cases suggests that a dispute does not exist, at least for the purposes of adjudication, until both sides have presented their case and had time to consider and/or debate their differences, it is suggested that a respondent is not limited in its response to a referral to adjudication to facts and arguments previously presented. The same is probably not true, or at least it is uncertain, where the other party in a reference to adjudication is the claimant. However, even if the other party were to seek to show that its action had been justified, for entirely different causes, it may be in the referring party's best interest to agree to the adjudicator's considering and deciding those matters by giving him an extended jurisdiction, by agreement. By extending the adjudicator's jurisdiction in this way the dispute will be decided, and the other party will have no reason to challenge the decision if found against it.

Replies and further details

Few adjudication procedures make provision for further submissions by the parties, and some adjudicators will seek to disallow further unsolicited replies. It is suggested that where the matter referred to adjudication includes a claim or claims by the other party, the referring party should be allowed a right to respond in any event, since this is strictly a response not a reply.

It is for the adjudicator, under section 108(2)(f) of the HGCRA, to 'take the initiative in asserting the facts and the law'. This undoubtedly gives the adjudicator the right and indeed the duty to ask the parties to respond to any questions he may have concerning the matters on which he has to decide. This will involve an early consideration by the adjudicator of the submissions made and for the adjudicator to request further statements and/or documents where he or she considers that there are issues that have not, or may not have been, correctly considered.

The adjudicator has to balance his duty against any inclination to consider a different line of argument not referred to him. Where he or she considers that the decision sought is correct but only justified for differing reasons, then an adjudicator should give the responding party the opportunity to present argument to the contrary.

The law, however, does recognise that the adjudicator has to balance his or her duty to the parties with the obligation to reach a decision within the restricted timetable, while at the same time making every effort to enable justice to be done.

Costs of adjudication

Unlike arbitration and litigation where the losing party will normally have to pay the legal costs incurred by the winning party, the normal arrangement in adjudication is that the parties pay their own costs. While on the one hand it appears unreasonable that a party that has been wronged should have to pay to have its wrong righted, on the other it means that a party seeking a decision over a relatively small sum is in control of the costs it incurs and is risking, and does not have any fear that the other side may employ an expensive team of lawyers for which the referring party may become liable if the decision goes against it.

However, although the standard forms of subcontract and the Scheme do not provide for cross-party costs, the parties may by agreement allow for the costs to be borne by one party entirely. One way in which this can be achieved, other than by express agreement, is for both parties in their submissions to the adjudicator to request that the adjudicator awards costs. Provided both parties seek such action, it has been held that there is an agreement for the adjudicator to do so. This was the situation considered by Judge Marshall Evans QC in the case of *John Cothliff Limited* v. *Allen Build (North West) Limited* [1999] LV9 22641, where, referring to paragraph 13 of the Scheme, at page 11 he said:

> 'I decide that the adjudicator has got power to award costs, at least where, as in this case, costs have been expressly sought in the application placed before the adjudicator, and where he has allowed representation, at least on behalf of the defendant by lawyers, and apparently on behalf of the claimant by a firm of dispute pursuing quantity surveyors, whom I am told are the leaders in that specialised field of extracting money from contractors up the line, or it may be denying it to contractors down the line.
>
> The Scheme is incorporated as applicable by an implied term in the contract by virtue of the section and subsection to which I have already referred (section 114, subsection 4 of the 1996 Act). When you look at the Scheme, the crucial parts appear to me to be paragraphs 13 and 16. Paragraph 13 says: "The adjudicator may take the initiative in ascertaining the facts and the law necessary to determine the disputes and shall decide on the procedure", those are the important words, "to be followed in the adjudication".
>
> I think it was intended and plainly intended to be a sweep-up clause to give the adjudicator general power to control, regulate and direct all matters to the procedure, its implementation, conduct, and the hearing and so on, and that it is wide enough to give the adjudicator a discretion as to whether it is appropriate in the circumstances of the particular case to make an award of costs.'

In a case such as the above, both parties have actively sought to extend the adjudicator's jurisdiction in this respect. Much more unreasonable is the subcontract which incorporates a clause requiring one party to be liable for the costs of both parties regardless of the outcome. Such arrangements may, for example, require the referring party to bear all the costs in the clear anticipation that it will normally be the subcontractor that will be the referring party. Unless it can be claimed that such a condition has not reasonably been brought to the subcontractor's notice, as in the case of *Interfoto* v. *Stiletto*, referred to in the section 'Incorporation of terms – general principles' in Chapter 2, such agreement will be upheld.

Such a term existed in the subcontract between *Bridgeway Construction Ltd* v. *Tolent Construction Ltd* [2000] LVO 99069, TC 14100. Judge Mackay held that, however unreasonable such a clause might be, the parties had agreed to it and it was not for the court to change their bargain and as a consequence it would not void the resultant adjudicator's decision. In his judgment he said, at page 9:

> 'It seems to me that contracting parties can contract how they like and it is unsatisfactory legally, if, at the end of the day, a disappointed party can come along and say, "Well, the contract was entirely wrong."
>
> Therefore, I find that the terms are not void and that the application to remove them and to alter the parties' position as a consequence is unsuccessful.
>
> Also, I take the view that as the claimants have argued matters before the adjudicator not on pounds, shillings and pence but on principle as to who should pay or who should not pay with regard to the contractual matters the subject matter of the adjudication, they are bound by the adjudication. They gave the adjudicator the right to determine such issues and they are bound by his determination.
>
> Therefore, it seems to me that the issues put forward by the claimants are decided in this way, namely that there was no voidness in the contractual terms; that the mere fact that the claimants must pay the costs of the other side in the adjudication is not in excess of jurisdiction; that with regard to the costs, expenses and charges, the adjudicator was entitled to find as he did. The mere fact that the parties did not argue how much should be the costs does not invalidate the activities of the adjudicator. In their submissions, the parties dealt with the matter fully enough for the adjudicator to give his findings, which he did.'

The decision

The decision should state clearly the conclusions reached by the adjudicator and these may be expressed in the form of directions to the parties. Such directions will normally state, where appropriate, the timetable for implementation, such as the timetable for payment of a sum found due.

The decision of the adjudicator, while strictly an interim or temporary decision, is nevertheless binding on the parties provided that the question answered in the decision is that referred to the adjudicator. Only in exceptional circumstances will the decision of the adjudicator not be enforced by the courts. Such circumstances include showing that the adjudicator had no jurisdiction in the matter referred, or as a result of the method of the adjudicator's appointment or exceptionally that the adjudicator, by his or her actions during the course of the adjudication, has demonstrated bias or a possible bias.

The adjudicator's decision will only be required to contain full reasons if the sub-contract so requires, or where requested to do so by either party. Many adjudicators will in any event give some outline of their thinking when writing their decision.

It is inevitable that one or both parties will disagree with and/or be dissatisfied with the decision. Generally the decision will be binding in its entirety, 'warts and all'. It will not be possible to claim that some parts of the decision will be accepted and other parts rejected. This is because, once appointed, the adjudicator is to decide the facts and the law. In the event that the adjudicator makes a mistake as to the facts or law of the matter or matters referred, this will be a mistake of fact or law and not a decision *ultra vires* his or her jurisdiction. As Judge

Humphrey Lloyd QC expressed the situation in his decision in *KNS* v. *Sindall*, at paragraph 24:

> 'There may be instances where an adjudicator's jurisdiction is in question and the decision can be severed so that the authorised can be saved and the unauthorised set aside. This is not such a case. There was only one dispute even though it embraced a number of claims or issues. KNS's present case is based on severing parts of the adjudicator's apparent conclusions from others. It is not entitled to do so. Adjudicators' decisions are intended to be provisional and in the nature of best shots on limited material. They are not to be issued as a launching pad for satellite litigation designed to obtain what is to be attained by other proceedings, namely the litigation or arbitration that must ensue if the parties cannot resolve their differences with the benefit of the adjudicator's opinion. KNS must therefore accept the whole of this decision and if it does not like it to seek a remedy elsewhere (in the absence of successful negotiation or some other form of ADR [alternative dispute resolution]). Furthermore I do not consider that it is right to try to dismantle and then to reconstruct this decision in the way suggested by KNS for that intrudes on the adjudicator's area of decision-making. Mr Mason [The adjudicator] understandably and properly said that he had not set out all his reasons. Had he done so he might well have explained why a certain course was not in accordance with his thinking. In addition, the parties have to accept the decision "warts and all"; they cannot come to the court to have a decision revised to excise what was unwanted and replace it with what was or was thought to be right, unless the court is the ultimate tribunal. This decision is a decision on whether Sindall was by March 2000 right not to pay KNS any more money. A party cannot pick amongst the reasons so as to characterise parts as unjustified and therefore made without jurisdiction. It is simplistic to say that a decision-maker is not authorised to make mistakes. It does not follow that every error or mistake falls outside the decision-maker's jurisdiction. Errors of fact and judgment (and even of law) are an inevitable risk of any decision-making process but if it is to be binding, even temporarily, the parties have to accept the risk of such errors. If the decision is erroneous but within the authority granted a party has to accept the result.'

Unlike an award in adjudication or a judgment at law, the decision of the adjudicator cannot be appealed, in the legal sense that the adjudicator erred in law. The only lawful option for the disappointed party is to commence the action afresh by way of an action in arbitration or at law, both of which are likely to be both time-consuming and expensive. Only in circumstances where there is a significant value involved, is the losing party likely to take the matter further.

For both parties, win or lose, the adjudication process with its early decision can be seen to bring the dispute to an end, with the associated saving of further expense in both management time and consultants' fees.

Enforcement

In the event that the losing party fails to comply with the adjudicator's decision, it will be necessary to seek an enforcement order from the courts. From the first the courts have implemented the intentions of Parliament, as seen by them, that adjudicator's decisions are to be complied with and at once.

There have now been an extensive number of cases, both reported and not reported, recording the decisions of judges to the effect that an adjudicator's decision is to complied with, whether right or wrong. In the judgment of Mr Justice Jackson in *Carillion Construction Limited* v. *Devonport Royal Dockyard* [2005] BLR 310, concerning the enforcement of an adjudicator's decision following a dispute over payment in connection with construction work at Devonport Royal Dockyard, the judge reviewed very fully the law, as it was at that time, at part 5 of his judgment:

> '58. The statutory provisions which I read out in part 1 of this judgment came into force in 1998. The first authoritative decision on the matter in which those provisions should be interpreted was given by Mr Justice Dyson in *Macob Civil Engineering Limited* v. *Morrison Construction Limited* [1999] BLR 93. In that case, Mr Justice Dyson enforced an adjudicator's decision and rejected a challenge on natural justice grounds.
>
> 59. At page 97, Mr Justice Dyson said this:
> "The intention of Parliament in enacting the Act was plain. It was to introduce a speedy mechanism for settling disputes in construction contracts on a provisional interim basis, and requiring the decisions of adjudicators to be enforced pending final determination of disputes by arbitration, litigation or agreement: see section 108(3) of the Act and paragraph 23.2 of Part 1 of the Scheme. The timetable for adjudications is very tight (see section 108 of the Act). Many would say unreasonably tight, and likely to result in injustice. Parliament must be taken to have been aware of this. So far as procedure is concerned, the adjudicator is given a fairly free hand. It is true (but hardly surprising) that he is required to act impartially (section 108(2)(e) of the Act and paragraph 12(a) of Part 1 of the Scheme). He is, however, permitted to take the initiative in ascertaining the facts and the law (section 108(2)(f) of the Act and paragraph 13 of Part 1 of the Scheme). He may, therefore, conduct an entirely inquisitorial process, or he may, as in the present case, invite representations from the parties. It is clear that Parliament intended that the adjudication should be conducted in a manner which those familiar with the grinding detail of the traditional approach to the resolution of construction disputes apparently find difficult to accept. But Parliament has not abolished arbitration and litigation of construction disputes. It has merely introduced an intervening provisional stage in the dispute resolution process. Crucially, it has made it clear that decisions of adjudicators are binding and are to be complied with until the dispute is finally resolved."
>
> 60. At page 98, Mr Justice Dyson said this:
> "The present case shows how easy it is to mount a challenge on an alleged breach of natural justice. I formed the strong provisional view that the challenge is hopeless. But the fact is that the challenge has been made and a dispute therefore exists between the parties in relation to it. Thus on Mr Furst's [Counsel for Devonport Dockyard's] argument, the party who is unsuccessful before the adjudicator has to do more than assert a breach of the rules of natural justice, or allege that the adjudicator acted partially, and he will be able to say that there has been no 'decision'.
>
> At first sight, it is difficult to see why a decision purportedly made by an adjudicator on the dispute that has been referred to him should not be a binding decision within the meaning of section 108(3) of the Act, paragraph 23(1) of the Scheme and clause 27 of the contract. If it had been intended to qualify the word 'decision' in some way, then this could have been done. Why not give the word its plain and ordinary meaning? I confess that I can think of no good reason for not so doing, and none was suggested to me in argument. If his decision on the issue referred to him is wrong, whether because he erred on the facts or the law, or because in reaching his decision he made a procedural error which invalidated the decision, it is still a decision on the issue. Different considerations may well apply if he purports to decide a dispute which was not referred to him at all."

61. The judgment of Mr Justice Dyson in *Macob* has subsequently been approved by the Court of Appeal and therefore it carries added authority. It should, however, be noted that this judgment has certain consequences. If an adjudicator makes an error of law, he is not complying with the second limb of paragraph 12(a) of the Scheme. Nonetheless, such non-compliance with the Scheme does not prevent his decision being valid and enforceable. Indeed, errors by an adjudicator may give rise to other breaches of the Scheme. For example, an adjudicator may wrongly decide that a piece of evidence is irrelevant and therefore he may fail to take that evidence into account as required by paragraph 17 of the Scheme. Nevertheless, such non-compliance does not deprive the adjudicator's decision of its temporarily binding force. These are consequences which flow from Mr Justice Dyson's reasoning in *Macob.*

62. In *Bouygues (UK) Limited* v. *Dahl-Jensen (UK) Limited* [2001] 1 All ER (comm) 1041 the adjudicator erroneously awarded to a subcontractor monies which should have been retained by the main contractor pending certificates of completion under the main contract. Both this court and the Court of Appeal held that the adjudicator's decision should be enforced. At paragraphs 14 to 15, Lord Justice Buxton said this:

> "Here, Mr Gard [the adjudicator] answered exactly the questions put to him. What went wrong was that in making the calculations to answer the question of whether the payments so far made under the subcontract represented an overpayment or an underpayment, he overlooked the fact that that assessment should be based on the contract sum presently due for payment, that is the contract sum less retention, rather than on the gross contract sum. That was an error, but an error made when he was acting within his jurisdiction. Provided that the adjudicator acts within that jurisdiction his award stands and is enforceable.
>
> 15. Bouygues contended that such an outcome was plainly unjust in a case where it was agreed that a mistake had been made, and particularly in a case, such as the present, where Dahl-Jensen was in insolvent liquidation, and therefore the eventual adjustment of the balance by way of arbitration will in practical terms be unenforceable on Bouygues's part. I respectfully consider that the judge was quite right when he pointed out that the possibility of such an outcome was inherent in the exceptional and summary procedure provided by the 1996 Act and the CIC [Construction Industry Council] adjudication procedure."

63. At paragraphs 27 to 28, Lord Justice Chadwick said this:

> "27. The first question raised by this appeal is whether the adjudicator's determination in the present case is binding on the parties – subject always to the limitation contained in section 108(3) and in paragraphs 4 and 31 of the Model Adjudication Procedure to which I have referred. The answer to that question turns on whether the adjudicator confined himself to a determination of the issues that were put before him by the parties. If he did so, then the parties are bound by his determination, notwithstanding that he may have fallen into error. As Mr Justice Knox put it in *Nikko Hotels (UK) Limited* v. *MEPC plc [1991] 2 EGLR 103 at 108*, in the passage cited by Lord Justice Buxton, if the adjudicator answered the right question in the wrong way, his decision will be binding. If he has answered the wrong question, his decision will be a nullity.
>
> 28. I am satisfied, for the reasons given by Lord Justice Buxton, that in the present case the adjudicator did confine himself to the determination of the issues put to him. This is not a case in which he can be said to have answered the wrong question. He answered the right question. But, as is accepted by both parties, he answered that question in the wrong way. That being so, notwithstanding that he appears to have made an error that is manifest on the face of his calculations, it is accepted that, subject to the limitation to which I have already referred, his determination is binding upon the parties."

64. Lord Justice Peter Gibson agreed with both judgments. The Court of Appeal decided *Bouygues* on 31 July 2000. Just over a week later, this court gave judgment in the first round of *Discain Project Services Limited* v. *Opecprime Development Limited* [2000] BLR 402. In this case, the adjudicator had oral and written communications with one party, from which the other party was excluded. Judge Bowsher QC held that this was a serious breach of the rules of natural justice, such that this court ought not to give summary judgment enforcing the award. Instead the judge gave leave to defend.

65. *Discain* proceeded to trial. Judge Browsher's judgment at the conclusion of the trial is reported at 80 Construction Law Reports 95. At trial, the judge heard oral evidence from the adjudicator and there was a much more extensive citation of authority than had been possible at the application for summary judgment. Judge Bowsher adhered to his original view that there had been a substantial breach of the rules of natural justice and he declined to enforce the adjudicator's decision.

66. In paragraph 34 of his judgment at trial, Judge Bowsher cited the passage from Mr Justice Dyson's at page 98 of *Macob* which I read out a few minutes ago. He explained that this passage did not mean that any breach of the rules of natural justice by an adjudicator, however serious, had no effect. At paragraph 39, Judge Bowsher cited with approval the following dictum of Judge Humphrey Lloyd in *Glencot Development & Design Co Limited* v. *Ben Barrett & Son (Contractors) Limited* [2001] 80 Con LR 14 at 31:

> "It is accepted that the adjudicator has to conduct the proceedings in accordance with the rules of natural justice or as fairly as the limitations imposed by Parliament permit . . ."

67. At paragraph 68 of his judgment, Judge Bowsher stated his conclusions in *Discain* in the following terms:

> "So the parties have entered into a compulsory agreement that the decision of the adjudicator is binding until the dispute is 'finally determined' by legal proceedings et cetera. Although I have heard a trial of action, I have not 'finally determined' the dispute that was before the adjudicator. This action is brought only to enforce the decision of the adjudicator and there has been no examination of the merits of what lay behind that decision. On the face of the 1996 Act and the Scheme, therefore, the decision is still binding on the parties. However, just as the court will decline to enforce contracts tainted by illegality, so I do not think it right that the court should enforce a decision reached after substantial breach of the rules of natural justice. I stress that an unsuccessful party in a case of this sort must do more than merely assert a breach of the rules of natural justice to defeat the claim. Any breach proved must be substantial and relevant."

68. In *C&B Scene Concept Design Limited* v. *Isobars Limited* [2002] BLR 93 the Court of Appeal held that an adjudicator's decision should be enforced, even though it might be based upon an error of law. Sir Murray Stuart-Smith (with whom Lord Justice Rix and Lord Justice Potter agreed) said this on pages 98 to 99:

> "24. In *Northern Developments (Cumbria) Limited* v. *J&J Nichols* [[2000] BLR 158], Judge Bowsher QC cited with approval the following formulation of principles stated by Judge Thornton QC *Sherwood* v. *Casson* [sic] [*Sherwood & Casson* v. *Mackenzie* (1999) 30 November, Judge Thornton QC, Case HT 99000188]:
>
> (i) A decision of an adjudicator whose validity is challenged as to its factual or legal conclusions or as to procedural error remains a decision that is both enforceable and should be enforced;
> (ii) A decision that is erroneous, even if the error is disclosed by the reasons, will still not ordinarily be capable of being challenged and should, ordinarily, still be enforced;

(iii) A decision may be challenged on the ground that the adjudicator was not empowered by the Act to make the decision, because there was no underlying construction contract between the parties or because he had gone outside his terms of reference.
(iv) The adjudication is intended to be a speedy process in which mistakes will inevitably occur. Thus, the court should guide against characterising a mistaken answer to an issue, which is within an adjudicator's jurisdiction, as being an excess of jurisdiction.
(v) An issue as to whether a construction contract ever came into existence, which is one challenging the jurisdiction of the adjudicator, so long as it is reasonably and clearly raised, must be determined by the court on the balance of probabilities with, if necessary, oral and documentary evidence.

25. I respectfully agree with this formulation. I would also add, as I have already pointed out, the provisional nature of the adjudication, which, though enforceable at the time can be reopened on the final determination.

26. Errors of procedure, fact or law are not sufficient to prevent enforcement of an adjudicator's decision by summary judgment. The case of *Bouygues (UK) Limited* v. *Dahl-Jensen (UK) Limited* [2000] BLR 522 is a striking example of this. The adjudicator had made an obvious and fundamental error, accepted by both sides to be such, which resulted in a balance being owed to the contractor, whereas in truth it had been overpaid. The Court of Appeal held that the adjudicator had not exceeded his jurisdiction, he had merely given a wrong answer to the question which was referred to him. And, were it not for the special circumstances that the claimant in that case was in liquidation, so that there could be no fair assessment on the final determination between the parties, summary judgment without a stay of execution would have been ordered . . .

29. But the adjudicator's jurisdiction is determined by and derives from the dispute that is referred to him. If he determines matters over and beyond the dispute he has no jurisdiction. But the scope of the dispute was agreed, namely as to the employer's obligation to make payment and the contractor's entitlement to receive payment following receipt by the employer of the contractor's applications for interim payment, numbers 4, 5 and 6 (see paragraph 12 above). In order to determine this dispute the adjudicator had to resolve as a matter of law whether clauses 30.3.3–6 applied or not, and if they did, what was the effect of failure to serve a timeous notice by the employer. Even if he was wrong on both these points, that did not affect his jurisdiction.

30. It is important that the enforcement of an adjudicator's decision by summary judgment should not be prevented by arguments that the adjudicator has made errors of law in reaching his decision, unless the adjudicator has purported to decide matters that are not referred to him. He must decide as a matter of construction of the referral, and therefore as a matter of law, what the dispute is that he has to decide. If he erroneously decides that the dispute referred to him is wider than it is, then, insofar as he has exceeded his jurisdiction, his decision cannot be enforced. But in the present case there was entire agreement as to the scope of the dispute, and the adjudicator's decision, albeit he may have made errors of law as to the relevant contractual provisions, it is still binding and enforceable until the matter is corrected in the final determination."

69. The Court of Appeal's judgment in *C&B Scene* was delivered on 31 January 2002. Some three months later, in *Balfour Beatty Construction Limited* v. *Lambeth LBC* [2002] BLR 288, Judge Humphrey Lloyd QC refused to enforce an adjudicator's decision for breach of natural justice. The facts of this case, however, were extreme. The dispute concerned a

claim for extension of time, together with loss and expense, on a Local Authority building contract which had overrun. The adjudicator used a different methodology to that which either party had put forward and made his own independent analysis of the critical path. The adjudicator did not invite either party to comment on this approach before issuing his decision.

70. On 22 January 2003, the Court of Appeal gave judgment in *Levolux AT Limited* v. *Ferson Contractors Limited* [2003] EWCA Civ 11; 86 CLR 98. The Court of Appeal upheld the judgment of this court, enforcing an adjudicator's decision. Lord Justice Mantell (with whom Lord Justice Ward and Lord Justice Longmore agreed) cited with approval the passage on page 97 of Mr Justice Dyson's judgment in *Macob*, which I read out a few minutes ago. Lord Justice Mantell then discussed the Court of Appeal's decision in *Bouygues*. At paragraph 9, Lord Justice Mantell said this:

> "The case of *Bouygues* is a good illustration of the Scheme put into practice. The adjudicator had made what was acknowledged to be an obvious and fundamental error which resulted in the contractor recovering monies from the building owner whereas in truth the contractor had been overpaid. The Court of Appeal held that since the adjudicator had not exceeded his jurisdiction but had simply arrived at an erroneous conclusion, the provisional award should stand."

71. Later in 2003, the Court of Appeal returned to the topic of enforcing adjudicators' decisions. In *Pegram Shopfitters Limited* v. *Tally Weijl (UK) Limited* [2003] EWCA Civ 1750; [2004], 1 All ER 818, the Court of Appeal held that there should not be a summary judgment enforcing an adjudicator's award, because it was arguable that there was no construction contract between the parties. Lord Justice May gave the leading judgment. He set out the historical background to the 1996 Act. At paragraph 9, Lord Justice May said this:

> "A number of first instance decisions in the Technology and Construction Court have striven to implement the policy of Parliament. Enforcement proceedings, as they are called, are brought using the CPR part 8 procedure and habitually there is a claim for summary judgment. Judges of the Technology and Construction Court have rightly been astute to examine technical defences to such applications with a degree of scepticism consonant with the policy of the 1996 Act, aptly described by Lord Justice Ward in *RJT Consulting Engineers Limited* v. *DM Engineering (Northern Ireland) Limited* [2002] EWCA Civ 270, (2002) 83 Con LR 99, [2002] 1 WLR 2344 as 'pay now, argue later'. There has been a number of appeals to this court. I understand anecdotally that this court may be regarded as less than entirely supportive of the policy of the 1996 Act. There certainly are cases in which this court has upheld challenges to the enforceability of decisions of adjudicators, but examination of the cases shows that this has occurred when legal principle has to prevail over broad-brush policy, as was the case in the *Gilbert-Ash* case."

72. In paragraph 12 Lord Justice May went on to note that, despite the general policy of the 1996 Act, the Court of Appeal would not uphold an adjudicator's award, if a respectable case had been made out for disputing the adjudicator's jurisdiction.

73. In *Gillies Ramsay Diamond and Others* v. *PJW Enterprises Limited* [2004] BLR 131 the Inner House of the Court of Session in Scotland dismissed certain challenges to the decision of an adjudicator. One of the unsuccessful challenges concerned the adequacy or inadequacy of the reasons given by the adjudicator. Lord Justice Clerk, (with whom Lord MacFadyen and Lord Caplan agreed) said this at paragraph 31:

> "31. In my opinion, a challenge to the intelligibility of stated reasons can succeed only if the reasons are so incoherent that it is impossible for the reasonable reader to make sense of them. In such a case, the decision is not supported by any reasons at all and

on that account is invalid (*Save Britain's Heritage* v. *No. 1 Poultry Limited*, supra [[1991] 1 WLR 153]). In my view, that cannot be said in this case. The adjudicator has understood what questions he had to answer. He has reached certain conclusions in law on those questions which, however erroneous, are at least comprehensible. Even if the question is one of the adequacy of the reasons, I am of the opinion that the reasons are sufficient to show that the adjudicator has dealt with the issues remitted to him and to show what his conclusions are on each (*Save Britain's Heritage* v. *No 1 Poultry Limited, supra*, at page 167)."

74. The last case which I must refer to in this series is *Amec Capital Projects Limited* v. *Whitefriars City Estates Limited* [2004] EWCA Civ 1418; [2005] BLR 1. In this case, the Court of Appeal reversed the decision of Judge Toulmin CMG QC that an adjudicator's decision should not be enforced for breaches of the rules of natural justice. Lord Justice Dyson in his judgment (with which Lord Justice Kennedy and Lord Justice Chadwick agreed) included the following important passages:

"14. The common law rules of natural justice or procedural fairness are two-fold. First, the person affected has the right to prior notice and an effective opportunity to make representations before a decision is made. Secondly, the person affected has the right to an unbiased tribunal. These two requirements are conceptually distinct. It is quite possible to have a decision from an unbiased tribunal which is unfair because the losing party was denied an effective opportunity of making representations. Conversely, it is possible for a tribunal to allow the leading party an effective opportunity to make representations, but be biased. In either event, the decision will be in breach of natural justice, and be liable to be quashed if susceptible to judicial review, or (in the world of private law) to be held to be invalid and unenforceable . . .

22. It is easy enough to make challenges of breach of natural justice against an adjudicator. The purpose of the Scheme of the 1996 Act is now well known. It is to provide a speedy mechanism for settling disputes in construction contracts on a provisional interim basis, and requiring the decisions of adjudicators to be enforced pending final determination of disputes by arbitration, litigation, or agreement. The intention of Parliament to achieve this purpose will be undermined if allegations of breach of natural justice are not examined critically when they are raised by parties who are seeking to avoid complying with adjudicator's decisions. It is only where the defendant has advanced a properly arguable objection based on apparent bias that he should be permitted to resist summary enforcement of the adjudicator's award on that ground . . .

41. A more fundamental question was raised as to whether adjudicators are in any event obliged to give parties the opportunity to make representations in relation to questions of jurisdiction. I respectfully disagree with the judge's view that the requirements of natural justice apply without distinction, whether the issue being considered by the adjudicator is his own jurisdiction or the merits of the dispute that has been referred to him for decision. The reason for the common law right to prior notice and an effective opportunity to make representations is to protect parties from the risk of decisions being reached unfairly. But it is only directed at decisions which can affect parties' rights. Procedural fairness does not require that parties should have the right to make representations in relation to decisions which do not affect their rights, still less in relation to 'decisions' which are nullities and which cannot affect their rights. Since the 'decision' of an adjudicator as to his jurisdiction is of no legal effect and cannot affect the rights of the parties, it is difficult to see the logical justification for a rule of law that an adjudicator can only make such a 'decision' after giving the parties an opportunity to make representations."

75. In preparing this judgment, I have looked at a large number of first instance judgments concerning the validity or enforceability of adjudicators' decisions. For the sake of brevity, I will not embark upon an analysis of every such case. Instead, let me set out my own observations.

76. Prior to 1998, if there was a dispute about payment within the construction sector, money would generally remain in the pocket of the paying party until final resolution of that dispute. This was a source of concern, for reasons set out in a number of reports including Sir Michael Latham's report, *Constructing the Team*, published in 1994. The statutory system of compulsory adjudication was set up to address this problem. The purpose of an adjudication was and is to determine who shall hold the disputed funds, and in what proportions, until such time as the dispute is finally resolved.

77. In order to achieve this objective, it is necessary that adjudication should be as speedy and inexpensive as circumstances permit. The adjudicator is not necessarily expected to arrive at the solution which will ultimately be held to be correct. That would be asking the impossible. The adjudicator is required to arrive at an interim resolution within strictly drawn constraints.

78. Over the past seven years, adjudication has been widely used in the construction industry. On many occasions, the parties have chosen to use the adjudicator's decision as, or as the basis for the final settlement of their disputes. This is a perfectly sensible and commercial approach. It has been remarked upon by the judges of this court. Nevertheless that perfectly sensible and commercial approach, which many parties choose to adopt, cannot change the juridical nature of adjudication or transform the legal duties which are imposed upon adjudicators by statute.

79. One can detect in the first instance cases over the last six years some slight differences in emphasis and approach. In borderline cases what one judge may regard as permissible error of law or procedure on the part of an adjudicator, another judge may characterise as excess of jurisdiction or a substantial breach of the rules of natural justice.

80. In my view, it is helpful to state or restate four basic principles:

1. The adjudication procedure does not involve the final determination of anybody's rights (unless all the parties so wish).
2. The Court of Appeal has repeatedly emphasised that adjudicators' decisions must be enforced, even if they result from errors of procedure, fact or law: see *Bouygues*, *C&B Scene* and *Levolux*.
3. Where an adjudicator has acted in excess of his jurisdiction or in serious breach of the rules of natural justice, the court will not enforce his decision: see *Discain*, *Balfour Beatty* and *Pegram Shopfitters*.
4. Judges must be astute to examine technical defences with a degree of scepticism consonant with the policy of the 1996 Act. Errors of law, fact or procedure by an adjudicator must be examined critically before the court accepts that such errors constitute excess of jurisdiction or serious breaches of the rules of natural justice: see *Pegram Shopfitters* and *Amec*.

81. May I now turn from general principles to five propositions which bear upon this case:

1. If an adjudicator declines to consider evidence which, on his analysis of the facts or law, is irrelevant, that is neither (a) a breach of the rules of natural justice nor (b) a failure to consider relevant material which undermines his decision on *Wednesbury* grounds or for breach of paragraph 17 of the Scheme. If the adjudicator's analysis of the facts or the law was erroneous, it may follow that he ought to have

considered the evidence in question. The possibility of such error is inherent in the adjudication system. It is not a ground for refusing to enforce the adjudicator's decision. I reach this conclusion on the basis of the Court of Appeal decisions mentioned earlier. This conclusion is also supported by the reasoning of Mr Justice Steyn in the context of arbitration in *Bill Biakh* v. *Hyundai Corporation* [1988] 1 Lloyds Rep 187.

2. On a careful reading of Judge Thornton's judgment in *Buxton Building Contractors Limited* v. *Governors of Durand Primary School* [2004] 1 BLR 474, I do not think this judgment is inconsistent with proposition 1. If, however, Mr Furst [Counsel for Devonport Dockyard] is right and if *Buxton* is inconsistent with proposition 1, then I consider that *Buxton* was wrongly decided and I decline to follow it.

3. It is often not practicable for an adjudicator to put to the parties his provisional conclusions for comment. Very often those provisional conclusions will represent some intermediate position, for which neither party was contending. It will only be in an exceptional case as *Balfour Beatty* v. *The London Borough of Lambeth* that an adjudicator's failure to put his provisional conclusions to the parties will constitute such a serious breach of the rules of natural justice that the court will decline to enforce his decision.

4. During argument, my attention has been drawn to certain decisions on the duty to give reasons in a planning context. See in particular *Save Britain's Heritage* v. *No. 1 Poultry Limited* [1991] 1 WLR 153 and *South Bucks DC and Another* v. *Porter (No. 2)* [2004] 1 WLR 1953. In my view, the principles stated in these cases are only of limited relevance to adjudicator's' decisions. I reach this conclusion for three reasons:
   (a) Adjudicators' decisions do not finally determine the rights of the parties (unless all parties so wish).
   (b) If reasons are given and they prove to be erroneous, that does not generally enable the adjudicator's decision to be challenged.
   (c) Adjudicators often are not required to give reasons at all.
5. If an adjudicator is requested to give reasons pursuant to paragraph 22 of the Scheme, in my view a brief statement of those reasons will suffice. The reasons should be sufficient to show that the adjudicator has dealt with the issues remitted to him and what his conclusions are on those issues. It will only be in extreme circumstances, such as those described by Lord Justice Clerk in *Gillies Ramsay*, that the court will decline to enforce an otherwise valid adjudicator's decision because of the inadequacy of the reasons given. The complainant would need to show that the reasons were absent or unintelligible and that, as a result, he had suffered substantial prejudice.'

In the earlier judgment of Lord Macfadyen, in the case of *The Construction Centre Group Limited* v. *The Highland Council* [2002] CA 127/02, the judge explained the nature of the adjudicator's decision and why it was to be complied with. This was not because the decision was correct but because the parties had agreed to comply with it. He put it in these words, in paragraph 9:

'It is, in my view, important to appreciate the nature of this action. In it, the pursuers do not ask the court to endorse the soundness of the adjudicator's decision on the merits of the dispute referred to him. Rather the pursuers merely ask the court to recognise that the parties have themselves contractually to implement the adjudicator's decision. The pursuers seek decree from the court, not because they are in the right of the dispute, but

> because they are contractually entitled to require the defenders to implement the adjudicator's provisional determination of the dispute, whether it be right or wrong.'

There has been debate as to whether an adjudicator's decision made after the lawful determination of a subcontract by the contractor could be overridden by the express provisions of the subcontract, allowing the contractor to make no further payments until the completion of the subcontract work and the contractor to be able to assess its losses arising out of that determination.

The Court of Appeal in the case of *Ferson Contractors Limited* v. *Levolux A. T. Limited* [2003] BLR 118, upheld the decision at first instance that the provision to pay the adjudicator's award overrode any other provisions of the subcontract. Although in that case the adjudicator had in effect found that the determination had been unlawful, Lord Justice Mantell, at paragraphs 29 and 30, found that the adjudicator's decision would be enforceable in any event:

> 'Here what was claimed, in opposition to the application for summary judgment, was a right to withhold payment following a valid determination of the contract. Rightly or wrongly, the adjudicator held that there had been no valid determination. So, even accepting that the logic of *Parsons Plastics* [*Parsons Plastics (Research and Development) Ltd* v. *Purac Ltd* [2002] BLR 334] applies in the circumstances of the present case, its application would not result in clauses 38A.7 and 38A.9 being overridden by clauses 29.8 and 29.9. And in any event this logic is insufficient to support Judge Thornton's conclusion expressed, as it is, in such broad terms.
>
> 30. But to my mind the answer to this appeal is the straightforward one provided by Judge Wilcox. The intended purpose of section 108 is plain. It is explained in those cases to which I have referred in an earlier part of this judgment. If Mr Collings [Counsel for Ferson] and Judge Thornton are right, that purpose would be defeated. The contract must be construed so as to give effect to the intention of Parliament rather than to defeat it. If that cannot be achieved by way of construction, then the offending clause must be struck down. I would suggest that it can be done without the need to strike out any particular clause and that is by the means adopted by Judge Wilcox. Clauses 29.8 and 29.9 must be read as not applying to monies due by reason of an adjudicator's decision.'

Such action requires the services of a solicitor and representation by a barrister but will normally be dealt with on a summary judgment basis and with the minimum of delay.

Chapter 19
Statutes

Introduction

It may be unnecessary to say that statutes will apply, as appropriate, whether or not they are specifically referred to in the Sub-Contract Documents. Where reference is made specifically to certain legislation, this will not normally alter or affect the duties and obligations between the parties. Thus, it is suggested, a clause requiring that the subcontractor comply with all necessary health and safety legislation and/or with specific regulations will not increase the statutory duty of the subcontractor either to the contractor or to its employees.

The same will not necessarily be so in the case of matters that arise from a statutory provision. If express restrictions have been imposed on the project, by powers such as those given to planning authorities, it may be that the imposing of such restrictions on the subcontractor by the contractor after finalisation of the Sub-Contract Agreement will constitute a variation to the subcontract where these were not made known to the subcontractor in the enquiry documents or during the pre-subcontract negotiations.

Since the minimum requirements of the statutes will apply to the subcontract regardless of any specific subcontract condition to that effect, such conditions are, it is suggested, therefore superfluous and unnecessary. However, where the contractor wishes to impose safety or other requirements or restrictions beyond those required by law, it may incorporate its own conditions within a Sub-Contract Agreement and such conditions will be binding on the subcontractor. For example, before the law made the wearing of safety helmets compulsory, many contractors made the wearing of helmets a requirement of all subcontractors' site operatives and such a condition would have been enforceable.

Statutes operate in differing ways. Some impose express obligations on certain parties under certain circumstances. Others will impose duties that require a certain level of performance but allow the party on whom the duty is imposed to decide the method by which it will satisfy that requirement. Further statutes will prevent certain actions under specific circumstances. In certain cases the law may make specific subcontract conditions ineffective.

Discussed below are three statutes of specific relevance to subcontractors in the construction industry.

The Housing Grants, Construction and Regeneration Act 1996, Part II

This Act became effective for all construction subcontracts entered into after 1 May 1998, following the approval of the statutory instrument known as the Scheme for

Construction Contracts (England and Wales) Regulations 1998 (the Scheme). This Act is certainly a most significant government-made regulation of and restriction on the right to freedom of contract between the parties to construction subcontracts. The HGCRA requires the parties to a construction subcontract to incorporate terms satisfying certain requirements into all such subcontracts with few exceptions, and such requirements will be implied through the Scheme where a subcontract fails to comply with those requirements.

The HGCRA can be broadly divided into two parts: dealing with payment, and giving a right for disputes to be decided by adjudication. That part dealing with payment can again be subdivided into:

1. The requirement to have an adequate mechanism for payment
2. The restriction of the right to withhold payment
3. The giving of the right to suspend performance in the event of non-payment
4. The making of conditional payment provisions generally ineffective.

The regulations for adjudication provide for the reference of any dispute to adjudication at any time, for the appointment and referral to an adjudicator within seven days of a notice of adjudication, that the adjudicator's decision is made within 28 days and that such decision is binding on the parties until decided by arbitration, litigation or agreement.

Part II of the HGCRA is the part that deals with, and is relevant to, construction subcontracts. This part starts at section 104, where a 'construction contract' is defined as follows:

> '104 – (1) In this Part a "construction contract" means an agreement with a person for any of the following –
>
> (a) The carrying out of construction operations;
> (b) Arranging for the carrying out of construction operations by others, whether under subcontract to him or otherwise;
> (c) Providing his own labour, or the labour of others, for the carrying out of construction operations.
>
> (2) References in this Part to a construction contract include an agreement –
>
> (a) to do architectural, design, or surveying work or
> (b) to provide advice on building, engineering, interior or exterior decoration or on the laying-out of landscape,
>
> in relation to construction operations.'

It can be seen from the above that construction subcontracts incorporate not only the physical construction work but also professional services and associated work such as landscaping. It is, however, limited to construction operations, which are defined in section 105:

> '105 – (1) In this Part "construction operations" means, subject as follows, operations of any of the following descriptions –
>
> (a) construction, alteration, repair, maintenance, extension, demolition or dismantling of buildings, or structures forming, or to form, part of the land (whether permanent or not);

(b) construction, alteration, repair, maintenance, extension, demolition or dismantling of works forming, or to form, part of the land, including (without prejudice to the foregoing) walls, roadworks, power-lines, telecommunication apparatus, aircraft runways, docks and harbours, railways, inland waterways, pipe-lines, reservoirs, water-mains, wells, sewers, industrial plant and installations for purposes of land drainage, coast protection or defence;
(c) installation in any building or structure of fittings forming part of the land, including (without prejudice to the foregoing) systems of heating, lighting, air-conditioning, ventilation, power supply, drainage, sanitation, water supply or fire protection, or security or communications systems;
(d) external or internal cleaning of buildings and structures, so far as carried out in the course of their construction, alteration, repair, extension or restoration;
(e) operations which form an integral part of, or preparatory to, or are for rendering complete, such operations as are previously described in this subsection, including site clearance, earthmoving, excavation, tunnelling and boring, laying of foundations, erection, maintenance or dismantling of scaffolding, site restoration, landscaping and the provision of roadways and other access works;
(f) painting or decorating the internal or external surfaces of any building or building structure.

(2) The following operations are not construction operations within the meaning of this Part –

(a) drilling for, or extraction of, oil or natural gas;
(b) extraction (whether by underground or surface working) of minerals; tunnelling or boring, or construction of underground works, for this purpose;
(c) assembly, installation or demolition of plant or machinery, or erection or demolition of steelwork for the purposes of supporting or providing access to plant or machinery, on a site where the primary activity is –
    (i) nuclear processing, power generation, or water or effluent treatment, or
    (ii) the production, transmission, processing or bulk storage (other than warehousing) of chemicals, pharmaceuticals, oil, gas, steel or food and drink;
(d) manufacture or delivery to site of –
    (i) building or engineering components or equipment,
    (ii) materials, plant or machinery, or
(e) components for systems of heating, lighting, air-conditioning, ventilation, power supply, drainage, sanitation, water supply or fire protection, or for security or communications systems except under a contract which also provides for their installation;
(f) the making, installation and repair of artistic works, being sculptures, murals and other works which are wholly artistic in nature.'

Despite the considerable detail given in this section its interpretation has led to several disputes and reported cases.

In the case of *Nottingham Community Housing Association Limited* v. *Powerminster Limited* [2000] BLR 759, the question, to be decided by the court, was whether a contract to carry out the annual servicing of gas appliances and to supply a responsive repair and breakdown service was a construction operation. Powerminster considered that the work fell within the meaning of section 105(1)(a) of the HGCRA. Nottingham claimed that since subsection (c) dealt only with installation it did not include maintenance and repair. At paragraph 17 of his judgment Mr Justice Dyson found that this work was a construction operation:

> 'I do not consider that the meaning of paragraph (a) is unclear. For the reasons that I have attempted to explain, it seems to me that the installation, alteration, repair and maintenance of heating systems etc. in buildings do fall within paragraph (a).'

From this judgment, it is clear that maintenance of parts of buildings, whether listed separately in section (c) or not, is a construction operation. Section (c) allows for installation within existing buildings or structures of systems for heating, lighting, air-conditioning, ventilation, power supply, drainage, sanitation, water supply or fire protection, or security or communications systems.

For construction work to be covered by the HGCRA it must not fall into one of the excluded operations listed in section 105(2). Sub-section (c) deals with plant or machinery on a site where the primary activity is one of those listed. In *Palmers Limited* v. *ABB Power Construction Limited* [1999] BLR 426, Judge Thornton QC had to decide if scaffolding erected for the construction of a boiler, and its supporting frame outside and away from any building, was a construction operation for the purposes of implying the adjudication provisions of the HGCRA into the contract. He found that it was:

> '22. Thus, structures and works forming part of the land (which are not confined to buildings but are clearly intended to refer to all structures and works of whatsoever type) are linked with this word. Moreover, power-lines, telecommunications apparatus and industrial plant are expressly included within the definition of "works" forming part of the land. Thus, it is clearly envisaged that the assembly and fixing to the land of industrial plant and similar features are included within the definition of construction operations and are also included in the definition of "construction".
>
> 23. The nature, size and method of fixing into position of the steel structure and the boiler itself clearly have the consequence that the boiler forms part of the land once assembled and fixed into position. Indeed, it would be hard to conceive a more rigid and permanent structure than the steelwork in question. The fact that much of the boiler is assembled on the site but away from its permanent resting place and then lifted into position cannot affect the conclusion that a construction activity is involved. Since much industrial plant is expressly included in the definition of a construction operation, the only reasonable conclusion is that ABB's work is a construction operation.'

It should be noted that what gave Palmers the right to adjudication was not that scaffolding was a construction operation, but that the construction of the boiler was. It is likely that scaffolding for a floating structure, for example, or constructed from such a structure would not be a construction operation under the Act.

In two cases, the court has been required to consider the meaning of the word 'plant'. In *Homer Burgess* v. *Chirex*, the judge had to decide whether the pipework joining items of plant was itself part of the plant. It was common ground that the primary activity on the defender's site was the production, processing or bulk storage of pharmaceuticals. It followed, therefore, that if the pipework was part of the plant, then its installation would not be a construction activity for the purposes of adjudication under the HGCRA. However, if the pipework was not part of the plant, then it probably was a construction activity.

The judge, Lord Macfadyen, had to decide whether the adjudicator had been in error in his consideration of the meaning of the word 'plant' in section 105(2)(c). He concluded, at page 135, that the adjudicator had been wrong and that the pipework was part of the plant:

> 'In these circumstances I am of opinion that the adjudicator did fall into error in his construction of the word "plant". Having regard to the general description of the pipework in question as forming the links between various pieces of machinery or equipment, by which ingredients and pharmaceuticals in process of manufacture are conveyed from one stage of the manufacturing process to another, I am of opinion that the pipework was clearly part of the plant being assembled or installed on the defenders' site. Without such pipework, the individual pieces of machinery or equipment would be unable to operate. The pipework is in a real sense part of the apparatus which, once it was installed, the defenders were going to use in order to carry on their business of manufacturing pharmaceuticals. The installation of pipework was in my opinion an operation which fell within the scope of the exception in section 105(2)(c)(ii), and was accordingly not a construction operation.'

It must be noted this was pipework connecting the process plant and was, no doubt, going to be used to convey the product being manufactured. The situation might have been different if the pipework in question had been connecting the production plant to a normal building service, such as water. In such cases, it is likely that a distinction will be made between pipework to the nearest isolation valve and that connecting that valve to the plant itself.

A very similar situation arose in the case of *ABB Power Construction Limited* v. *Norwest Holst Engineering Limited* (2000) 77 Con LR 20. In this case, while there were secondary issues, the primary question was whether insulation was part of the boiler installation. Since the work was for "repowering" in an existing power station, and the extent of the insulation was limited to "within three metres of the boilers and did not require work to any existing pipework with which the new pipe work connected", the insulation was part of the plant.

The evidence of the defendant was contained in a statement quoted at paragraph 6:

> '. . . the insulation or cladding was for "boilers, ducting, silencers, pipework, drums and tanks" without which "there would effectively be no plant at all". He then states what is obvious but which has formally to be found, namely that without insulation to provide protection against internal heat of about 530°C (with a design maximum of 630°C) the boiler casings would disintegrate within days and would be too hot for anybody to approach safely and that insulation of all the equipment for the plant was necessary to ensure that it and that the process in its entirety were safe, workable, and efficient.'

The judge found, in paragraph 15, that:

> 'The evidence is that the provision of insulation is an integral part of the construction of pipework, boilers and the like which are required so that power may be generated. Hence if the test were: Does the insulation perform a plant-like function? Then the answer is undoubtedly: Yes. Without insulation the pipework, boilers etc would not function as they are designed to perform, nor could the plant be operated safely and efficiently.'

In the case of *ABB Zantingh Limited* v. *Zedal Building Services Limited* [2001] BLR 66, the judge had to decide whether the site of the installation of a power generation plant was or was not a construction site within the definition of construction under the HGCRA. To decide this point it was necessary for the judge to determine the extent of the site. If the site were found to be confined to the area within which the generators were to be positioned, then the site would probably have been a site where the primary activity was power generation and consequently excluded under section 105(1)(c)(i). However, if the site were considered to be the wider area of the whole factory, then the primary activity would be printing and the construction work would fall under the requirements of the HGCRA.

The judge considered the issues under four headings, the last of which was 'site'.

> '28. It seems to me that this is the central issue between the parties. If the site is defined as the small areas on which the generators stood in Oldham and Watford, surrounded by a security fence, then the primary activity of the sites must be power generation, because the only activity of those sites is power generation. That must be so even though the activity (as shown by the planning applications) was intended to be merely temporary. If the site is defined as the whole areas occupied by MCP [Mirror Colour Print Group] at Oldham and Watford, then it cannot conceivably be said that the primary activity of those sites is power generation. Taking those sites as a whole, power generation can only be regarded as ancillary to the primary activity of printing colour magazines whether or not excess power might be sold to others.
>
> 32. The main contract between MCP and SSE [Scottish and Southern Electricity plc] for Oldham refers to the site as Mirror Colour Print (Oldham) Ltd Hollingwood Avenue, Oldham. Mr Raeside [Counsel for ABB Zantingh] submits that that is not a definition of 'the site' but is merely 'an address'. The main contract with SSE incorporated the standard terms of MF/1(rev3) including:
>
> "Site means the actual place or places, provided or made available by the purchaser, to which plant is to be delivered or at which work is to be done by the contractor, together with so much of the area surrounding the same as the contractor shall with the consent of the purchaser actually use in the connection with the works otherwise than merely for the purposes of access."
>
> 33. That definition, made in a contract to which neither Zedal nor ABB were party, might be thought to support the case presented by ABB. However, I do not accept that what some other parties defined as the site is the same as what was envisaged by Parliament for different purposes particularly when in the same contract they refer to the site in more general terms.
>
> 34. When Parliament refers in section 105(2) to "a site where the primary activity is" the reference must be to a place broader than a generator surrounded by a security fence. To make any sense of the Act, one has to look to the nature of the whole site and ask what is the primary purpose of the whole site. Is the primary purpose power generation, or, in this case, printing?'

The judge concluded that the work did not fall within any of the exceptions provided by section 105(2) of the HGCRA.

In section 106 of the HGCRA, it states that the requirements of the Act do not apply to contracts with a residential occupier.

> '106. – (1) This Part does not apply –
>
> (a) to a construction contract with a residential occupier (see below), or

> (b) to any other description of construction contract excluded from the operation of this Part by order of the Secretary of State.
>
> (2) A construction contract with a residential occupier means a construction contract which principally relates to operations on a dwelling which one of the parties to the contract occupies, or intends to occupy, as his residence.
>
> In this section "dwelling" means a dwelling-house or a flat; and for this purpose – "dwelling-house" does not include a building containing a flat; and "flat" means separate and self-contained premises constructed or adapted for use for residential purposes and forming part of a building from some other part of which the premises are divided horizontally.'

In practice, subcontractors by definition will not themselves contract with a residential occupier, since their subcontract will be with the contractor, whose contract will be subject to the restrictions of the HGCRA.

While these sections 104, 105 and 107 set out the nature of work that will fall under the provisions of the HGCRA and that which will not, it is always possible for a subcontractor to undertake work not covered by the HGCRA and, by expressly including a clause or clauses, make the Act effective as though it were a construction subcontract to which the HGCRA applied.

Section 107 makes the HGCRA applicable only to contracts in writing.

> '107. – (1) The provisions of this Part apply only where the construction contract is in writing, and any other agreement between the parties as to any matter is effective for the purposes of this Part only if in writing.
>
> The expressions "agreement" "agree" and "agreed" shall be construed accordingly.
>
> (2) There is an agreement in writing –
>
> if the agreement is made in writing (whether or not it is signed by the parties),
>
> (a) if the agreement is made by exchange of communications in writing, or
>
> (b) if the agreement is evidenced in writing.
>
> (3) Where parties agree otherwise than in writing by reference to terms which are in writing, they make an agreement in writing.
>
> (4) An agreement is evidenced in writing if an agreement made otherwise than in writing is recorded by one of the parties, or by a third party, with the authority of the parties to the agreement.
>
> (5) An exchange of written submissions in adjudication proceedings, or in arbitral or legal proceedings in which the existence of an agreement otherwise than in writing is alleged by one party against another party and not denied by the other party in his response constitutes as between those parties an agreement in writing to the effect alleged.
>
> (6) References in this Part to anything being written or in writing include its being recorded by any means.'

It can be seen from these definitions that a contract in writing, for the purposes of the HGCRA, can be made after the work which is the subject of the contract is complete. The easiest way to bring this about is for one party to write to the other, setting out its understanding of the terms of the contract, which unless dissented from by the other party should satisfy subsection (2) of section 107 of HGCRA.

However, the words of section 107(4) have been the subject of different interpretations as to whether the requirement of the words 'with the authority of the parties to the agreement' applies equally to the agreement being recorded 'by one of the parties' or 'by a third party'. In the view of the writer, the use of the comma after the phrase 'one of the parties' separates that option for recording from the second option comprising the remainder of the sentence. Thus the phrase 'with the authority of the parties to the agreement' relates only to the reference to 'a third party' and not, more generally, to the recording by one of the parties.

This difference of interpretation was discussed in the case of *Millers Specialist Joinery Company Limited* v. *Nobles Construction Limited* [2001] TCC 64/00, where Judge Gilliland QC concluded that where one party confirmed the agreement, it did not require the authority of the other party for such confirmation to form a contract in writing for the purposes of subsection (4). At paragraph 13 of his judgment he stated:

> 'Mr Pennifer [Counsel for Nobles Construction] next submitted that because section 107(4) had not been satisfied, the agreement was not "evidenced in writing" for the purposes of section 107(2)(d). I do not accept that submission. It involves the view that subsection (4) provides an exhaustive definition of what is meant in subsection (2)(d) by the words "evidenced in writing". Subsection (4) however is not expressed to be a definition setting out what was meant by the words "evidenced in writing". It merely states that an agreement will be evidenced in writing where it has been recorded by a person with the authority of the parties. It does not state that this is the only way in which an agreement may be evidenced in writing. In the present case both parties accept that Mr Dalton's [the quantity surveyor for Nobles'] letter of 12 January is a correct record of what had then been agreed. That in my experience is a classic example of when an agreement can in the ordinary and accepted signification of the words be described as being "evidenced in writing". It would in my judgment be a remarkable result if a document which the parties accept is a correct record of what had been agreed was not to be regarded as written evidence of what they had agreed. There does not appear to be any rational justification for excluding such a document from the definition of a construction contract under Part II of the 1996 Act and insisting on a requirement that authority to record the terms must have been given before an agreement can be said to be evidenced in writing and thus be a construction contract for the purposes of Part II. Authority is only relevant where the document is to be treated as binding or effective without any further step or assent being necessary. In my judgment the words "evidenced in writing" in sub-section (2)(c) are used in their ordinary sense as referring to a written document which sets out or refers to the relevant terms of the agreement and sub-section (4) is not intended to restrict the application of sub-section (2)(c). Sub-section (4), like sub-sections (3) and (5), rather is a provision which widens or extends the ambit of what is to be regarded as an agreement in writing for the purposes of Part II of the 1996 Act. It is probably directed to the situation where at or after a meeting it has been agreed that someone should prepare minutes of what had been agreed and the effect of the provision is to make clear that the minutes themselves are to be treated as written evidence of the agreement even if it cannot be shown that the minutes have actually been assented to by all the parties.'

In any event, where one party confirms the agreement reached between the parties, and the other party, without objecting, acts on that confirmation, it is probable that it will be regarded as having consented to that record of the agreement by its subsequent conduct.

Subsection (5) makes the exchange of written submissions in adjudication proceedings, in which the existence of an agreement other than in writing is asserted by one party and not denied by the other, an agreement in writing for the purposes of the HGCRA.

There have been few reported challenges that a subcontract was not a contract in writing for the purposes of the HGCRA. In *A & D Maintenance & Construction Limited* v. *Pagehurst Construction Services Limited* (2000) 16 CLJ 199, Judge Bowsher, among other matters, had to decide if there was a contract in writing for the purposes of the HGCRA. He found there was because the respondents had made responses in the adjudication proceedings:

> '15. In the course of lengthy submissions, both parties made reference to the subcontract. Whilst the parties do not agree upon the precise terms evidenced by the subcontract confirmation form, nonetheless there is sufficient, in my judgment, to warrant finding that those exchanges in the reply and response complied with Section 107(5) of the Act. In other words there was a further and alternative basis for holding that there was an agreement in writing under the Act to which the Scheme applied.
>
> 16. The defendants participated in the adjudication. They could have challenged the adjudicator as they threatened to do so, or sought an immediate ruling by the adjudicator as to his jurisdiction which could have been the subject of an immediate challenge. They did not do so.'

The Late Payment of Commercial Debts (Interest) Act 1998

The Late Payment Act has removed the traditional prejudice of the law against the payment of interest on debts, which was based on the medieval concept that usury was unchristian and therefore wrong. Parliament, by way of the Late Payment Act, has acknowledged the commercial realities that cash flow is essential to the economic well-being of companies and that late payment is a significant cause of insolvency, particularly with small and/or newly established businesses.

The Late Payment Act has been brought into force progressively, by way of ministerial orders and regulations. The Late Payment of Commercial Debts Regulations 2002 brought the Late Payment Act into force for all commercial contracts entered into after 7 August 2002, by repealing previous regulations which had limited the effect of the Late Payment Act to certain sized businesses for contracts entered into after 1 November 1998.

Earlier regulations had restricted the Late Payment Act to certain categories of contract by differentiating between contracts between small businesses, defined as ones employing less than 50 employees, and contracts with other organisations. Commencement Order No. 1 at section 3 introduced the effects of the Late Payment Act into contracts made after 1 November 1998, as follows:

(a) contracts made between a small business supplier and a purchaser who is a United Kingdom public authority; or
(b) contracts made between a small business supplier and a large business purchaser.

The schedule attached to this order gives details as to how the size of a business is to be calculated.

The rate of statutory interest was defined, in the Rate of Interest (No. 2) Order 1998, at section 4:

> 'The rate of statutory interest for the purposes of the Late Payment of Commercial Debts (Interest) Act 1998 shall be 8 per cent over the official dealing rate per annum.'

This rate applies to qualifying debts occurring after 1 November 1998 and before 7 August 2002. Qualifying debts after that date are subject to the Rate of Interest (No 3) Order, which, at section 4, bands interest rates in twice-yearly increments as follows:

> 'For the purposes of this Order the official dealing rate to be used is that in force on 30th June or 31st December in any year. This rate will apply as the official dealing rate for the following six month period, namely 1st July to 31st December or 1st January to 30th June respectively.'

In addition, the Late Payment of Commercial Debts Regulations 2002 provides for an additional fixed sum payment, for each qualifying debt:

> '5A – (1) Once statutory interest begins to run in relation to a qualifying debt, the supplier shall be entitled to a fixed sum (in addition to statutory interest on the debt).
> (2) That sum shall be –
>
> (a) for a debt less than £1,000, the sum of £40;
> (b) for a debt of £1,000 or more, but less than £10,000, the sum of £70;
> (c) for a debt of £10,000 or more, the sum of £100.
>
> (3) The obligation to pay an additional fixed sum under this section in respect of a qualifying debt shall be treated as part of the term implied by section 1(1) in the contract creating the debt.'

The effect of this regulation, where subcontracts allow for periodic payments under the HGCRA or otherwise, is that potentially, where a contractor is regularly late with payment, a lump sum will become payable over and above each payment as due.

The Late Payment Act, like the HGCRA, works by implying into subcontracts to which it applies the requirements of the Act, which can be summarised as being that any qualifying debt will carry interest, referred to as statutory interest, as set out in the Late Payment Act.

Section 4 of the Late Payment Act defines when statutory interest begins to run:

> '4. – (1) Statutory interest runs in relation to a qualifying debt in accordance with this section (unless section 5 applies).
> (2) Statutory interest starts to run on the day after the relevant day for the debt, at the rate prevailing under section 6 at the end of the relevant day.
> (3) Where the supplier and the purchaser agree a date for payment of the debt (that is, the day on which the debt is to be created by the contract), that is the relevant day unless the debt relates to an obligation to make an advance payment.
> A date so agreed may be a fixed one or may depend on the happening of an event or failure of an event to happen.
> (4) Where the debt relates to an obligation to make an advance payment, the relevant day is the day on which the debt is treated by section 11 as having been created.
> (5) In any other case, the relevant day is the last day of the period of 30 days beginning with –

(a) the day on which the obligation of the supplier to which the debt relates is performed; or
(b) the day on which the purchaser has notice of the amount of the debt or (where that amount is unascertained) the sum which the supplier claims is the amount of the debt,

whichever is the later.
(6) Where the debt is created by virtue of an obligation to pay a sum due in respect of a period of hire of goods, subsection (5)(a) has effect as if it referred to the last day of that period.
(7) Statutory interest ceases to run when the interest would cease to run if it were carried under an express contract term.
(8) In this section "advanced payment" has the same meaning as in section 11.'

There may, in many subcontract situations, be some confusion as to the date on which the debt is created. There may be express payment provisions, within the subcontract, or, in the absence of express provisions, the provisions of the Scheme will be implied where relevant. In all other cases the provisions of section 4(5) will apply, which effectively gives a payment period of 30 days from the date of performance where the contract sum is established, or from the date of notice to the contractor where it is not.

Section 7(2) of the Late Payment Act allows the parties after a debt is created 'to agree terms dealing with the debt'. The provisions of the Late Payment Act may therefore be changed, by agreement, in regard to any existing debt. Such agreement might, for example, allow for payment by instalments or relief from some future obligation, such as latent defects.

Part II of the Late Payment Act allows that in certain circumstances statutory interest may be ousted by express terms of the subcontract. Section 8 sets out the circumstances by which the statutory provisions may be ousted:

'8. – (1) Any contract terms are void to the extent that they purport to exclude the right to statutory interest in relation to the debt, unless there is a substantial contractual remedy for late payment of the debt.
(2) Where the parties agree a contractual remedy for late payment of the debt that is a substantial remedy, statutory interest is not carried by the debt (unless they agree otherwise).
(3) The parties may not agree to vary the right to statutory interest in relation to the debt unless either the right to statutory interest is varied or the overall remedy for late payment of the debt is a substantial remedy.
(4) Any contract terms are void to the extent that they purport to –

(a) confer a contractual right to interest that is not a substantial remedy for late payment of the debt, or
(b) vary the right to statutory interest so as to provide for a right to statutory interest that is not a substantial remedy for late payment of the debt,

unless the overall remedy for late payment of the debt is a substantial remedy.
(5) Subject to this section, the parties are free to agree contract terms which deal with the consequences of late payment of the debt.'

This section produces a number of difficulties, which include the definition of a substantial remedy.

Section 8(1) refers to contractual terms which purport to exclude the right to statutory interest. It is suggested that in order for any contractual provisions for interest to oust the right to statutory interest, they must expressly seek to oust the right to statutory interest in relation to such debts. Where such terms exist in standard forms and do not specifically seek to oust the statutory provisions, they may be considered to apply only to situations where the statutory provisions do not apply.

It is suggested, but is by no means certain, that the provision of a simple rate of interest, less than that set as the statutory rate, would not satisfy the requirement for an alternative substantial remedy. This is even more likely to be the case now that the statutory provisions include a lump sum payment in addition. A substantial remedy might be a higher lump sum with a lower rate of interest or a variable rate of interest, increasing with the period of delay in payment, provided that the higher rates are significantly above the statutory rate.

Section 8(4)(a) expressly excludes a contractual right to interest that is not a substantial remedy, or varying the right to statutory interest so as to provide for a right to statutory interest that is not a substantial remedy for late payment. Both of these provisions prohibit the right to vary the rate of statutory interest, unless it is substantial or the overall remedy is substantial. To argue that a lesser rate is substantial begs the question, when would such a lesser rate cease to be substantial? This would leave the matter open for the arbitrary decision of the courts. It is suggested that Parliament, by way of defining the statutory rate, have defined what is substantial.

Section 9 seeks to assist in defining 'substantial remedy' as follows:

> '9. – (1) A remedy for the late payment of the debt shall be regarded as a substantial remedy unless –
>
> (a) the remedy is insufficient either for the purpose of compensating the supplier for late payment or for deterring late payment; and
> (b) it would not be fair or reasonable to allow the remedy to be relied on to oust or (as the case may be) to vary the right to statutory interest that would otherwise apply in relation to the debt.
>
> (2) In determining whether a remedy is not a substantial remedy, regard shall be had to all the relevant circumstances at the time the terms in question are agreed.
> (3) In determining whether subsection (1)(b) applies, regard shall be had (without prejudice to the generality of subsection (2)) to the following matters –
>
> (a) the benefits of commercial certainty;
> (b) the strength of the bargaining positions of the parties relative to each other;
> (c) whether the term was imposed by one party to the detriment of the other (whether by use of standard terms or otherwise); and
> (d) whether the supplier received an inducement to agree to the term.'

Section 9(1)(a) suggests that a remedy is not substantial if it is insufficient for 'deterring late payment'. On this basis, a remedy providing less than the statutory rate will only be substantial if there are never any late payments. Certainly, if there are more than very occasional late payments, such a rate will have proved itself not to be sufficient to satisfy this test. Thus in order to justify a lower rate of interest, a contractor will need to show that such lower rate has, in fact, been a deterrent to its making payments late.

Other significant requirements for accepting an alternative remedy as substantial can be found at section 9(3). This section lists four factors to be considered:

'(a) Which envisages a situation where an alternative remedy makes more certain the rights of the subcontractor than can be found in the CDIA [Late Payment Act].
(b) Considers the relative bargaining positions of the parties, which for most subcontractors will act in their favour in resisting lesser rates of interest imposed upon them by the contractor.
(c) It is considered that it will be difficult for a contractor to claim that a straight reduction in interest rate will not be other than to the detriment of the subcontractor. Further it must be noted that this applies even when "standard terms" are being used.
(d) However, the CDIA acknowledges that a subcontractor may be offered an inducement to agree a change in terms, which is for a remedy that on its face is not substantial. The inducement may be considered as part of the remedy. Such an inducement might, for example, be a very short payment period creating the debt at a much earlier date.'

Contracts (Rights of Third Parties) Act 1999

Traditionally the law has strictly limited rights under a subcontract to the parties to that subcontract. Where it has been considered necessary or desirable for a third party to have rights and/or obligations related to a subcontract, then it has been necessary to arrange for a collateral contract in the form necessary, such as a warranty between the supplier of goods and/or services and the ultimate user.

The Contracts (Rights of Third Parties) Act 1999, at section 1(1), expressly provides for a third party to be able to enforce a term of the subcontract where either there is an express provision for it to do so or the term purports to confer a benefit on it:

'1. – (1) Subject to the provisions of this Act, a person who is not a party to a contract (a "third party") may in his own right enforce a term of the contract if –

(a) the contract expressly provides that he may, or
(b) subject to subsection (2), the term purports to confer a benefit on him.'

Subsection 1(2) allows for the parties not to intend for such a term or terms to be enforceable:

'1. – (2) Subsection (1)(b) does not apply if on a proper construction of the contract it appears that the parties did not intend the term to be enforceable by the third party.'

Subsection 1(3) requires that the third party be expressly identified, in the subcontract, by name or as a member of a class:

'1. – (3) The third party must be expressly identified in the contract by name, as a member of a class or as answering a particular description but need not be in existence when the contract is entered into.'

The following subsections, 4–6, limit and allow the above rights to those within the terms of the subcontract.

Many, if not most, standard forms of construction contracts expressly exclude the rights of third parties as allowed for in section 1(2) of the Contracts (Rights of Third Parties) Act 1999. SBCSub/C, at clause 1.6, provides for such exclusion by the words:

> '1.6 Notwithstanding any other provision of this Subcontract, nothing in this Subcontract confers or is intended to confer any right to enforce any of its terms on any person who is not a party to it.'

However, many subcontracts are entered into, either by accepted offer or under various contractors' bespoke terms and conditions, when it will rarely be apparent that the parties did not intend, that either any express term or the subcontract as a whole, not to be enforceable by a third party.

On the face of it, subsection 1(1) of the Contracts (Rights of Third Parties) Act 1999 requires express words in the subcontract or alternatively a very clear indication that a benefit is to be created for the third party, and that in the absence of such an express provision no right will exist. However, subcontracts frequently incorporate clauses requiring the provisions of the main contract to apply to the subcontract. While the interpretation of such clauses is complex and is discussed more fully in the section 'Incorporation of terms – general principles' in Chapter 2, it must be at least possible that such a general clause would grant rights to a third party named in the main contract.

The circumstances where this Act is likely to be applied could be where one of the parties ceases to exist, for example, because of insolvency or as the result of a takeover. Where a contractor becomes insolvent, it is possible that the employer may be able to enforce terms of the subcontract by way of provisions of this Act. Similarly, provided subsection 1(2) does not apply, a subcontractor could rely, where appropriate, on terms of the main contract to seek to enforce a right under that contract.

The general reaction of the construction industry has been to take advantage of the provisions for excluding the rights of third parties from subcontracts, and to ignore the potential benefits of the Contracts (Rights of Third Parties) Act 1999 to avoid the need for separate collateral contracts and warranties. As a consequence, since no issues or disputes arising from this Act are known to have been tried by the courts, the potential effect of the Act is unknown.

Table of Cases

A *v.* B [2002] Scot CS 325 315–17
A & D Maintenance & Construction Limited *v.* Pagehurst Construction Services Limited (2000) 16 CLJ 199 522
Abbey Developments Limited *v.* PP Brickwork Limited [2003] EWHC 1987 (Technology) 203–204, 365
ABB Power Construction Limited *v.* Norwest Holst Engineering Limited (2000) 77 Con LR 20 518
ABB Zantingh Ltd *v.* Zedal Building Services Ltd [2001] BLR 66 519
AC Controls Limited *v.* British Broadcasting Corporation (2002) 89 Con LR 52 37–38
Agip SpA *v.* Navigazione Alta Italia SpA (The Nai Genova and Nai Superba) [1984] 1 Lloyds Rep 353 87
Ailsa Craig Fishing Co Ltd *v.* Malvern Fishing Co Ltd [1983] 1 Lloyds Rep 183; [1983] 1 WLR 964 443
A. J. Brenton t/a Manton Electrical Components *v.* Jack Palmer [2001] HT 00/436 494
Alfred C. Toepfer *v.* Peter Cremer [1975] 2 Lloyds Rep 118 61
Alfred McAlpine Homes North Ltd *v.* Property & Land Contractors Ltd (1995) 76 BLR 59 . . . . 229, 244, 436, 441–2
Allen Wilson Shopfitters *v.* Mr Anthony Buckingham [2005] EWHC 1165 (TC) 307
Alstom Signalling Limited (t/a Alstom Transport Information Solutions) *v.* Jarvis Facilities Limited (No. 1) (2004) 95 Con LR 55 . . . . 74–5, 81
Altec Electric Ltd *v.* J. V. Driver Projects Ltd (2002) 15 CLR (3rd) 199 52–3
Amec Capital Projects Limited *v.* Whitefriars City Estates Limited [2004] EWCA Civ 1418 510
Amec Process & Energy Limited *v.* Stork Engineers & Contractors BV [2002] 1997 ORE 659 310–11
Anisminic Ltd *v.* Foreign Compensation Commission [1969] 2 AC 147 500
Aqua Design & Play International Ltd (in liquidation t/a Aqua Design) and Fenlock Hansen Ltd (t/a Fendor Hansen) *v.* Kier Regional Ltd (t/a French Kier Anglia) (2002) 82 Con LR 117 46–7
Ascon Contracting Limited *v.* Alfred McAlpine Construction Isle of Man Limited (1999) 66 Con LR 119; [1999] 1998-ORB-315 . . . . 115, 238, 403–404, 415–18
Aughton Ltd (formally Aughton Group Ltd) *v.* M. F. Kent Services Ltd (1991) 57 BLR 1 . . . . 49, 53
Aurum Investments Limited and Avonforce Limited (in liquidation) *v.* Knapp Hicks & Partners and Advanced Underpinning Limited (2000) 78 Con LR 114 . . . . 154–5, 196–7
Avintour Ltd *v.* Ryder Airline Services Ltd [1994] SC 270 73

Babcock Energy Ltd *v.* Lodge Sturtevant Ltd (1994) 41 Con LR 45 . . . . . 406–407, 439–9
Baden *v.* Société Génerale [1993] 1 WLR 509 84
Balfour Beatty Building Ltd *v.* Chestermount Properties Ltd (1993) 62 BLR 1 199
Balfour Beatty Construction Limited *v.* London Borough of Lambeth [2002] BLR 288 . . . . . 512, 588
Barque Quilpué Ltd *v.* Brown [1904] 2 KB 261 386
Belgravia Property Company Limited *v.* (1) S & R (London) Limited (2) Taylor Woodrow Management Limited [2001] BLR 424 462–7
Bellefield Computer Services and Others *v.* E. Turner & Sons Limited and Others [2002] EWCA Civ 1823 166
Bentley Construction Limited *v.* Somerfield Property Company Limited and Somerfield Stores Limited (2002) 82 Con LR 163 225
BHP Petroleum Ltd and Others *v.* British Steel plc and Dalmine SpA [2000] 2 Lloyds Rep 277 443
Biggin & Co Ltd *v.* Permanite Ltd and Others [1951] 2 KB 314 407
Bill Biakh *v.* Hyundai Corporation [1988] 1 Lloyds Rep 187 512
Blackpool & Fylde Aero Club Ltd *v.* Blackpool Borough Council [1990] 3 All ER 25 2
Blyth & Blyth Limited *v.* Carillion Construction Limited (2001) 79 Con LR 142 . . . . . 96–7, 167
Bouygues (UK) Limited *v.* Dahl-Jensen (UK) Limited [2000] BLR 522; [2001] 1 All ER (comm) 1041 . . . . . 506, 508
Bovis Lend Lease Limited (formerly Bovis Construction Limited) *v.* R. D. Fire Protection Limited and (1) Huthco Limited (2) Baris UK Limited (2003) 89 Con LR 169 407–10
Boynton and Another *v.* Willers [2003] EWCA Civ 904 12
Bracken and Another *v.* Billinghurst [2003] EWHC 1333 (TCC) 290, 291
Bridgeway Construction Ltd *v.* Tolent Construction Ltd [2000] LVO 99069, TC 14100 502–503
Brightside Kilpatrick Engineering Services *v.* Mitchell Construction (1973) Ltd (1975) 1 BLR 62 . . . . . 46, 47–48
British Bank for Foreign Trade Ltd *v.* Novintex Ltd [1994] 1 KB 623 73
British Steel Corporation *v.* Cleveland Bridge & Engineering Co Ltd (1981) 24 BLR 94 . . . . . 27, 245, 428
British Westinghouse Electric & Manufacturing Co Ltd *v.* Underground Electric Railways Co of London Ltd [1912] AC 673 421
Brontex Knitting Works Ltd *v.* St John's Grange [1944] WN 85 444
Butler Machine Tool Co Ltd *v.* Ex-Cell-O Corporation (England) Ltd [1979] 1 WLR 401 20
Buxton Building Contractors Limited *v.* Governors of Durand Primary School [2004] 1 BLR 474 512
C & B Scene Concept Design Limited *v.* Isobars Limited [2002] BLR 93 . . . . . 507, 508
Cammell Laird & Co Ltd *v.* Manganese Bronze & Brass Co Ltd [1934] AC 402 338
Canada Steamship Lines Ltd *v.* The King [1952] 1 Lloyds Rep 1 443

Carillion Construction Limited (trading as Crown House Engineering) *v.* Ballast Plc (formerly Ballast Wiltshire plc) [2001] EWCA Civ 1098 . . . . 13, 28
Carillion Construction Limited *v.* Devonport Royal Dockyard [2005] BLR 310 505–12
Carillion Construction Ltd *v.* Felix (UK) Ltd [2001] BLR 1 292–3
Carr *v.* J. A. Berriman Pty Ltd (1953) 89 CLR 327 366
Cegelec Projects Limited *v.* Pirelli Construction Company Limited [1998] No. 1997 ORB 646 . . . . 51, 54, 56–7
Chandler Bros Ltd *v.* Boswell [1936] CA 179 391
Charles Rickards Ltd *v.* Oppenheim [1950] 1 All ER 420 . . . . 61–2, 389–90
Chatbrown Ltd *v.* Alfred McAlpine Construction (Southern) Ltd (1986) 35 BLR 44 431
Chichester Joinery Ltd *v.* John Mowlem & Co plc (1987) 42 BLR 100 22
Chiemgauer Membran Und Zeltbau GmbH (formerly Koch Hightex GmbH) *v.* The New Millennium Experience Company Limited (formerly Millennium Central Limited) [2000] Ch-1998-K-No 3692 446
Circle Freight International Ltd (t/a Mogul Air) *v.* Medeast Gulf Exports Ltd (t/a Gulf Export) [1988] 2 Lloyds Rep 427 64–6
C. J. Elvin Building Services Limited *v.* (1) Peter Noble (2) Alexa Noble [2003] EWHC 837 (TCC) 399
Clark Contracts *v.* The Burrell Co [2002] SLT 103 259
Clarke & Sons *v.* ACT Construction (2002) 85 Con LR 1 . . . . 63–4, 66, 67
Clydebank Engineering & Shipbuilding Co *v.* Don Jose Ramos Yzqinerdo y Castaneda [1905] AC 6 424
Commission for New Towns *v.* Cooper (Great Britain) Ltd [1995] Ch 259 . . . . 84, 87
Commissioner for Main Roads *v.* Reed & Stuart Pty Ltd and Another (1974) 12 BLR 55 365–6
Comorex Ltd *v.* Costelloe Tunnelling (London) Ltd [1995] SLT 1217 48
Comyn Ching Limited *v.* Radius plc [1994] ORB 728 26
Connex South Eastern Ltd *v.* MJ Building Services Group plc [2004] BLR 333 19
The Construction Centre Group Limited *v.* The Highland Council [2002] CA127/02 . . . . 316, 317–8, 512–3
Co-operative Insurance Society Limited *v.* (1) Henry Boot Scotland Limited (2) Henry Boot plc (formerly known as Henry Boot & Sons plc) (3) Crouch Hogg Waterman Limited (2002) 84 Con LR 164 152
Co-operative Wholesale Society Ltd *v.* Birse Construction Ltd (1996) 46 Con LR 110 467
Cory Bros & Co Ltd *v.* The Mecca [1897] AC 286 294
Costain Building & Civil Engineering Limited *v.* Scottish Rugby Union PLC [1993] SC 650 427
Costain Civil Engineering and Tarmac Construction *v.* Zanen Dredging & Contracting Company (1996) 85 BLR 77 . . . . 144, 179, 245
Courtney and Fairbairn Ltd *v.* Tolaini Brothers (Hotels) Ltd and The Thatched Barn Ltd (1974) 2 BLR 97 69–71
Croudace Construction Ltd *v.* Cawoods Concrete Products Ltd (1978) 8 BLR 20 443
Crown House Engineering Ltd *v.* Amec Projects Ltd (1989) 48 BLR 32 428

CR Taylor (Wholesale) Ltd *v.* Hepworths Ltd [1977] 1 WLR 659 421
D & C Builders *v.* Rees [1965] 3 All ER 837 . . . . 289, 291–2
David McLean Housing Contractors Ltd *v.* Swansea Housing Association [2002] BLR 125 313–15
Dawber Williamson Roofing Ltd *v.* Humberside County Council (1979) 14 BLR 70 377
Dawnays Ltd *v.* F. G. Minter Ltd and Trollope & Colls Ltd (1971) 1 BLR 16; [1974] AC 689 284–5
Day *v.* McLea (1899) 22 QBD 610 291
Dean & Dyball *v.* Kenneth Grubb Associates (2003) 100 Con LR 92 478
Decro-Wall International SA *v.* Practitioners in Marketing Ltd [1971] 1 WLR 361 . . . . 384, 392
Didymi Corp *v.* Atlantic Lines & Navigation Co Inc [1989] 2 Lloyds Rep 108 78
Discain Project Services Limited *v.* Opecprime Development Limited [2000] BLR 402 507
Dodd *v.* Churton [1897] 1 QB 562 . . . . 153, 199–200
D. R. Bradley (Cable Jointing) Limited *v.* Jefco Mechanical Services Limited [1988] ORB 1986-D-959 389
Duncan *v.* Blundell (1820) 3 Stark 6 337
Dunlop Pneumatic Tyre Company Limited *v.* New Garage & Motor Company Limited [1915] AC 79 423–4
Durabella Ltd *v.* J. Jarvis & Sons Ltd (2001) 83 Con LR 145 . . . . 25, 299
East Ham Corporation *v.* Bernard Sunley & Sons Ltd [1966] AC 406 421
Edmund Nuttall Limited *v.* R. G. Carter Limited [2002] BLR 312 476
Ellis-Don Ltd *v.* The Parking Authority of Toronto (1978) 28 BLR 98 226–7
Emcor Drake & Scull Ltd *v.* Sir Robert McAlpine Ltd [2004] EWHC 1017 (TCC) 38–9, 259–60
Emson Eastern Ltd (in receivership) *v.* E M E Developments Ltd (1991) 55 BLR 114 324
Erith Contractors Ltd *v.* Costain Civil Engineering [1994] A&DRLJ 123 486
Euro Pools PLC *v.* Clydeside Steel Fabrications Limited (17 January 2003) unreported 437–8
F&G Sykes (Wessex) Ltd *v.* Fine Fare Ltd [1967] 1 Lloyds Rep 53 . . . . . 73, 76–81
Fastrack Contractors Limited *v.* Morrison Construction Limited [2000] BLR 168 . . . . 477, 480, 493
Felthouse *v.* Bindley (1862) 11 CBNS 869 24
Ferson Contractors Limited *v.* Levolux A. T. Limited [2003] BLR 118 . . . . 379–80, 513
F. G. Minter *v.* Welsh Health Technical Services Organisation (1980) 13 BLR 1 . . . . 230–31, 302, 308–309
Fraser Williams *v.* Prudential Holborn Ltd (1993) 64 BLR 1 27
Galliford (UK) Ltd *v.* Aldi Stores (8 March 2000) unreported 221–2
George Trollope & Sons and Colls & Sons Limited *v.* Washington Merritt Grant Singer (1913) Hudson's Building Contracts, 4th Ed, Vol. 2, p. 849 232–3
George Wimpey UK Limited (formerly Wimpey Homes Holdings Limited) *v.* V I Components Limited [2004] EWHC 1374 (ch) 85–7

Gilbert-Ash (Northern) Ltd *v.* Modern Engineering (Bristol) Ltd (1973) 1 BLR 73 . . . . 278, 284, 285, 286, 430
Gillies Ramsay Diamond and Others *v.* PJW Enterprises Limited [2004] BLR 131 509
GLC *v.* The Cleveland Bridge & Engineering Co Ltd (1986) 34 BLR 72 416
Glencot Development & Design Co Limited *v.* Ben Barrett & Son (Contractors) Limited [2001] 80 Con LR 14 507
Goodwin & Sons Ltd *v.* Fawcett (1965) 195 EG 27 371
Gordon's Executors *v.* Gordon [1918] 1 SLT 407 74
Greater London Council *v.* Cleveland Bridge & Engineering Co Ltd (1986) 8 Con LR 30 112–13
Guinness plc *v.* CMD Property Developments Ltd (formerly Central Merchant Developments Ltd) and Others (1995) 76 BLR 40 . . . . 362, 400
Hackwood Ltd *v.* Areen Design Services Ltd [2005] EWHC 2322 (TCC) 41–2
Hadley *v.* Baxendale (1854) 9 Exch 341 . . . . 189, 226, 310, 443
Hall & Tawse South Limited *v.* Ivory Gate Limited (1996) 62 Con LR 117 . . . . 19, 27, 33
Hallamshire Ltd *v.* South Holland Council (2004) 93 Con LR 103 68–9
Hammond & Co *v.* Bussey (1887) 20 QBD 79 408
Hardwick Game Farm *v.* Suffolk Agricultural & Poultry Producers Association [1968] 1 Lloyds Rep 547; [1969] 2 AC 31 . . . . 65, 338–9
Harmon CFEM Facades (UK) Ltd *v.* The Corporate Officer of the House of Commons (2000) 67 Con LR 1 3
Harvey Shopfitters Ltd *v.* ADI Ltd (6 March 2003) unreported 29–31, 62
Henry Boot Construction *v.* Alstom Combined Cycles Ltd [1999] BLR 123; [2000] BLR 247; (2005) 101 Con LR 52 . . . . 219, 305
Hillas & Co Ltd *v.* Arcos Ltd (1932) 38 Com Cas 23 71
Hoenig *v.* Isaacs [1952] 2 All ER 176 . . . . 63, 252, 272, 319–20, 322
Holland Dredging (UK) Ltd *v.* The Dredging and Construction Co Ltd and Imperial Chemical Industries plc (Third Party) (1987) 37 BLR 1 59–60
Holland Hannen and Cubitts *v.* WHTSO (1983) 18 BLR 117 190
Holme *v.* Guppy (1838) 3 M&W 387 200
Homer Burgess Limited *v.* Chirex (Annan) Limited [2000] BLR 124 . . . . 499–50, 517
How Engineering Services Limited *v.* Lindner Ceiling Floors Partitions plc (1999) 64 Con LR 67 229
Hurst Stores & Interiors Ltd *v.* M. L. Europe Property Ltd (2004) 94 Con LR 66 83–5
Hussey *v.* Horne-Payne (1879) 4 App Cas 311 75
IJS Contractors Limited *v.* Dew Construction Limited (2000) 85 Con LR 48 394–5
Independent Broadcasting Authority *v.* EMI Electronics Ltd and BICC Construction Ltd (1980) 14 BLR 1 150–52
Inland Revenue Commissioners *v.* Barbara Fry, Lawtel, 30 November 2001 291
Interfoto Picture Library Ltd *v.* Stiletto Visual Programmes Ltd [1988] 1 All ER 348 . . . . 56, 502
Jacob and Youngs *v.* Kent (1921) 129 NE 889 422

James Longley & Co Ltd *v.* South West Regional Health Authority (1983) 25 BLR 56 249
James Miller & Partners Ltd *v.* Whitworth Street Estates (Manchester) Ltd [1970] AC 583 41
The Jardine Engineering Corporation Ltd and Others *v.* The Shimizu Corporation (1992) 63 BLR 96 233
Jarvis Facilities Limited *v.* Alstom Signalling Limited (trading as Alstom Transport Information Solutions) (No. 2) [2004] EWHC 1285 (TCC) 296–7
Jeancharm Limited t/a Beaver International *v.* Barnet Football Club Limited (2003) 92 Con LR 26 425
J. J. Fee Ltd *v.* The Express Lift Company Limited (1992) 34 Con LR 147 . . . . . 185–6, 188
J. M. Hill & Sons Ltd *v.* London Borough of Camden (1982) 18 BLR 31 371
John Barker Construction Ltd *v.* London Portman Hotel Ltd (1996) 83 BLR 31 203
John Cothliff Limited *v.* Allen Build (North West) Limited [1999] LV9 22641 502
John Martin Hoyes Directional Drilling Ltd *v.* R. E. Docwra Ltd [2000] 1999-TCC-117 49–50
Joinery Plus Limited (in administration) *v.* Laing Limited [2003] BLR 184 290–91
Jones *v.* St John's College (1870) LR 6 QB 115 200
J. Spurling Ltd *v.* Bradshaw [1956] 1 WLR 461 . . . . . 42, 444, 445
Karl Construction Ltd *v.* Palisade Properties plc [2002] CA 199/01 . . . . . 302, 427
Keeton Sons & Co v. Carl Prior Ltd (14 March 1985) unreported 65
Kemble *v.* Farren (1829) 6 Bing 141 424
KNS Industrial Services (Birmingham) Limited *v.* Sindall Limited (2000) 75 Con LR 71 . . . . . 284, 286, 480, 496, 504
Lafarge Redland Aggregates Limited *v.* Shephard Hill Civil Engineering Limited [2000] BLR 385 485–7
Laing Management (Scotland) Limited *v.* John Doyle Construction Limited [2004] BLR 295 240–42
Leslie & Co Limited *v.* The Managers of the Metropolitan Asylums District [1904] J of P, Vol. LXVIII 86, CA 277
Lester Williams *v.* Roffey Brothers & Nicholls (Contractors) Ltd (1989) 48 BLR 69 . . . . . 118, 328, 363
L'Estrange *v.* F. Graucob Limited [1934] 2 KB 394 44–5
Levolux AT Limited *v.* Ferson Contractors Limited [2002] BLR 341; [2003] EWCA Civ 11 . . . . . 286, 509
Lintest Builders Ltd *v.* Roberts (1980) 13 BLR 38 354
London Borough of Merton *v.* Stanley Hugh Leach Ltd (1985) 32 BLR 51 . . . . . 189–90, 386
Lord Elphinstone *v.* Monkland Iron & Coal Co (1886) 11 App Cas 332 424
The Lord Mayor, Aldermen & Citizens of the City of Westminster *v.* J. Jarvis & Sons Ltd and Another (1970) 7 BLR 64 . . . . . 322–3, 352–3
Love and Stewart *v.* S. Instone & Co Ltd (1917) 33 TLR 475 25
L. Schuler AG *v.* Wickman Machine Tool Sales Ltd [1974] AC 235 41
McCutcheon *v.* David MacBrayne Ltd [1964] 1 Lloyds Rep 16; 1 WLR 125 65

Mackay *v.* Dick (1881) 6 App Cas 251 190
Machenair Limited *v.* Gill & Wilkinson Limited [2005] EWHC 445 (TCC) 43
Macob Civil Engineering Limited *v.* Morrison Construction Limited [1999] BLR 93 505
Mamidoil-Jetoil Greek Petroleum Co SA *v.* Okta Crude Oil Refinery [2001] EWCA Civ 406; [2001] 2 Lloyds Rep 76 78
Maxi Construction Management Limited *v.* Mortons Rolls Limited [2001] CA39/01 257–8
May and Butcher Ltd *v.* The King [1943] 2 KB 17n 76
Mersey Steel & Iron Co *v.* Naylor, Benson & Co (1884) 9 App Cas 434 445
Michael A. Johnson *v.* W. H. Brown Construction (Dundee) Ltd [2000] BLR 243 . . . . . 362, 432–3, 433–4
Millar's Machinery Co Ltd *v.* David Way & Son (1934) 40 Com Cas 204 444
Miller Construction Ltd *v.* Trent Concrete Cladding Ltd (4 August 1995) unreported 14
Millers Specialist Joinery Company Limited *v.* Nobles Construction Limited [2001] TCC 64/00 521
Modern Building Wales Ltd *v.* Limmer & Trinidad Co Ltd (1975) 14 BLR 101 49
Monk Construction Ltd *v.* Norwich Union Life Assurance Society (1992) 62 BLR 107 . . . . . 26, 28, 31, 35–6, 244–5, 280
Monmouthshire County Council *v.* Costello & Kemple Ltd (1965) 5 BLR 83 476
The Moorcock (1889) 14 PD 64, CA 471–2
Morgan *v.* Perry (1974) 229 Estates Gazette 1737 434
Mowlem PLC (trading as Mowlem Marine) *v.* Stena Line Ports Limited [2004] EWHC 2206 (TCC) 40
Mowlem (Scotland) Limited *v.* Inverclyde Council [2003] XA 29/02 103
Moyarget Developments Limited *v.* Mrs Rove Mathis and Others [2005] CSOH 136 73–4
Muir Construction Ltd *v.* Hambly Ltd [1990] SLT 830 371
Murray Building Services *v.* Spree Developments [2004] TCC 4804 81–3
Myers *v.* Brent Cross Service Co [1934] 1 KB 46 337
Newton Woodhouse *v.* Trevor Toys Ltd [1991] OBENF/1130/90 and OBENF/91/0684 385
Nikko Hotels (UK) Limited *v.* MEPC plc (1991) 2 EGLR 103 506
Northern Construction Co Ltd *v.* Gloge Heating & Plumbing Ltd (1984) 6 DLR (4th) 450 4
Northern Developments (Cumbria) Limited *v.* J. & J. Nicols [2000] BLR 158 . . . . . 286, 507
Norwest Holst Construction Ltd *v.* Co-operative Wholesale Society Ltd [1998] All ER (D) 61 . . . . . 229, 404, 441
Norwest Holst Construction Limited *v.* Renfrewshire Council (20 November 1996) unreported 146, 180–81
Nottingham Community Housing Association Limited *v.* Powerminster Limited [2000] BLR 759 516–17
P & M Kaye Ltd *v.* Hosier & Dickinson Ltd [1972] 1 All ER 144 . . . . . 353–41, 358, 361
Pagan SpA *v.* Feed Products Limited [1987] 2 Lloyds Rep 601 74

Palmers Limited *v.* ABB Power Construction Limited [1999] BLR 426 517
Panamena Europea Navigation *v.* Frederick Leyland [1947] AC 428 273
Pantland Hick *v.* Raymond & Reid [1893] HL 22 . . . . 342–3, 403, 404
Parsons Plastics (Research and Development) Ltd *v.* Purac Ltd [2002] BLR 334 513
Pegram Shopfitters Limited *v.* Tally Weijl (UK) Limited [2003] EWCA Civ 1750; [2004] 1 All ER 818 509
Philips *v.* Ward [1956] 1 WLR 471 434
Pigott Foundations Ltd *v.* Shepherd Construction Ltd (1993) 67 BLR 49 . . . . 105, 106, 345, 352, 386–7, 416
Pitchmastic plc *v.* Birse Construction Limited (19 May 2000) unreported 273
Plant Construction plc *v.* Clive Adams Associates, JMH Construction Services Ltd [2000] BLR 137 196
Poseidon Freight Forwarding Co Ltd *v.* Davies Turner Southern Ltd and Another [1996] 2 Lloyds Rep 388 . . . . 43–4, 45
Public Works Commissioners *v.* Hills [1906] AC 368 424
The Queen in right of Ontario *et al.* *v.* Ron Engineering & Construction Ltd [1981] 1 SCR III 4
Queensland Electricity Generating Board *v.* New Hope Collieries Pty Ltd [1988] 1 Lloyds Rep 205 78
Rees & Kirby Ltd *v.* Swansea City Council (1985) 30 BLR 1 309–10
R. G. Carter Limited *v.* Edmund Nuttall Limited [2000] HT-00-230 57–8
Ritchie Brothers (PWC) Limited *v.* David Philip (Commercials) Limited [2005] BLR 384 498
RJT Consulting Engineers Ltd *v.* DM Engineering (Northern Ireland) Ltd [2002] BLR 217 . . . . 71–2, 509
Roberts *v.* Bury Commissioners [1870] LR 5 CP 310 273
Robertson Group (Construction) Limited *v.* (First) Amey-Miller (Edinburgh) Joint Venture; (Second) Amey Programme Management Limited and (Third) Miller Construction (UK) Limited [2005] BLR 491 . . . . 31–3, 226, 227–8, 446
Robinson *v.* Harman (1848) 1 Exch 850 421
Roche Products Ltd and Celltech Therapeutics Ltd *v.* Freeman Process Systems Ltd and Haden MacLellan Holdings plc, Black Country Development Corporation *v.* Kier Construction Ltd (1996) 80 BLR 102 . . . . 54–5, 57
Roe *v.* Naylor (No. 2) (1919) 87 LJ (KB) 958 45
Rossiter *v.* Miller (1878) 3 App Cas 1124 74
Rotherham Metropolitan Borough Council *v.* Frank Haslam Milan & Co Ltd and MJ Gleeson (Northern) Ltd (1996) 78 BLR 1 337–8
The Rugby Group Limited *v.* ProForce Recruit Limited [2005] EWHC 70 (QB) 34–5
Rupert Morgan Building Services (LCC) Limited *v.* David Jervis and Harriet Jervis [2004] BLR 18, CA 259
Ruxley Electronics & Construction Ltd *v.* Forsyth (1995) 73 BLR 1 420–22
Save Britain's Heritage *v.* No. 1 Poultry Limited [1991] 1 WLR 153 . . . . . 510, 512
S. C. Taverner and Co Limited *v.* Glamorgan County Council (1941) 57 TLR 243 192
Shanks *v.* Gray [1977] SLTR Notes 26 . . . . 432, 434

Shawton Engineering Limited *v.* (1) DGP International Limited (t/a Design Group Partnership) (2) DGP Limited (formerly known as DGP (Consulting Engineers) Limited [2005] TCC 44/01 344–5
Shepherd Construction Ltd *v.* Mecright Ltd [2000] BLR 489 294
Sherwood & Casson *v.* Mackenzie [1999] HT 99000188 507
Shirlaw *v.* Southern Foundries (1926) Ltd [1939] 2 KB 206 472
Sindall Limited *v.* Solland (unreported) 15 June 2001 477
Six Continents Retail Ltd *v.* Carford Catering Ltd and R. Bristoll Ltd [2003] EWCA Civ 1790 198–9
Skanska Construction UK Limited (formerly Kvaerner Construction Limited) *v.* Egger (Barony) Limited [2004] EWHC 1748 (TCC) 243, 249
SL Timber Systems Limited *v.* Carillion Construction Limited [2001] BLR 516 283
Smith *v.* UMB Chrysler (Scotland) Ltd [1978] SC (HL) 1 443
Solland International Limited *v.* Darayden Holdings Limited TCC 15 February 2002 286
South Bucks DC and Another *v.* Porter (No. 2) [2004] 1 WLR 1953 512
Stent Foundations Limited *v.* Carillion Construction (Contracts) Limited (2000) 78 Con LR 199 20
Stickney *v.* Keeble [1915] AC 386 390
Stour Valley Builders *v.* Stuart [1974] 2 Lloyds Rep 13 .... 290, 291
Strathmore Building Services Limited *v.* Colin Scott Greig t/a Hestia Fireside Design [2000] CA 18/00 282
Sudbrook Trading Estate Limited *v.* Eggleton [1983] AC 444 77
Surrey Heath Borough Council *v.* Lovell Construction Ltd and Haden Young Ltd (Third Party) (1988) 42 BLR 25 355
Swan Hunter and Wigham Richardson Ltd *v.* France Fenwick Tyne & Wear Co Ltd, The Albion [1953] 1 WLR 1026 445
Sweatfield Limited (formally JT Design Build Limited) *v.* Hathaway Roofing Limited (31 January 1996) unreported 394
Tate & Lyle Food Distribution Limited and Silvertown Services Lighterage Limited *v.* Greater London Council and Port of London Authority (22 May 1981) unreported 242–3, 439
Team Services plc *v.* Kier Management & Design Ltd (1993) 63 BLR 76 ... 275–6
Thomas Bates & Son *v.* Wyndam's (Lingerie) Ltd [1981] 1 WLR 505 86
Tim Butler Contractors Limited *v.* Merewood Homes Limited [2000] TCC 10/00 500
Tinghamgrange Ltd *v.* Dew Group Ltd and North West Water Ltd (1995) 47 Con LR 105 220–21
Tito *v.* Waddell (No. 2) [1977] Ch 106, 332C 422
Total M & E Services Limited *v.* ABB Building Technologies Limited (formerly ABB Steward Limited) (2002) 87 Con LR 154 .... 191–2, 420
Trustees of the Stratfield Saye Estate *v.* AHL Construction Limited [2004] EWHC 3286 (TCC) 72
Turner *v.* Garland & Christopher (1853) Hudson's Building Contracts, 4th edn, Vol 2, p. 1 151
T. W. Thomas & Co Ltd *v.* Portsea Steamship Co Ltd [1912] AC 1 .... 49, 53
VHE Construction plc v. RBSTB Trust Company Limited [2000] BLR 187 286

Walford *v.* Miles [1992] 2 AC 128 75–6
Walker and Another *v.* The London & North Western Railway Company [1876] Common Pleas 518 123–4
Wallis *v.* Smith (1882) 21 Ch D 243 424
Webster *v.* Bosanquet [1912] AC 394 424
Weldmarc Site Services Limited *v.* Cubitt Building & Interiors Limited [2002] HT-01–408 . . . . . 244, 383, 435
Weldon Plant Limited *v.* The Commission for New Towns [2000] BLR 496 216–17
Wells *v.* Army & Navy Co-operative Society (1903) Hudson's Building Contracts, 4th edn, Vol. 2 p. 346 . . . . . 352, 386
Wescol Structures Limited *v.* Miller Construction Limited (8 May 1998) unreported 51–2, 276–7
Whittal Builders Company Limited *v.* Chester-le-Street District Council [1985] No. 1980 W 3060 235
William Lacey (Hounslow) Ltd *v.* Davis [1957] 1 WLR 932 7
William Tomkinson & Sons Limited *v.* The Parochial Church Council of St Michael-in-the-Hamlet with St Andrew Toxteth in the City and Diocese of Liverpool and Holford Associates [1990] OR No. 21/87 . . . . . 355, 360–61, 399–400
William Verry (Glazing Systems) *v.* Furlong Homes Ltd [2005] EWCH 138 (TCC) 479–80
William Verry Ltd *v.* North West London Communal Mikvah [2004] BLR 308 497–8
Williams *v.* Fitzmaurice (1858) 3 H & N 844 . . . . . 7, 16, 33–4
Woolmer *v.* Delmer Price Ltd [1955] 1 QB 291 445
Young & Marten Ltd *v.* McManus Childs Ltd (1968) 9 BLR 77; [1969] 1 AC 454 . . . . . 155, 336–7

Index